Major
Technological
Risk

Other Pergamon Titles of Interest

Pergamon Related Journals
Free Specimen Copy Gladly Sent on Request

Major Technological Risk

An Assessment of Industrial Disasters

PATRICK LAGADEC
Doctor of Political Science

Translated from the French by
H. OSTWALD

Technical Editor
J. C. CHICKEN

PERGAMON PRESS

OXFORD · NEW YORK · TORONTO · SYDNEY · PARIS · FRANKFURT

U.K.	Pergamon Press Ltd., Headington Hill Hall, Oxford OX3 0BW, England
U.S.A.	Pergamon Press Inc., Maxwell House, Fairview Park, Elmsford, New York 10523, U.S.A.
CANADA	Pergamon Press Canada Ltd., Suite 104, 150 Consumers Road, Willowdale, Ontario M2J 1P9, Canada
AUSTRALIA	Pergamon Press (Aust.) Pty. Ltd., P.O. Box 544, Potts Point, N.S.W. 2011, Australia
FRANCE	Pergamon Press SARL, 24 rue des Ecoles, 75240 Paris, Cedex 05, France
FEDERAL REPUBLIC OF GERMANY	Pergamon Press GmbH, 6242 Kronberg-Taunus, Hammerweg 6, Federal Republic of Germany

First edition 1982

Library of Congress Cataloging in Publication Data

Lagadec, Patrick.
Major technological risk.
Translation of: Le risque technologique majeur.
Includes bibliographies and index.
1. Industrial accidents. 2. Industrial safety.
3. Risk management. I. Title.
HD7262.L2413 1982   363.1'1   82-3678
ISBN 0-08-028913-4

British Library Cataloguing in Publication Data

Lagadec, Patrick
Major technological risk.
1. Technological assessment 2. Risk
I. Title II. Le risque technologique
majeur. *English*
363.1   T174.5
ISBN 0-08-028913-4

In order to make this volume available as economically and as rapidly as possible the typescript has been reproduced in its original form. This method unfortunately has its typographical limitations but it is hoped that they in no way distract the reader.

This book is a translation of Le Risque Technologique Majeur, published by Pergamon Press France S.A., 1981

Printed in Great Britain by The Anchor Press Ltd Tiptree, Essex

PREFACE

This work is — with only some slight alterations — the publication of the doctoral thesis in political science presented and defended publicly on November 14, 1980 by Patrick Lagadec at the Institute of Political Studies at the University of Political Sciences of Grenoble.

The jury consisted of the following persons:

Messrs L. Nizard, Professor of political science at the University of Grenoble II, President

F. d'Arcy, Professor of political science at the University of Grenoble II

J. Capdevielle, Research fellow at the National Foundation for Political Science

J. -F. di Chiara, Subdirector, deputy to the Director of Civil Defence, in charge of operational services

C. Henry, Professor at the Polytechnic

J. M. Martin, Research Director at the C.N.R.S.

J. -J. Salomon, Professor at the C.N.A.M, chair of Technology and Society

P. Vesseron, Chief Mining Engineer, head of the service of the industrial environment at the Ministry of the Environment and Cadre de Vie.

ACKNOWLEDGEMENTS

We wish to express our gratitude to the members of the jury, in particular to L. Nizard who presided.

J. F. di Chiara, J. J. Salomon and Ph. Vesseron have given us invaluable assistance. We are aware of all we owe to Cl. Henry who welcomed us to the Econometric Laboratory in September 1977. To the Polytechnic and to the Laboratory team go thanks for having provided us with the means for conducting this research.

Without the support of a great many specialists — whom, unfortunately, we cannot all mention by name — this work could not have been accomplished. We are indebted to many for information, advice and corrections which above all permitted us a series of monographs, summaries and finally this last work which has been twice recast.

In the first place we thank our foreign friends. Our studies on Italy would never have seen the light of day without the help of Laura Conti, well known for her competence and courage; we shall not forget the very long hours spent under her guidance on learning the complexities of Italian reality, the (discouraging!) meanderings of the case of Seveso. Cl. Rise and C. de Rosa of Milan also gave us valuable assistance. In the wake of our first studies of that country we were received by the board members of Hoffmann-La-Roche in Basle to whom go thanks for the time they gave us, the documents given to us and the corrections suggested to us. We have been particularly impressed by our long meeting with G. Reggiani who is equally well known for his competence and courage in the Seveso affair.

For the British case we have benefited from the remarkable reception given to us by the board of the Health and Safety Executive: J. H. Dunster (Deputy Director), A. V. Cohen, H. E. Lewis; J. R. H. Shenkel, F. R. Farmer and J. H. Bowen from the Safety and Reliability Directorate (UK Atomic Energy Authority) were prepared to discuss delicate issues with us. T. A. Kletz, safety specialist at I.C.I., made it possible for us to understand more clearly the point of view of industry in this matter. We owe much to L. McGinty, News Editor at the monthly *New Scientist*, Sir Bernard Braine, conservative MP for Essex Southeast, received us several times at Westminster and is cordially thanked on this occasion. P. Haslam from the Association

for the Defence of Canvey Island made us realise another aspect of the problem
of major risk: we remember the discussions at the methane terminal of British
Gas amongst 2,000 caravans strangely placed in the impressive setting of
Canvey and the search for essential documents among hundreds of papers in
possession of the Association. We shall also never forget the support of
R. Johnston, K. Green (University of Manchester), B. Wynne (University of
Lancaster), G. Atherley (University of Brimingham), D. Napier (University of
of London), Sir F. Warner (Consultants Cremer & Warner), S. McKechnie (Trade
Union ASTMS).

   Still abroad, we had support from Dorothy Nelkin (Cornell University, USA),
Helga Novotny (Centre Européen de formation et de recherche en action sociale,
United Nations, Vienna), H. Otway (I.I.A.S.A. — I.A.E.A. Vienna, later Ispra,
Italy), J. Ravetz (University of Leeds), W. D. Rowe (former administrator of
the EPA, Washington); various encounters at Bielefeld, Berlin, Frankfurt,
Oslo have provided very stimulating work. Some of the meetings would not have
been possible without the assistance from the EEC Commission. We shall not
forget our correspondents from across the Atlantic, especially, R. Kasperson
and R. Kates (Clark University); J. Burton (University of Toronto) gave us
essential support in the study of the rail disaster of Toronto.

   In France, very many organisations and persons have given us active co-
operation. To them all we express here our gratitude. In particular to:
A. Pradinaud from the Service of the Industrial Environment and F. Perier,
until recently Mining Engineer at the Interdepartmental Directorate of
Industry and Mines at Lyons; P. Tanguy (Director of the Institute for
Protection and Nuclear Safety at the Atomic Energy Commission) and R.
Andurand (Department of Nuclear Safety — C.E.A.), Messrs Chastel and Jeudi
(Directorate of Civil Safety); Messrs Macart, Jeanette, Goubet (Ministry of
Industry), Messrs Tribuillois (Plenary Assembly of Insurance Companies),
Hure (Société Commerciale de Réassurance) and Deprimoz (Technical Accident
Group) as well as K. Konrad (Muenchner Rueck), Messrs Deschanels and
Lavedrine (aérospace); Mr Choquet (I.N.2.P.3.); Mr Moyen (I.N.R.S.), Messrs
Bachmann, Desse and Bourrée safety specialists at the Rhone-Poulenc group),
Mr Rainelli (I.N.R.A., Rennes), Mr Martin-Bouyer (I.N.S.E.R.M.).

   We also owe much to those who have helped with the publication of a
synthesis of this work in *La Recherche*, Martine Barrère, as well as for the
collective work of which we were in charge and which was published by
Futuribles, also in November 1979: H. de Jouvenel, L. Puiseux, Martine
Rémond-Gouilloud, R. Crisp, C. Lepage-Jessua, Nicole Nowicki, F. Fagnani and
J. F. Belhoste, Y. Stourdzé, Ph. Roqueplo, P. Blanquart were among those who
helped to clarify our ideas while work was in progress and especially in its
final phase.

   The trade unions (particularly the C.F.D.T. of the Atomic Energy) and the
Scientists Group for information on Nuclear Energy (G.S.I.E.N.) have provided
us with valuable complementary information on the nuclear aspects of our
research.

   Obviously, no organisation and no persons shall be responsible for this
paper: we take exclusive responsibility for it.

   Marie-Louise Pouderous has shouldered the heavy duty of designing the
presentation of this paper and to a large part its typing. Thanks go to her
for her competence, her speed and her always stimulating advice.

CONTENTS

NOTICE

The following pages will show the diversity and the sharpness of the challenges posed by major technological risk. The scientific research work too will not escape those called into question. It will not be possible to follow the rule demanding that all references (sources) are clearly stated: a large number of interesting documents in our field of major risk are confidential. We have therefore been constrained to break this basic rule sometimes. To refuse recognition of this confidentiality would have made the examination of major technological risk a forbidden task. This does, of course, not cover any involuntary omissions.

To all those who have helped me with their competence, their demands, their respect and their friendship.

GENERAL INTRODUCTION

A political approach to major technological risk

"To the limit of what we can do it has all the seriousness of our thought."

Karl Jaspers
The Atom Bomb and the Future of Man
Paris, Buchet-Chastel 1963

GENERAL INTRODUCTION

THE POLITICAL APPROACH TO RISK

1. A NEW FIELD OF QUESTIONS ON THE VERY FOUNDATIONS OF OUR
INDUSTRIAL CIVILISATION

Major technological risk: three apparently simple words which nevertheless
raise incalculable numbers of questions; social questions, challenges to
political conscience and action in our time which are quite formidable.
Three disruptive words which are promptly abandoned if they are somewhat
emphasized: "Life means risk; without risk no activity, no dynamism but
certainly famine, epidemics, and at the end of the road: confrontation and
devastating war." This solid common sense, which is neither wrong nor quite
true, does not, however, succeed in obscuring the issue. It is quite true
that risk is a dimension indissolubly linked with life; that man has torn
himself away from his primal condition by manufacturing tools, the risk
makers; that risk has taken on a large size and new forms since tools have
become transformed into machines and the craftsman into the workman: the
railway, the spinning mills, the mines have caused a large number of deaths.
It is quite correct: risk is nothing new. Let us not forget: on March 10,
1906 — and this is no exception in our industrial countries — an enormous
explosion wrecked the French mine of Courrières: 1,100 dead, many hundreds
made invalids for life; 562 widows, 1,133 orphans; hunger and cold; 50,000
strikers, troubles. Clémenceau declared martial law in the area; 25,000
soldiers criss-crossed the miners' dwellings, forbade free traffic; terror,
clandestine activity, 'conspiracy', arrest of responsible trade unionists;
resumption of political life and work. Mourning, misery and bitterness! (1).

Yet, technological risk*poses a new problem today. Since the first third
of this century the means of production developed by man have changed scale,
there are for instance the large industrial complexes, and the risk has risen
quantitatively. For several decades science has also uniquely transformed

---

*
The term 'technological' will be used in a very broad sense: proper
development of techniques but also the industrial development associated
with the implementation of these techniques. See p. 8

3

industrial activities: numerous processes, new products, have been introduced.
With them risks appeared which had been unknown in the past, and this just
when massive support from science was strengthening the pretentions of
contemporary homo faber to control more and more, to perfection, it was
believed, all dangers incurred. The enthusiasm of conquest, the arrogance
which sometimes inflates success, also the painful feeling — after Hiroshima
— of necessary redemption in the civilian world after one had 'succeeded' so
much in the military one, were to leave the risk issues in the dark for a
long time. At present, after certain setbacks have appeared, the interest
in long term effects has been revived, the global approach to development
imposes itself almost side by side with sectoral growth studies, and there
is now more decided interest in this issue which has never been far from the
minds of the most experienced scientists. The sciences and techniques,
perhaps already as such but certainly when regarded as social processes,
show their ambivalence more and more clearly in the same measure as they
develop, model and transform our life styles and our natural environment.

   Certain events have heightened the attention paid to this ambivalence.
There is the emergence of social movements or more or less social manifes-
tations of some breadth which call into question the choices made in the
name of technical progress. There are also as immediate starters a series
of events which have taken up a permanent column in the press: the disasters.
These extraordinary phenomena which are always called 'extremely rare' but
ever more urgent, begin to cause broader reflections on the issue of major
technological risk.

   We now come to a series of views about the nature of risk. We have the
beginnings of an experience in the field of civilian use of nuclear energy
which, as a first, has benefited from some attention to the problems of
safety. In fact, one finds oneself rather at a loss in the face of our
three words: major technological risk. The good thing at the beginning of
the 1980s is this: the issue is about to be recognised. How does one
approach it? There is an unpleasant surprise in store: the classical
measuring sticks appear to be failing. Dividing up the difficulties and
tackling them one by one, relying on a law of accumulation is no longer an
operable method. The phenomena at issue impose themselves from the outset
in their global characteristics. Understanding them requires not only good
sectoral, 'vertical', knowledge in which the specialists trained since World
War II excel but also 'horizontal' and 'diagonal' measures which permit the
updating of general logic. One is reluctant in the face of such perspectives:
on the one hand because of difficulties — due to lack of practical experience
— because one has been used to division, to partial, functional logic; on
the other hand because of the calling into question what would result from
a transgression of the established rules of exclusively localised
comprehension. One hesitates because the questions to be put on the
analysing table are not soothing. A disaster is not the result of a single
technical or human failure; it is the product of a collection of factors
which stem from the mechanisms, the management, the development of the area,
up to the great technological options and the choice of life style ... . In
short, the approach to the issue of major technological risk requires very
often that one goes back to the very foundations of our industrial
civilisation, its technical trump cards, its highest values. The 'burning'
nature of the issue puts it beyond begging the question and beyond the
rhetoric of good manners. It is not astonishing in such circumstances that
the phenomenon still largely escapes intellectual grasp and that the
discourse is difficult to link up with reality.

   This is from where the strange 'perfidy' of our apparently quite innocent

words stems. For some years and in a seldom disputed manner they tend to
turn seminars that are organised to deal with this topic into a Tower of
Babel. The difficulty and the fear of understanding are tall obstacles.
From the outset, the speaker has already lost a good part of his audience
who are confused by his choice of definitions, by his way of describing the
subject or by his approach.

On the strength of this basic statement and in order to take a further
step let us remember for a moment the way in which the issue has arisen in
order to help with the approach to it. Three events in particular support
the raising of the question. Seveso: the escape of dioxin, a product of
extreme toxicity, probably mutagenic and teratogenic. The *Amoco-Cadiz:* the
wreck of one of those ocean-going giants on March 16, 1978 which affected
large coastal areas. Three Mile island: the nuclear accident which raised
the fear of mass evacuation of the population on March 28, 1979 in
Pennsylvania, USA.

Basing our thoughts on such events, which in large measure still remain
just danger signals and have not been taken as seriously as they should be,
we shall subsequently uphold the following central idea which for the sake
of clarity we define thus: there is a clear discontinuity between today's
major technological risk and the dangers, of the same origin, that were
known in the past. On the one hand, the size of the phenomenon has changed:
for instance those who had the responsibility in July 1976 could ask themselves secretly whether the accident of Seveso would not force the
evacuation of Milan, Italy's economic capital, for an indefinite period of
time. On the other hand, the nature of dangers has changed: in the first
place because of quantitative growth — not just deaths but collective
slaughter — and then the quality of the factors involved which let the
danger hang over the heads of not only the present generation but also over
those of their descendants: it is now the totality of life that is called
into question.

One can no longer hide the truth from oneself: large agglomerations, huge
areas are nowadays exposed to the menace of serious destruction, contamination and evacuation. The populations concerned in each case run into
hundreds of thousands, in fact into millions. The causes of disasters are
certainly multiple even if they are still poorly accounted for and little
known.

These exactly are the menaces and their possible translation into
catastrophic events which we call here 'major technological risk'. These
same risks concern directly the industrial societies, their lives, their
territories, their organisations, their option, their future, their
reproduction. Examination of the questions linked with this new situation
is the objective of the present study. How does such a situation challenge
the conscience and the political actions of our time? What are the
challenges that must be clarified and dealt with? How can collective
reason, freedom and goodwill regain hold of some of the leverage which seems
to be slipping away from us? How can the often absent and sometimes totally
routed understanding of the phenomena retrieve its place in the decision
making process on matters of development: one realises cruelly in times of
disaster that all discussions turn on themselves, fatalism and belief in
miracles take over! There are so many questions which the body politic can
no longer ignore.

The kind of thinking that is required is no doubt new. Until now the
accident constituted, in a manner of speaking, part of the labour contract,

no matter what the objections to this situation may have been. Regulatory
social systems were actually set up to 'manage' the prolongation of the
working activity. With the possibility of civilian slaughters, not
'manageable' in the classical sense of the term, a new state of facts
appeared on the scene. However it is in another field that it has already
become necessary to recognise a solution of analoguous continuity. This is
the military field, the question which developed with the arrival of nuclear
forces. To be sure: on that occasion too it took a good decade to recognise
against all common sense, i.e. against 'experience', because Dresden and
Hamburg counted more victims than Hiroshima and Nagasaki, that 'THE' bomb
was no longer just 'a' bomb. One had to admit that Clausewitz's principle
no longer worked so well: when battle signifies the end of all it is no
longer a continuation of politics. The available means have changed the
nature of the issue. The same is true here, with industrial activity and
its menaces. This is why research such as that by Karl Jaspers in his work
'*The Atom Bomb and the Future of Man*' (2) is at least as a point of
departure so enlightening for us. In this book he presents us with a
reflection which takes up position precisely 'at the limits', at the edge of
what had been called 'the unthinkable' and which must provoke thought rather
than limit reflection. As far as the key points of this author are
concerned let us review the routes to avoid in a similar problematic
situation which he has pointed out to us.

In borderline situations all is called into question and not just a
marginal element of the system. On account of this, pigeon hole thinking is
no longer adequate; 'common sense' leads to escapism inasmuch as the main
rules and references are no longer applicable. The 'realists', the
'specialists' and the 'law-abiding' are always caught short by the events.
The big organisations with their concern for regularity and orderliness excel
in the management of what is in vogue: they will be the ones to abut borderline situations the most frequently on account of the nature of the latter.
In borderline situations the demand for knowledge is imperative. However,
they call for more than just a simple examination of symptoms; or, if one
takes up Hegel's most quoted statement: truth is allied with reality. The
important thing is not just the bomb but war, says Jaspers, which means for
us not just the disaster but the organisation and management of economic
activity. This type of examination immediately comes up against blackmail:
"Shut up, you are causing panic!" Jaspers retorts:
    "In Germany, in 1933, they said: above all, no civil war!" (p. 656). He
    continues; "Fear is useful only if it transforms itself into a force of
    action (p. 656); if panic terror tumbles into unreason, enlightened fear
    leads to liberation through reason" (p. 658).

In borderline situations the body politic must not give in to the illusion
of a necessary return to equilibrium: there are situations of real
disruption. The ultimate resort must not be taken for granted. There is
no natural necessity; the future devolves upon man and his responsibility.

In borderline situations first thoughts, common sense, the desire to rub
out the fundamental questions evoked by the new situation are not at all
helpful. In situations of disruption the foremost task is to retrieve
'sense'. For this perspication is needed, for instance of the kind that
Themistocles had, which means the ability to free oneself from the false
'necessities' of the moment, to retrieve the principles of the intelligence
of the whole and to subordinate human activity to a new political direction.
Without this ability to take up the issue as it emerges and to respond to it
with a project, a policy remains crippled.

In borderline situations it is man's freedom that is challenged. They are therefore not a matter for prediction but rather for preparedness: not to come up with cut and dried thoughts but to construct from the situation what will not bar man from his calling to freedom. When all is at stake freedom cannot be freedom just for some. One must therefore beware of expecting prophets because "then the Fuehrer arrives" (2, p. 646) or to throw into despair the sages who have always said that man is too foolish to expect anything of him: the enlightened despot has hardly ever been better than his subjects. The means to that searched-for freedom is democracy, as reality or as project, to be equally redefined since the emerging disruption affects it as much as it affects every other part of the situation. Democracy will therefore have to be something other than a collection of formal institutions. Freedom and democracy challenge planning and the body politic directs it. In the face of the borderline event it is neither a question of drawing up a master plan to create order nor a simple reorganisation of activities but the opening of new possibilities for human activity, knowing that it is impossible to foresee what man will do with an act of freedom. That in the end is possible only at the price of a change of the collective will.

We have to consider two obstacles mentioned by Jaspers which are particularly acute in our field. These are on the one hand the will to ignore and on the other to renunciation. The first attitude manifests itself as a constant one: we shall have to examine it closely in the light of experience and try to understand its fundamental resilience. The second one also occurs frequently. In the military field one hears all the time: "Considering what man is, there is no hope." In our field as well, but in a more moderate fashion, we have been told: "Considering what is nowadays the supreme value in our society, the economy in its narrowest sense, there is no room for manoeuvre. If the choice is between a factory and a large city one does not hesitate for a second: one keeps the factory and rebuilds the agglomeration elsewhere if need be and permits expansion".*

If one follows that line one has to visualise the situation as follows: We are on the threshold of the metastable, water is still liquid at -10° and the first disturbance will turn everything to ice on the crest of the wave which is about to break. In that case there is indeed nothing more to be done. There remains meditation on the finite nature, not of humanity (which applies to the military) but more or less of the Western World which draws its strength from its industrial apparatus and its technological abilities. We do not share this evaluation. We maintain that hope remains possible, hope but not certainty; that through research and experimentation new fields for reflection and collective action can be opened up at various levels and thanks to a better knowledge of it all in its totality and also in its local differences.

Obviously all this work requires the delimitation of a particular field of examination. The preceding pages show already a number of choices which we had to make. We now go back to these options in order to make them explicit.

2. PRIORITIES FOR OUR EXAMINATION

Various types of risk could be qualified as 'major'. Here we are interested

---

*These ideas do not come from someone on the edge of society or someone stretched to the limits of decision.

in technological risk, giving this term a very broad application: the tool as
well as its application, the complex, the project which has become possible
because of the development in science and technology, as a fruit of human
ingenuity. In order to include the British (3) or European (4) application
of the term we shall examine the menaces which due to industrial activity
weigh down no longer just on persons and assets inside the walls where they
work but quite particularly on what is outside those walls, priority being
given to the menaces which result in sudden brutal events.

We leave largely outside our field of analysis that which results from the
military domain. We therefore do not examine the scenarios which flourish
nowadays in the USA where everybody, thanks to a little pocket computer, can
know the dangers he faces if that nuclear bomb explodes at such and such a
distance from where he is and at such and such an altitude. However, as has
already been seen in the preceding pages this military world will not be kept
totally in the background. In fact, there exists a continuum, and if it
comes to the crunch the priorities chosen for our work shall decide and not
the specific nature of the phenomena. The military domain can suggest
interesting reflections for our examination of industrial risks; in terms of
national defence, civil defence, sometimes in the magnitude of events there
exists a continuity which must not be ignored; the same is true between
demonstrations, riots, acts of foul play, acts of war or again with regard
to the 'state of siege' as declared in the wake of the Courrières disaster
or because of major internal upheaval. We are quite aware of the dangers to
the quality of research threatening from too hasty conclusions. Let us
confirm then that those factors linked with the military shall be considered
when necessary but they shall not be given priority in our reflections.
Incidentally, as it has been frequently pointed out, let us make it clear
that we reject the idea according to which no risk other than nuclear war may
claim attention as long as universal peace is not ascertained: such ideas
paralyse and cheaply justify inaction. The terms of the argument have to be
reversed: inasmuch as certain advances have been made in fields like ours the
approach to the military issue, which on the evidence is much more serious
and complex, can be facilitated.

A second attempt at restricting our field of investigation will be made on
similar principles. The field of natural disasters shall also largely be
kept in the background. Thus we shall not for instance deal with such serious
phenomena as the earthquake of Tan Shan which has cost 650,000 Chinese lives
on July 27, 1976; or the flooding which is considered the worst reported
natural disaster in history and which in July/August 1959 caused the death of
two million people in northern China (5); or the seismic tidal waves in
Pakistan, the earthquakes in Central and South America, the volcanic
eruptions in South East Asia etc. The very frequent and often confused
discussions on connections between so-called 'natural' disasters and disasters
of industrial origin justify dwelling on this point for a moment. Here again
we have an obvious continuum and that along several lines of analysis. From
one extreme to the other one notices three alignment pickets: from the
seismic tidal wave which ravages an area in Alaska to the foreseen and
expected earthquake which may destroy deposits of gas and hydrocarbon, which
in turn may ravage the neighbouring area when passing through a densely
populated but poorly protected zone. From one extreme to the other human
responsibility increases. Let us dwell on this to avoid all misunderstanding.
As Jean-Jacques Salomon reminds us in his foreword to a recent issue of the
magazine *Futuribles* which is devoted to this topic (6), Rousseau has already
called Voltaire to task for the latter's interpretation of the Lisbon
disaster: "It was not nature which assembled 20,000 six and seven story
houses; if the inhabitants had spread out and lived in less crammed conditions

one would have found them the next day twenty leagues from there and quite as happy as if nothing had happened" (7). Let us finally go back to certain cases in which man showed stupefying irresponsibility in his choices of development, given the menaces of natural origin prevailing at the chosen location. If one follows the specialists of the world's largest reinsurance company, the Muenchner Rueckversicherungs-Gesellschaft, Tokyo stands as a symbol for the whole planet: an extreme earthquake probability in the order of one per cent per year (10^{-2}/year); damage valued at two hundred billion US dollars (8); but the bay continues to receive methane carriers, Kawasaki still shelters hydrocarbon depots which cannot withstand an earthquake of any significance (9). What then is our choice? We shall concentrate exclusively on cases where human responsibility is heavily involved. For simplicity's sake and also to facilitate the social learning process on this issue we shall even tend to concentrate primarily on those cases in which man is totally responsible for the situations that have been created so as to avoid the all too tempting flight into the invocation of the scourges of nature. It should be understood that this does not prevent the marginal inclusion of other situations in which man, in order not to be completely responsible for the phenomena, has appropriately managed all that would relieve him of his responsibilities. As in the military domain, the so-called natural disasters are full of lessons to be learned: issues of size of phenomena, prevention, evacuation, rescue plans etc. To repeat: these events will not be ignored but they will not be given priority in the analysis.

A third choice of priority must be explained. The analysis is not directed towards the professional risks with which the industrial worker is familiar; not that he or it are unimportant, there are more than two thousand deaths at work every year in France, but again for the sake of simplification. We concentrate mainly on the phenomena which can affect the population outside factories. We recognise the dangers of such limitation and the comments from the British trade unions for instance who are sometimes surprised at the concern shown for the citizen (i.e. the voter) when the worker faces considerable dangers every day (10). Here too we reply in terms of the learning process: inasmuch as even very large scale risks, which are sensitive in terms of electoral impact, remain undealt with or insufficiently taken into account there is little chance for professional risks to be controlled with the necessary determination. What conscience and collective action will be able to achieve in the still more sensitive field of collective risk may become an important gain for the worker who (which should not be forgotten) is also a citizen and often lives in very exposed urban areas. There again the connecting links between the 'worker' and the 'citizen' are numerous: between the 'classical accident' and the 'major disaster': a disaster is often nothing but an incident that has found a fatal sound box, nothing but a series of incidents that have amplified each other, nothing but an outcome of minor difficulties etc. We therefore do not neglect the lessons to be learned from the observation of more classical industrial accidents.

A fourth priority has also been observed. Between the two spheres which are usually distinguished nowadays, safety and protection, we have largely chosen the first one. In doing so we have adopted the British application of the term 'major hazard': the risk of a sudden brutal event (reaching beyond the perimeter of an installation). Not that the events with which safety deals are more serious: Minamata is no less serious than Feyzin. The same type of pollution of waterways in Canada is no less worrying than the railway accident in Toronto. The six tonnes of cosmetic cream containing lead which are sold in France each year are perhaps no less dangerous than events of spectacular appearance. The massive release of carbon dioxide into

the atmosphere, the continuous release of highly toxic mutagenic products are
issues of equal seriousness. In short, it is not exclusively the seriousness
of the phenomena which has guided our choice but again in the first place the
concern with the learning process. If some additional steps can be taken
under the pressure from shock events as key factors for social change for a
better policy on matters of safety we shall have a positive experience which
will be useful so as to enter into their multiple aspects of diffuse risks.
When, by contrast, one is still capable of denying — even when confronted
with rigorous analysis — the most formidable evidence as is the case today
("Petrol evaporates", "Nothing went wrong at Harrisburg", "Seveso is a
pollution of minds") then there is little chance that the delicate questions
of slow and diffuse risks will be adequately approached or even just
recognised: whence we have chosen our priority. Obviously no tight partition
will be drawn artificially between the two types of reality called to mind
which again are only extremes within a context that is: phenomena which
sometimes present problems both of safety and of protection.

Finally, as a last element in our grading system, we have conducted our
research mainly in the developed countries of western Europe and North
America. But even though these areas are the most representative, the other
industrialised countries are also part of our research. As elsewhere, and
this is of some importance, the countries of the Third World import western
technology. Let us consider this point. The poor countries are already
those most affected by natural calamities; but now there are the additional
risks of technological origin and of large scale. During summer 1979 a dam
burst northwest of Bombay: it is estimated that there were more than 25,000
victims. The concentration of people, which is typical for these countries,
makes the menace that lies in wait for them even more serious; the context
into which our high technologies are transferred will make the menaces linked
with the use of these tools grow still further. This is a very important
field for research to cover at a time when 'redeployment', the 'North-South
Dialogue' are being discussed.

In this field which is thus differentially illuminated: what have been the
particular perspectives of our undertaking? We shall present them in the
following pages.

3. RESEARCH PERSPECTIVES: MAJOR TECHNOLOGICAL RISK AND THE EXERCISE
   OF COLLECTIVE CHOICE

Being interested in the issue of risk the journalist collects events,
reports on debates and reports and, always in search of the facts which
emerge from the usual 'noise', pays general attention to disastrous events
and willingly devotes his headlines to them. The technician with the aid of
the mathematician (probabilist) and the statistician tests his tools, studies
his control circuits and safety mechanisms. The administrator asks himself
about the vulnerability of his establishments. The economist evaluates the
cost of prevention, of damage, the insurance markets and, far removed from
this world, the historian tries to put the present time and the experience of
the past in perspective which is a bit more deeply rooted; equally, the
sociologist undertakes to examine how a particular way of social functioning
can be the cause of the phenomena observed, giving the events their depth,
their social reality which the rapid readings that are the proper domain of
the exact sciences are incapable of demonstrating. The examining magistrate
must be mentioned here as well as the enquiry commission whose function it
is to unearth a bit more of the truth about what has happened. In these
events the thickest veils are always drawn over what might cast a shadow on

the image of a brand name, a corporation, vested interests, local or national pride. The social psychologist may in the same manner examine the various psychological mechanisms to which individuals or groups resort when suddenly plunged into a frightening situation on account of some information or event.

Why then take up the more global aspect outlined previously, a political aspect, when according to some a few good technical studies would take care of the issue perfectly. Why, as one also hears all the time that the problem of major risk is no business of the body politic inasmuch as there is quite simply no room for manoeuvre in terms of techno-economic development, and the less fuss made of these issues the better society will be served?

Two thoughts have nevertheless determined our decision. One is general: the incessant quest, particularly noticeable in the countries from which we have information, for power in matters of development policy. For about fifteen years a number of voices including some from the highest ranks in the state have been raised to wish, request, insist on a kind of development planning that would work on a very much enlarged basis: enlarged in space, time, choice of variables and ultimate goals — and be under closer control of citizens or their representatives.*

Can one imagine that this aspiration should have died by now, an aspiration which has induced the French parliament to vote unanimously for a law that provides for the implementation of and a certain publicity for studies of the impact of large development projects, which has also pledged us to change existing legislation on listed installations. Even if one wants to fight against it with all one's energy, that is still better than to turn away in horror and obscure the issue. This aspiration finds a perfect field of projection in the issue of major technological risk. What kind of risk? Risk for what purpose? Risk for whose benefit? Risk suffered mainly by whom? Risks started, chosen, legitimised by whom? How? Even if not all these questions are as yet acute in their explicit form nothing forbids, but rather the contrary, the asking of questions before we reach situations in which they might be experimentally blocked in our industrial societies because the events occur like lightning in matters of major risk, and its effects can be quick and formidable. Let us not forget: Clémenceau declared martial law in a disaster situation which, while serious, had nothing in common with present-day menaces. Can the body politic afford to remain indifferent to this?

Here we have a further thought which must induce us to maintain our analytical point of view. Maurice Strong, while Secretary General of the UN Environment Programme, gave a warning in August 1974 which must not be ignored: "Pollution will engender intolerable social and political tensions which will express themselves in outbreaks of violence and acts of desperate individuals and groups of the population" (13). What is true of pollution is true, to a much higher degree, of risk. Within hours there will be a rush, disarray, panic, an unmanageable situation on a regional level and consequently the danger of largescale uprisings, emergency legislation and all that may ensue from it. Can this be ignored? The body politic,

---

*For instance G. Pompidou: "In a field on which the everyday life of man depends directly, control by the citizens and their effective participation in the appointment of their framework of existence is required more than anywhere else ... A kind of environmental ethics must be created and spread which imposes some elementary rules without which it will become impossible to breathe in this world." (Speech made by the Président de la Republique in Chicago, February 28, 1970 (12, p. 25)).

challenged by technology and its application, in such brutal and pointed
fashion cannot lag decades behind the times and debate things in the manner
of sophists with nothing more to feed on than some evasive and grandiloquent
references to the competence of the experts. Political reflection has not
got the measure of present dangers, of the ignorance of our industrial
society of what significant breakdowns may mean or have in store for it from
one moment to the next, being more or less forewarned by alarms of sorts but
incapable of carrying the burden of enlightenment: so, it remains only to
play the game of hide and seek, with possible bitter surprises in the case of
serious events, if the mechanism of semblance comes to an end; radical
reactions on the part of citizens when, after a disaster, they find a scapegoat (which might be just that technology but equally science in general, the
form of government, labour ...) and make it pay for the fault of having won
their approval. Can this be ignored by the body politic?

The reasons for the reluctance to take an interest in this field of
investigation must be well understood, and we therefore come back to them.
Let us mention here only the following: the intuitive fear that there might
be no solution; the feeling that existing scientific education is not
sufficient to tackle such a problem; the confused idea that our mental
schemes, which are unrivalled when it comes to getting the maximum out of a
functional-partial logic, are invalidated when it becomes necessary to think
in terms of a totality and still there can be no question of returning to
'primitive thinking'. Part of the difficulty arises from the crucial
economic pressures attached to the argument and the resistance of all-
powerful organisations. To put this difficulty into perspective with the
urgency and seriousness of the dangers makes the rigorous and elaborate
investigation of our fundamental search still more necessary. Faced with
these realities, already existent or in suspense, how can a political
approach be construed? How can that fundamental quest of communities in
all times: the mastery of life and reproduction, still be exercised?

This is the line of our research. Before we outline the way the present
paper is constructed, which can only be an indicator in this vast research,
it will be useful to present the itinerary which has permitted us to reach
this platform.

4. ITINERARY

The study of great risk appears as a scientific discipline of rather broad
scope. The first stage of the paper was therefore devoted to gathering the
largest number possible of texts which have been circulating among
specialists, mainly from large organisations that work on energy issues.
These 'papers' abound. They deal with specific risks: radiation, certainly,
but also with asbestos, pesticides, genetic engineering, gases, dangerous
products of all sorts etc. They try to define the methodologies which would
permit identification of policies. They measure behaviour, psychological
attitudes. They study the reaction of the masses in cases of disaster. They
examine insurance problems etc. What has been missing most of the time was
what we are researching: an account of the fundamental choices when investment
decisions are made; a statement on the general circumstances that have lead
to such and such a disaster: the levers by means of which choice can still
be exercised; the margins of freedom that remain; the dangers that exist in
a more general way etc. All this needed to be brought up to date or rather
to be structured into an interpretation that would permit a clarification of
all that remained obscure because of disinterest or, in addition, on account

of interests to the contrary or on account of still deeper ineptitude. The
political approach demanded that one did not accept as satisfactory the
reasoning along the lines of: the calculations show what is 'acceptable';
there is no other solution; there has only been a banal incident ... Only
a series of rigorous confrontations between observers, texts, debates and
experiences permitted us to get beyond the commonly accepted evidence which
very quickly blocks all reflection of which the object is collective choice,
the ability of a society to conduct its development and its own transform-
ations in a clear and open manner.

There was a specially suitable way of tackling this task: the study of
cases that had been very precisely reported in well explained contexts.
Such study permits the most rigorous cross-referencing and the tightest
comparisons. This methodological programme had to make possible a reflection
born of a 'battle' with reality which never asserts itself from just the
evidence offered. This examination of multiple facets of reality could lead
to better founded questions; it could provide better safeguards against the
ease with which intellectual speculation, which avoids continuous confrontation with the concrete data, may be tempted to yield.

We have undertaken our research in several countries, and the organisational and cultural differences were always stimulating and a source of
very valuable investigation. Our attention was immediately attracted to
Great Britain, a country that has devoted itself to the study of industrial
risk for a decade and which has established a Committee of Enquiry for this
purpose presided over by Lord Robens (1970-1972) and which has recently given
itself legislation and apparatus conceived on completely new schemes of
thinking. The situation in Italy appeared equally interesting for study
because of its opposite scenario: nothing new has been forthcoming there for
rather a long time that would truly have helped in facing major risk. We
thus had two contrasting scenarios to examine and we have done this taking
into account in each case the law and regulations in force.

Rather than making a listing of reports it seemed more fruitful to us to
grasp the dynamics of the ongoing processes with regard to major risk in the
two countries. These were easily accessible in the case of England where the
arrangements in this field had been in full swing for some years. In this
case the thing to do was to look at the progress, the results, the slant
between objectives and results achieved, the pressures exercised by the
various social forces taking part in the debate. As for Italy, nothing of
the kind existed. Another way of approaching, in a dynamic fashion, the
social and institutional responses to the challenge presented by risk was to
examine, with a magnifying glass, what happened in crisis situations, in
great disasters. This is what we have done.

In Italy a disaster came up for analysis: Seveso. The disaster as it has
sometimes been called mainly because of the special nature of the dioxin,
its stability, which reduces the rescue services largely to impotence. In
Great Britain the great incident was Flixborough. In this disaster in 1974,
a factory was wiped out by the combustion of a gas cloud. These two cases
have been examined in depth, and in order to systematize the analysis a
series of other accidents were studied in each of the two countries. This
work has already permitted the sketching out of a possible scenario (which
will be submitted for criticism): the British one; and a scenario which must
be called unacceptable: the Italian one. France was studied, but for the
time being has been left to one side so that the commenced investigation and
the learning process should not be encumbered or even blocked by the evident
proximity to the reported data. In the same spirit we have approached the

nuclear issues with the greatest caution as they have become taboo in the
very sense of this term.

On this first basis we have broadened our reflection thanks to a series of
meetings both abroad and in France in order better to mark out the big stakes,
better to clarify the possible progress in this complex field of great
industrial risk. It was found useful always to observe a hierarchy of
concerns, the pecking order of information, the grid of analyses, the
disciplines called upon to throw light on the situation because the major
difficulty for this reflection is perhaps to be found at source, i.e. on the
level of proper thought organisation. If it were too strongly focused on a
specific experience, on a particular line of activity, it would drift and
mistake means for the end, restoring again a functional-partial logic where
the question is decidely global, directed at the whole and not only at its
parts. How many times has it been suggested to us to stick to a specific
question: D.D.T., transport ... and to avoid a multiple topic of enquiry!
This would certainly have been easier but would have made us bypass the
essential: the challenge of major risk presents itself to our industrial
society in a global fashion, in the actual state of science, the techniques
and their use.

Alternatively being too widely distributed to treat the differences between
the constituent parts of the whole with finesse our thinking would run the
risk of getting lost in hasty and vain generalisations that might be tempting
(in the guise of facile but hardly honest appearance) to have recourse to
excessive theorising, to a globalising discourse or too accommodating an
ideology. We have therefore worked on two fronts to try and avoid the snags
just mentioned. There was on the one hand the sharp confrontation of ideas,
issues, practices of others in this field: the proving of results submitted
to specialists in other fields but broken off at the examination of complex
social questions. This should permit a better focusing on the basic
hypothesis, the deduction, the grids of reading, the points of entry into the
analysis of the phenomena etc; in short: to get hold of a tool for general
reflection that would be operative and not blinding; to take a few steps
towards a theory and not just to theorise a practice from a few observations
which one may always be tempted to hold up as 'facts'.*

On the other front we have continued our work 'on site' with new case
studies: the aftermath of some disasters and particularly the study of risks
presented by large industrial agglomerations, Canvey Island specifically, a
high risk zone in the Thames estuary. In this case we have met the parties
concerned, studied the reports, the clashes in parliament between the local
MP and the successive governments since the start of the public discussion on
the 'acceptability' of the situation.

These efforts have permitted us to approach more easily the French problems
and the nuclear issues; there again a number of meetings and the study of the
accident at Three Mile Island were essential departure points.

---

*Let us remember a word from Goethe: "Every fact is already a theory" (2,
p. 85-86). Along the same line the Vocabulaire de la Philosophie precise
has: "Fact: That which is or which happens inasmuch as it is held to be a
real datum of experience on which thinking can depend". "The notion of fact
when defined comes to an affirmative judgment of exterior reality"
(Seignobos and Langlois, Introduction aux études historiques, 156). This
term has therefore essentially the character of a value judgment (14, p. 337).

Among other problems there remained the task of putting this information and this reflection in a historical perspective. We have therefore investigated the past in order to clarify the continuities and disruptions which exist in the matter of major risk and this again in order to avoid all complacency which hasty generalisations or plain common sense might suggest.

This itinerary is thus at the basis of the present attempt. We shall now present its general layout.

5. LAYOUT OF THE THESIS

1. Part One: the record, approached under the aspect of responsibility

As we have stressed, reflection and political action can only be exercised within the narrow reference of effective reality. It was therefore important, for the first part, to present the 'record' of technological risk.

In Chapter 1 we enter immediately into the core of the concrete difficulty: we shall present a series of cases of serious accidents that are well known in their outlines but much less in their essential structures and their complexity which in order to be understood require a large detour by means of scrupulous examination.

Chapter 2 responds to the necessity of placing this experience of major risk in a historical perspective. To raise the question of what is new today it is necessary to ask oneself about the continuities and disruptions in this field; it is also necessary to place industrial risk in the more general context of multiple risks with which our industrial societies are familiar. This has been done in three subchapters:

the first one examines the rise of technological risk during the period of the industrial revolution and until the second world war;

the second studies the continuities and the innovations observed during the post-war period in connection with large-scale industry;

the third investigates the threats which exist at present and about which the cases examined in the previous chapters represent so many warnings.

We have tried, for each period, as we have said, to locate the general context in which the problem of major technological risk presents itself. This effort leads, of course, to supplementary developments which, however, have seemed indispensable to us: reflections on technological risk have a very strong tendency to refer to undefined 'others' (natural disasters, wars, famine ...); it was necessary to reintegrate these references which are all too often used to beg the question.

This first part involves a number of cases for illustration. The barrenness of synthetic documentation on this matter has lead us not to neglect this initial contribution. These elements have been presented in a spirit which is rather different from that of other papers on which we were able to draw; it will hardly be a question of 'spinning out' the scenes of disasters in detail. What we are concerned with are the great lines which determine the action, the exercise of responsibility: how has such and such a disaster been analysed, foreseen, remedied?

This factual examination is therefore not a simple compilation of data but already a political reflection. What interests us, then, for example in the

case of the wreck of the *Titanic*, is not the highlife on board the liner or
the piercing cries of the passengers in the night ... but mainly the fact
that no notice had been taken of the series of warnings that had been
received. In the same way we should hardly have given our attention to the
deaths in the fire at the Opéra Comique in 1887 if there had not been a
debate in parliament thirteen days before the disaster and if the minister
had not on this occasion admitted the fact that there had to be a fire at
the Opéra (some time), that there would then be hundreds of victims and that
one could only hope that this event would not happen soon ... while he was
waiting for a response from his colleague at the Ministry of Finance. This
kind of behaviour, brought to light with examples from the past, could not be
ignored; because if reality has changed, and today's risks are no longer
those of yesteryear, the organisational, social and cultural attitudes
doubtless take longer to change.

2. Part Two: The technical management of major risk

The layout of this second part is easily explained. The industrial
societies have established a considerable number of institutional, administrative and legal measures to cope with — prevent, combat and remedy —
industrial risk. We shall examine these three points in Chapter 3 which
will also include a study of what has already brought about the recourse to
science and to advanced techniques in the field of safety.

Meanwhile, technical management of technological risk faces difficulties.
In Chapter 4 the limitations of this management will be studied in outlines;
this will be done in two parts since we shall see first a series of
difficulties which we consider to be only relative: adjustments will permit
them to be removed; then there are some much more serious limitations, some
of them quasi-absolute, about which earnest questions are asked of the body
politic.

3. Part Three: Social regulations concerning major risk

If the utensils are insufficient, are they at least being used? Do sound
social regulations permit the mitigation of the difficulties encountered on
the level of management? Chapter 5 examines by turns the practices of the
three main 'actors' involved: the operator, the authorities, the citizen.
The serious imperfections brought to light by these examinations are, however, not yet a complete account.

In Chapter 6 we shall go beyond an analysis that has been centred around
a specific 'actor'. We shall examine the network of 'actors' who shape
'situations' that seem also pregnant with a disagreeable aftertaste in the
matter of major risk.

4. Part Four: Politics

Throughout this factual, technical and social examination the more
immediately political reflection which must bear upon two interrelated
aspects will be articulated: the practice of rationality, which is necessary
for the conduct of a coherent action, and the practice of democracy, which
is necessary at least up to a point, as even the most reticent will admit,
so that the projects can take shape, live and let a community live. This
political reflection is the object of the fourth and last part of this thesis.
What are the interactions between risk, politics and the process of development? We have devoted three chapters to their study.

The political approach to risk. In Chapter 7 we examine the scenario
within which the body politic, anxious to maintain existing practices and

projects, tries to brush aside the difficulties presented by major risk. The latter is referred to a technical commission. The experts will find the answers, good ones obviously, and the citizen will have to accept, or more or less tolerate, the choices made.

A second scenario will be studied in Chapter 8. Here the body politic agrees to take major technological risk into account. On the one hand, on the level of rationality, it agrees to change the rules of action rather substantially; on the other hand, on the level of democracy, it agrees to open up the decision making process considerably.

We thus have two types of response, but they remain just immediate ones. Major risk brings up a more serious question. Major technological risk subverts the body politic: it calls into question reason and democracy at the same time and also the relationship between reason and democracy, the social power relationships which have developed on the basis of a certain link between reason and democracy. It will be seen that this raises deeper questions than those examined in the two preceding chapters: they will be dealt with at last in Chapter 9.

We shall therefore see, in a word, the Western World shaken in its very foundations. There remains: not to despair; not to sing the praises of the upright, not to brush the question aside or leave it to others but to undertake the real task: to examine the elements of reflection which must be brought together in order to face the unprecedented.

REFERENCES

(1) D. COOPER-RICHET, Drame à la mine. *Le Monde*, 25-26 novembre 1979.

(2) K. JASPERS, *La bombe atomique et l'avenir de l'homme*. Buchet-Chastel, Paris, 1963 (708 pages) (Die Atombombe und die Zukunft des Menschen, Piper, München, 1958).

(3) Advisory Committee on Major Hazards: First Report Health and Safety Commission, H.M.S.L., London 1976.

(4) Proposition de directive du conseil sur les risques d'accidents majeurs de certaines activités industrielles, présentée au Conseil le 19 juillet 1979 (no C 212/4 à 14). *Journal Officiel des Communautés Européennes* du 24 juillet 1979.

(5) F. FAGNANI et J. F. BELHOSTE, Le risque: essai de mise en perspective. *Futuribles*, no 28, novembre 1977, pp. 107-125.

(6) J. J. SALOMON, De Lisbonne (1755) à Harrisburg (1979). *Futuribles*, no 28, novembre 1977, pp. 5-10.

(7) J. J. ROUSSEAU, Lettre no 300, 18 août 1756. *Correspondance générale*. vol. II, Ed. Dufour-Colin, 1924, p. 306.

(8) *World Map of Natural Hazards*. Münchener Rückversicherungs Gesellschaft, 1978.

(9) *Le Monde*, 11 août 1979.

(10) J. GRAYSON et Ch. GODARD, Industrial safety and the trade union
     movement. Studies for Trade Unionists, vol. 1, no 4.

(11) S. HARGOUS, Les Indiens du Canada sont contaminés par le mercure.
     Le Monde, 25-26 novembre 1979.

(12) Robert POUJADE (Ministre de l'Environnement), Lettre du 10 février 1971.
     Nuisances et Environnement, février 1971, p. 25.

(13) C. M. VADROT, Mort de la Méditerranée. *Le Seuil,* Paris, 1977, p. 213.

(14) A. LALANDE, *Vocabulaire technique et critique de la philosophie.*
     P.U.F., Paris, 1976.

PART ONE

A change in the scale
and the nature of major risk

CHAPTER ONE

A Series of Grave Warnings

1. 1st June 1974: Flixborough
2. 10th July 1976: Seveso
3. 18th March 1978: The *Amoco-Cadiz*
4. 16th March 1979: Three Mile Island
5. 10th November 1979: Toronto

"Despite the terror it has experienced the world wants,
this time again, to feel reassured up to the moment when
it becomes evident that the extent and the nature of the
means employed ... have gone beyond all expectations".

Karl Jaspers
The Atom Bomb and the Future of Man

"Pangloss explained to him how everything was at its (for
the) best. Jack was not of this opinion. All this was
necessary, explained the shady (?) doctor and the
particular misfortunes contribute to the common weal inas-
much as the more particular misfortunes there are the more
all is well."
While he was arguing the sky darkened, the wind blew from
all four quarters of the world, and the ship was assailed
by the most horrible storm ...
Half the passengers who were already weakened ... did not
even have the strength to worry about the danger. The
other half let off shrieks and prayed; the sails were torn,
the masts broken, the ship filled with water. Trying as
best they could, nobody understood anybody else, nobody
gave orders."

Voltaire
Candide or Optimism

I. SATURDAY, 1st JUNE 1974: FLIXBOROUGH

1. A FACTORY DESTROYED, 2,450 HOUSES DAMAGED

On Saturday at 4.53 pm the chemical works of Nypro Ltd., at Flixborough, a
small rural community 260 km north of London, was almost completely erased
by an explosion which made one think of an act of war as the Commission of
Enquiry later had to report (1, p. 1).

The explosion had been caused by the ignition of a cloud of 40-50 tonnes of
cyclohexane, a highly inflammable product that dilutes greatly in the air and
in the temperature surrounding a hot spot. A pipe had leaked, letting gas

escape at a temperature of 155°C and a pressure of 8.8 bar. Within thirty
seconds a cloud two hundred metres in diameter and 100 metres in height
formed; driven by the wind at 25 km/h it ignited when it came into contact
with the discharge tower of the hydrogen unit 100 metres from the point where
the leak had occurred. An explosion ensued: it was heard at a distance of
50 km; it devastated the 24 hectares of factory site. Even though comparisons
of the effect of the explosion are difficult to establish, some have spoken
of the equivalent of 16 (±2) tonnes of TNT. The fire raged with flames
rising up to 70-100 metres.

Among the seventy two people present at the site twenty-eight died (of whom
nineteen were in the control room), thirty-six others were injured. Outside
the factory fifty-three injured were counted; hundreds more suffered minor
injuries which were not officially registered.

The material damages estimated at tens of millions of dollars, more than
180 million dollars for the reconstruction of the factory alone, covered a
vast area. All buildings within a radius of 600 metres were destroyed and
more than 2,450 houses were damaged in the vicinity. Windows were shattered
within a radius of 13 km. The instant increase of air pressure at the
epicentre was more than 2 bar. All fixed fire extinguishing installations
were immediately destroyed. It took two and a half days to get to the
principal sources of the fire (2-5). Some people, after having visited the
devastated site, compared the disaster to what might have been caused by a
minor atom bomb (6).

If this comparison is exaggerated it shows at least the shock which the
disaster caused among the British population. The chemical industry, this
was a factory which produced caprolactam, an interim product in the manufacture of nylon, showed itself capable of endangering very seriously its
workers and, a newly discovered fact, the safety of the local population. On
the available evidence the drama would have been of a different magnitude if
it had happened on a working day: the factory usually employed five hundred
and fifty people. No exact count was taken but everybody realised that one
had come very close to a disaster of enormous scale, one that could not be
compared with anything that had been known in this branch of industry until
then. In fact, the administration buildings, the technical offices, the
control room, laboratories and maintenance workshops had been completely
destroyed.

Some further thoughts occurred. The factory was located at a distance from
urban centres, set among agricultural fields. The two nearest villages were
Flixborough and Amscott, both about 800 metres away; at 3.5 km distance there
was Burton, at 5 km the urban centre of Scunthorpe. A few figures will show
what the explosion could have caused if the factory had been differently
located: 72 houses out of the 73 at Flixborough, 73 out of 79 at Amscott,
644 out of the 756 at Burton were damaged to various degrees. In other words:
the deflagration had affected 90 per cent of the buildings within a 3.5 km
radius and remained very dangerous up to a distance of 5 km and even beyond
(the projection of a large piece of equipment found at 6 km from the factory
is a further proof of this). (7)

The potential risks connected with the operations of the chemical industry
had thus been measured. Henceforth one had to take into account extramural
deflagration of gas clouds which could annihilate urban centres since not all
factories are located in the countryside like Nypro at Flixborough.

It was clear to everybody that this type of event did not permit any

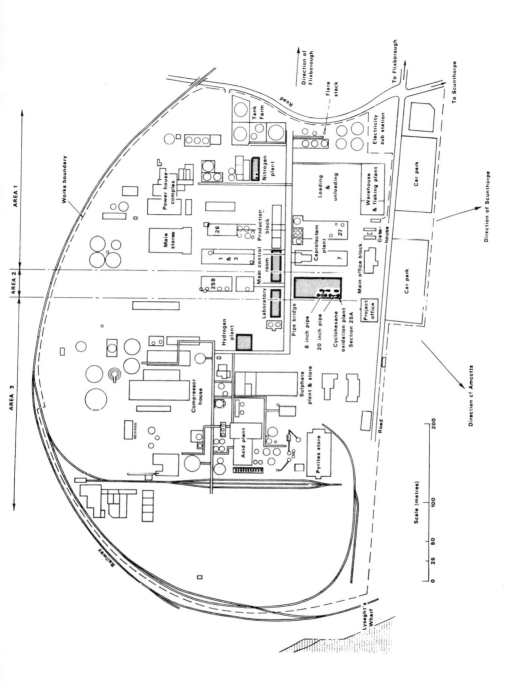

Fig. 1: Simplified plan of the Nypro factory.
(Source: 1, annex)

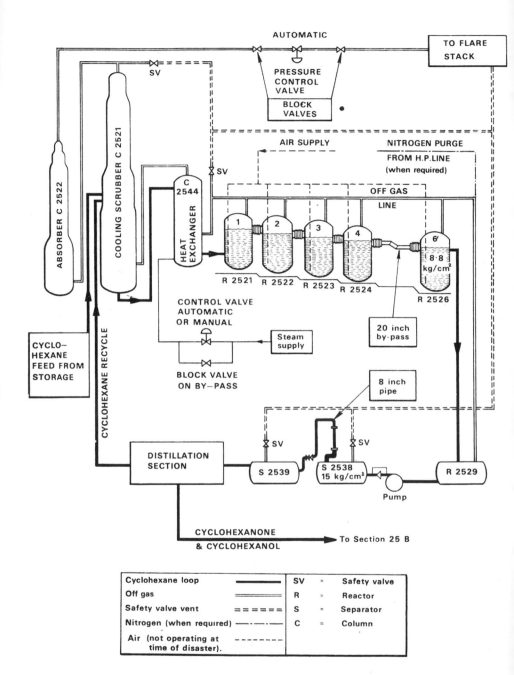

Fig. 2: Installation for the oxydation of cyclohexane. Section 25A.
Simplified diagram (not drawn to scale).
(Source: 1, annex)

countermeasure to be taken between the moment of alert and the start of the drama as was strongly pointed out in a first report two years later by a working group set up by the British government (8).

In one sentence the local Member of Parliament told the House of Commons the whole chagrin of the inhabitants: "My fellow citizens can now sleep in peace. The factory is destroyed, the harm is done." (9)

2. THE ANALYSES OF THE COMMISSION OF ENQUIRY

The public authorities acted rather quickly after the drama. On June 27, 1974 the Secretary of State for Employment, Michael Foot, ordered two enquiries. One, carried out by a Commission of Enquiry, was to examine the causes of the accident in order to establish the responsibilities and to determine guilt. The other was to be carried out by a committee of experts who would advise the government on what measures to take to get the workings of chemical factories which might cause other 'Flixboroughs' under control. Set up in 1975, this committee (consultative committee on major risk) had to get to grips with work of a broader scope. In order to study the case of Flixborough one has therefore to direct oneself to the report by the Commission of Enquiry which, even though it is sometimes rather limited,* appears instructive in many respects.

1st: The company and its factory**

a) The Nypro company and the Flixborough factory. Initially the factory was a subsidiary of Fisons Ltd; it had been established in 1938 for the manufacture of fertilizer. In 1964 it passed to the Nypro company which had been formed with the participation of Dutch State Mines (DSM) with a view to the manufacture of caprolactam, an intermediary product in the manufacture of nylon.

In 1967 Nypro was reorganised with the following participants: DSM (45%), British National Coal Board (45%), Fisons Ltd. (10%).

In August 1967 a first unit of 20,000 tonnes/year of caprolactam made from phenol was put on stream.

In 1972 the caprolactam capacity was increased to 70,000 tonnes/year by the addition of a new unit which employs a process based on cyclohexane. In 1974 the Nypro company was the sole manufacturer of caprolactam in Great Britain.

b) The installation involved in the explosion. The installation concerned (25A on the map) is a unit for the oxidation of cyclohexane by the use of air and includes six reactors in sequence, each unit having a capacity of 45 m^3 and made of mild steel (13mm) with rustproof plating (3mm) internally. The safety valves being calibrated at 11 bar.

The reactors are equipped with a central stirring rod. The oxidizing reaction of the cyclohexane is accomplished with a catalyst at 155°C under

---

*The Commission tried in fact to deal only with those elements which explain directly the sequence of events in the accident. Besides, this report has evidently one acute concern: not to blame Nypro or the chemical industry in general. (We shall come back to this point later on.)

**According to the account given by E. Bachmann (5).

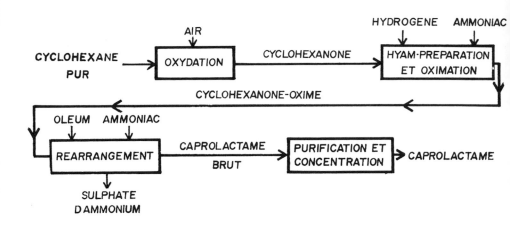

Fig. 3:   The manufacture of caprolactam.
(Source: 5, p. 2)

8.8 bars of pressure by means of air injection with the help of a perforated
gradient. Each reactor contains 25 m^3 of liquid. The throughput circulating
from one reactor to the next through piping systems of 28inch diameter is
250-300 m^{3}h.

2nd: The sequence of events leading up to the accident (January-June 1974).

a) The absence of a competent mechanical engineer. At the beginning of 1974
the maintenance engineer left the factory for personal reasons, and by June
1974 the company had not yet found a replacement. None of the other
engineers, even though they were graduates, had special competence in
mechanics.

The duties of the maintenance engineer, especially coordination, were given
provisionally to a subordinate (a foreman of sorts who had a technician's
diploma and who had completed his training). This technician had spent ten
years of his career in the public electricity supply service and four years
in maintenance. His qualification was insufficient for the job given to him
temporarily and also insufficient for the detection of certain design
anomalies in connection with important modifications of the equipment.

These notes taken from E. Bachmann (5, p. 1) will later be completed by
observations made by the Commission. They suffice here to locate the
incident which occurred on March 25, 1974 when the following organisation
chart is added:

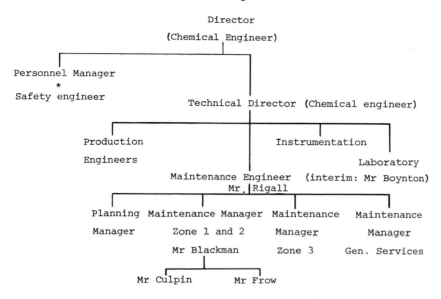

Fig. 4: Organisation Chart

b) The discovery of an escape of cyclohexane at reactor No. 5 on March 27, 1974. On the morning of March 27 it was discovered that cyclohexane was escaping from reactor No. 5. Investigation showed a vertical crack in the outer casing of the reactor; a small quantity of cyclohexane escaped from this crack; this indicated that the internal casing was also defective. The production engineer on duty telephoned the director for zone 2 and they agreed that the installation would have to be closed down, depressurised and cooled while a complete inspection was to take place**(Par. 33).

c) The desire to restart production as soon as possible (March 28). The following morning, March 28, the director inspected the crack and found that it was about two metres long. This indicated a serious situation and the morning was spent deciding what had to be done (Par. 54).

During this meeting it was decided that reactor No. 5 would be closed down for inspection; that it would be possible to continue oxidation with the remaining 5 reactors; that a by-pass had to be built to link reactors No. 4 and 6 and that when this by-pass had been put in place the factory would go back on stream (Par. 55).

So, a by-pass was to be installed between reactors No. 4 and 6, more precisely: between their 28inch diameter expansion fans; but since the

---

*Organisational link assumed.

**The lines in italics are extracts from the report of the Commission of Enquiry; the respective paragraphs are given in brackets.

factory had only 20inch pipes the by-pass had to be fitted to each fan with
the interposition of a plate and strap. This was not the only anomaly: In
fact, the whole of the 'repair work' must be considered: it was very poor
patchwork which nobody had thought about thoroughly.

It seemed clear to us, wrote the Commission of Enquiry, that:

"Nobody in the meeting — apart from Mr Blackman — was quite aware of the
problem presented by the restarting of operations; without establishing the
cause of the crack in reactor No. 5, without disassembling and inspecting the
five other reactors to find out whether one of those might not have the same
defects even though they had not yet developed to the point where they would
cause an escape.*

*Nobody seems to have considered that the link between reactor No. 5 and 6
implied a major technical problem or that it was more than a routine plumbing
job, nor were possible design problems and alternatives discussed.*

*Even the fact that the access and the exit of the by-pass were at different
levels was not brought up in the meeting.*

*The main point at the meeting was to restart the oxydation process with a
minimum of delay (Par. 56).*

*d) Haste and incompetence are stressed by the Commission. We entirely
absolve all persons of the blame that their desire to restart production could
have led them knowingly**to go on stream with a dangerous process without
paying attention to the safety of those who were to work on it.*

*We have, however, no doubt that it was indeed this desire which led them
to neglect the fact that it was potentially dangerous to restart production
without having examined the remaining reactors and without having determined
the cause of the crack in the fifth reactor. Equally we have no doubt that
the error of judgment concerning the problem presented by the linking of
reactor No. 4 to No. 6 was largely due to this same desire.*

*In the case of Mr Blackman we feel that two supplementary factors came into
play which made him neglect the difficulty of linking reactors Nos 4 and 6.
In the first place he had the worry about the cause of the crack in reactor
No. 5; secondly there was the new and difficult problem of disconnecting
reactor No. 5. These were his main preoccupations which led him not to think
of taking the appropriate measures concerning the construction, the tests,
the installation and the elements of support for the assembly of the link.
(Par. 57).*

*If there had been a suitably qualified engineer present at that moment,
with sufficient status and the authority to impose his views, he would in our
opinion have insisted that there was to be no restart before the other
reactors had been completely inspected and the cause of the crack in reactor
No. 5 determined ... (Par. 58).*

---

*The engineer attached to the maintenance engineer, Mr Rigall.

**'Knowingly' it is in fact lucky that the irresponsibility had not been
consciously planned.

No doubt, the fact that the error had been committed to attend only to the operations being carried out did not cause or contribute directly to the disaster. Indirectly, however, we feel it played a part. If it had been decided to dismantle and inspect the other reactors and to wait for a report on the causes of the crack in reactor No. 5*the factory would have had to stay out of operation for several days.

The design and the construction of the whole link should not have been carried out in a hurry as it was in this case. There should have been time to consider what problems would arise and how they could be suitably dealt with. On account of this fact we do not feel certain that at least some of the problems would not have been identified and measures taken that would have prevented the disaster. (Par. 59).

e) *The implementation of the decisions of March 28: incompetence and no respect for standards.* The repair work was done and on April 1, 1974, without those responsible having tried to understand the cause of the fissure in reactor No. 5 or to make sure that the other reactors were in good order the installation was put back on stream. And yet, as the report on the inquiry indicates, it could only cause concern:

Nobody gave thought to the fact that the whole installation, once pressurised, would be subject to a torque reaction that would shear the fans which had not been designed for this.

Nor did anybody take into account the fact that the strong hydraulic pressure on the fans (some 38 tonnes at working pressure) would tend to buckle the by-pass at the joints.

No calculation was made to check whether the fans or the piping could take the load.

No reference was made to the respective rules of British Standards or to other applicable standards.

No reference was made to the guidelines for the user published by the manufacturer of the fans.

No piping layout was made beyond a chalk drawing on the ground.

No pressure test, either of the piping or of the whole unit, was made before it was fixed on ... (Par. 62).

The result was a unit the stressing of which was unknown and did not correspond to the requirements of British Standards or to those of the guidelines ... It is certain that if the engineers of Nypro had read the guidelines of the designer they would have seen that the whole pipe and fan assembly was not safe (Par. 63).

No pillar or other means was used either to support the piping from underneath or to prevent lateral movement. The four pillars which had been put

---

*The later enquiry attributed this crack to a fissuring corrosion caused by nitrates carried in the water with which in the past small escapes of cyclohexane had been sprinkled. This water had penetrated into the insulation and when it evaporated had deposited nitrates on the steel of the apparatus (5, p. 6).

*up were designed mainly as a support during the assembly work in order to
prevent the weight of the assembly from pulling on the fans. For that
purpose they probably were adequate. They were totally inadequate in
operating conditions. This was not surprising because since no attention had
been paid to the supports in operating conditions, apart from Mr Blackman who
at the time of designing gave his assistant a sketch of the supports ... .
These were, however, not put up, and Mr Blackman took no steps to insist on
the installation of these supports. (Par. 68).*

By April 1 the system was set up, after trial and necessary modifications,
to eliminate an escape (Par. 69). But the Commission adds:

*There had been neither planning nor control of the design, of the construc-
tion, of the trial or the adjustment of the unit; there was no checking
either on the way the work had been carried out (Par. 17).*

*f) From April 1 to May 29, 1974: the repair holds up. A unit had been
installed the design of which had not been calculated, which neither
conformed to the rules of British Standards nor to the recommendations of the
manufacturer of the fan which latter was subject to rotary movement when
under pressure; which was not, in aggregate, held in place from above and
inadequately secured from below. As a result the fans were subjected to
forces for which they were not designed (Par. 72).*

*This assembly, even though it had been tested pneumatically at 9 kg/cm^3,
had not been tested at safety pressure i.e. the pressure at the valve, i.e.
11 kg/cm^3 (Par. 72). Such a test would almost certainly have caused the
rupture of the pipe-fan assembly, and the disaster would have been avoided.
The tests that were made were not tests of the strength of the assembly but
tests for leakage. (Par. 73).*

*One or several of the reactors which were still in use while not leaking
yet or not sufficiently for any leakage to be detected could already have
had substantial fissures ... Such fissures could have spread and caused
a serious rupture of the boiler ... (Par. 72).*

Until May 29 the system must have functioned normally:

*The assembly did not cause any problem. It was never closely checked but
was looked at in passing on many occasions by a large number of witnesses
... (Par. 74).*

*g) May 29 - June 1: difficulties. The four days preceding the disaster were
full of difficulties. On Wednesday, May 29, a leakage was discovered which
forced a stoppage. The process was restarted in the early hours of Saturday
after repairs and escape tests. At 4 am a new leakage occurred; others were
discovered; the process was stopped.*

*Subsequently it was found that these leakages had righted themselves, and
at about 5 am operations were restarted ... (Par. 78).*

Shortly afterwards the process was stopped again because of a leakage.
Repairs could not be carried out immediately because the necessary special
tools were not available. The Commission remarks:

*A delay in order to obtain the necessary tools might have led to the
development of a favourable situation i.e. one without danger in serious
conditions which contained risks (Par. 79).*

The process was restarted at 7 am on Saturday morning; it lasted till 3 pm. Difficulties arose again as regards temperatures and pressures (pressures which were disquieting without being outright alarming); there was not enough nitrogen (Par. 82), a substance which was, however, essential for the safety of the process (Par. 211). One cannot establish exactly what happened during the final process because the explosion killed everybody in the control room and destroyed all instruments.

h) 4.53 pm: the disaster. Despite observations made by certain experts*the Commission upholds the hypothesis according to which at 4.53 pm the provisional 20inch pipe broke. The two fans broke, the piping whipped and broke free and fell to the ground. Through two 28inch openings (exits from reactors 4 and 6) hot cyclohexane escaped under pressure in massive quantities. Between 25 and 35 seconds later the combustion occurred, followed by fire.

3. Beyond the actual event, an unpleasant context from the point of view of safety

a) Deficient organisation. Let us sum up some of the observations made by the Commission of Enquiry:

The maintenance engineer left the company at the beginning of the year (1974) and had not yet been replaced at the time of the disaster (Par. 19).

There was a reorganisation of the process going on that was to become fully effective on July 1, 1974 (Par. 19).

There was at least a safety engineer whose precise position in the organis- ation chart seemed a bit uncertain who, however, considered himself responsible to the personnel director even though he had the right of direct access to the director general (Par. 23).

Even though steps had been taken to replace the maintenance engineer the position had not been filled; a coordinating function was exercised, however, by Mr Boynton ... In our opinion he was not qualified to act as a coordinator of the engineering department at an installation such as Flixborough, and the exercise of this function should not have been demanded of him, not even for a short period (Par. 24).

For his functions in zone 2 (where section 25A was located) Mr Blackman (the engineer attached to the maintenance engineer) had under him as assistant engineers Mr Culpin and Mr C. G. Frow, the supervisor of section 25A and three other sections. None of them was professionally qualified as a mechanical engineer even though all of them had some technical qualifi- cations and a certain technical experience. Mr Blackman in particular is, in our opinion, a reliable and devoted man of practical sense even though as it appeared later had been subjected to an excessive work load which led him into error (Par. 26).

The engineering section had a weak structure as the company recognised. For this reason the company called on Mr J. F. Hughes of the National Coal Board in 1974 to obtain advice on reorganisation. Subsequently the situation

---

*According to whom the final cause of the disaster has to be looked for in the rupture of an 8inch diameter pipe that was involved and secondarily of a 20inch diameter pipe.

*got worse with the departure of Mr Rigall. From that time on there was no
qualified mechanical engineer with a status of sufficient authority to deal
with complex or novel engineering problems or to demand that necessary
measures were taken.*

*This was also recognised by Nypro because as from the departure of
Mr Rigall, Mr Boynton and the other engineers were told that if there were
problems they could call on Mr Hughes for assistance. Mr Hughes was on the
site only sporadically but it was possible to communicate with him even
though this could entail some delay because he had quite a few other important responsibilities ...*

*This weakness of the engineering section was made so much more serious
since the director and the technical director were both chemical engineers
without any training or qualification in mechanics (Par. 27).*

b. Serious infringements in the field of stocking dangerous materials.   The
enquiry report is precise (Par. 194):

On June 1, 1974 Nypro stocked: 330,000 gallons of cyclohexane, 66,000
gallons of naphta, 11,000 gallons of methyl benzene, 26,400 gallons of
benzene, 450 gallons of gasoline. The stocking of these potentially
dangerous substances is explicitly put under the control of the local
authority which is in charge of issuing the licence provided for in the
Petroleum (Consolidation) Act of 1928.

In fact, the only licences that had been issued authorised: 7,000 gallons
of naphta, 1,500 gallons of gasoline.

3. MORE GENERAL SOCIO-ECONOMIC INVESTIGATIONS BEYOND THE ANALYSES
   BY THE COMMISSION OF ENQUIRY

The following remarks, made by non-official obervers, can be assigned to
the dossier as a complement to the enquiry report.

1. The nonexistence of public control

Nowhere in the court report is the role of the Factory Inspectorate, the
public administration responsible for industrial safety, discussed or even
mentioned. Should the Inspectorate have been alerted to the matter of the
temporary piping? What did it make of the fact, which was admitted by the
Court of Enquiry, that Nypro stocked more than 400,000 gallons of dangerous
products while it had a licence for only 7,000 gallons? (7, p. 5).

2. The economic difficulties of the industrial group

Why, at the meeting of March 28, did those responsible and the engineers
of Nypro rush into a job without giving much consideration to safety? Were
they a wretched lot of people? A small number of economic facts may help us
to understand. The caprolactam factory of Nypro was programmed for a
production of 70,000 tonnes p.a. In reality it produced only 47,000 tonnes
p.a. at the time of the accident. Dutch State Mines as well as the National
Coal Board lost money in the operation. They had requested the government's
Price Commission to authorise a 48 per cent increase in the price of
caprolactam. This authorisation was refused. In other words, Nypro was
subject to serious economic and commercial pressure. This surely explains
the undue and risk-laden haste on March 28 (7, p. 51).

3. Strong competition

An even more important question: why was the factory built for this particular technological process? It was in fact a supplier of caprolactam to two important fibre manufacturers, Courtauld and British Enkalon. These were in direct competition with the other big nylon manufacturers, ICI and Dupont. These two latter companies held patents (on a process for the manufacture of caprolactam) which most of the experts considered safer than the one used at Flixborough. Once again economic competition forced the construction of a dangerous factory. The same commercial pressure was in force when nothing better than a dangerous repair job was done (7, p. 5).

4. CONCLUSION: THE WARNING SHOT OF FLIXBOROUGH

The explosion on June 1, 1974 strongly shook the British population. It caused consternation: within thirty seconds a whole area could be devastated by an accident at a chemical factory. One could no longer ignore the danger, knowing full well that one had not been altogether unlucky since the 'demonstration' took place in open country.

The incident of Flixborough, even if it was not different from explosions of gas clouds that had occurred elsewhere in the world, was unique in that it provided, for the industry and for the British rescue services, the first direct experience of the consequences of such an event. In a perfectly clear manner it demonstrated the necessity to postulate the possibility of a massive escape of gas and the formation of clouds from containers holding inflammable liquids kept under pressure and at temperatures above their boiling point (10, p. 217).

Flixborough: the factory after the disaster.

REFERENCES

(1) The Flixborough Disaster. Department of Employment. Report of the
Court of Inquiry. Her Majesty's Stationery Office (H.M.S.O.), London
1975 (56 pages).

(2) P. TANGUY et J. F. GUYONNET, La prévision rationnelle des grands risques
Le Progrès Technique, A.N.R.T. no 11-12, décembre 1978, pp. 33-40.

(3) La catastrophe de Flixborough. Produits Chimiques Ugine Kuhlmann
(P.C.U.K.), note interne, 9 janvier 1976, (10 pages).

(4) Les leçons de la catastrophe de Flixborough. Rhône-Poulenc, note interne,
12 janvier 1976, (8 pages).

(5) E. BACHMAN, Les leçons de la catastrophe de Flixborough. Rhône-Poulenc,
(9 pages).

(6) H. D. TAYLOR, Flixborough: the implication for management. A Keith
Shipton Developments Special Study, June 1975.

(7) L. MC GUINTY, Contribution pour une séance de travail organisée par le
Laboratoire d'Econométrie de l'Ecole Polytechnique, mai 1978, (11 pages).

(8) Advisory Committee on Major Hazards: First report Health and Safety
Commission, H.M.S.O., London 1976.

(9) J. ELLIS, député de Brigg et Scunthorpe, Déclaration à la Chambre des
Communes House of Commons, Official Report, Parliamentary debates,
Tuesday 18 June 1974, Vol. 875, no 5, col. 255.

(10) W. M. DIGGLE, Major emergencies in petrochemicals complex: planning
action by emergency services. Imperial Chemical Industries Ltd (I.C.I.).
I. Chem. E. Symposium, series no 47, pp. 217-223.

II. SATURDAY JULY 10, 1976: SEVESO

On that day, when production had been finished and most of the workers of
the Icmesa*factory at Meda (in Lombardy, Italy) already enjoying their weekend, the temperature of a reactor which had been left in a cooling phase
suddenly rose for unknown reasons; the 'safety' disc loosened and permitted
a reddish cloud to escape into the atmosphere. Children noticed it; the
cloud disappeared. The scene occurred in the northern outskirts of Milan,
18 km from the economic capital of Italy**.

Later it will be shown that in fact tetrachlorodibenzodioxine (TCDD) or
'dioxin' had thus spread around in unknown quantities. This is one of the
most violent, most dangerous, most difficult to combat poisons which human
intelligence has ever succeeded in manufacturing.

In the case previously examined, Flixborough, the disaster occurred within
thirty seconds, and the destruction of the factory, the seventy-metre high
flames, the smashed roofs of houses left no doubt about the reality of the
disaster. Here, everything was different: the drama took on a different
appearance. The cloud dissipated, everything was normal in appearance. But
uncertainty was omnipresent, fear took the place of stupor. Perhaps death
was there, very near, ready to strike or to install itself for years to
consummate its victory. This was the spectre of dioxin.

1. DIOXIN, A FORMIDABLE POISON

The aerosol***pushed into the atmosphere on July 10, 1976 came from a
factory load of trichlorophenol (TCF). This substance was sold to Givaudan
for the synthesis of hexachlorophene****.

Hexachlorophene is among a number of chlorinated derivates of which
increasing use has been made over the last twenty five years for civilian
and military purposes. These products enter into the manufacture of insecticides, herbicides, bactericides etc. Especially well known is the
2-4-5T used in forestry for clearing undergrowth, the hexachlorophene used
for disinfection in hospitals (see Figs. 8 and 9 infra). The chlorinated

---

*The Icmesa factory, situated at Meda (a community adjoining Seveso which
was more severely struck by the accident and which on account of this gave
its name to the accident) belongs to the Swiss company Givaudan, itself a
subsidiary of the Swiss Hoffmann-la-Roche group which is well known as one
of the world's pharmaceutical leaders. It is known that Givaudan had
recently been at the centre of the hexachlorophene affair (still recalled
as the "Morhange talc" affair).

**See Figs. 5, 6, 7, p. 37). There was apparently nothing extraordinary
going on, given the constant pollution one is used to in the area. However, this was an extremely serious accident.

***The cloud consisted of polyethylene-glycol, of di-ethylene, of soda and
trichlorophenol, containing a lot of dioxin (9).

****And only for that as the industrialist (10) who was suspected of having
manufactured it for military uses stressed.

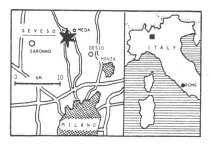

Fig. 5:   Seveso, 20 km from Milan
(Source: 7, p. 162)

Fig. 6:   The communities concerned
(Source: 7, p. 175)

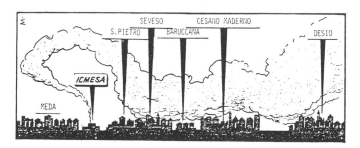

Fig. 7:   Exhibit of the possible drift of the cloud
(Source: 7, p. 162)

derivates, because of their antivegetative properties, are also used as
additives in numerous products: varnishes, paints, inks, textile fibres ...
Dioxin is still present in trichlorophenol in quantities that vary according
to the use intended for the product: extremely weak if it is destined for
medical use (less than 0.1 ppm = parts per million), less weak for other
applications, especially when they are military; this was the case with the
'orange agent' widely used in Vietnam (the concentration of dioxin could go
as high as 50 ppm, i.e. i.e. up to 500 times higher than the upper mark
permitted for herbicides in France — decree of July 29, 1975). (3;4, p. 3;
5, p. 40); 6, p. 11; 7, p. 1).

Dioxin can form up in large quantities if there is an accidental increase
of temperature and pressure in the reactor in the course of production of
trichlorophenol. This is precisely what happened at the Icmesa factory on
July 10, 1976.

1st: Sharp toxicity, astonishing stability

Dioxin is a very toxic substance and is generally compared with products
already considered extremely dangerous in order to explain the degree of
toxicity of TCDD: One of the most violent poisons, 500 times more toxic than
strychnin, 10,000 times more than cyanide (8). The DL (lethal dose 50) i.e.
that which kills 50 per cent of experimental animals, is 5 micrograms/kg for
the rabbit and 29 micrograms/kg for the rat (9).

This product has remarkable stability; it can be eliminated only in
negligible quantities; it accumulates in the liver, in nerve and fatty
tissues. The toxic dose may be absorbed all at one time or fractionally in
repeated doses. For primates no experiments have been attempted, given the
toxicity of the product.

2nd: Probably very grave but yet little known deferred effects

Pathology presents mainly the following features (12):

- Chlorine acne i.e. inflammation of the sebaceous glands of the skin
  (cysts, boils),
- Changes in the liver, renal, thyroid, pancreatic functions; gastro-
  intestinal lesions;
- Reduction of libido and sexual potency;
- Changes in the central nervous system (which express themselves in
  deficiency of memory, degradation of social relations, sleep problems,
  emotional instability ...).

This pathology gives rise to fear of the following effects:

- Immunosuppressive effect i.e. reduced resistance to infectious diseases;
- Teratogenic effect; TCDD is a powerful embryotoxin for all types of
  animals (including horses). It has therefore the property of causing
  teratogenic effects: death of the embryo or change in the foetal development with malformation at birth from very small doses. The toxic effect
  on the foetus of the rat manifests itself also if the mother has been
  intoxicated during the last phase of pregnancy; it is also transmitted
  by sucking. The distinctness of these experiences is such that it
  appears extremely improbable that such phenomena would not also be
  produced in human beings (12, p. 133).
- Mutagenic effect: it can carry delayed cancer or even hereditary malformation. This effect has not been proved. However, as S. Zedda writes
  "the first lessons from Vietnam and certain experimental hypotheses
  constitute more than (just) a warning (12, p. 34)".

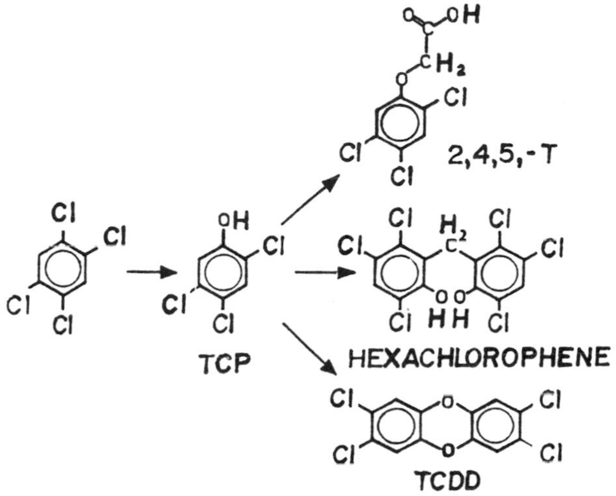

Fig. 8: Hexachlorophene and '2, 4, 5T' are prepared from the same composite: trichlorophenol. In the synthesis of this composite from tetra-
chloro-benzene an undesirable secondary reaction produces dioxin in very
small quantities if the temperature is held down to no more than 200°C,
in large quantities if the reaction is not controlled because the latter
is exothermic.

(Source: 7, p. 163)

- Enzymatic induction and inhibition effect: certain enzymes are considerably
  induced while others are inhibited: the synthesis of DNA seems also
  weakened. The consequences of the effect remain largely unknown.

The existing uncertainties concerning the exact effects of the substance
must be stressed, especially with human beings and here particularly those
concerning the mutagenic and teratogenic effects. The director of the Hygiene
Laboratory of Lombardy said that since it has never been seriously studied
TCDD remains partly a mystery (13).

The effects are real if one follows Dr Ton That Thut of the Hanoi hospital
who had to treat victims of the discharge of American defoliants: 60,000
tonnes of defoliants with trichlorophenol containing nearly 20 tonnes of
dioxin between 1965 and 1972. These statements are disquieting as 30 per
cent of the people affected have died and, years later, some of those
poisoned continue to die from tumours of the liver and the mortality among
their newly born is abnormally high (14). The international Commission of
Enquiry, appointed at the request of the Americans has shown itself more
reserved and did not establish a clear correlation between the accidents

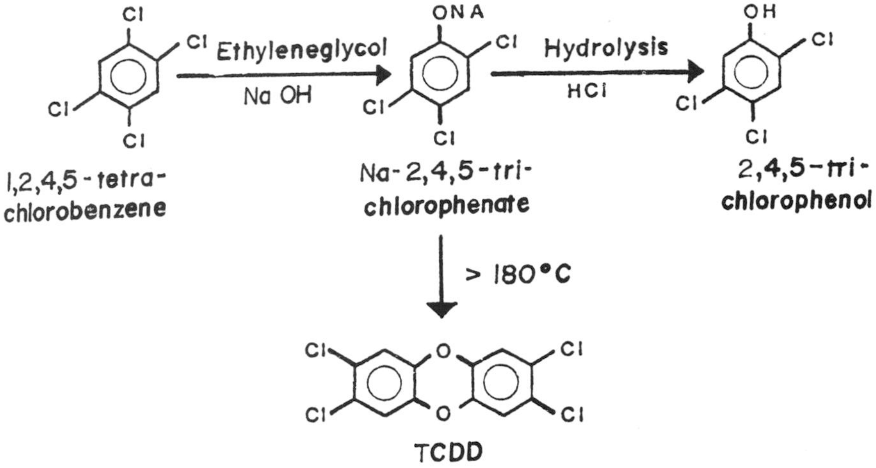

Fig. 9: The transformation of tetra-chloro-benzene into
         trichlorophenol
                    (Source: 4, p. 2)

observed and the discharge of defoliant (9, 15); in the same sense and
concerning the Laboratory Professor Tuchmann-Duplessis noted in a report:
the experimental results while showing the great harmfulness of dioxin also
suggest that there could be important differences between the reactions of
rodents and those of primates (15, p. 6).

If they do not remove the fundamental ambiguities about the affects of the
product the consequences of the accidents that occurred in chemical factories
which produce trichlorophenol (500 victims, 15, p. 5) nevertheless make clear
the dangerous nature of chlorinated derivates of dioxin. The recall of these
events shows yet another characteristic of dioxin: its stability.

3rd: A substance difficult to eliminate — as the precedents have shown

Accidents have occurred more particularly on November 17, 1953 at BASF,
Ludwigshafen (West Germany); at Dow Chemical Corp. at Midlands (Michigan) in
1961; at Philips in Amsterdam in 1966; in England on August 23, 1968 at the
Coalite and Chemical Products Ltd. (Bolsover, Derbyshire); in 1968 in
France near Grenoble (3; 12, p. 25- 16).

The greatest discretion surrounded these dramas. However, after nineteen
years of silence one learns that the German accident had caused forty two
serious cases of chlorine acne; there were also fourteen victims who had
damage to their liver and kidneys, to the cardio-vascular system, to the
nervous systems of seven of them. Identical symptoms appeared with the wives
and children of exposed workers. Some cases of chlorine acne were still
under treatment fifteen years after the accident. The resistance of TCDD was
demonstrated on that occasion: two years after the accident when an attempt
was made to use the premises again new cases of chlorine acne were reported

Substance	Molecular weight	Minimum lethal dose, moles/kg	Minimum lethal dose, ug/kg
Botulinus toxin A	9×10^{-5}	3.3×10^{-17}	0.00003
Tetanus toxin	1×10^{-5}	1×10^{-15}	0.0001
Diphtheria toxin	7.2×10^{-4}	4.2×10^{-12}	0.3
TCDD	322	3.1×10^{-9}	1
Saxitoxin	372	2.4×10^{-8}	9
Tetrodotoxin	319	2.5×10^{-8}	8–20
Bufotoxin	757	5.2×10^{-7}	390
Curare	696	7.2×10^{-7}	500
Strychnine	334	1.5×10^{-6}	500
Muscarin	210	5.2×10^{-6}	1.100
Diisopropylfluorophosphate	184	1.6×10^{-5}	3.100
Sodium cyanide	49	2.0×10^{-4}	10.000

Table 1: Comparative toxicity of various substances
(Source: 7, p. 164)

Species	LD_{50} ug/kg	Time in days between application and death
Guinea pig	0.5–2	5–34
Rabbit	10–115	6–39
Rat	22–100	9–48
Chicken	25–50	12–21
Rhesus Monkey	< 70	28–47
Dog	> 30–300	8–15
Mouse	114–284	20–25

Table 2: LD 50 for various animal species
(Source: 7, p. 164)

among the workers. Everything that was combustible was burned and everything
that was not was cast in blocks of concrete for submersion in the Atlantic
ocean*. Work continued with the use of diving suits and oxygen masks.

In 1973 it was learned that the English accident had caused seventy nine
cases of chlorine acne; some appeared still three years after the explosion
of the reactor with workers of a factory under construction who had not been
exposed before. The locations had to be sanitized again meticulously; the
contaminated material was buried at a depth of forty five metres in an
abandoned coal mine*. (See previous footnote).

Fifty workers were poisoned, of whom ten rather seriously in the Dutch
accident. During the two years that followed the escape of dioxin, which
was thought to have been between 200 and 300 grams, intestinal sarcoma were
registered, serious liver diseases which caused the deaths of four workers.
All attempts at depollution failed and after ten years it was decided to
demolish everything for submersion in the Atlantic ocean*(see previous footnote). In the French accident twenty one cases of chlorine acne were registered with loss of weight, anorexia etc.

*The reports do not specify whether the various 'dustbins' (mines, ocean)
that were used can be so used without danger.

The case of a Czechoslovakian factory must be added which, between 1965 and 1968 (the year the establishment was closed down) was responsible for seventy eight cases of chlorine acne, practically all the factory's workers; there were two deaths, eleven cases of liver and metabolic troubles and psychic troubles in the majority of cases.

To these accidents, at the level of production, one can add those which have occurred in other links of the chain. In Missouri, in May 1971, a training track for horses was sprinkled with oil with the aim of stopping the dust flying about. This caused numerous victims among the animals. A 31-33 ppm contamination with dioxin in the soil was found. The soil was twice replaced (October 1971 and April 1972) but the horses continued to die until January 19' among eighty five horses that had used the track during that period fifty eight fell ill and forty three died; there were twenty six abortions, numerous deaths at birth, and six congenital malformations were counted. There were also four cases of human contamination of which one was particularly grave. Three weeks later the same oil recovery company sprinkled two other tracks with the same results. Three years later it was found that the oil in question came from a factory which from 1969 to 1971 had produced trichloro-phenol which contained lots of dioxin: the production was stopped at the end of the Vietnam war but no apparatus for the elmination of residue was installed (12, p. 28; 5).

Among the numerous other cases of intoxication due to chlorinated derivates the one which occurred in Japan in 1969 needs to be specially mentioned. A vegetable cooking oil caused serious trouble to hundreds of people. Two years after the time of the accident no improvement was found in those intoxicated. In this case there was a teratogenic effect (15, p. 3).

2. THE FACTORY AT MEDA: SERIOUS INSUFFICIENCIES

1st: The operation as seen by the Hoffmann-la-Roche group

a) The production of trichlorophenol. In the past trichlorophenol, used for the synthesis of hexachlorophene in the Givaudan group, had been bought out-side. Since the end of the 1960s it became, however, more and more difficult for Givaudan to procure trichlorophenol in the quantities wanted at the degree of purity required. Trichlorophenol was used in large quantities by the manufacturers of herbicides. The raw materials crisis of 1974 brought with it the depletion of this intermediary product which called for the decision to manufacture it.

Already during the years 1970-1972 pilot runs had been undertaken with a view to autonomous production. Eighty seven tonnes of trichlorophenol were manufactured during these trial runs. The production was interrupted in 1973 and 1974 and requirements covered by purchases. Production, correctly speaking, started in 1975 at Meda at the rate of 105 tonnes for that year. In 1976, 130 tonnes had been synthesised up to the day of the accident. All loads were manufactured without incident.

The trichlorophenol manufactured being destined exclusively for medical and cosmetic uses, the requirements were particularly high. Quality control had demanded the development of highly sensitive methods of analysis which could, at short notice, only be applied at Givaudan. This explains why the analyses could not be realised on the spot but on the contrary the samples had to be sent to Switzerland (10).

b) Safety. The optimal temperature for this reaction is 170° and the heating
method of this apparatus did not permit to go beyond 190°, well below the
critical temperature (230°). If there was no alarm bell installed it was not
for financial reasons, its cost is extremely low, nor out of negligence but
simply because there was no reason to expect a sudden increase of temperature
(17).

A safety valve was installed on the apparatus but its purpose was not to
anticipate the effects of an exothermic reaction but to serve safety in an
operation at the start of reaction (17).

c) A factory in the process of modernisation. From 1970 onwards the programme
of modernisation and restructuring which was in progress for the whole
Givaudan group was also applied at Icmesa. Manufacturing was rationalised,
quality control strengthened, equipment modernised while a new installation
for sewerage treatment which met the requirements was to go into operation
from autumn 1976. In the space of five years Givaudan had to invest 16
million Swiss francs in this small company, an amount that greatly exceeded
the possibilities of the old owners. This amount represents an investment of
100,000 Swiss francs per working place. A good part of these improvements
had been achieved at the time when the accident occurred. The manufacture of
trichlorophenol was part of this modernisation programme (10).

d) The choice of locality. When the need for proper manufacturing became
imperative its location was discussed. Several of Givaudan's and Roche's
factories were considered. The choice finally fell on Icmesa for the following reasons:

Icmesa had always been cut out as an important supplier of chemical
products for the Givaudan group, and this role remained in the long term
restructuring plan.

The manufacturing rationalisation programme for the old Icmesa factory
demanded the allocation of new products to that company in order to
maintain employment in the long term and permit the factory to run at least
without loss. The manufacture of trichlorophenol tallied also perfectly
with this allocation of functions from the point of view of chemistry
inasmuch as the existing apparatus and installations lent themselves to the
intended manufacture.

Greater freedom in matters of safety and protection of the environment
played no part (10).

e) A formal denial: the installation had no links at all with the military.
Never, neither at Icmesa, nor at Givaudan, nor at Roche have toxic substances
been manufactured or supplied for military use (10).

2nd: Some radical criticisms from Italian disparagers of Hoffmann-la-Roche

a) Safety. In a reaction in which very precise control of temperature is
fundamental not only for the avoidance of tragedies but also for the achievement of a pure final product no thought had been given to the introduction
of automatic temperature control, not even to that of a relay that would
ring an alarm bell at the bottom of the ladder (12, p. 30).

Knowing that the reactor could explode they solved the problem by
constructing a safety valve which would lead off directly into the atmosphere
without any precaution, and this in a densely populated area. They did not

even approach the problem of the terrible parasitic reaction which transforms
TCF into dioxin and which instead of an explosion could provoke the diffusion
of the product with serious danger for the workers and the population living
nearby (12, p. 30).

b) The choice of location. Roche had chosen Italy to manufacture trichlorophenol also because ... of the scientific underdevelopment in that country,
because of the absence of restrictive regulations and the weakness of controls
(12, p. 30), a country where the health authorities stifle all recourse,
where municipal councils, engulfed to their necks in real estate scandals
became vulnerable to blackmail because, for instance, they built residential
quarters in the place allocated for a hospital. In addition, the salaries
were low: 'understanding' trade unions. It is not through dishonesty, it is
because of the myth of 'industrialisation' equals "progress was too fast, it
was sustained with too much infatuation by the entire Italian Left that the
unions might present dangerous interlocutors to the lieutenants from Basle".
(18, p. 167).

c) An additional suspicion. Trichlorophenol was manufactured in Italy by
Icmesa with the proviso that the factory did not directly produce dioxin as
a strategic weapon for the USA and that it sustained certain technicians who
had studied the installations of the Meda factory (18, p. 66).

Fritz Moeri, the actual builder of Icmesa's reactor, has expressed doubts
in an interview given to Pierpaolo Bollani (*Tempo*, 8 August, 1976) about the
fact that during the accident the factory was producing trichlorophenol (18,
p. 68).

3rd: Observations by the Commission of Enquiry*

a) Process control. It was not automatic; it was worked manually and therefore discontinually (the continual system was in the process of being
installed (p. 63)).

b) Cooling system. This too was operated manually. Those responsible at
Icmesa have always maintained in their depositions that this manual cooling
system was valid (p. 63). The director of Givaudan, G. Waldvogel, has
declared on the subject that the temperature must not exceed a certain level
and that the necessary valves had to be opened (p. 64).

The Commission retorts:

This logic renders the responsibility even heavier because it is quite
evident that if the merely manual controls were considered adequate the
continuous presence of people who are capable of applying them is an
absolute necessity (p. 65).

c) There was not even an automatic signal or switching-off system.
Registering this information which was given by P. Paoletti (production
coordinator at Icmesa) the commission notes that the system did have its
usefulness (p. 66).

----

*All references given under this heading, unless mentioned otherwise, are
relative to the report by the parliamentary Commission of Enquiry (No. 16).
Only the page numbers are therefore given.

Coming back to the accident of July 10, the Commission insists:

> The speed of intervention by Mr Galante, the workshop manager who was the
> first to realise that there had been an accident, may in fact have been due
> to chance (chance that he was in the vicinity of the establishment at the
> time of the accident) ... Recognising that something abnormal was happen-
> ing he intervened and avoided a still more serious disaster (p. 66).

d) Staff training. The staff was not aware of the risks connected with the
production of trichlorophenol which is serious not only as concerns major
accidents but also for the more normal functioning of the company: on various
occasions production residues had escaped from containers or pipes (p. 62,
deposition by Mr Paoletti). The staff was not qualified to deal with these
products.

e) The process used. In the examination of the causes of the accident of
July 10, 1976 it must be taken into consideration that those responsible at
Icmesa and at Givaudan have argued the absolute impossibility of foreseeing
such an event and the negative character of experiences they have had subsequently in the laboratory (deposition by Guy Waldvogel, director general
of Givaudan and Joerg Sambeth, technical director of Givaudan).

However, in the scientific literature between 1971 and 1974 one finds the
description of other accidents in the production of TCF which were followed
by the formation of dioxin; and the Commission considers it totally improbable
that the technical directors of Givaudan and Icmesa could have been unaware
of this. They themselves have, in other declarations made to the Commissions
(Sambeth, von Zwehl, in charge of technical services at Icmesa; Paoletti,
production coordinator at Icmesa) confirmed that they knew the work by Milnes
(*Nature*, Vol. 232, 1971, p. 395) before July 10, 1976. This author had
pointed out since 1971 that in a mixture of caustic soda and ethylenic glycol
an exothermic reaction could develop ... that it could develop rapidly and
out of control up to 410° and would then release large quantities of gaseous
products.

In the case of Icmesa the Commission learned from a deposition (by an
Icmesa technician) that at the time of the accident the reactor temperature
which was not at all controlled was between 450 and 500° which signifies that
the temperature had been considerably above the safety threshold and that
conditions for the formation of a significant quantity of dioxin had
developed, be it because the solvent had evaporated, be it that other substances had formed in an uncontrolled manner (p. 66).

The Givaudan patent provides that the distillation of the solvent occurs
after the acidification of the trichlorophenol; at the Icmesa factory the
inverse process had been used. If this inversion had been avoided the prolonged contact at high temperature between ethylenic glycol and caustic soda
(a contact which involves risk factors) would have been avoided and
consequently the distillation of the solvent would not have occurred in a
basic but in an acid environment. To this must be added that in the Icmesa
process the diminution of the solvent, a gradual diminution, had as its
necessary consequence the continuous reduction of the thermic head of vapour
and favoured the conditions of danger of which Milnes had spoken. Finally,
the change made in the molar ratio of the initial concentrations between the
tetrachlorobenzene, the caustic soda and the ethylenic glycol must be
stressed. Whereas the proportion in the Givaudan patent was 1:2:11.5 it was

1:3:5.5 in the Icmesa process*(p. 69).

These statements we found in the conclusion of this technical article:

Ethylene-glycol is expensive; stocking it is expensive; the transfer of
ethylene-glycol is expensive and so is its handling. Whereas if one uses
a smaller quantity one reduces staff, working time, energy consumption and
the production services (19).

These modifications have as an effect on the one hand a noticeable
variation (change) of production cost, on the other an increased risk of
TCDD forming and exothermic reaction (p. 70).

f) Numerous infringements.   Infringements in relations with the mayor.

The commission examined the placement of the factory and its activity the
nature of which had changed with the lapse of time:

Icmesa had stated that the establishment was destined for the manufacture
of pharmaceutical products ... It was obvious that any change in the type
of production which could have required the inclusion of the establishment
in one of the two categories foreseen in Article 216 (of the Testo Unico**)
would have obliged the company to notify the mayor fifteen days before the
start of the new production. This never happened. Icmesa asked only
permission for the enlargement of the factory (p. 47).

Mainly on account of this fact the housing plan of Icmesa was approved (on
June 30, 1973) without inspection of the Icmesa factory since it had been
designed for the manufacture of pharmaceutical products. However, during the
years 1969/70 the establishment had in fact been modified for the manufacture
of TCF; so it started and reached a maximum level in 1975 (6,361; 33,000, :
40,350; 38,400; 105,000; 142,000 kg during the years 1970/71/72; 1974/75/76
(pp. 49-50).

Even in 1972, in a report on atmospheric pollution requested by the mayor
following disquiet about the function of the installation Icmesa made no
specific mention of the manufacture of TCF. It confined itself to alluding
to an incinerator installation for organic residues and residues of phenol.
(p. 77).

Infringement in relation to INAIL

Article 12 of the Testo Unico law concerning the obligatory insurance
against work accidents obliges industrialists to declare to the National
Institute of Work Accidents (INAIL) the kind of work which might cause work
related disease carried out fifteen days before the start of its operations.
There exists a list (annex 4 of the Testo Unico) of work related diseases
which are to be insured. In this list the diseases caused by phenol and
glycol can be found under numbers 15 and 22.

---

*Much more complete technical explanations may be found in an issue of the
 review Sepere which is devoted to the case of Seveso.

**Basic law on sanitary matters of 1934, is the main piece of Italian
  legislation on matters of the environment, nowadays largely outdated; however, it has not yet been suitably recast (as we shall see later).

At the time the insurance report to INAIL was prepared (1947) Icmesa declared that the Meda factory manufactured chemical and pharmaceutical products; subsequently it never declared modifications undertaken that involved risk level and type of manufacture. (p. 75).

Infringement in relation to labour law

Icmesa did not instal the commission on working conditions provided for in Article 9 of the law of May 20, 1970, No. 300; this commission has to examine the conditions of noxiousness (harmful conditions) (p. 76).

Infringement in relation to the provincial administration

The competent provincial administration on matters of air and water pollution had forbidden Icmesa to release industrial refuse into public waters (June 27, 1957). But the situation remained alarming. In a report of October 18, 1969 from the Provincial Laboratory for Hygiene and Prophylaxis one reads: Multiple, persistent nauseous odours, continuous and constant danger for the ground water level and the nearby stream. At last, in January 1972, Icmesa presented the installation project for water purification and incineration of residual dirt from the treatment of phenol (pp. 78/79).

Infringement in relation to the ANCC*

An industrial establishment in terms of the law (of July 26, 1965), No. 966 Article 2) has to obtain a certificate of prevention from the Vigili del Fuoco. It has to request control visits; the initiative is incumbent upon it and not on the fire police. Icmesa, after an interruption of its production in 1973, had to request a new authorisation (pp. 81-82).

When after a year it restarted production of TCF it requested renewal of the certificate of prevention. The Vigili del Fuoco carried out their visit but did not issue the certificate because they had reported deficiencies in the documentation. The commander of the Vigili del Fuoco notified the mayor of Meda and Icmesa that the renewal of the certificate of prevention was subject to the presentation ... of a technical report on the manufacture and the substances used. Icmesa resumed production of trichlorophenol until July 10, 1976 without the certificate of prevention (p. 82).

Infringement in relation to the Works Inspectorate

The inspector of works in Milan should also have been notified of the change of manufacture when Icmesa started manufacturing TCF. However they did not do that (p. 83).

4th The deficiency of the public authorities in matters of preventive control

a) Virulent criticism of the passivity of the administration. G. Pecorella, lawyer for the plaintiffs is specific: the drama would not have happened if certain public organisations had not shut their eyes on what was happening in the establishment at Meda. The CRIA**knew since 1972 that Icmesa handled phenol:

---

*National Association for the Control of Combustions, an organisation charged with the control of apparatus working with pressure, from pressure cookers to reactor tanks in chemical industry installations

**Regional committee in charge of the control of the release of pollutants into the atmosphere (CRIAL = CRIA for Lombardy).

In fact, on February 2 of that year a request had been filed by Icmesa for an authorisation to instal an incinerator for phenol residues. CRIAL then requested a technical report on the industrial installation used from Icmesa in order to be able to determine the quantity and composition of the incinerated substances which were to be released into the air. This report was filed on March 7, 1972. In its reply of June 27, 1972 CRIAL expressed doubt about the completeness of the report and seemed to think that not all the ongoing operations had been mentioned as the necessity of setting up an incinerator would lead one to assume. It requested a new report to be filed within 30 days which was to include a description of all operations carried out, of all raw materials handled and the quantities of the substances which were to be released into the atmosphere.

The reply took thirty months without CRIAL ever making any move to get back at Icmesa. The report of March 28, 1975 speaks of the production of trichlorophenol for which a burner for phenolised water had been set up but it insists nevertheless that "the production of trichlorophenol has meanwhile been stopped but might be restarted very soon"; finally the report states that all the reactors "have a ... direct outlet into the atmosphere with a view to letting out possible accidental excess pressure (safety discs, low pressure valve)". Even at that time CRIAL was in no hurry at all to inform the mayor who would have had to impose the necessary protection measures on Icmesa (22, p. 108).

It must be pointed out that the members of CRIA include the chairman of the regional council, the regional director for ecology, the provincial medical officer, the health authorities, the chief of the regional works inspectorate, the representatives of the departments of the region, the president of the chamber of commerce etc. The authorities were therefore in the picture about what was going on at Meda.

The virulent criticisms become understandable if one considers for example the following table which shows how a quarter of a century had not been sufficient to get respect for the law (42):

1957: An enquiry shows that Icmesa poisons the water of a nearby stream.

1958: Those responsible at the factory announce that they have installed a purification system.

1959: A control shows that the water remains toxic.

1962: After two reminders from the provincial authorities Icmesa announces that it has put a new installation into operation.

1965: The quality of the water is still found unacceptable.

1969/75: New disposal systems for refuse are installed; the reduction of toxicity does not seem sufficient. A new process is introduced; it remains ineffective.

1975: A new dosśsier is sent to the Director of Public Prosecutions.

b) *Moderate criticism from the Commission of Enquiry: the public authorities have not shown much zeal but they have not committed serious mistakes*. The parliamentary commission analysed the responsible authorities one by one. Summed up, its analysis is as follows:

The Inspector of Works had relied on the workers for information instead of requesting it from the management of Icmesa. Since the workers had no information from their company the inspector found himself blocked as he had not made use of his rights towards the company (16, p. 76).

ANCC did not investigate in depth; it has the excuse of industrial secrecy;
only the inspector was entitled to know the production process. ANCC did
not think of questioning the inspector (16, p. 98).

The fire brigade had no knowledge either of the industrial secrets. Only
the inspector ... But they did not ask the inspector; neither did they
alert the police commissioner who would have had the possibility of
exposing the situation (16, pp. 95-96)

INAIL had no right to know the production process either, not even in the
execution of its functions. But it did not occur to them, as it appears,
to get in touch with the inspector (16, p. 100).

All the persons mentioned in this list are more or less part of CRIA. It
appears that there has not been a more global approach to the issues (16,
pp. 94-95).

As the Commission remarks, everything was done in the most 'sclerotic'
manner; everybody tended to work within the narrowest limits of their
authority without using all their powers, without bothering about what was
happening at Icmesa*.

3. THE CALENDAR OF IMPOTENCE

1st: July 10-24: Dioxin takes over control of the area; the industrial
owners keep quiet; the bureaucracy reasserts itself

Saturday, July 10:

At 12.37 h, as we have said, the safety disc of the reactor in block B of
the Icmesa factory slackened following a sudden increase in temperature and
pressure; the industrial owners will say that the reason for this was unknown
and inexplicable:

The manufacturing process, properly speaking, had been completed at 6.00 h
and the night shift left the factory after having cut off all energy supply
to the apparatus (10).

The Commission of Enquiry**for its part commented:

The last cycle for the week had started at 16.00 h on Friday 9th, i.e. ten
hours late compared to normal conditions. Based on working hours, those
responsible at Icmesa knew what would happen at the time of the interruption
(16, p. 69).

---

*The situation must be seen in the general context of the control of instal-
lations in Italy. The case of Meda is in no way exceptional.

**In the same sense S. Zedda thinks he can maintain: on Friday, July 9, at
7.00 h in the evening a new cycle of reaction and distillation which
normally lasts for fifteen hours was imposed on the workers, in full knowledge of the fact that on the following morning at 6.00 h the workers would
leave and the weekend would start. Well, on that Saturday, July 10, something did not work in the reaction of distillation and smothering in the
sodium-trichlorophenate water. However, the reactor was left to itself
without an automatic alarm signal until the cloud escaping from the valve
indicated at 12.40 h that the drama had started (12, p. 30).

At 12.37 h an employee was in the vicinity of the establishment and, as has been mentioned earlier, intervened. This was a lucky coincidence.

Outside, children saw a cloud for a moment.

Mr Galante alerted Mr Paoletti who stood in for the man in charge of production who was on holiday. To our knowledge no other intervention by any of those responsible at Icmesa took place until the next day (16, p. 105).

Sunday, July 11:

The first effects of the accident were noticed: vegetation burned, animals taken by disease; some twenty children had sores on their arms, red spots on their faces, some sort of burns on their bodies, high fever, intestinal troubles. Police inquired (20, p. 12).

The man responsible in the company (the engineer von Zwehl), also on holiday, was joined by Mr Paoletti: he asked for samples of the burned vegetation to be sent to the Givaudan laboratories in Switzerland for analysis. This was done on Sunday evening. A little earlier, at 17.45 h, two representatives from Icmesa informed the commander of the carabinieri that a cloud of herbicide had been spread over the area around the factory. The mayor of Meda and the health officer of the community were alerted (16, p. 106).

But there is an immense difference between a 'herbicide' and dioxin. However, since 14.15 h the technical director of Givaudan (Dr Sambeth) who had been given the news had established the hypothesis of an escape of dioxin if one follows his deposition at the Commission of Enquiry:

We had heard of accidents of a similar kind and I thought of this possibility; I thought at that moment, and I still think, that there was a very high concentration of dioxin around the safety disc and a smaller concentration elsewhere. I could not think at that time that the dioxin could have expanded over a very large area (16, p. 107).

Monday, July 12:

The industrial owners did not close the factory: work was resumed normally on Monday. On that day Icmesa confirmed by letter to the local health authority that an incident had occurred at its factory on Saturday and that measures of precaution had been suggested to the neighbours; but once more there was only an allusion to 'herbicides':

Meda, 12 July 1976
For the att. of the Health Officer
Office for Health and Hygiene
20050 Seveso

Following our previous conversations we confirm that on Saturday, July 10, 1976 an incident occurred inside our establishment. The factory was at a standstill, as is usual on a Saturday, which is a non-working day. We are still studying the causes of the accident ... At the moment we can only assume that an inexplicable exothermic reaction has occurred in a reactor that had been left in a cooling phase. (There were the necessary substances for the production of raw trichlorophenol in the reactor: tetrachlorobenzene, caustic soda etc.). At the end of the normal working hours (06.00 h on Saturday) the reactor containing the raw product was left in a non-operating

state ... as usual. We do not know for what reason a rupture of the safety
disc occurred at 12.40 h which permitted a steam cloud to escape which after
hitting vegetation inside our establishment moved in a southeasterly direction,
driven by the wind, and within a short time dissolved.

Not being able to evaluate the nature of the substances carried by those
vapours and their exact effects we have intervened with neighbours asking
them not to consume garden prdoucts, knowing that the final product is also
used in herbicides (21).

This was then the first official document on the accident. The lawyer for
the plaintiffs, G. Pecorella, called it "a perfect example of criminal
hypocrisy" (22, p. 106).

Tuesday, July 13:

The health authorities sent this letter to the mayors of Meda and Seveso.
They added their own evaluation:

After enquiries undertaken, no danger to persons living in the surrounding
areas is to be feared (22, p. 106).

Wednesday, July 14:

The analyses carried out at the Givaudan laboratories at Duebendorf
(Switzerland) showed that dioxin was present (deposition Sambeth, Commission
of Enquiry, 16, p. 108).

On site, the deaths of a large number of animals in the area adjacent to
the factory were reported (16, p. 113).

Thursday, July 15:

Serious cases of poisoning were reported among the population (16, p. 110).
The mayors announced by means of posters that precautions must be taken in
the affected area (ban on the consumption of garden vegetables) and met the
industrial owners: the latter made no mention of the presence of dioxin (16,
p. 113).

Friday, July 16:

Fifteen children, four of which were in a grave condition, were admitted
to hospital; but nobody knew what treatment to apply. A strike was called;
the inhabitants insisted that the authorities give them some exact information (20, p. 13; 23, p. 13).

The Italians on their part took samples for analysis (16, p. 114).

Saturday, July 17:

The mayors of Meda and Seveso added emphasis to the health advice given;
they ordered the burning of the polluted garden products, the killing and
burning of the affected animals. On the same day the director of the
provincial chemical laboratory also established the hypothesis: there could
have been an escape of dioxin (16, p. 114).

Sunday, July 18:

The mayor of Meda ordered the closure of the factory; seals were affixed
to the doors of the accident block (16, p. 114). The director of the
provincial chemical laboratory in Milan, in a statement to the technicians of

Icmesa, declared the possibility of the presence of dioxin in the toxic cloud (16, p. 108).

Monday, July 19:

While five more children were hospitalised, the director of the provincial chemical laboratory learned during a visit at Givaudan's that the industrial owners too knew that there had been a formation of dioxin (16, p. 111).

Tuesday, July 20:

Upon the return from Switzerland of the responsible Italian persons the health directorate knew therefore for certain the seriousness of what had happened on July 10; the mayors were informed (16, p. 115).

In the area, animals died within a radius of 3 km from the zone originally declared endangered (23, p. 14; 22, p. 107).

Wednesday/Thursday, July 21/22:

Meeting at the mayor's office in Seveso; no decision was taken. Under pressure from the Regional Council*supplementary protective measures for the citizens were adopted (prohibition to eat (meat from) animals from the area; closure of certain establishments, on the spot medical checks etc.). The multiplication of pathological facts and the expanse of the affected area provoked the local health authorities to demand from the police commissioner the declaration of a state of emergency by the Provincial Health Council. The scientific literature on this strange product called dioxin, which did not exist in the files of the anti-poison centre in Milan, was gathered (23, pp. 14-15).

From the most official quarters the tone was only just one of disquiet:

> Police Commissioner of Milan
> Official Communique
> Milan, July 22, 1976

In connection with the Icmesa ... accident ... the Police Commissioner of Milan has received the provincial medical officer, Professor Eboli, the director of the chemical laboratory, Laboratory of Hygiene and Prophylaxis, Dr Cavallaro, the health officer of the community of Seveso, Professor Ghetti ...

There is general agreement on stating that contrary to what has been suggested there exists at this time no toxic gas cloud. No extensions of the phenomenon beyond the communities mentioned below has been reported.

As a precaution, the Police Commissioner advises not to eat produce from the area ...

The Provincial Health Council has been called into session for tomorrow. (23, pp. 15-16).

Friday, July 23:

A large meeting gathered together the medical science experts at the Police Commissioner's office in Milan. The meeting went on all afternoon. In the evening, a terse communique minimised the seriousness of the situation (20, p. 13):

---

*Regional parliament, regional legislature. A law of 1976 about the regions very strongly decentralised powers in Italy.

The Provincial Health Council in session at the Police Commissioner's office at midday for the examination of the events at Seveso and Meda ... confirms the validity of the measures taken by the Region of Lombardy and the initiatives by the local authorities concerning the prevention of possibly damaging effects on the population of the communities concerned.

The meeting has concluded that it is not necessary to suggest civil defence measures.

The university representatives who took part in the meeting have unanimously stressed that further measures need not be considered necessary or urgent (23, p. 16).

In the television news, Vittorio Rivolta, the Director General of the Health Service confirmed: Everything is under control (20, p. 13). A few hours later these communiques looked ridiculous. The Director of the Medical Research Centre of Roche in Basle, Giuseppe Reggiani, confirmed: The situation is very serious; draconian measures are necessary; 20 cm of earth surface must be removed, the factory must be buried, the houses destroyed (20, p. 14).

To support his statement he presented summary charts of pollution drawn up by Swiss technicians (20, p. 14)

Saturday, July 24:

One last effort was still made to throw out the spectre of a disaster. The regional health director implicated G. Reggiani in the *Corriere d'Informazione:*

This person was dumped on us; nobody expected him, and nobody expected such severe statements. To my knowledge, he is not an official representative of the company and I shall today request to know on whose behalf he speaks. I have made clear to him the seriousness of what he says. I have the impression that this person is bluffing. And this person will have to answer for his statements (23, p. 18).

However G. Reggiani received his 'official' recognition: "a doctor who acts as our consultant", asserted the director general of Givaudan in a letter to the medical authorities of Meda and Seveso (4, p. 9a).

These events brought about a substantial change of scenery. A large gathering of high-ranking medical personalities, politicians and administrators was held at the health directorate. At the end of the afternoon the verdict was pronounced with embarrassment:

One hundred and seventy nine people will have to abandon their houses within 24 hours; their dwellings are in an area which is too highly polluted (20, p. 14).

In the meeting it was also spelled out that people must eat absolutely no produce from the area (vegetables, eggs, meat, milk ...) which they had done over the last two weeks because of lack of sufficiently exact information (except for the immediate neighbours of the factory). The mayors found a population in uproar as they left the meeting. The harm was at last recognised, more or less, the existence of the tip of the iceberg was no longer denied.

TCDD - EXPERIMENTAL TOXICITY	DOSE [1]
- LD_{50} FOR LABORATORY ANIMALS	1 (0.6-115) µG/KG
- SYSTEMIC LESIONS (LIVER, BLOOD, SKIN ETC.) AND CLINICAL CHEMISTRY ABNORMAL FINDINGS AT CHRONIC ADMINISTRATION	0.1 "
- EMBRYOTOXICITY, FOETOTOXICITY, POSTNATAL GROWTH RETARDATION	0.25 "
- REPRODUCTION, FERTILITY	0.01 "
- IMMUNODEFICIENCY (ATROPHY OF LYMPHOID TISSUE)	0.1 "
- ENZYM INDUCTION	1 "
- CARCINOGENICITY	0.1 "
- MUTAGENICITY (?)	2 µG/ML
[1] INDICATIVE VALUES	

Table 3:   Experimental studies on animals
(Source: 4, p. 9)

TOXIC EFFECTS OF TCDD IN MAN	
DERMATOLOGICAL:	CHLORACNE
	PORPHYRIA CUTANEA TARDA
	HYPERPIGMENTATION AND HIRSUTISM
INTERNAL:	LIVER DAMAGE (MILD FIBROSIS, FATTY CHANGES, HAEMOFUSCIN DEPOSITION AND PARENCHYMAL-CELL DEGENERATION)
	RAISED SERUM HEPATIC ENZYME LEVELS
	DISORDERS OF FAT METABOLISM
	DISORDERS OF CARBOHYDRATE METABOLISM
	CARDIOVASCULAR DISORDERS
	URINARY TRACT DISORDERS
	RESPIRATORY TRACT DISORDERS
	PANCREATIC DISORDERS
NEUROLOGICAL:	POLYNEUROPATHIES
	SENSORIAL IMPAIRMENTS (SIGHT, HEARING, SMELL, TASTE)
PSYCHIATRIC:	NEURASTHENIC OR DEPRESSIVE SYNDROMES

Table 4:   Toxic effects on man
(Source: 4, p. 9)

2nd: July 25 - August 30: some measures against the dioxin; much effort to save the institutions on the spot

From Sunday 25th to Thursday 29th:

The army encircled 12 hectares of contaminated area with barbed wire on the 25th. On Monday 26th, the evacuation of the area designated by the authorities started at 11.15 h for 225 people. They were allowed to take with them those of their clothes which on that day were in wardrobes a ruling which can only be reported with astonishment: a piece of linen that was drying in the open air on the 10th or 11th could be taken along on the 25th. In fact, the inhabitants carried with them much more: food, various objects, before starting 'proper' housemoving on a large scale (20, pp. 14-15; 16, p. 229).

A first scientific report (worked out by NATO) was sent to the regional government authority; it gave draconian limits for the acceptable threshold of dioxin: 0.0125 micrograms/m^2.

That was in any case a concentration not measurable with the available apparatus. What was it good for, then? The NATO report will be forgotten (23, pp. 38-41). It will be kept secret but its dislosure on August 24 had the effect of a bombshell: the threshold established by the official services was in fact 400 times higher than the figure of 0.0125 micrograms/m^2 (20, p. 18).

Major confusion ensued. On July 24 Rome had let it be known that on the strength of a decree of January 14, 1972 the Region had full authority and responsibility in matters of health (23, p. 22). Two days later, on the 26th, the mayor of Seveso let it be known that as from July 16 there had been agreement between him and the authorities to proceed with the evacuation of the neighbourhood of Icmesa but that a series of events had intervened to hold up this evacuation (23, pp. 38-41). On the 27th the Minister of Health (at national level) accused the Region for its slowness in informing Rome (23, p. 21).

Also on the 27th, Vittorio Rivolta, regional minister of health, was given authority by the government of Lombardy to take matters in hand (24). The army cordoned off an additional 15 hectares. A total of 227 people had been evacuated by that date.

The following day the disquiet increased. These lines from a correspondent of *Le Monde* show clearly the fear that can grow within a population that has been affected, little by little, as if by chance, by a mysterious invisible evil and which has frustrated the technicians and those responsible:

There is concern for the inhabitants of Baruccana and Cesano even though they are located several kilometers from the company, because a new contaminated area has been disclosed in those parts. As in Seveso, for some ten days chicken, rabbits and dogs began to die. Now, there are perhaps 15,000 inhabitants and not just a few hundred in danger, because the symptoms of intoxication with dioxin are very little known and take a long time to show themselves. Skin lesions, the most obvious symptoms, appear only after several days. Gastro-enteritis is feared and even very long-term effects, genetic effects.

And Marc-Ambrose Rendu notes also:

The men from the 3rd artillery regiment in Milan who have surrounded the factory and 30 hectares of neighbouring land with barbed wire have worked with bare hands, without special precautions. Only yesterday were they given rubber boots. Must they be put under medical surveillance? (25).

During the night a bomb exploded in Rome, in front of the offices of the Italian subsidiary of Hoffmann-la-Roche. The outrage caused considerable damage but there were no casualties (13)

Friday, July 30 - Sunday, August 1:

Vittorio Rivolta launched an appeal to the population on July 30. Laura Conti, communist party member of the regional parliament, a medical doctor, secretary of the Health-Ecology Commission of the Regional Council, reported this appeal and criticised it:

- it defines dioxin as an "unrecognised gas", which is wrong;
- it states that "the dioxin has struck a limited, isolated and evacuated area": a large area is, however, neither isolated nor evacuated;
- it says "that no danger exists outside the evacuated area, that there are only hygienic measures to be observed"; that is inaccuarate;
- it says finally "that in order to understand the phenomenon properly one can say that the polluted area is comparable to an area struck by fire where the fire has been brought under control"; which is an enormous psychological error. V. Rivolta indicates the end of danger while there remain deferred effects (23, pp. 31-32).

- This kind of declaration, classical in cases of disaster, could not check the negative development which was taking shape.

The economic effects of the toxic cloud were beginning to be felt: closure of restaurants, refusal to accept delivery of orders already shipped by furniture manufacturers; hotel owners in some holiday centres refusing accommodation to people whose identity cards show Seveso as their domicile.

Therapeutic abortion on account of the possible teratogenic effect of dioxin on the psychical health of the mother*was authorised by the Health Minister, L. DalFalco. The Christian Democrats, the movement "Communion and Liberation", the clergy of the province denied the dioxin danger. V. Rivolta, a Christian Democrat, was accused of "abortionism" for having admitted that there were risks for the women and the new-born. The polemics about abortion raged as other groups, who appeared to have come from Rome, pleaded in opposition to "Communion and Liberation" (20, p. 15).

On that Friday, July 30, the first map**of the polluted area was at last ready. There were distinguished:

- a zone A of strong concentration of endioxin, up to 5,000 micrograms/m^2 and more on a surface of 115 hectares with 700 inhabitants;
- a zone B, much less affected, up to 15 micrograms/m^2 and above the threshold of 5 micrograms/m^2, stretching over 205 hectares with 4,280 inhabitants (3, p. 15, 26).

At the end of July more than fifty people were counted as receiving hospital treatment in the region. The citizens of Seveso, very disquieted, went for examination at the rate of 600 a day at a centre installed in a school in town (24).

Monday, 2nd - Sunday, 8th August:

On August 2, 511 inhabitants were dislodged, and during the night a great bustle descended on the office of the mayor at Desio. The fact was that dioxin in larger quantities than those found at Seveso and Meda had been spotted at Cesano, Maderno and Desio. Must there be more evacuations? The The representatives of the Region recoiled: "Here, 7,000 people must leave.

---

*Only the health of the mother permits therapeutic abortion according to Italian law. Since dioxin has no proven impact on the physical health of the pregnant woman there remained only the aspect of psychical health to obtain legal interruption of pregnancy.

**See map.

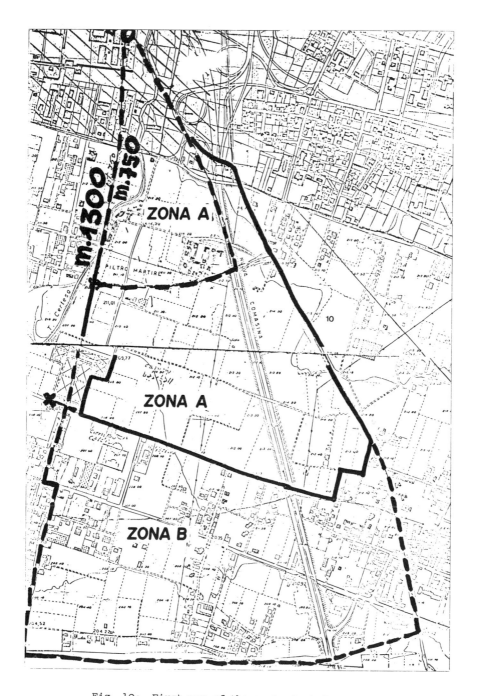

Fig. 10: First map of the contaminated area

We are aware that the evacuation of such a large number of people would cause
a great 'shock'. The population was advised to use discretion: the women and
children were removed; people were asked not to procreate during the coming
months (20, pp. 15-16, p. 229).

The authorities were presented with diverse offers of 'solutions' for the
treatment of the dioxin. As in similar cases, the market of ideas was packed
but it remained difficult to discern what was valid from what was useless,
even dangerous, patchwork.

Since the molecules of TCDD crumble at temperatures above 800° one thought
of the flamethrowers used by the military specialists of "nuclear, bacterio-
logical, chemical" warfare (13). The sowing of the oil with bacteriae that
would destroy the dioxin molecules biologically was also proposed; or perhaps
the release of ozone in order to increase the destructive effect of the sun
rays; or the decontamination of people, animals and objects with household
soap which would dissolve the dioxin molecules and thus facilitate the attack
on them by the sun rays; for the soil it was proposed to spread a mixture of
vegetable oil and animal fat (28).

Faced with this variety of proposed solutions the authorities were impotent.
As L. Conti pointed out, it was not for V. Rivolta to compare these different
scientific hypotheses. There existed an 'ecology' service but that was not
taken into account. In fact, there was a strong rivalry between it and the
'health' service (23, p. 43).

The uncertainty about the quantity of dioxin spread about weighed heavily:
500 grams, 2 kilograms? Or, according to a British expert, D. F. Lee, who
arrived on the spot, quite a different figure:

The estimated quantity of 2 kilograms of escaped dioxin appears understated.
According to my theory, which I hope is wrong, there could in the end have
been 130 kilograms of TCDD (23, p. 42).

Since the reactor was sealed off it was difficult to choose between all
these hypotheses. Even in September Hoffmann-la-Roche still noted:

Dioxin escaped in a quantity which can still not be exactly determined.
(10, p. 4).

Monday, August 9:

The medico-epidemiological commission set up by the Region after having
considered the available literature approved a document confirming that
dioxin is teratogenic for animals but that there are no data on man; that it
is nevertheless reasonable to admit the danger; and that the first three
months of pregnancy should be considred in zones A and B (23, p. 79).

Laura Conti (23) remarked that this document had been awaited with extreme
impatience, as if it meant to her alone a decision. In any case, it meant
only a sign of awareness of the dangers of dioxin; it permitted in no way
therapeutic abortion, on the grounds of teratogenic risk, because the law
concerns itself only with the health of the mother. Proving that there is
danger to the foetus is useless where the law is concerned. The document of
August 9 may perhaps help to plead psychical traumatism of the mother.
Giovanni Cerutti saw there at last, according to him, "a loophole for therapeutic abortion" (20, p. 16).

The Catholic hierarchy remained vigilant. Monsignor Giovanni Colombo, Cardinal-Archbishop of Milan, reacted briskly:

> While many have felt a duty to help relieve the difficulties we deplore that so many negative positions have been taken up such as the organisation of a campaign for abortion and the spreading of new and often unfounded alarm ... And the archbishop stressed the generous offer by some couples who had declared their willingness to adopt a child that was born deformed. "We invite those who are prepared to do this to make themselves known" (29, p. 100,; 30).

On the spot, a clinic for the medical follow-up on the population and the centralisation of observations was set up; financial aid of 240 million ffrs was released. There were still eleven people in hospital out of twenty four registered admissions; 220 pregnant women of which 117 in their first three months of pregancy had been examined; malformations were feared (31)

Tuesday, August 10

The local disquiet was revived by the state of health of the mayor of Cesano Maderno, one of the first persons to have visited the contaminated fields after the accident. The medical tests showed that the mayor had an excessive amount of white corpuscles. Could this be due to an ingestion of dioxin? (32). In the same vein, there was the death under suspicious circumstances of a woman of 35 in July: the examining magistrate had ordered an autopsy (25)

A ghetto feeling developed in the area as it felt itself rejected:

> The Swiss authorities banned the import of fruit and vegetables coming from the Milan area and air samples were taken along the border (i.e. some 30 km from the polluted area) in order to discover any contamination that might reach Swiss territory. A shipment of furniture of Seveso manufacture was stopped at the German-Swiss border because the addressee, a company in Cologne, had refused to take delivery, being afraid that it might have been impregnated by the dioxin cloud (32).

Scientific opinions are not designed to appease fear. The weekly paper *Tempo* published an interview with Ton That Thut from the Hanoi hospital:

- out of every 1,000 people intoxicated 300 deaths were registered;

- the dioxin had caused serious liver damages;

- malformations and an extraordinarily high number of spontaneous abortions had been noted.

Situations such as those that followed must be seen in this mental climate. It shows the disarray of a population in the grip of a problem that was largely beyond the capacity of the authorities. The life-buoy for these people was science. Science had to know.

Laura Conti, stationed at a secondary school in Seveso, found herself, therefore, facing a population at a loose end and divided into two camps: one favourable, one hostile to therapeutic abortion. After her discrete intervention she had to face a militant from "Communion and Liberation" asking the following question:

- Wouldn't it be better, if one wanted a woman to abort, to do so on the basis of a test which would show whether she has been contaminated by dioxin?

- There is no such test ...

People resume the attack at the end of the gathering:

- You have to give a figure!

- You speak of danger, but danger must be expressed in a figure!

- A figure: you have it in your head; if not you wouldn't say the women should abort!

- I can't get out of this thinks the person involved ... The only figure available ... but the Vietnamese situation is not the same ...

- Tell us that figure! We know that it's not identical, but one wants to have an idea!

- In Vietnam, among those who had hepatitis, 30 per cent of those who had had it, had cancer of the liver ...

- 30 per cent, 30 per cent, said the people (23, pp. 76-77).

On August 10 a second map*of the polluted area was ready. It showed:

- a zone A of 108 hectares, evacuated (730 people)**

- a zone B of 169.4 hectares; the children from that zone as well as the pregnant women were evacuated during the day;

- a zone R (R standing for caution) surrounding the first two, of 1,430 hectares.

Wednesday, August 11:

The scientific commission***appointed by the central government (on August 4) issued its verdict: Everything in the Seveso area must be destroyed, even the houses (20, p. 16).

Professor Cimmino declared: We shall need months, perhaps years, to understand the situation in depth (34).

The president of the Region, Cesare Golfari, rejected this opinion flatly:

No initiative for the sanitation of the terrain shall be taken; all decisions shall be taken by the Regional Council (22).

The Region argued that the destruction might raise polluted dust; V. Rivolta accused the Cimmino commission of pessimism; he thought that improvement was possible (declaration of August 14): We have well-founded hopes of recovering at least four fifths of the houses of zone A ... The improvement will be possible without pulling down the houses or removing the soil (19).

On that Wednesday, August 11, it was decided to call the Regional Council

---

*See map

**The exact borders of the zone had changed at least three times (16, pp. 119-120).

***Cimmino Commission.

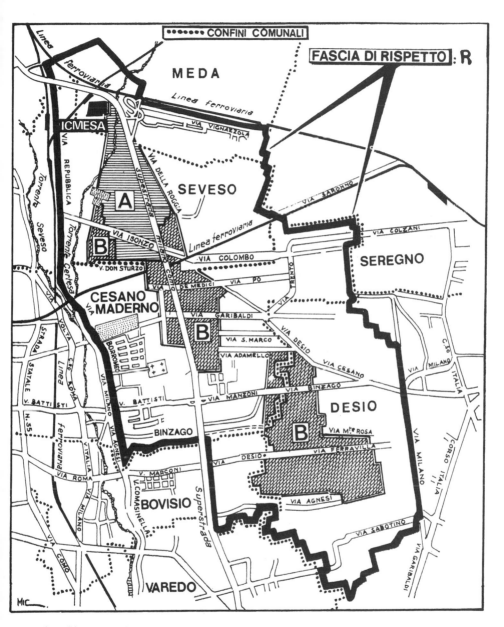

Fig. 11: Second contamination map.   Populations: see table 5, page 62
(Source: 39)

TOWNSHIP	ZONE A	ZONE B	ZONE R	OUTSIDE	TOTAL
SEVESO	681	626	7 945	7 738	16 990
MEDA	55	-	4 017	15 493	19 565
CESANO M.	-	2 731	14 991	16 111	33 833
DESIO	-	1 342	4 608	27 061	33 011
BARLASSINA	-	-	72	5 559	5 631
BOVISIO	-	-	167	11 058	11 225
LENTATE	-	-	-	13 037	13 037
MUGGIO	-	-	-	18 690	18 690
NOVA M.	-	-	-	18 467	18 467
SEREGNO	-	-	-	36 838	36 838
VAREDO	-	-	-	11 841	11 841
TOTAL	736	4 699	31 800	181 893	219 128

Table 5: Population of the different communities in the area
concerned and in the zones of contamination (A, B,
R) which have been established.
(Source: 4, p. 11)

into session. The date was fixed for ... August 24, 1976.

The technical director of Icmesa was arrested and charged with causing
"disaster by carelessness" and "omission of measures against work accidents"
(26).

During a press conference called at Basle by the directors of Hoffmann-la-
Roche, Mr Guy Waldvogel, director of Givaudan, indicated that he knew in
advance the risk of an appearance of toxic products at the Seveso factory
but that he had never imagined that "such a disaster" could occur. This was
why no emergency plan had been worked out with the local authorities. Mr
Alfred Jann, president of Hoffmann-la-Roche, said: We shall pay the damages.
We have sufficient financial reserves for that (26).

A young woman from the contaminated area had a stillborn child (26).
Whether or not this had anything to do with the accident of July 10, the event
did not improve the tense atmosphere in the area.

Thursday, 12th - Monday, August 23rd

Confusion grew. Therefore:

- the polemics about abortion continued, the Pope joined in the condemnation
  (20, p. 17).

- the controversy between the Cimmino commission and the Region carried on;
  between those who favoured the destruction of the area and those who
  strained to reassure the people;

- the evacuated inhabitants began to protest (20, p. 17);

- charges were brought against the mayor of Meda and against two regional representatives who had delayed the decontamination measures (20, p. 17).

Tuesday, August 24:

The Regional Council could at last meet to discuss the action taken by the regional government. After preliminary polemics the political parties (except for the extreme Left) reached agreement and approved the action taken by the regional government. The agreement was, nevertheless, matched by reservation: It will be necessary to recheck and verify the real presence of dioxin in the soil and in the subsoil, a woman socialist declared in the name of the health commission (20, pp. 17-18).

Very complex elements intervened on that day. What would be the attitude of the opposition, in particular that of the Italian communist party? Would the regional government be overthrown? During the morning session Laura Conti for the Italian communist party drew up the indictment of the action carried out, an action from which the Council had been kept away since July 10; emotions and tensions were at their highest pitch: the audience heard the description of a reality which the officials had systematically wanted to cover up and to cover themselves since the beginning of the affair. The meeting dispersed, paled or overtaken by controlled feelings. Was that the end of the regional government?

In the afternoon, the government's action received a positive judgement. Why? Several factors came into play. On the one hand, at national level, the Christian Democrats let it be known that if there were problems in Milan the historic compromise would be endangered in Rome: this was conveyed to the communists of Milan before the meeting resumed. On the other hand, the condemnation of the regional government at a time when it was in a difficult position might perhaps have signalled the appointment of a government commissioner. The opposition was not in favour of this. Finally, not all the actions of the Region were condemned: the attitude of V. Rivolta, a Christian Democrat, towards abortion was judged dignified and courageous (23, pp. 93-100, Discussions).

The session of August 24 had at least permitted to save the institutions. The real battle against the dioxin had not started; that had to wait ...

3rd: September - October: in search of a politically, economically and socially acceptable dioxin; nature would do the rest

Rains on the confusion. Torrential rains fell on the region during the first week of September, and it was feared that it might disperse the dioxin. Decontamination had started, a procedure had at last been set up by Givaudan and was accepted, but according to the industrial owners' own statement it was rather late. The summer sun which had been useful for the degradation of the poison no longer shone; autumn with its rain was already there, the leaves were falling. Was the dioxin already in the plain of the river Po? For its part, the regional executive did not exactly show wild determination in the battle against the dioxin which strongly embarassed Hoffmann-la-Roche and did not fail to raise criticism.

Laura Conti noted that every decision was overthrown the moment precise plans were forthcoming which had been devised with the most sophisticated techniques. They fell victim to the fetishism of precision. Everything was measured... in order to have a clear conscience. All this fervour was aimed at overthrowing all decisions. (23, p. 109). Laura Conti juxtaposed this

inertia with a phrase of V. Rivolta pronounced after a meeting on August 24:
If within three months experience shows negative results we shall let nature
run its course (23, p. 100).

The group from the Italian communist party had at the time talks with the
president of the regional government. It got the impression that the cause
of the inertia was to be found in "the confident expectation of miraculous
solutions that would ameliorate the soil; solutions that would permit leaving
everyting as it was." (23, p. 114). This inertia was taken by the population
as proof of the harmlessness of dioxin. In public gatherings the secretary
of the Health-Ecology Commission frequently noticed this negation of danger:
You are allright, aren't you? We, we are allright! (23, p. 116).

Shame and fear about abortion. The battle raged on in the field of abortion.
The women found themselves tossed from one hospital to another, thrown out,
like this one for example:

We don't do political abortions here. Your child is in excellent shape.
So, there are no objective causes that could involve psychic trouble (35).

They had to prove before colleges of forensic psychiatrists that they were
in grave psychic danger. For the women who had lived through the traumatic
experience of the toxic cloud the interruption of pregnancy, often desired,
has been a drama of violence and shame, writes Marisa Fumagalli (29) who
quotes the witness of a psychiatrist in *Unita* of September 23:

The five hospitalised women were spared nothing; from the intolerable
psychological intimidation exercised continuously and subtly by the nursing
and auxiliary staff of the gynaecological section to the niggardly attitude
of the psychiatrist of the Desio hospital who subjected these women to further
examinations which were unnecessary from a scientific point of view and
shameful at human level: he began the conversation by exhibiting false certifi-
cates which attested to the non-pollution of the dwellings of these same women
(29, p. 101).

Driven back by the counter-attack from the Milan doctors the hospitals of
the area had to give in finally. The women of Seveso were shown no respect.
It had even been insinuated that some of them, taking advantage of the
dramatic situation had feigned non-existant psychic troubles (29, p. 102).

One could not get out of this abortion mire. Thus, on October 30, *Avvenire*
(a conservative Catholic journal from Bologna, close to the Catholic hierarchy)
published, with a lot of noise, a "document on the effects of dioxin" which
originated in the provincial health office of the Democracia Christiana in
Milan:

Small doses of dioxin such as those that may have been ingested by the
inhabitants of the area are probably quite harmless for in order to make
dioxin toxic it must necessarily be present in the human body, like any other
poison, in certain concentrations below which it is not dangerous ...

It has been said that dioxin is teratogenic; this is true for certain
animals but there is an infinite number of substances that one comes across
every day which are teratogenic for certain animals but not at all for man.

It seems quite improbable that teratogenic effects could exist with the
women who have not previously shown functional lesions of the liver or the
kidneys because of the dioxin.

It is true that doubts exist concerning malformations: their probability
would be increased by 50 per cent. What does this mean? If in normal
circumstances four out of every one hundred children are born with some more
or less serious malformations, 50 per cent more would mean six out of every
hundred (23, p. 120).

First official enquiries. After weeks of self-satisfaction, of assurances
and smug optimism the regional authorities began to show some reservations,
to accept some questions. For instance the president of the regional
executive, Mr Cesare Golfari:

In these sectors (sector B) life can return to normal. Elsewhere one will
have to wait till vegetation is removed, the soil stripped and the houses
cleaned. For zone A, on the other hand, especially in the sectors adjoining
the factory, no early return to normal activity can objectively be foreseen.
This will require much time and doubtless very different methods.

We have been to the United States, and we have questioned Vietnamese,
English, German and Swedish scholars. Nobody has been in a position to supply
us with an exact decontamination technique. The only means is photosynthesis.
For dioxin to disappear completely under the influence of the sun, years are
needed. It remains to be seen how the process can be accelerated. Diverse
methods have been suggested to us for this purpose (35).

Here again is V. Rivolta leaning on the legislative and administrative
staff for a denunciation:

"The legislative disorder, the dispersion of competences, the lack of funds
for the public administration to have existing standards enforced which
latter are anyway fragmentary and incoherent." According to him, legislation had to be revised, but without waiting any longer "the services for
prevention, control and assistance must be improved taking into account
that the Lombardy region does not even have a medical officer for every
province".

But it is very late. After so many denials of the noxiousness of the product
these enquiries and these very belated measures met only with the weariness
of an exhausted population. Its main aim was: to forget.

Dioxin? Sure ... But nobody has died, and in any case we are not going to
live like this, in a state of siege, until the end of time!

Scepticism became widespread among these people who had been torn away from
their houses, objects of the headlines of world news, these women in the
centre of politico-moral conflicts about abortion, condemned to give birth to
malformed children or to try to be considered insane in order to get an
abortion which basically they were ashamed of; and all that because of an
invisible poison, often claimed to be 'under control' by the top people.

In the end, these people did not trust science any more (35), Robert Solé
remarked and wrote:

After having had much fear, the inhabitants of Seveso, Meda, Desio and
Cesano Maderno are now very weary.

That is understandable: this dioxin: after all, nobody has seen it. Never
has a disaster zone shown such a normal face. Even if tomorrow a new
yellowish cloud, loaded with the worst poison were to soar over the roofs:

who would guard against it? Here, the sky is cloudy for eight months of the year. A thousand factories sprinkle it permanently with fumes: it means work. Work, 20 km north of Milan, is hallowed. Merchants, artisans and small industrialists have not understood that one lets them live there, lets them come and go but that one closes their businesses. Some of them cheat: they work at night. Big red and white posters had to be put up on the walls to threaten them with sanctions.

The weariness of some does not let one forget the persistent disquiet of others. They were waiting for certainties: they had been given an avalanche of scientific formulae, as obscure as they were contradictory. Obviously, the scholars were groping in the dark, and the politicians had suspended their judgement (35).

One must also remember that the majority of people who were evacuated from zone A owned their houses: they found only temporary shelters, though luxurious ones.

There was also the wrong that had been inflicted on all those small businesses of the province: Two hundred and seventeen businesses had been closed in zones A and B since July and their inventories blocked. As for the 5,000 others, the authorities had to buy full page advertisements to confirm that they were 'sound' and that their products may quite safely be bought (35).

At the end of three months no method of sanitation seemed to be working. The discontent of the inhabitants of the communities concerned (more than 100,000)grew from day to day

October 7: Meeting of the Regional Council — the wager of the 5 micrograms/m^2

This day was chosen for the second meeting of the Regional Council after the accident. The stake at this meeting was not a small one: it had to be decided at what scale evacuations had to be carried out. There had been 800 people in zone A; there were 4,800 in zone B; the watermark would rise to 12,000 if one counted the people who very regularly visited the disaster zone. Still greater precautions could have involved 20,000 people and more*. The field was rather large for a decision on evacuation.

Complicating the 'technical' issue there was the political issue of V. Rivolta's tendency which was in danger of being overthrown by other tendencies within the Democracia Christiana. These difficulties explain a certain confusion in the behaviour of the assembly.

The council started by approving the report by the socialist Scevarolli (president of the Health-Ecology commission) which begins with very hard criticisms of the regional government: Seveso is a tragedy. Until now the whole matter had been dominated by uncertainty and unwise optimism (20, p. 18).

Subsequently, the approach suggested by V. Rivolta i.e. continuation with what had been done so far was approved. The most serious question, however, was that of the threshold to be observed concerning the dioxin. In his report the president of the Health-Ecology commission reminded the assembly of the two opposing demands: the theoretical and the practical:

---

*The communities in the neighbourhood of Icmesa had 220,000 inhabitants.

Our duty is to conduct ourselves as if the most serious risks (resulting from dioxin) had been proved. This would mean, theoretically, to reject any safety threshold if there were not the practical and valid requirement to establish a conventional threshold which in accordance with the recommendations by the epidemiological commission was established at 0.01 micrograms/m^2 for areas of the inhabitable zones and 5 micrograms/m^2 for open spaces (36, p. 92).

The 'Proletarian Democracy' (party) with Mario Capanna wanted to maintain strictness in practical measures:

There is only one way of being sure not to suffer the attacks of dioxin and that is to ascertain that its level is zero i.e. that it does not exist, that it cannot be measured even with the most sophisticated apparatus (36, p. 92).

All parties other than Proletarian Democracy sided with the proposals made by the Region: in the end it was decided to institute an "acceptable" level of pollution: a choice of "unheard of" gravity according to Mario Capanna, a necessary choice according to Laura Conti who sided with the idea of this threshold while knowing the ambiguity of the measure: the cup had to be drained to the dregs; one should not have produced a substance that involved the risk of leading to such a profound drama faced with which one was helpless. Now that it had happened one could not get out of it easily, not even with radical measures. The "socially acceptable" still has to be looked at:

Are we wrong or are we right? It is difficult to tell. Nevertheless, if someone wants to criticise us the following has to be taken into account: if we had chosen another path, for instance considering the dioxin too dangerous to allow acceptance of even the most minimal concentrations measurable by apparatus (between ten and five hundred times lower than those established) then we would have had to envisage the evacuation of about 12,000 people. Where would we have accommodated them? If after three months one can establish that the people suffer from being uprooted even if they are put up in luxurious dwellings, near their usual place of habitation, what kind of existence would it have meant for 12,000 people in a tent village, far from where they used to live? What difficulties of adaptation would have emerged? What problems for the education of the children?

We accepted the threshold of 1 part in 10 billion (5 micrograms/m^2). I would do it again if it had to be done again. But is is frightening to take such a decision. It was frightening because we knew very well that a 'safety threshold' did not exist. There is no quantity of dioxin, no matter how small, of which one could be certain that it carried no danger to the receiving organism. This happens with all 'mutagenic' substances, i.e. those capable of causing mutations in the inheritable cell substance (patrimoine heriditaire). These mutations, if they occur in the cells of the ovaries of the testicles, can cause either sterility or the birth of diseased children; if they occur in the cells of the bone marrow they can cause leukaemia, even after many years.

If we had known that dioxin is definitely mutagenic for the human species, I do not know what our conduct would have been. However, the mutagenic property of dioxin has been proved only on bacteriae, and consequently the substance is considered to be probably mutagenic or carcinogenic in man. We found ourselves, in a manner of speaking, in the presence of a 'probability to the second power' (square). The organism which lives on polluted soil does not have a certain but only a probable expectation of swallowing a substance of which we do not know for sure whether it can be carcinogenic for man.

Choosing the solution of leaving 12,000 people in zone B i.e. on polluted soil meant choosing a remote probability that some child might one day be attacked by leukaemia rather than put 2,000 children definitely in a situation of disorientation, of psychological and affective bewilderment.

If one day I am told that there is a child in zone B that has leukaemia — then, perhaps, a painful feeling of having been wrong may grow up within me: terribly, irreparably wrong, and I shall have to carry this wrong with me for the rest of my life (37, pp. 47-48).

On that day everybody rallied around the proposition of 5 micrograms/m^2. The limits for zone B were subsequently reduced which led to the adoption of the second map. That map, by the way, looks rather arbitrary: it seemed to be based much rather on community limits or geographical facilities (roads, rail tracks) or again on economic 'understanding' (factories and workshops left outside the zone) than on the presence of dioxin. This accounted for the number of factors that had caused the incredulity of the inhabitants of the zone. These, who had been the object of the headlines of the world news, subjected to often contradictory information and measures and spectators of the impotence of the authorities vacillated between weariness, fear and dejection.

October 10: the 'mutiny' of incredulity and refusal

On October 10 the protests of some 700 evacuees, who despite promises had not yet been rehoused near their normal dwelling places (but were staying at the Agip motel, luxuriously but not satisfactorily) blew up in dramatic fashion. During the first hours of the morning of that Sunday, October 10, the disaster victims took their cars and went off on the road to Seveso, forced the barbed wire fences and retook possession of their houses. For several hours the evacuated zone was to be the unreal scene of a gigantic theatrical performance: the actors in the drama played a particularly black comedy; they played to a real world in this universe which showed no sign of disaster: the houses, the gardens, the area, the greenery all appeared so welcoming. They invited each other for meals, for picnics. It needed the police, the constabulary, the provincial and regional authorities to stop such 'true' performance. The regional authorities undertook the task of informing the disaster victims yet again and of starting sanitation with the help of volunteers.

At the end of October, three months after the disaster, the decision was taken to remove the contaminated soil. At the beginning of November one had not got beyond this decision: it remained to be organised and put into action. As regards the treatment of the contaminated equipment, four months after the escape of the poison no decision had yet been taken. The political crisis that developed in Lombardy was not such as to favour a solution of the problem: once more one had enough to do with the institutional and political issues in order not to be able to take the necessary interest in the dioxin problems.

4th: 1976 - 1980: the burden of Seveso. Those responsible chose to deal with people's troubles rather than with the danger. The people demanded compensation, silence and oblivion.

The month of November 1976 appears as a turning point for the population of the contaminated area: they reduced the problem to a matter of compensation; and reducing the drama to a private and personal dimension contributed to Seveso being locked into its solitude.

The first results, after five months, were luckily not as severe as one might have feared. If there were still tens of persons under treatment all those who had been hospitalised had been discharged. Even if this first experience had not been reassuring for the future it permitted a feeling of relief as to the effect of the contamination: there had been no deaths; there had been no slaughter. But what might the future hold as regards malformations at birth and cancers? One did not know yet. The danger remained. The dioxin was still there. What to do?

On December 5 a big meeting was organised to establish a decontamination policy; the agreement reached shows a certain abandon; the improvement should be realised within an expected period of nine months provided that no circumstances intervened that would impede implementation (23, p. 158).

Thus a demand for sacrifices which would inevitably have gone with improvements was avoided: a step backwards was taken. Choosing between danger and discomfort, discomfort was tackled.

During the first three months of 1977 the situation got worse; psychoses, diverse alerts, confusion, as the secretary of the Health-Ecology Commission notes again:

Chloracne was found where no presence of dioxin had been proved and even where, on the basis of previous searches, its presence had been excluded. This shook the population again. Rightly so: dioxin can act in minimal quantities, and the cases of chloracne were a denial of the misleading assurances by the public authorities. The type of information that had been given previously turned public opinion not against the authorities but against science and the scientists; such was the fear and distrust of all scientific achievements (23, pp. 165-166).

The confusion continued into the second quarter of 1977 with births of malformed children; with denied information; with retorts to the lack of responsibility of the analyses carried out and even to the refusal to put them to work etc. In short: the disinfection of the soil and the buildings in zone B went on; economic activity in Seveso was slowed down, the population no longer knew what to think of it all. It was to down tools and, disguised as an anniversary celebration, close the books on the memory of the drama:

A year had passed at Seveso: one preferred to forget. The inhabitants did not want to celebrate the first anniversary of the dioxin cloud which had poisoned their lives, in every sense of the word, since July 10, 1976. This day of remembrance was rather the day of voluntary oblivion. The mayor was on holiday, many of his citizens had gone fishing or gone to cafes. The curate in his sermon took care not to make the least reference to the disaster. Nevertheless, the situation had not changed: one still did not know how to fight the dioxin pollution and one did not know the long term effects (38).

In October one hundred and twenty people (out of the eight hundred evacuees) were able to reoccupy*their houses; others were readmitted in December. However, there was still no question of restarting agriculture or of playing in the gardens: the children had to be taught accordingly.

---

*Zone A was subdivided into 7 sectors: sectors A6 and A7, the least polluted, were reopened.

It was getting close to the second anniversary. The industrial owners and official circles were relieved by the results achieved, especially in May 1978 by the study of Professor Tuchmann-Duplessis. One had expected the worst, and in the end:

The contamination had been relatively moderate because with the exception of the cases of chloracne found in a small number of children no important pathological changes were discovered ... The frequency of abortions in the contaminated area remained clearly below the figures usually registered in Europe ... The number of malformations, while higher for the year 1977 (1.36%) than for the preceding year (0.13%), remained clearly below the frequency of 2.5 to 3 per cent which are generally found in western countries. The number of malformations registered in 1977 does not reflect the real increase in prenatal development troubles but an improvement in medical enquiries ... The postnatal development of the children seemed normal (15, pp. 5, 8, 11).

Others, such as a people's committee close to the Italian communist party, showed more scepticism and accused the authorities of having discouraged the collection of pertinent data. The population had chosen, once again, to forget, as Joelle Kuentz reports:

You, do you see the poison?

There were no deaths and no abnormal births.

You would say, wouldn't you, that all this was very exaggerated?

Just look how green everything is around here. You have worried us, you have predicted the end of Seveso, you said the women would have to abort if they didn't want to give birth to monsters. The feminists came from Rome to spread their claptrap. Then came the priests and took the matter up. They told us to accept the monsters as the will of God. But basically all these things were political manoeuvres, and they continue. So, we have had enough of journalists and politicians! Excuse me, I have to go.

Fear, yes: it's still present but one must forget it. Look how beautiful this garden is. And look on the other side of the fence which surrounds the most polluted area: it's just as green as here. Well, one really asks oneself where the dioxin is. My sister, she continues to worry herself. She preaches every day to her children: don't do this, don't do that! What is it good for? You know, there are also many things in this story which one doesn't understand. Take the cemetery, for instance. It was in zone A, the most polluted. That means one wasn't allowed to enter it without a mask and special clothing. Well, today "they" have given free access: without having done anything. It seems that dioxin has been discovered at Meda, on the other side of the factory. Now, the community of Meda refused to be put within the poisoned zone because it didn't suit the people there. You realise something from that, do you?

... before remarking:

It's true: one is astonished at the map. The outlines too often follow the communal borders, too often leave out certain roads and railway lines so that one cannot discover from it much more than the fantasy forms of the toxic cloud, the reflection of a myriad of interests which don't have much to do with the rigour of scientific experiments ...

The women in zone B who were not allowed to breast-feed received a 'milk allocation'. The children between three and fourteen years of age in zone B and the so-called 'suspected' zone (where the quantities of dioxin had not been experimentally measurable) received the equivalent of 600 ffrs.

per annum if their parents moved them from Seveso for twenty days. On the other hand, the children who lived outside these zones, even within a metre of the demarcation line and who went to school in the polluted zones received nothing. In the same way the families of the B zones benefited from receiving an amount corresponding to their expenses for fruit and vegetables since they could no longer cultivate their gardens while households located in the suspected zone who no longer had the right of using their vegetable gardens received nothing.

In the end nobody believed in the presence of dioxin any longer (39). However, nobody could tell whether the dioxin had left the scene. During the whole year of 1977 there was controversy about the registered malformations. To do justice to all, let us say the margin of uncertainty was perhaps very large but that in any case on the one hand there was no slaughter of the civilian population, on the other hand it would still be unwise to close the files. This is, by the way, the decision of the court: to judge on the basis of the facts known today with this first phase not prejudicing at all what might be found subsequently. With a product like dioxin there is no final point; at best the outcome is in suspense ...

4. BALANCE SHEET

1st: Health

The Hoffmann-la-Roche company drew up the following balance sheet in 1978; it was to include these encouraging points which were proven during the following months:

- The skin affections were benign; they affected only the most sensitive fraction of the population i.e. the children and the adolescents. In the majority of cases they disappeared without leaving a trace.

- The peripheral nervous system was not affected in its functions; the contamination with dioxin caused no repercussions in this respect.

- The livers of the people affected never produced attacks such as result from hepatic insufficiency or, for that matter, any other affection.

- No anomaly was found in the blood, the organic functions and the metabolic process in the cases examined.

- Pregnancy, embryonic development as well as the development of the children did not show any disturbances.

- Examination of the cellular structures responsible for heredity did not reveal any anomaly (40).

As concerns malformations there has been great controversy between the authorities and the "people's committee" which comprised citizens close to the Italian communist party who read the official statistics differently. Here are the two statistics (41):

	1974	1975	1976	1977	1978
Authorities	—	—	4	38	53
People's committee	2	9	34	122	157

SEVESO: The inconsequence of the measures taken after the
accident. Translation of the poster: Polluted zone.
Close windows and air ducts.

   The Hoffmann-la-Roche company explained that neither set of figures is
wrong: there were just different methods of analysis and greater attention
given to the cases of malformations after the accident than usual in similar
circumstances. The opponents contested the figures supplied and noted that
the studies made covered the whole area of the eleven communities near
Icmesa — i.e. 220,000 people — and not only the area most affected by the
accident; due to this there was no significant difference from the usual
figures for malformation; it would have been necessary to study the affected
areas separately. The Democracia Christiana, conclude these opponents,
wanted foremost to let the drama be forgotten. And since Italy was strongly
up against the economic crisis ...

   On other subjects — spontaneous abortions, birth-rate, death-rate — the
same remarks can be made. There had not been a disaster. But, based on the

figures made available, there is divergence between the official publications
which while recommending a follow-up on the situation do not give alarming
figures and the opponents who, like Laura Conti, say that the necessary
researches have been shirked.

2nd: Territory

Zones B and R had never been evacuated. Zone A, divided into seven sub-
zones, could be partly recovered (sub-zones A6 and A7). Thus, out of the 95
hectares that were most affected 60 could be reopened to the population which
means for five hundred and eleven people out of seven hundred and thirty six
evacuees (152 families out of 212). This was achieved in autumn 1977. For
zones A1, A2, A3, A4 and A5 the prognoses are given with some reservations
and range from optimism to most severe pessimism: some ascertain that the
most contaminated territory must be erased from the map (16, pp. 238-240).

3rd: Economy

- Agriculture: 61 farms; 4,000 family vegetable gardens; damage to cultures;
  slaughtering of animals; operation forbidden.

- Industry: 2 businesses in zone A (10 in zone B).

- Artisans: 37 businesses in zone A (121 in zone B).

- Population: help given to disaster victims.

- Sanitation plan

In 1978 the Commission of Enquiry gave an initial sum claimed by the region:
27 billion Lire (136 million ffrs.) remarking, however, that no final figure
could be given before court procedures were concluded. Still, this is just
an approximate estimate. If one takes into account all that the Region had
to set up in the way of funds for dealing with the issue (researches, diverse
interventions) one arrives at a figure of 120 billion Lire (600 million Ffrs.).
And, of course this figure does not include what might appear in the future
since the evil has not altogether disappeared (16, pp. 258-264).

5. SEVESO — TO AVOID OBLIVION

So, there has been no slaughter in Lombardy, neither in 1976 nor afterwards.
This observation does, by the way, not constitute a recognition of the
quality of the established balance sheets. But it is true that there has
been no great disaster as had been feared at the time the evil was diagnosed.

Some derive from this an argument to claim that in this affair there has
only been exaggeration, pollution of minds, phobia entertained by the press
and some marginal groups. One cannot let such blinding of the public take
place easily. The actors in the drama know it well: even if they are not
quite prepared to follow Dr Reggiani who is respected by everybody for his
courage and honesty and who made this remark: for nearly two weeks one did
not know in Basle whether one had to demand — and for how many years? —
the evacuation of Milan. Whence perhaps the caution of the industrial
owners in summer 1976 (2).

Whence also the impossibility of smiling when Seveso is invoked: on
Saturday, July 10, 1976 "something happened" truly at Meda; something that
concerns all of the world's chemical industry and doubtless also all people
who might one day be affected by the factories of this industry.

REFERENCES

(1) P. LAGADEC, Développement, environnement et politique vis-à-vis du risque: le cas de l'Italie — Seveso. Laboratoire d'Econométrie, Ecole Polytechnique, mars 1979 (280 pages).

(2) P. LAGADEC, Développement, environnement et politique vis-à-vis du risque: le cas de l'Italie — Seveso. Additif no 1. Laboratoire d'Econométrie, Ecole Polytechnique, octobre 1979 (9 pages).

(3) O.E.C.D., Report on the dioxin contamination at Seveso. 12th Meeting of the Chemicals Group, October 20th 1976, Paris (21 pages).

(4) E. HOMBERGER, G. REGGIANI, J. SAMBETH, H. WIPF, The Seveso accident: its nature, extent and consequences. Givaudan, Hoffmann-La-Roche et Icmesa, version provisoire (71 pages).

(5) G. REGGIANI, Human Exposure. Localized contamination with TCDD (Seveso, Missouri and other areas). *Topics in Environmental Health — Halogenated Biphenyls, Terphenyls, Naphthalenes, Dibenzodioxins and Related Compounds.* (Chapter II) Version provisoire (108 pages), pour parution chez Elsevier Biomedical Press, Amsterdam.

(6) G. REGGIANI, Toxic Effects of TCDD in man. Nato — Workshop on Ecotoxicology. July-August 1977, Guildford (England), (17 pages).

(7) G. REGGIANI, Medical Problems Raised by the TCDD Contamination in Seveso, Italy. *Archives of Toxicology,* 40, 1978, pp. 161-188.

(8) T. NGUYEN-DAM, *Le Monde,* 11 août 1976.

(9) J. MOORE, Un accident d'origine chimique: la catastrophe de Seveso due à la dioxine. Sécurité civile et industrielle. *France-Selection,* no 285, avril 1979, pp. 14-23.

(10) Roche Nachrichten, septembre 1976.

(11) Seveso: l'usine Icmesa aurait vendu de la dioxine bien avant l'explosion. *La Tribune de Genève,* 25 août 1976, p. 3.

(12) S. ZEDDA, La leçon de chloracné. *Survivre à Seveso,* F. Maspero/Presses Universitaires de Grenoble, 1976, pp. 21-44.

(13) *Le Monde,* 30 juillet 1976.

(14) *Le Monde,* 30 août 1976.

(15) H. TUCHMANN-DUPLESSIS, Pollution de l'environnement et descendance. A propos de l'accident de Seveso. Laboratoire d'Embryologie, Université René Descartes, mai 1978 (11 pages).

(16) Camera dei Deputati VII Legislatura. Commissione Parlamentare di inchiesta sulla fuga di sostanze tossi che avvenuta il 10 luglio 1976 rello stabilimento Icmesa e sui rischi potenziali per la salute e per l'ambiante derivanti da attivita' industriali (Legge 16 giugno 1977, n. 357). Juillet 1978 (470 pages) — Notre traduction.

(17) Lettre de la société Hoffmann-La-Roche à Madame J. KÜNTZ.

(18) Cl. RISE, V. BETTINI, C. CERDERNA, Derrière l'Icmesa. *Survivre à Seveso, op. cit.*, pp. 61-68.

(19) G. MAZZA, V. SCATTURIN et Gruppo P.I.A. (Gruppo di prevenzione e di igiene ambientale del consiglio di fabricca Montedison, Castellanza). Icmesa: Come et per ché? *Sapere*, novembre décembre 1976, pp. 10-36.

(20) G. CERRUTI, Cent jours à la dioxine. *Survivre à Seveso, op. cit.*, pp. 9-20.

(21) E. de ROSA, Scienza et Societa. *Scientific American*, édition italienne, mars 1977.

(22) G. PECORELLA, Qui va payer? *Survivre à Seveso, op. cit.*, pp. 105-117.

(23) L. CONTI, Visto da Seveso. Feltrinelli, Milano 1977.

(24) *Le Monde*, 31 juillet 1976.

(25) *Le Monde*, 29 juillet 1976.

(26) *Le Monde*, 13 août 1976.

(27) *Le Monde*, 1-2 août 1976.

(28) *Le Monde*, 3 août 1976.

(29) M. FUMAGALLI, Avorter à Seveso. *Survivre à Seveso, op. cit.*, pp. 95-102).

(30) *Le Monde*, 10 août 1976.

(31) *Le Monde*, 12 août 1976.

(32) *Le Monde*, 11 août 1976.

(33) *Tempo*, no 31, 8 août 1976.

(34) *Le Monde*, 17 août 1976.

(35) *Le Monde*, 5 août 1976.

(36) M. CAPANNA, Un nuage sur l'institution. *Survivre à Seveso, op. cit.*, pp. 85-94.

(37) L. CONTI, Trop d'échéances manquées. *Survivre à Seveso, op. cit.*, pp. 54-58.

(38) *Le Monde*, 12 juillet 1977.

(39) *Le Matin*, 13 juin 1978.

(40) Roche Nachrichten, juin 1978.

(41) *La Stampa*, 11 mars 1979. Corriere della Sera, 26 février 1979.

(42) *Le Monde*, 28 octobre 1976.

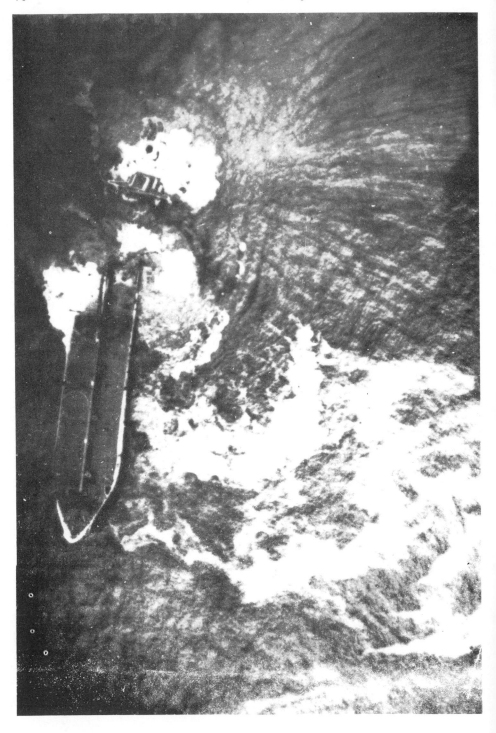

III. THURSDAY, MARCH 16, 1978: THE AMOCO-CADIZ

On March 16, 1978, the *Amoco-Cadiz*, an oil tanker of more than 230,000 gross weight tonnes, flying the Liberian flag, ran aground on the shallows about four miles*from the small Brittany port of Portsall on the coast of Nord-Finistère. Its cargo of 230,000 tonnes of crude oil spread on the sea causing the pollution of more than 400 km of beaches from Pointe Saint-Mathieu in the West to the island of Bréhat in the east.

This pollution, exceptional as to the volume of oil spilled as well as to the stretch of coastline affected, struck the Brittany coast which had already, for eleven years, been victim to disasters of the same kind. But those, despite their seriousness, had not so far reached a comparable scale (1, p. 5).

Those who have not seen the shoals of drifting fish, those oil coated birds in their agony, those rocks looking sad in the slimy setting that enveloped them, those oyster banks affected right down into their sediments, those who have not touched the sand soiled right into the bowels of the beaches and the seaweed stained and abandoned in blackness, those who have not admired the dignified and courageous Brittany people fighting, sometimes up to their waists in that mire of despair — they can hardly get the measure of such a disaster.

In order to understand one must also have seen the spectacle of that steel monster, broken but still proud of her prow pointing towards the coastline; the defeated monster that flouted the soldiers and all those of good will who went there, from Pointe Saint-Mathieu to the Isle de Bréhat to extirpate untiringly, drop for drop, the harm which the next wave would bring back upon them.

There was a magnificent spirit of solidarity which transformed the cry of alarm and impotence of a civilisation which had been overcome by the creation of its own genius (2, p. 4).

1. THE STRANDING**

1st: The facts as they seem to have occurred out at sea

08.00 h:***

The *Amoco-Cadiz* which belonged to the Amoco company whose registered office is in Chicago, flying the Liberian flag and skippered by the Italian Captain P. Bardari, was to the southwest of the island of Oessant. Coming from the Persian Gulf it was going to Rotterdam via Lyme Bay (GB). It went up the northward lane of traffic**** at a speed of 9.5 knots.

---

*The distance is doubtless somewhat overstated in this report.

**We take up — summing up sometimes — the attempt at reconstruction of the facts presented in the report by the Senate Commission of Enquiry (1, pp. 15-52).

***The hours are given at GMT i.e. one hour after the official French time at the occurrence.

****Defined by the Intergovernmental Maritime Consultative Organisation (IMCO), this device is more commonly known as the 'rail'

The tug *Pacific*, an ocean-going salvage vessel equipped with 10,000 HP, registered in Hamburg and skippered by Captain H. Weinert, a West German, left the port of Brest at 08.24 h. The weather forecast was bad, predicting southwesterly or westerly winds force seven gusting to force 9 and perhaps to gale force 10 later.

09.45 h:

Running at 14.5 knots the *Pacific* was rounding the headland of Saint-Mathieu and approaching the channel of Le Four.

The *Amoco-Cadiz* was about 7.5 miles north of Ouessant. The helmsman told the captain that the vessel was off course: There was a breakdown of the steering system, it was blocked in a position that steered the vessel to portside.

Captain Bardari had the engines stopped and decided to call for assistance. He drafted a so-called safety TTT*message saying that the *Amoco-Cadiz* "was no longer manoeuvrable" and asking other vessels to stand by.

10.20 h:

The message was sent by telegraph (09.50 h) and by radio (10.00 h). The station Le Conquet-Radio picked up the telegraph message at 10.20 h.

11.05 h:

The *Amoco-Cadiz* made contact with Radio-Conquet; it enquired about the nearest tug station. Captain Bardari unable to contact the ship's owners in Chicago because of the time difference tried to alert the Amoco company's representatives in Genoa, this via Radio-Conquet (11.15 h). But he did not succeed.

11.20 h:

Attempts at repairing the blockage having been unsuccessful, Captain Bardari requested tug assistance; this request was transmitted by Radio-Conquet.

11.28 h:

The *Pacific*, about thirteen miles away, turned round and made contact with the tanker; it offered its services on the basis of the assistance type contract**(Lloyd's open form). The *Amoco-Cadiz* asked them to wait.

12.08 h:

The chief officer of the *Pacific* asked the *Amoco-Cadiz* again whether she accepted the Lloyd's open form. All he received for an answer was the address of the ship's owners in Chicago. In fact, Captain Bardari was trying to inform the ship's owners of the situation.

12.20 h:

Since its call was received the *Amoco-Cadiz* had already drifted about two

---

*First degree in the scale of accident messages (see following footnote).

**This form of arrangement permits avoidance of negotiations prior to assistance operations; in case of successful salvage the amount of remuneration is fixed by arbitration, usually in London (1, p. 168).

miles to the southeast. It had crossed the southern limit of the northward
lane. It rolled heavily under the influence of winds that gusted to force
8-10. Its captain made contact with the tug and informed it of a "breakdown
of steering, engines in good order" and "his intention to be tugged".

The tug put itself to within about 400 metres starboard of the *Amoco-Cadiz*.
Captain Weinert talked "to somebody from the tanker who by (his) accent
seemed to be English".

Despite the bad weather the island of Ouessant was clearly visible.

13.15 h:

This was the very beginning of preparatory operations for tugging. The
two ships were about 5 miles north-northwest of Ouessant; the wind blew from
the northwest; the seas were high; a weak current ran towards the island.
The *Amoco-Cadiz* had still not accepted the offer of assistance. Shortly after
14.00 h the *Pacific* began to pull.

The *Amoco-Cadiz* then refused a re-newed offer of assistance; this did not
improve relations between the two captains and for a moment seemed to affect
the tugging operation, according to the skipper of the *Amoco-Cadiz*. Nevertheless, the tugging continued, but not without some difficulty: the master
of the *Pacific* confirmed (at 15.00 h) that he did not know the position of the
tanker's rudder or whether the engines of the *Amoco-Cadiz* were working. A
second tug, the *Simson*, however far away, took course towards the two vessels.
vessels.

15.15 h:

A new contract proposal made by the tug was strongly rejected by the
tanker's skipper.

16.00 h:

The *Amoco-Cadiz*, having received approval from Chicago, accepted the
Lloyd's open form. On the basis of the statements made by the tug's skipper,
four and a half hours had passed since the first offer of service.

The Captain of the *Amoco-Cadiz* for his part maintained that he only wanted
a tugging contract and that the captain of the tug had changed the nature of
the contract during the tugging (to be precise: at 14.35 h).

16.05 h:

Captain Weinert was informed that the *Simson* expected to arrive by 23.00 h.
He in turn informed Captain Bardari.

According to the tug's skipper this was the only occasion during the whole
day when he was directly in contact with his counterpart from the tanker. He
learned then that the engines of the tanker were in working order but he
still had not received an answer as to the position of the rudder.

16.15 h:

The tugging chain snapped at the tanker end. The wind was straight west,
force 8, pushing 9-10. There was a heavy swell. The waves were close to 8
metres (high, deep). The current began to return (it now bore west-southwest).
The tug stopped its engines and began to recover its tow. It informed the
tanker that it would make a new attempt as quickly as possible and suggested
the tanker should reverse its engines. It received no answer. Captain
Bardari had the engines reversed.

<u>17.05 h</u>:

The crew of the *Pacific* began preparations for a second attempt at tugging. The *Pacific* was stopped and pitched heavily. Two sailors were injured.

Preparations continued (17.35 h) and conversations with the tanker took place (16.23, 17.15, 17.59 h).

<u>18.20 h</u>:

Contact between the *Amoco-Cadiz* and Chicago via Le Conquet-Radio. Captain Bardari informed his interlocutor of the situation. He told him in particular that "if all other methods fail then the ship will drift towards the coast and distress measures will have to be taken".

<u>19.06 h</u>:

The tug approached the tanker. The wind which had veered to northwest had become stronger; the swell was strong; the current got stronger. Three attempts to launch a line failed: 19.10, 19.15, 19.20 h. A mechanical accident led to a fourth failure. The *Amoco-Cadiz* dropped its portside anchor (20.04 h). A fifth rocket was launched, and the tug could at last make fast (20.55 h) which had proved difficult because of another mechanical accident (20.28 h).

<u>20.55 h</u>:

The *Amoco-Cadiz* requested the tug to pull softly; the tanker could not lift its anchor. After a two minute breakdown of the rudder the *Pacific* began to pull slowly (20.57 h).

<u>21.04 h</u>:

The tanker ran aground astern for the first time. Its engines were flooded. The lighting was cut and the radio contact broke off. Captain Weinert reported this to Hamburg (21.13 h).

<u>21.39 h</u>:

The stern of the *Amoco-Cadiz* hit the seabed a second time.

<u>21.43 h</u>:

Weinert called Hamburg and put "all engines full ahead". The tanker fired red flares. It began to lose oil. The black tide had started.

<u>21.50 h</u>:

The *Pacific* requested a helicopter to evacuate the crew.

<u>21.55 - 22.00 h</u>:

The engines of the *Pacific* were stepped up to full strength. The tanker had now definitely run aground.

<u>22.12 h</u>:

The tug line snapped.

<u>22.30 h</u>:

The *Simson* arrived on the scene.

Midnight:

Helicopters from the Navy arrived and began to evacuate the crew with a winch.

01.45 h:

Forty two people out of forty four had been rescued. Captain Bardari and one officer stayed on board. They were evacuated at 05.03 h.

2nd: The facts as they seem to have been perceived on land

a) The perception of events at Radio-Conquet

Traditional functions of the station. Like all maritime radio stations along the French coasts Radio-Conquet had two principal functions: to ascertain a communications service between ships and the land and between ships; to ascertain a lookout service for the safety of human life at sea (this on coded frequencies and according to coded procedures*.

The XXXs called "emergency messages" constitute the second degree in the coded scale of incidents or accidents which vessels may transmit by radio. A XXX signifies that something serious is happening on board but that neither the safety of the vessel nor that of its crew is involved. No XXX was sent by the *Amoco-Cadiz*.

The distress messages which alone have the effect of starting interventions constitute the third and last degree in the scale of accidents which vessels in difficulty may transmit by radio. No distress message was received from the *Amoco-Cadiz* on March 16 before 23.18 h (1, p. 31)

Information received by the station:

Radio-Conquet picked up the end of the TTT message (see preceding footnote) indicating a breakdown of the steering system; despite calls for repetition of the message the transmitting vessel whose identity was not known did not answer.

11.05 h:

The *Amoco-Cadiz* made contact with the station, then made a request for assistance which was to be relayed by Radio-Conquet (11.20 h).

11.15 - 18.41 h:

During these hours the station put the *Amoco-Cadiz* in touch with Genoa (11.15 h) and twice with Chicago (for fourteen minutes at 11.30 h and for six minutes at 18.35 h). It also put the Pacific in touch with Hamburg seven times (altogether over forty minutes of communications) and transmitted messages between the tug and the tanker, the *Pacific* and the *Simson*.

19.50 h:

The second in command of the Portsall lifeboat telephoned Radio-Conquet to

---

*For the messages there is an established grading system: A TTT is a so-called 'safety' message. A TTT constitutes the first degree in the coded scale of incidents or accidents which vessels may transmit by radio. The purpose of the TTT is to inform other vessels of an incident on board that may present a danger to the vessels cruising in the vicinity. The TTTs are frequent and are not considered alarming in themselves.

advise that he had noticed "a light at about two miles from the rocks of Portsall".

20.00 h:

Le Conquet transmitted this information to CROSSMA*who replied: OK, well received. But is the light on land or is it a distress signal? Le Conquet replied: No, no distress signals but the two vessels are very close. CROSSMA replied: OK, navigation lights. Le Conquet then gave the telephone number of the second in command of the Portsall lifeboat to CROSSMA.

20.16 h:

CROSSMA alerted Radio-Conquet that the sub-master of the station of the National Lifeboat Society at Portsall "after a patrol had stated that there was a ship being towed by another".

21.00 h:

Questioned by the Operational Centre of the Navy (COM) Le Conquet confirmed that the *Pacific* was towing the *Amoco-Cadiz* and did not indicate that the latter was in difficulty.

21.13 h:

The *Pacific* called Hamburg and talked for eight minutes: at 21.43 h the *Pacific* called Hamburg again and talked for five minutes.

21.50 h:

The *Pacific* requested a helicopter for the *Amoco-Cadiz*; Radio-Conquet transmitted the message to CROSSMA (22.00 h).

22:00 h:

Radio-Conquet established a connection between the *Amoco-Cadiz* and Chicago (connection relayed by the *Pacific*); the tanker explained its distress situation and the fact that it was losing oil but did not issue a distress signal.

22.34 h:

The *Pacific* requested a call to Hamburg and talked for five minutes.

23.18 h:

The *Amoco-Cadiz* sent out a SOS call which Radio-Conquet transmitted to CROSSMA (23.22 h).

b) The impression of the events by the Navy

Duties of the Navy. Among the duties of the Navy the Senate Commission mentioned the following two responsibilities besides the more classical ones of policing, assistance and rescue:

In the terms of Article 16 of the law of July 7, 1976 concerning the prevention and repression of maritime pollution the state and therefore the competent authorities, among them the Navy, are authorised to intervene

---

*Regional Operational Surveillance and Salvage Centre for the Channel, located at Cotentin (Joburg).

for the purpose of prevention in cases of breakdown at sea occasioned by a vessel that could create serious and imminent danger likely to affect the coastline.

In the terms of a more recent law (however preceding the accident of the *Amoco-Cadiz*), the decree of March 9, 1978 concerning the organisation of actions by the state at sea, the port-admiral had conferred upon himself a very general responsibility as representative of the Prime Minister and of every minister. He has general administrative police authority at sea and is invested with a general responsibility in all fields in which the state acts.

The look-out facilities of the Navy are mainly the signal stations: there were four of them in the area concerned with the *Amoco-Cadiz*. The Operational Centre of the Navy (COM) is in charge of carrying out military operational missions and missions of "public service"; it has to watch permanently "the situation of vessels in the Atlantic area" (Order of April 30, 1974).

Information received by the Navy

09.23 h:

The signal station at Stiff (Ouessant) intercepted a message which it could hardly understand and the origin of which it could not make out. "Engine ... keep ... clear ...". It did not report to COM. At 11.15 h the signal stations communicated between themselves about the communications from the Pacific and calls in English. A look-out was set up.

13.16 h:

The signal station at Stiff, having heard the *Pacific*, asked it what was happening. The Pacific replied that it was towing the Liberian tanker *Amoco-Cadiz*. The signal station informed COM immediately and requested the position of the vessels. The deputy officer on duty at COM did not consider this information to be alarming and did not react.

13.20 h:

The signal stations at Stiff and Molène heard the beginning of a communication in English but the interlocutors were using a frequency which the signal stations had neither the technical means nor the duty to survey.

14.00 to 18.30 h:

The signal stations followed the positions of the *Amoco-Cadiz*. The station at Molène requested information from the one at Creac'h (Ouessant) when it saw the convoy at 15.00 h. Creac'h answered: "tanker in tow, no reason for alarm". Neither Stiff nor Creac'h considered it necessary to inform COM which was unaware of the course of the convoy.

18.30 h:

In accordance with regulations the signal stations ceased their look-out at sunset.

20.34 h:

The signal station at Saint-Mathieu which keeps look-out during the night was informed by a private party that a convoy consisting of a tanker and a tug seemed to be immobile, very near the coast, off Portsall. The signal station alerted COM. COM was also alerted at 20.35 h by the look-out of the Molène station who, having finished his duty normally, nevertheless noticed lights rather near the coast and called COM.

20.40 h:

COM took a number of measures (calls to vessels, alert, and later rescue operations).

3rd: Impression of events received by CROSSMA (Ministry of Transport)

a) Duties of CROSSMA. This administration is mainly in charge of the prevention of accidents at sea, maritime assistance, maritime rescue, functions which are not immediately connected with a problem like that of the running aground of a tanker; however, Article 16 of the law July 7, 1976 concerning the prevention and repression of maritime pollution confers on the state, and thus on services like this one, the right of summons in cases of damage at sea which might cause serious and imminent danger and affect the coastline.

b) Information received by CROSSMA. Informed by the second in command of the life-boat at Portsall of the presence of unusual lights two miles from the rocks of Portsall, the station Radio-Conquet transmitted this information to CROSSMA Joburg. CROSSMA asked "whether the lights are on land or whether they are distress signals". Le Conquet explained that they were not distress signals but that the two vessels were very close to the rocks. CROSSMA replied: "OK, they are navigation lights". Le Conquet gave CROSSMA the phone number of the second in command of the life-boat at Portsall from whom the information came. CROSSMA then requested additonal information from the rescue station at Portsall. It received the answer that those responsible had gone out for a while to watch the situation. At 20.10 h the rescue station at Portsall called CROSSMA back and indicated "that it is a false alert; the lights are those of a tanker under tow by the tug *Pacific*".

20.16 h:

CROSSMA for its part reassured the station Radio-Conquet by advising that "the second in command of the life-boat at Portsall after a patrol along the coast has stated that there was a ship being towed by another".

21.42 h:

COM, informed by the signal station at Aber Wrac'h that a red signal rocket had been launched by the tanker, told CROSSMA that a tanker of 230,000 tonnes was about to run aground.

22.48 h:

The rescue station at Portsall advised that a fishing boat was going to the site where the vessel had run aground.

23.10 h:

CROSSMA alerted the sea rescue station of Ouessant and then the one at Molène.

23.38 h:

CROSSMA was informed that Polmar-Sea had been put into operation.

2. SEARCH FOR EXPLANATIONS

The Senate Commission underlines before giving its analysis:

Rather than looking for "scapegoats" it intends to carry out a useful, calm

analysis but without complacency towards the organisation and actual
functioning of the administrations concerned.

1st: Search for explanations about the events at sea

The Commission of Enquiry gave the following facts (1, pp. 54-59).

- Difficult atmospheric conditions: waves, frequently coming up to 9, even
  10 metres; the strongest gusts of wind were registered at the most
  critical moments of the towing operation.

- At the outset, worrying but not disastrous technical conditions: the
  tanker's engines were in working order; the tug was near the *Amoco-Cadiz*
  when the latter asked for assistance.

- A laborious communication between the two skippers: at least one of them
  had difficulties expressing himself in English; they were only rarely in
  direct contact. The technical information on operations in connection
  with the towing went rather badly. There was, among other things,
  divergence between the two skippers on the tactics to be adopted for
  carrying out the towing successfully. On the other hand, the respective
  skippers kept mostly in touch with their ship's owners.

- At no time did either the captain of the *Pacific* or that of the *Amoco-
  Cadiz* think it necessary to ask for help. At 21.43 h, more than an hour
  and a half after running aground, the tanker launched red signal rockets.
  The first request stating serious difficulties was made around 21.50 -
  22.00 h. It was then that the *Pacific* asked Radio-Conquet for helicopter
  assistance to evacuate the tanker.

2nd: Search for explanations about the part the authorities concerned played or could have played

a) Radio-Conquet. The anlaysis of the senate report on the enquiry reads as
follows:

By means of radio messages which they heard or received, transmitted or
sent during the whole of March 16 those responsible at the maritime radio
station of Le Conquet, which comes under the direction of the international
telecommunications network, were in a position to know fairly fully what
was happening at sea (1, p. 30).

- The station was in a position to know that the Amoco-Cadiz was being
  damaged on account of the TTT message overheard at 11.20 h. This message
  confirmed after all the significance of the incomplete message of which
  the origin could not be identified and which had been received at 10.20 h.

- The station was informed that an attempt to assist the damaged vessel was
  under way.

- Those responsible at the station seemed to be in a position to appreciate
  that the attempt at giving assistance was not taking place in the best of
  circumstances. This evaluation would appear to have been possible without
  violating the secrecy of private communications. In fact, the station
  gathered in transit a number of unusual communications coming from
  the convoy formed by the *Amoco-Cadiz* and the *Pacific*. They are here
  recalled to memory: message from the *Amoco-Cadiz* to Genoa at 11.15 h; to
  Chicago at 11.30 h (eighteen minutes), at 18.35 h (six minutes), at 22.00 h
  (twenty one minutes); messages between the *Pacific* and the *Simson* at 13.07 h;

from the Pacific to Hamburg at 11.35 h, 13.40 h, 15.04 h, 15.55 h, 16.20 h,
17.15 h, 18.03 h, 21.13 h, 21.43 h, 22.34 h. In addition, certain information was brought directly to the attention of those responsible at Le
Conquet. Thus at 13.07 h the *Simson* after having asked Le Conquet to
advise the *Pacific* that it wanted to make contact, specified that it was
on its way to the *Pacific* at full speed. At 19.56 h the second in command
of the life-boat alerted Le Conquet about a disquieting fact in this
context: the presence of a light two miles off the rocks at Portsall (1,
p. 60).

So, Radio-Conquet had knowledge throughout the day on March 16, 1978 of a
large number of facts, in themselves of various degrees of seriousness but
the convergence of which was of a disquieting nature. In addition, those
responsible at the station were not unaware of the weather situation which
was hardly propitious to the execution of an assistance, even in good
conditions, to be given to a ship of more than 200,000 tonnes, fully loaded,
by a single tug even though it was publicly known to be the most powerful in
the area (1, p. 61).

The information that was spreading about after 20.00 h did not give rise to
any doubts. Still at 21.00 h when COM addressed a request to Radio-Conquet
the reply was just that the *Pacific* was towing the tanker *Amoco-Cadiz* without
any hint about the difficulties of the tanker.

It appeared that in applying a very recent regulation Radio-Conquet like
all exterior services or public establishments who had competence at sea
seemed to have a duty to inform the port admiral. This duty to inform seems
to have been capable of being interpreted rather vaguely. Article 5 of the
decree Nr. 78-272 of March 9, 1978 concerning the organisation of actions of
the state at sea stipulates in fact that "the authorities on land, the
exterior services and public establishments of the state that have competence
at sea shall keep the port-admiral informed of matters likely to be of
importance at sea and shall communicate to him all the relevant information
on regulations in force and on decisions taken".

All these details lead one to be astonished at the prolonged silence of
Radio-Conquet towards the diverse administrative authorities with responsibility at sea. From the communications which were passed about during the
day on March 16, 1978 the maritime radio centre of Le Conquet seemed in fact
to have been in a position to conclude that something serious was happening
offshore (1, pp. 61-62).

However, in order to understand the attitude of Radio-Conquet we must draw
attention to four more aspects:

- The extremely large number of communications channelled through the station
  this necessarily led those responsible at Radio-Conquet to minimise the
  seriousness of events of which they had knowledge.

- The absence of any distress message on the basis of regulations in force,
  until the decree of March 9, 1978, the station was not held to alert the
  authorities.

- The customary rule of responsibility and freedom of decision of all ship's
  masters: the skipper of a ship is the only judge of the situation, and
  inasmuch as he does not officially request assistance or inasmuch as he
  does not communicate the difficulties he is faced with he is assumed not
  to need any particular help.

- The private and confidential nature of the communications which transit through the station.

The whole question, the senate enquiry report concludes, is to see whether respect for the principle of secrecy of private communications could have overruled the application of the principle to inform the port-admiral which was established in the decree of March 9, 1978 and of which it is after all not certain whether the directions for its application had been brought to the knowledge of those responsible at the maritime radio station of Le Conquet by circular letter or instruction before March 16, 1978 (1, p. 63).

b) The Navy. The Navy for its part received only little information, did not have a sufficient structure to deal with such non-military information and in any case did not have the proper means to foresee the running aground. Even if Article 16 of the law of July 7, 1976 concerning the prevention and repression of maritime pollution and the recent decree of March 9, 1978 conferred upon it responsibilities for summons and assistance, the Navy could hardly carry out such a mission successfully.

A few explanatory facts may be added (1, pp. 66-68):

- The Navy was never informed by the vessels themselves.

- On account of the weather situation (which at the same time increased the the risks and difficulties of navigation or assistance) and of other priority duties no vessel or plane patrolled the sector involved during March 16; the communications between the tanker and the tug were therefore not picked up.

- On account of the gale the indirect sources of information of the Navy (fishing boats, boats from other administrations) could not play their usual part.

- Only the station of Radio-Conquet had solid information on the situation but this was not transmitted to COM.

- All through the afternoon of the 16th the signal stations registered information that could have seemed abnormal as to the course pursued by the *Amoco-Cadiz* and the *Pacific*. They were actually in a position to tell that the two vessels moved only very slowly and were outside the shipping lane, on the land side.

- Being responsible for the application of the regulation concerning the passage of vessels the Navy could have been worried following receipt of information at 13.35 h coming from the signal station at Le Stiff: the convoy was in an abnormal position. A tolerance might be admissible for a small vessel but not for a tanker of more than 300 metres length. Rule 10 (par. d) of an order of June 30, 1977 was clear: "the coastal shipping lanes must not normally be used by through-traffic. "The size of the vessel in difficulty, its position, the weather conditions, the nature of the assistance operation should have been taken into account and consequently have caused concern.

However, (1, pp. 68-69):

- There was an error in the assessment of the speed and direction of the convoy which information had been transmitted to COM; this may largely be explained by the fact that they had been arrived at from observations

registered by the very rudimentary means then at the disposal of the
signal station.

- The non-reaction to the information transmitted by the signal station at
  Le Stiff (13.50 h) may partly be explained by the fact that at the time
  there was at COM no group charged with the surveillance of commercial
  traffic. COM was mainly an echelon for the centralisation of military
  type information.

- The order of June 30, 1977 spelled out that the signal station had to
  report infringments to the surveillance vessel; on March 16, 1978 no
  vessel was on surveillance duty in the shipping lane.

But in addition and lastly (1, pp. 69-71):

- Even if the Navy could and should have intervened offshore in the area
  concerned on that day, particularly in view of Article 16 of the law of
  July 7, 1976 (there certainly was serious and imminent danger for the
  coastline in the sense of Article 11-4 of the Brussels Convention of 1969),
  it did not have the tugging facilities that would have been adequate to
  the requirements of the *Amoco-Cadiz*.

c) CROSSMA. They were informed even more belatedly than the Navy and had,
according to them, no means of intervention that would have been adequate to
the situation that existed on March 16, 1978, no matter what duties seemed to
have been given them, mainly those of the decree of April 30, 1974 (1, p. 74;
48).

3. POLLUTION

The essential points to note concerning the pollution*caused by the *Amoco-
Cadiz* are the following:

- Light oil containing one third aromatic substances which evaporate but
  also spread easily in sea-water with contents of high toxicity for the
  living environment.

- A very short distance between the location of the wreck and the coast
  (less than two miles).

- A massive quantity of hydrocarbon spilled within a few days: 223,000
  tonnes in less than two weeks, i.e. a discharge of 18,000 to 20,00 m^3 per
  day which had a disastrous effect on the environment (1, p. 107).

From the time of the stranding, oil flowed towards Portsall and its
vicinity, pushed by the winds which blew at gale force from the north-northwest and covered the coast in oil. On March 17, a circular area around the
wreck with a radius of four miles was polluted. On the 18th, a north-north-
easterly wind pushed the slick southwards: it reached the cape of Saint-
Mathieu and threatened the offshore area of Brest. On the 20th the wind
changed from northeast to west and pushed the oil slick to the northeast,
sparing the offshore area of Brest; the oil reached Roscoff during the night.
In the afternoon of the 21st it arrived in the bay of Lanniou; the gale
increased again in force after a relative lull. On the 23rd the north coast
of Brittany was affected up to Sillon de Talbert in the east: two thirds of
the cargo had already been discharged into the sea.

---

*See map of maximum spread of the oil slick p. 89 (3, p. 16 or 4, p. 21).

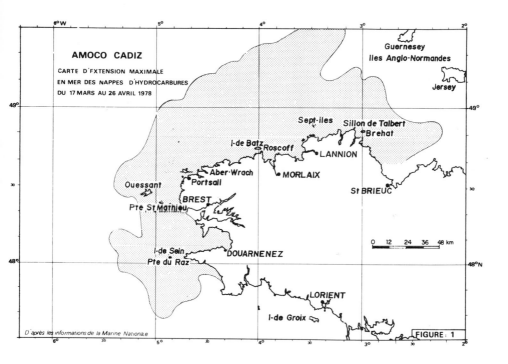

Fig. 12: Map of maximum extension of hydrocarbon slicks on the sea.
(Source: National Centre for the exploitation of the sea)

Then (March 25-26) the equinoctial tide came and carried the slicks high up
on the rocks and beaches, to places which only another equinoctial tide could
again reach. The estuaries were particularly affected. Offshore very large
areas of oil slick drifted east and for a while threatened the bay of Mont
St. Michel, Jersey and Cotentin.

On March 29 the winds changed again; some places, such as Perros Guirec
so far spared, were polluted. The oil slicks were pushed offshore by the
southwesterly wind. The largest of them (75 miles long) drifted west afterwards under the effect of a north-northeast wind. A small part of it passed
the cape of Saint-Mathieu; it broke up and veered south; on April 19 some
oil slicks rounded the cape of La Chèvre and touched the bottom of the bay of
Douarnenez on April 21; some traces of hydrocarbon reached the bay of
Audierne in May.

At the beginning of May evaporation, dissolution and the battle against
the oil slicks which had been attacked with dispersants and precipitants led
to a residual state that excluded noticeable new arrivals on land. But the
wind had already slackened considerably: the great tides which follow the
equinoctial period could no longer affect the expected cleaning process.

The polluted area extended therefore from Le Conquet to Le Sillon de Talbert,

some sites outside this area having suffered light attacks as e.g. the bay
of Douarnenez. Inside the area very few sites had been spared (1, pp. 107;
2, pp. 206-209; 3, 4 and 5).

4. THE BATTLE AGAINST THE POLLUTION

1st: The Polmar plan

à) 1970: A battle plan designed following the black tide of the Torrey Canyon

The battle against maritime hydrocarbon pollution had been the subject of a
ministerial order issued on December 23, 1970 in response to the stranding
of the *Torrey Canyon* on March 18, 1967. In short, the purpose of this order
is to define the responsibilities of the administrations concerned and to
coordinate their interventions. The order is explicit in a document
entitled "Pollution of the sea and the coasts by hydrocarbons — Polmar
Plan". The plan specifies the allocation of responsibilities to the various
ministerial departments; in an annex there is especially attached a telephone directory and a list of materials and equipment for fighting pollution (1, pp. 85-87, 2, pp. 189-204).

b) 1976: An organisation redefined after recent black tides

Following the disasters of the *Olympic Brevery* and the *Boehlen* in 1976
certain modifications were considered indispensable; decentralisation of
the launching of the Polmar plan (where the 1970 law had given this
responsibility to the Prime Minister); launching from the moment when there
was a threat of pollution (without waiting for actual pollution), making
available an intervention fund (to respond to requirements without delay),
increased battle facilities and preparations.

These new lines of organisation were contained in a report from GICAMA*
(spring 1977) and were adopted by the Council of Ministers on May 25, 1977
(6, p. 33).

c) 1978: The reform undertaken in 1976 is not achieved

The reform of the Polmar plan was undertaken. But it had not been achieved
to any great extent, when the *Amoco-Cadiz* ran aground. On three capital
points the system was found to be lacking (1, pp. 191-199; 2, pp. 200-204):

General organisation

The instructions were in the process of being worked out and the defects
brought to light by the report of the government presented to parliament
eighteen months earlier (during the discussion of the law of July 7, 1976)
were still there (2, p. 202).

The circuit of decisions may still appear too complex inasmuch as the
measures to be taken require swiftness, singleness of responsibility and
simplicity in the administrative rules (2, p. 202).

---

*Interministerial Group for the coordination of actions at sea by the
administrations, created by a decree of April 19, 1972 for the purpose of
coordinating the actions of some fourteen administrations involved
(6, p. 33).

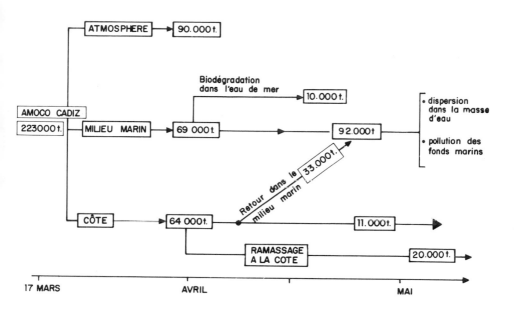

Fig. 13:   "What became of the 223,000 tonnes"

The means available

The report submitted by GICAMA in 1977 remained valid; it established that all the equipment available would at best permit the containment and elimination of an accidental discharge of 5,000 to 10,000 cubic metres of hydrocarbon on the hypothesis that weather conditions and local currents permitted it.   Now, since 1972 the creation of the means to combat the discharge of 30,000 tonnes of hydrocarbon had been defined as an objective. This effort of reflection had been made on a par with the procurement of equipment and material to cope with pollution (1, p. 140).

Preparation for action.
The parliamentary commission indicates:

No actual, complete and combined sea and land, exercise had been organised, neither under the rule of the order of 1970 nor since the reform of Polmar had started.   Apart from this, all the envisaged hypotheses were based on limited discharges (30,000 tonnes for the calculation of means, 15,000 tonnes for the exercise of mobilisation in 1975) (2, p. 203).

It was with this equipment for battle, duly analysed for its insufficiencies in 1975, that the situation had to be faced:

The reforms undertaken since 1976 and the decisions taken in 1977 had theoretically permitted a better adaptation of these tools to their task.

However, the slowness of realisation of these reforms constituted a major
obstacle (2, p. 204).

Summing up it appears therefore that in March 1978 we did not have suitable
equipment to combat pollution.   In addition, it is obvious that in the
aggregate the equipment was not up to the size of the disaster.

2nd: Application of the Polmar plan

a) Setting up the battle apparatus.   The Polmar plan was put into operation
by the Port-Admiral (sea operations) and the police commissioner of Finistère
(land operations) on March 16, 1978 at 23.45 h.   The Operational Centre of
the Civil Defence Directorate (in Paris) held a first meeting two hours later
at 01.45 h.   The following day the interministerial commission for the battle
against pollution by hydrocarbons designed the strategy to be followed; supply
of all barriers available in France to the disaster area; alerting all
European maritime states; an attempt at "lightening" the tanker as soon
as the gale had blown over without setting the vessel on fire which was considered dangerous and of little avail (the inflammable parts of the cargo
being the most volatile, no significant gain would have been achieved).

On March 17, while the plan Polmar-Land was directed from Quimper by the
police commissioner an advanced command post (PC) was set up in the disaster
area at Ploudalmézeau; on March 19, the police commissioner for the northern
coasts also put the Polmar plan into operation, three days before the first
oil slicks arrived.

While there were some difficulties at the start the coordination with the
mayors could be better established on Monday, March 20.   The coordination
between administrations, which was delicate because of the duality of the
Polmar plan, could be more affectively ascertained from March 24 onwards by
the appointment of Mr Marc Becam, Secretary of State at the Interior Ministry
(and mayor of Quimper) as coordinator.   From March 26 on, a representative of
the Port-Admiral took part in the meetings held at the command post Polmar-
Land at Ploudalmézeau.   Mr Becam held a daily press conference (1, pp. 85-92,
2, pp. 206-224).

b) The battle at sea.   The idea of setting the vessel on fire was dropped as
we have said.   There remained the pumping of oil on to other tankers.   However, in view of the weather, the scale of the disaster and the excessive
time required for bringing pumping gear to the site all efforts to master the
pollution at the vessel ended in deadlock (1, p. 95).

Action on the oil slicks could be envisaged in three ways:

Pumping by mechanical means, but the seas were too high to permit the use
of the available equipment;

Use of precipitants, but sending the oil to the bottom of the sea was no
acceptable solution, given the eventual dangers of such a 'disappearance'.

Finally the use of dispersants.   These products had been considerably
improved since their massive utilisation in 1967 on the occasion of the
stranding of the *Torrey Canyon* (the 10,000 tonnes used on the Cornish coast
were responsible, rather than the oil itself, for the death-rate suffered
by the maritime fauna (6, p. 23).   They are now less toxic.   Nevertheless,
their use still necessitates precautions: the dispersed oil may become
temporarily toxic; the toxicity of the dispersants can combine with the

toxicity of the hydrocarbon and produce a toxicity 4 to 5 times higher than that due to simple addition (synergic effect). This is why dispersants were banned at depths of less than 50 metres in sensitive areas (1, pp. 93-101, 2, pp. 226-239, 7).

c) The battle on land. The protection of sensitive areas is effected by barriers. The insufficiency of stocks, the absence of necessary instructions for their erection, the absence of transportation equipment and especially the weather conditions have often made this first line of defence inoperable.

There remained the pumping of the coast implemented, at first by individual farmers who had usable barrels. By the end of May 65,000 tonnes of products had been pumped (containing about 30,000 tonnes of oil). These products had to be taken to degasification stations at Brest, Nantes, Saint-Nazaire and Le Havre.

In addition, some 185,000 tonnes of solid waste were collected containing 10-15 per cent hydrocarbons. After heat sterilisation they had to be deposited in the port zone of Brest and in a cove near Trégastel (for the northern coasts). Between the beaches and the places of treatment or deposit there were intermediate storage pits arranged, mainly located outside the ecologically sensitive areas.

Finally there remained the clearing of the coastline or more or less of the most accessible parts, the most sensitive from the point of view of the tourist trade.

All this required an unprecedented effort on the part of the multiple administrations involved, the local representatives, the farmers and people living on the coast, so many times affected by an oil disaster (1, pp. 102-105, 2, pp. 240-256, 7).

3rd: Critical observations in the parliamentary reports

a) General organisation. The general opinion was that on the organisational level the coordination between administrations and between the different decision making bodies seems to have caused the most difficulties

It was paradoxical to find that France had more powerful means for intervention in far away theatres of operation (the necessary instruments transportable by air, replacement of bridges etc) than those, even though they are more elementary, required for the installation of an adequate command post in a situation like this. The creation of a mobile command unit equipped with the means of communication and for accommodation seems imperative (2, p. 215).

While in the case of minor maritime pollutions the implementation of the Polmar plan is not called for, the role devolved on the local communities and in particular on parishes is important while it seems to be non-existent in the case of implementation of the plan if one refers to the decree of 1970. Above all, the Polmar plan does not determine the role devolved on the local communities.

Based on the facts as far as they can be established when analysing the setting up of the whole organisation a certain irresolution is noticed during the first three or four days and this despite the immediate implementation of the Polmar plans and the effective mobilisation of all administrations involved (2, p. 221).

b) The battle at sea. In addition to the criticism of the absence of
available means and of delays in forwarding equipment (1, p. 95) the rule of
50 metres depths of sea for the use of dispersants was criticised by the
senators:

From the start of operations all the diverse services involved had agreed
on the necessity not to employ dispersants where the sea was not at least
50 metres deep as witness the joint communique of March 24 by the representatives from the Ministeries of the Environment, Interior (Directorate of
Civil Defence), Navy, IFP, ISTMM and CNEXO.

Technical directives to this effect were sent the following day to those
responsible for the various cleaning-up sites. However, the Commission has
been informed from several sources (delegations that went to Brittany,
diverse people questioned by the Commission) that this principle was actually
not respected either at the start of operations or at this present time when
the concentrated dispersants added to water are still used to clean rocky areas
on the coast with hoses.

The Commission is surprised that such hazards should have overruled the
principles which had been established as untouchable. The competent
authorities have shown themselves legitimately circumspect on the consequences
of the use of such substances on the coast but they did not see their way to
translating their reservations into actions (1, pp. 100-101).

c) The battle on land. Once more the lack of means, the lack of organisation
must be put on record. The case of the floating barriers is taken as a
typical example: the administrative allocations for the purchase, the forwarding, the stocking, the installation could only lead to inefficiency (1,
p. 180).

The case of the barriers underlines also the need for effective training.
The term barrier is really rather badly chosen and develops an attitude
which evokes rather the image of a "Maginot Line" which was associated with
this work tool. A barrier must be used to channel the discharge of an oil
slick towards a less sensitive area where in addition its recuperation will
be easier; it is not a static fortification. What with the tides and their
inversion every six hours it must be known how to place this instrument
judiciously and manage it consistently (2, p. 242). This requires knowledge
which cannot be conferred by a notice, no matter how well written.

Let us still register this observation concerning the choice of location
for the intermediate storage pits:

Studies made in recent years under the authority of the Ministry of the
Environment with the approval of the BRGM tended to record the sensitive
areas where such storage facilities were to be prescribed so as to localise
the proper place(s) for such use. Those responsible at the departmental
directorate(s) of agriculture from each of the two departments involved
did not seem to have such documents in their possession when they had to
determine these sites (1, p. 104).

5. THE BALANCE SHEET OF THE BLACK TIDE

1st: Ecological impact

Being an environment of very great riches, of great diversity, fragile
and with an unstable equilibrium, attacked by multiple forces in either

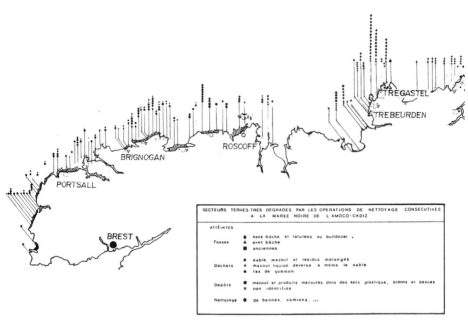

(Source : J. C. LEVEUVRE et M. LE DEMEZET, communication personnelle)

Fig. 14: Damaged land sectors

chronic or sudden fashion*, the coastline this time had to absorb the shock
of a massive discharge of hydrocarbon**.

The most affected areas were the estuaries and the semi-enclosed bays. A
massive and crushing mortality rate was registered among rock and beach
animals within a radius of five kilometres around the wreck and at accumulation points up to 100 kilometres away. The waters have generally regained
their 'normal' content of hydrocarbon at the end of three months except in
estuaries and bays.

---

*According to world averages which were reexamined for Britanny, pollution
by hydrocarbons occurs in a ration of 1.7 and 11 for, respectively oil
tanker accidents, degasification of tankers at sea, telluric fallout
(6, p. 125).

**See figure on page 91 (3, p. 34).

If the first effects of the black tide were spectacular, millions of razor-fish thrown up by the sea, more than 4,000 birds gathered, some ten million fish killed, one could draw up a provisional balance sheet of moderate serious-ness during the months following the disaster. The mortality rates registered are localised, selective, incomplete, in the order of 3 per cent of the total fauna, less than 5 per cent of the flora (8, p. 337). The high seas in the area of the stranding, the moderate and selective use of detergents, the gathering of the spilled oil have been favourable factors (3, p. 46, 8).

The situation can be described by quoting some points taken from the works of Cl. Chasse (6, p. 127):

- The very fertile field of large algae, some 400 km^2, "three times more productive per hectare than our best forests on land", is practically intact.

- The benthic animals have occasioned uneven losses. A third of the coast-line lost more than 50 per cent of its biomass while another third lost less than two per cent. The total loss of maritime animals is estimated at 260,000 tonnes of gross fresh weight.

- Plankton of which there is still little in March was not much affected.

- The very mobile fauna of fish and of large shellfish were altogether very little affected: it took to flight. High mortality rates, very localised, were confined to species of little economic importance.

- The number of birds killed is estimated at between 15,000 and 20,000.

In addition to this 'mortality section' of the balance sheet the problems of pathology and change of ecological equilibrium need to be examined:

- The animals that survived were contaminated at rates of 200, even 500 and sometimes 1,500 ppm (above 100 ppm a species is uneatable on account of its taste). With the sanitation of the area, decontamination worked within a few weeks but in areas where sediments remained strongly affected by hydrocarbons no restitution of this kind was possible.

- A reduction of vitality, emaciation, reduced resistance of some species was also noticed; species which disappeared did not come back; necroses were observed, tumours affected certain fish.

- The ecological equilibrium was affected: proliferation reduction, concentration*of certain species (6, p. 127).

Hydrocarbons had been trapped in substantial quantities in mudbanks, deposited on sand before being covered again by new layers of sand (carried mainly by the equinoctial tide of March 25/26, 1978), infiltrated the sedi-ments in depth. On account of this no final balance sheet can be drawn up. After March 17, 1978 it became necessary to establish a programme of studies over several years to follow the ecological impact of the disaster (9).

The specialists stress the need for this ecological follow-up over a long

---

*Whence unusual fishing catches, rose-coloured shrimps for instance which do not indicate a proliferation of the animal (and may bode ill for future catches).

period of time (because of the slowness of the return to equilibrium), the
importance of uncertainties that remain (problems of reproduction, recolonis-
ation, reutilisation of the most affected areas which were sometimes the most
valuable). One remembers these lines from a CNEXCO report which show the
difficulty of the task and the tenacity required for the success of the
attempt if one considers the still higher risk of a new black tide:

The deficit caused among several species will balance off over the years
if the environment is not again polluted. It is tempting to try and remedy
this state of affairs by going ahead with repopulating, introducing larvae or
young shellfish, flat fish and bivalves produced in enclosures. However, it
is necessary to keep in mind the fact that alimentary support (invertebrates)
constitutes a limiting factor the equilibrium of which cannot be artificially
restored. It seems illusionary, for many species, to go ahead with massive
repopulation with either adults or young ones without first evaluating the
profitability of such actions based on existing economic and biological data.
An experimental study of this kind with the help of a computer model is in
progress for flat fish and oyster beds. In all cases it would be useful if
the follow-up on the evolution of the haliotic stocks were continued until
1983 (3, p. 51).

2nd: Effects on human health

Two groups of the population were involved: the inhabitants (by respiratory
contact) and people working on the beaches (respiratory and skin contact).
The conclusions reached by specialists agree (10, 11, 12) with regard to short
and medium term effects:

During the acute phase an important number of troubles have been observed
which can be qualified as minor, even if their, rapidly receding, existence
was a genuine nuisance to people affected. On medium term, consulting
medical practitioners did not find new pathological effects. In the long
term, however, we must point out that these results give no indication on
future consequences of this pollution (10, p. 15).

For the long term effect precisely it does not seem that sufficient
financial means have been allocated which would permit the desired analysis;
rightly or wrongly, some people deplore that useful scientific work in this
field has not been particularly encouraged (6, p. 135, 12, p. 40).

3rd: The economic consequences

The various studies carried out and published, in particular*the one by
CODAFF**of 1979 (13) and the synthesis advanced in 1980 by the magazine
OCTANT (published by INSEE, 14, p. 45) permit the drawing up of the following
balance sheet:

a) Sea economy.

- The pollution of ports and fishing ports has resulted in a halt of all
  activities. Five hundred to six hundred professional fishermen out of a
  total of 1,800 (for the locations of Brest, Morlaix and Paimpol) were
  affected by the consequences of the black tide. After March 16 most of

---

*Various sources indicated hereafter: 1, pp. 115-122: 2, pp. 265-268; 6,
pp. 163-166, 181-191; 13; 14, p. 45; 15 pp. 1-9).

**Departmental Committee on Development and Housing for Finistere.

the fishermen gave up their activities. The resumption occurred only gradually after the end of April; it was generally resumed at the beginning of June. Much of the fishing tackle had been damaged, often made unusable.

- The marketing of cockles and shellfish involving about a hundred occasional fishermen who drew additional income from this had become impossible on account of the bad taste of the species which had been affected by hydrocarbons.

- The breeding grounds, the coast north of Finistère has the largest area of breeding grounds in France, had been polluted in their submerged parts (which required restoration work since the walls had been soaked with hydrocarbons); for the non-submerged parts the supply of seawater required delivery by road tankers.

- Oyster farming was the sector most affected. A large part of the stocks had to be destroyed in the oyster beds (the total number of oysters which could not be supplied to the market: 1,500 tonnes of which 250 tonnes from punts) and in the bay of Morlaix (500 tonnes destroyed the first time, 4,600 tonnes thereafter). Installations and equipment were damaged in the area of Wrac'h and particularly in the area of Benoit; the cultures had to be relocated (30,000 ffrs/hectare). The environment, sometimes still more polluted, made the resumption of activities difficult*.

- The harvest of algae impeded, at a time, could nevertheless be accomplished as far as Laminaria were concerned. The loss figures show no serious disruption for 1978. For this there are two reasons: the mechanisation of the boats and the fact that in 1978 too the jiggers absorbed the algae green while discharging the drying-out agents and making them ready for harvesting. For the lichens, the fucus harvested at much lesser depth than the Laminaria, the balance sheet looks much worse.

- The black tide has therefore had a general impact on the economy of the sea.

Employment in oyster fisheries was particularly affected. If the measures of indemnification have permitted to suspend lay-offs for forty nine out of a total of about three hundred (13, p. 2) short-time working was nevertheless introduced in the most severely affected sectors (two hundred and eighty nine on June 9 out of a total of three hundred and eighty two). As a consequence, business located upstream and downstream suffered the backlash of the crisis; naval repairers, suppliers of fishing equipment, the fish trade, fish transport. The halt of business reduced income to zero while fixed expenses remained.

Comparing the periods March to August in August 1977 and 1978 one can measure the average loss registered (in weight and in value): fish minus 4 per cent and minus 30 per cent, shellfish minus 32 per cent and minus 26 per cent, oysters — minus 80 per cent and minus 60 per cent.

In addition to the immediate losses which no doubt indemnification will deal with there are the medium term problems and the uncertainty about the future. This is mainly a question of reestablishing equilibrium as far as

---

*A burned-out factory can be rebuilt; an eco-system as rich as an oyster bed cannot be so easily rehabilitated.

eggs, larvae and young fish, which are much more vulnerable than fully grown ones are concerned. There is also the question of decontamination in the depths of the most affected areas, which are in some instances the most productive.

b) The tourist trade. The losses in the hotel industry were heavy (100 million ffrs)*. A considerable drop in the number of foreign visitors was noticed (90 per cent for certain periods). Camping sites were equally affected, operating at only 50 per cent of their capacity in July. Furnished apartments faced a drop of 60 per cent in June and 20 per cent in July. True, the poor weather conditions during the first half of July may have caused a negative effect but they could not disguise the essential cause of this drop in business.

c) Consequences which need evaluation in depth and in the long term. The black tide from the *Amoco-Cadiz* has hit a fragile economic entity (85,000 unemployed registered with the local employment agencies of Brest and Morlaix of which 45 per cent were aged under twenty-five). What will be the lasting effect of the shock? What will be the long term effects on economic life? What will be the attitude of the investors in respect of a sector of the economy which depends strongly on the quality of the environment?

4th: The financial aspect — Compensation

For the immediate battle against the pollution (staff, supply of services, ships, planes, helicopters, equipment purchases) total expense exceeded 415 million ffrs (2, p. 281). To this must be added the amounts provided for the redevelopment of sites (Decision by the Interministerial Council for the Development of the Area of July 18, 1978): 17.35 million ffrs.

Compensation has been paid or provided for to those affected and to the tune of about 45 million ffrs: the sea fishermen received a provisional compensation of 1,244 ffrs per fortnight of their idle time (15 March to 30 July 1978) and afterwards a complementary compensation for the rest of the year. The hotel owners apparently succeeded in making an arrangement with the authorities. The situation is more delicate with regard to the oyster farmers; the amounts suggested are in fact based only on part of the losses, destruction of stocks, cleaning up, keeping on staff while they could not work, transfer of oysters. Other items must be considered such as the reestablishment of the concession areas**, equipment maintenance, losses due to default in the growth of the oysters, loss of business during the partial lay-offs etc. People in the trade estimate their overall losses at about double the amount so far established by the authorities.

5th: Legal action

As regards penalties, an investigation opened by the examining magistrate at Brest led to the indictment of the skippers of both vessels (tanker and tug); the decision was upheld in an appeal hearing on October 27, 1979.

In civil law the French government has started proceedings against the companies (Amoco International and Standard Oil of Indiana) who owned the vessel in the court of Chicago. It claims 460 million ffrs in damages and

---

*Tourist bookings 1978 in percentage of those for 1977: April/May 30-40 per cent; June 50-55 per cent; first half of July 60-70 percent; second half 80-90 per cent (13, p. 16).

**Initial reinstatement is programmed and in the process of implementation.

interest. The court has declared itself competent on September 22, 1979 and
has refused the two companies in litigation the right to limit their
responsibility. This legal action by the state has been joined by the districts of Finistere and Côtes-du-Nord, fifty communities in Finistère and
hotel owners' associations. Such lawsuits are extremely costly, extremely
drawn-out and doubtless hazardous; these difficulties are strongly felt by
the plaintiffs (16). These latter ones have nothing like the financial
strength of their adversaries who are hardly bothered (quite on the contrary,
as some of the plaintiffs point out) by the prospect of spending large
amounts on court proceedings.

6. CONCLUSION: THE ABSENCE OF A SYSTEM OF PROTECTION

In order to understand the incapacity of the various parties involved in
the inexorable drift of the giant oil tanker one must get down to the
fundamental causes without being hindered by the search for some scapegoat.
With the Senate Commission of Enquiry we shall hold on to four essential
factors which have caused this impotence.

1st: The mentality of seafaring people

Here we are facing time-honoured traditions. If at sea there is self-
sacrifice when a human life is at stake the seafaring man, on the other hand,
being solitary, dignified and brave, does not call for help until there is
danger to human life (2, p. 94). The rule "after God, the skipper is the
only master" remains very much alive and powerful. The traditions at sea
can therefore still be summed up like this: absolute solidarity in case of
danger to human life; no intervention in case of distress (2, p. 95). If
one applies this to the events of March 16, 1978, the members of the senate
note, many of the attitudes find their explanation in this. "The law of the
sea means confidence in the skipper and not defiance ... No distress call,
no intervention" (2, p. 95).

2nd: The hardly responsible use of the maritime environment

If mentalities are still impregnated by the "chivalrous" spirit invoked
above behaviour at sea is also guided by other, much less noble, notions;
"the constant search for private and instant profit", the Senate Commission
of Enquiry spells out (1, p. 157). The sea, "common heritage of mankind" as
the United Nations wanted to define it? Or the sea looked at rather as a
"res nullius", a simple support for ships, justifying its users in their
search for reduced cost of investment and exploitation, for multiple
financial and fiscal advantages, weak or even non-existent regulations, so
many factors found in states which complacently lend their flag (without
these latter having a complete monopoly in this business).

From the point of view that the sea is a heritage one can readily explain
the phrase by the French Minister of the Interior, Ch. Bonnet, evoking the
inexorable drift of the tanker and its tug: "Negotiations which I shall not
hesitate to call sordid" (17). From the point of view that the sea is
simply a liquid mass permitting the traffic of ships and the discharge of
cumbersome refuse one can better understand an event like that of March 16,
1978. More so still as the systems of prevention and combat largely failed,
as the Commissions of Enquiry emphasised.

3rd: Administrative insufficiency of government action at sea

As the report by the delegates stresses this must be understood in depth
without going for an easy search for a guilty party. In other words: the

problem which arose on March 16, 1978 sat still on the fence without blame
being attributed. Thus, as for Radio-Conquet, one must not accuse unjustly:

Considering the principle of secrecy of the communications and of the
private and purely commercial nature of the assistance relations, it was not
exactly within the competence of Radio-Conquet, not even, as it seems, within
its possibilities and in any case not at all within its habits to evaluate
the seriousness of a situation from telephone conversations or, as a general
rule, to take an interest in the safety of shipping; at least not until the
publication of the decree of March 9, 1978 which placed responsibility on all
land-based authorities and on all exterior services and establishments of the
state who have competence at sea to keep the Port-Admiral informed of all
business likely to be of particular importance at sea, no law submitted the
maritime radio stations to a general information duty on events of which they
could have knowledge.

However, on March 16, the decree had been published in the *Journal Officiel*
of the 11th, no particular instruction to this effect had as yet been received
(2, pp. 95-96).

In general, the delegates indicate:

What we are dealing with is a complicated system in which information is
shared between diverse agents who ignore each other more or less, in which an
information which is somewhere cut into pieces, circulates badly, and so
causes finally and paradoxically the ignorance of the authority which has the
competence to intervene. A sometimes incoherent system, always marked by a
pretended coordination which has to replace the unity of command which is
indispensable in the face of danger, first potential, then actual. A system
in which one administration which has powers but no means is called upon to
request these from another administration which evaluates the opportunity to
grant these means and, conversely, an administration which has the means but
not the information which would cause it to put them to work or the power to
use them. Altogether a divided system deprived of all synthetic function.

In this respect it must be recognised that all the information collected,
and which agrees on the respective parts assigned to the Navy, the merchant
navy, the post and telecommunication authorities concerning life at sea and
its problems show at which point reform is needed (1, p. 223).

4th: The laxity of measures for surveillance of shipping and the deficiencies of the means of intervention

On March 16 no ship was on duty controlling the area of Ouessant when the
under-equipment of fixed radio and radar installations made this completely
hazardous from the coast.

Finally, supposing that the information would have made intervention on the
Amoco-Cadiz possible, the National Defence Forces had neither specialised
crews to attempt a repair of the damage nor tugs in a nearby position and with
sufficient power to prevent the stranding (2, p. 97).

To overcome these difficulties, and above all the first one, is an arduous
task. To do this the mentality at sea needs to be reversed. And to do that
one needs precise laws which spell out the responsibility of the seafaring
people. Firm orders are needed which permit the circulation of information
between the men on land and the commanding authority and mobilises the
government services at all levels. They must be given the means required
(2, p. 97).

Article 16 of the law of July 7, 1976 aimed exactly at these difficulties and deficiencies; it provides:

In the case of damage or accident at sea incurred by any vessel, aircraft, fishing tackle or platform carrying or having aboard noxious substances or hydrocarbons and being capable of creating serious and imminent dangers likely to affect the coastline or connected interests in the sense of Article 11-4 of the Brussels Convention of November 29, 1969 on the intervention on the high seas in case of an accident involving or capable of involving a pollution by hydrocarbons, the owner of said vessel, aircraft, fishing tackle or platform may be summoned to undertake all necessary measures to end these dangers.

In case such summons has no effect or not the desired effect within the time allowed or as a matter of course in an emergency the government may take the necessary measures at the expense of the owners or recover the cost from the latter (2, p. 93).

True, in the beginning the parliamentary amendment was much more striking than the text proposed by the government; the simple case of damage (without there being necessarily "grave and imminent danger") justified the implementation of the emergency intervention procedure (2, pp. 93-94). However, the 1976 law permitted already to expect better accident prevention.

On March 16 the laws had not yet been revised; the orders had not been given; the means were not available. The decree which sanctions the general police authority of the Port-Admiral and establishes a duty of supplying information for all services had been published five days before the drama; as for the decree which obliges the skippers of all vessels carrying hydrocarbon to report any damage likely to create a serious and imminent danger within the 50 mile zone and obliges on the other hand the tugs to inform the authorities of any request for help and assistance — that was published ten days after the accident.

For lack of decrees of application, lack of precise instructions, lack of means, Article 16 of the law of July 7, 1976 remained a dead letter; the exercise of the power to intervene could not become effective (2, pp. 97-98).

5th A general situation that could only lead to deadlock

All these considerations lead the Senate Enquiry Commission to these concluding words: they underline that some tactical errors do not appreciably aggravate a situation when the basic determinants of it are far too negative:

The lack of reaction by administrations which had responsibilities at sea has had no influence on the stranding of the tanker *Amoco-Cadiz* on March 16, 1978 (1, p. 81).

The Polmar plan was already well recognised for its insufficiencies: Even before the disaster of the *Amoco-Cadiz* occurred the authorities were perfectly well aware of the maladjustment of the means that we could mobilise. The report prepared for the government by GICAMA in 1977 spells it out quite explicitly as we have indicated earlier on (1, p. 196).

The Polmar plan could, therefore, provide no hope for a more honourable follow-up after the deadlock of prevention.

REFERENCES

(1) Rapport de la Commission d'Enquête du Sénat, présenté par A. Colin.
    Seconde session ordinaire 1977-1978, juin 1978, no 486 (289 pages).

(2) Rapport de la Commission d'Enquête de l'Assemblée Nationale présenté par
    H. Baudoin. Première session ordinaire 1978-1979, novembre 1978, no 665,
    Tome 1 (333 pages); tome 2 (94 pages).

(3) M. MARCHAND, G. CONAN et L. d'OZOUVILLE, Bilan écologique de la pollution
    de l'*Amoco-Cadiz*. Centre National pour l'Exploitation des Océans
    (C.N.E.X.O.) Rapports scientifiques et techniques, no 40, février 1979
    (57 pages).

(4) S. BERNE et L. d'OZOUVILLE, Cartographie des apports polluants et des
    zones contaminées. C.N.E.X.O., mai 1979 (175 pages).

(5) *Amoco-Cadiz*. Premières observations sur la pollution par les hydro-
    carbures. Colloque du C.N.E.X.O., 7 juin 1978, no 6, 1978 (239 pages).

(6) La pollution marine par les hydrocarbures. Union des Villes du Littoral
    Ouest-Européen (U.V.L.O.E.) Colloque de l'U.V.L.O.E., Brest, 28-29-30
    mars 1979, (227 pages).

(7) P. BELLIER, Lutte contre les pollutions marines accidentelles par les
    hydrocarbures. L'expérience de l'*Amoco-Cadiz*. Direction des ports et
    de la navigation maritime, mars-sept. 1978 (193 pages).

(8) Penn Ar Bed. Bulletin de la Société pour l'Etude et la Protection de la
    Nature en Bretagne, no 93, juin 1978.

(9) P. NOUNOU, La pollution pétrolière des océans. *La Recherche*, vol. no 97,
    février 1979, pp. 147-156.

(10) Conséquences médicales du naufrage de l'*Amoco-Cadiz* sur la population
     côtière et les volontaires civils. Bilan du 17 mars au 31 décembre 1978.
     Faculté de Médecine de Brest.

(11) G. LE MENN, Synthèse chimique et biologique. Conséquences médicales du
     naufrage de l'*Amoco-Cadiz* sur la population côtières et les volontaires
     civils. Bilan du 17 mars au 31 décembre 1978. Faculté de Médecine de
     Brest.

(12) S. SALAÜN, Un mode d'intoxication rare par les hydrocarbures. Les
     marées noires. Thèse de Doctorat d'Etat en Médecine, 17 avril 1980,
     Université de Bordeaux III (58 pages).

(13) Incidences économiques de la catastrophe de l'*Amoco-Cadiz*. Note de
     synthèse du Comité Départemental du Développement et d'Aménagement du
     Finistère (CORDAF), janvier 1979 (23 pages).

(14) OCTANT (Insee), Cahiers statistiques de la Bretagne, no 0, janvier 1980.

(15) F. BONNIEUX et P. RAINELLI, Le tourisme sur le littoral breton: son
     importance et son recul en 1978. Octant, cahiers statistiques de la
     Bretagne, no 2, 1980.

(16) *Presse-Environnement*, no 356, 14 décembre 1979.

(17) *Europe* 1, 21 mars 1978.

IV. WEDNESDAY, MARCH 28, 1979: THE NUCLEAR ACCIDENT AT
    THREE MILE ISLAND

On Wednesday, March 28, 1979, thirty-six seconds after 04.00 h, several
water supply pumps broke down in unit No. 2 of the nuclear centre at Three
Mile Island (TMI) which is ten miles southeast of Harrisburg in Pennsylvania.
Thus begins the accident of TMI. During the minutes, the hours, the days
that followed a series of events, equipment failure, inappropriate
procedures, human error and ignorance were to turn the accident into a
major crisis, the worst that the nuclear industry had ever experienced.

The accident focused national and international attention on the nuclear
installation of TMI and became the main subject of reflection for hundreds
of thousands of people. For people living in the communities of Royalton,
Goldsboro, Middletown ... and Harrisburg the rumours, the contradictory
official statements, a deficiency of knowledge about the subject of radioactive emissions, the constant possibility of mass evacuation and the fear
that a hydrogen bubble in the nuclear reactor might explode were effective
and immediate realities.

With these sentences the Commission of Enquiry set up by President Carter
two weeks after the accident introduced its detailed report on the event.

1. TMI — A NUCLEAR CENTRE, AN AMERICAN ENVIRONMENT

1st: Technical characteristics

Some simple technical data may be useful for the comprehension of what
follows: (according to 1, pp. 81-89 and 2, pp. 76-78). The centre of TMI
consists of two units: Reactor No. 1 with a capacity of 800 megawatts (MWe),
put into operation in 1974, had supplied, at the time of the accident in
reactor No. 2, 20 billion kWh (equivalent to six months electricity consumption of New York) and its rating of reliability had been excellent; the
second reactor, the one which was to have the accident, had a capacity of
900 MWe and had only been in commission for three months. These reactors
belong in the category called pressurised light water reactors (PWR), the
most frequently used in the world*.

The general functional principle of a nuclear centre is similar to that of
all thermal power centres and is concentrated around the following elements:
a heat source, in this case a nuclear reactor, permits the evaporation of
water; the steam thus generated drives a turbine which is coupled to an
alternator that produces electricity. The steam passes its condensation
heat on to the water circulating in the piping system which at TMI is cooled
by air in cooling towers. The installation comprises mainly**:

- a reactor building consisting of a surrounding concrete wall called a
  containment wall because of its protective function;

- a building containing the nuclear auxiliaries;

- a building containing the electricity generating installation.

---

*The French centres in particular are of this PWR category but under
 Westinghouse, not Babcock and Wilcox, licence like the generators at TMI.
**See Figure 16, page 110.

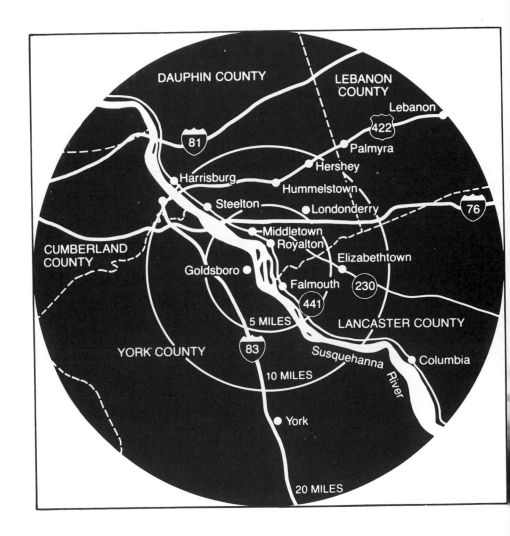

Fig. 15: The centre of TMI is located on land belonging to
the community Middletown in the state of Pennsylvania (USA)
on the bank of the Susquehanna River, 15 km from the state
capital, Harrisburg (90,000 inhabitants).
                    (Source: 1, p. 122)

Within the containment wall is the installation for the generation of
'nuclear heat'. At a centre like the one at TMI one finds within this wall:

- the reactor vessel containing the core where the fission of the atoms of
Uranium 235 occurs which releases the heat; the core is cooled with
ordinary water which circulates outside the vessel in two 'loops' comprising primary piping and in each of the loops a heat changer, then two
primary pumps in parallel. On this circuit is the pressuriser which
permits control of the pressure in the circuit; the pressuriser is
protected by discharge valves which open when the primary pressure reaches
a reading fixed at 156 bar, thus permitting the discharge of excess steam
into a so-called discharge tank.

- 'Steam Generators' or GV, where the primary water transmits its heat to
the so-called 'secondary circuit' while vaporising the water sent along by
supply pumps.

Outside the containment wall is mainly the electricity generating installation. The steam produced by the generators is sent to the turbine by means
of a piping system which runs through the containment wall. This steam
expands in the turbine where the transformation of thermal energy into
mechanical energy, electrical energy, takes place. The steam, when leaving
the turbine, condenses in the condensor where it transmits its condensation
heat to the water from the cold source. It is subsequently sent back to the
supply pumps.

The safety of the installation is ensured by a number of screens called
'barriers' between the fuel and the public and by a certain number of devices
which intervene in the case of malfunction. In PWR reactors like the ones
at TMI there are three barriers in sequence:

- The first one is made up of the casing which surrounds the fuel tubes.
This casing consists of Zircaloy which has a number of advantages from a
thermodynamic point of view but has the disadvantage of reacting to water
at temperatures above $1,200^{\circ}C$ and gives off hydrogen.

- The second one consists of the primary circuit: vessel, loops (rustproof
steel of very high quality, a very large number of small diameter) steam
generator and primary pump pipes and the pressuriser with its valves and
the discharge tank.

- The third one consists of the concrete wall which makes up the reactor
building.

In the context of the accident on March 28, 1979 the safety systems
involved were:

- The emergency switch off system consisting of an assembly of rods the
function of which is to stop the fissioning reaction*and the associated
control system.

- The auxiliary supply system of the steam generators employed in case of
loss of normal water supply.

---

*The reactor of TMI-2 had been in service only for a short while: this limited
the quantity of residual heat produced and was therefore a minimising factor
in the accident process.

- The safety injection system the function of which is to inject boronated
  water (neutron poison) into the primary circuit in case of depressurisation
  of that circuit. This function is generally achieved by three categories
  of systems: injection at high, medium and low pressure.

- The wall insulation system which permits, upon an indication of high
  pressure (+ 0.3 bar of relative pressure) to close all auxiliary pipes
  through which primary water runs and which comes out of the surrounding
  wall.

- The wall's cooling and ventilation system designed to reduce the temperature and pressure inside the third barrier in order to ensure its intact
  survival in case of an accident.

2nd: Institutional data

The centre at TMI is operated by Metropolitan Edison, a private company
which is a part-owner of the installation; as is the rule in the United
States, TMI belongs to a private group. In order to understand the institutional game that will unfold on the occasion of the accident the responsibilities of the various public authorities must also be spelled out.

The drawing up and the implementation of rescue plans are the responsibility
of each state. Accidents to do with radiation are dealt with specifically,
as the case may be, by the Health Service or by Civil Defence (Emergency
Service) as was the case in Pennsylvania, the agency being called 'Pennsylvania
Emergency Management Agency (PEMA)'. In cases of crisis PEMA coordinates
action under the responsibility of a council made up of the governor and the
lieutenant governor of the state, four members of the state parliament
(legislature) and the directors of the various agencies likely to intervene in
emergency situations. This council meets in an operation centre (Emergency
Operation Centre (EOC)), well equipped and well protected. In the counties
there exist local plans which spell out the responsibilities and procedures
to be followed in case of an accident, more particularly a nuclear accident;
a civil defence director activates an operation centre at state level.

At federal level the pilot organisation is the Nuclear Regulatory Commission
(NRC). Other administrations join in for support like EPA (Environmental
Protection Agency), DHEW (Department of Health, Education and Welfare), FEMA
(Federal Emergency Management Agency). All these agencies have means which
they can put at the disposal of the governors in cases of serious accidents
while the states remain responsible for the operations.

Before the accident at TMI the NRC recommended the establishment of two
zones of emergency intervention around the centres: the first one within a
radius of ten miles covering risks of radiation from the cloud of radioactive
material; the second comprising between ten and fifty miles deals with risks
of contamination by ingestion. The precise limits of these zones must be
determined by those responsible in each state. At the time of the accident
at TMI the general emergency plan of the state of Pennsylvania existed but
its annex, the plan for the protection against radiation for the state and
those for the counties, had not yet been approved by the NRC (complements and
more details had been requested for the criteria and the conditions for
evacuation of people in the neighbourhood of TMI). On the other hand, the
internal emergency plan at the centre had been established and approved. A
direct telephone line connected the operator with the civil defence command
post (EOC) at the seat of the governor in Harrisburg (2, pp. 10-11).

3rd: The overall situation in March 1979

At the beginning of 1979 there were seventy four nuclear power stations operating in the United States. The American nuclear programme had undergone successive adjustments — downwards. Objections to nuclear power were rather strong in the country and came also from non-marginal scientists; referenda had been organised in various states. Quite recently, on March 15, 1979 the NRC had ordered the closing down of centres which had faulty designs (insufficiency at the level of seismic risks). This decision had been the subject of a number of comments. Important decisions were expected from the White House concerning energy production and more particularly a relaunch of the nuclear programme.

Finally, a fiction film, *The Chinese Syndrome*, with Jane Fonda in the leading part, beat the box office records for several weeks in the USA: an incident in a nuclear power station, operators taken aback in front of control screens that gave false indications, an industry that is faster at hiding facts than at explaining them frankly*: so there was something to stir up the emotions of the inhabitants of Harrisburg in particular where the film had been showing on March 28, 1979 (2, pp. 6-7).

2. FIVE DAYS OF TECHNICAL UNCERTAINTY, POLITICAL CONFUSION AND SOCIAL DISQUIET

1st: Technical uncertainties

The Kemeny report, named after the president of the Commission set up at the request of President Carter, retraces the unfolding of the accident (1, pp. 90-142) and makes clear the succession of technical events which occurred and the parallel development of what the actors in the drama understood and decided to do, that is to say: the operators, then the engineers of the centre and finally those responsible at the NRC. The article written by P. Tanguy (3) spells out the scenario and draws up a balance sheet from it.

We shall still refer back to other reports in order to study the case of TMI (3-13). We shall take up the essential parts even if this entails being incomplete and insufficiently precise.

a) The first moments experienced by the operators in the control room (the first two hours). It all started on Wednesday, March 28, 1979 at 04.00 h 36 seconds while section two at the centre at TMI was running at 97 per cent of its nominal capacity. The initiating incident was the start of two water pumps supplying the steam generator, subsequent to a failure in a common up-stream circuit. This caused, two seconds later, the start of the turbo alternator group, five seconds later the opening of the discharge sluice of the pressure unit**and eight seconds later the emergency stop of the reactor which is effected by the dropping of absorbant rods into the core.

Nothing abnormal up to that moment. It was an incident, serious enough but not of an exceptional nature. However, it resulted in a cascade of alerts, one hundred of them within a few minutes. The operators' reaction as reported by Kemeny is understandable:

---

*So many intervening factors, in reality, in the event that happened on Wednesday, March 28.

**Since the pressure in the primary circuit rose following the increase in the temperature of the primary water that was caused by the stop of the secondary water supply.

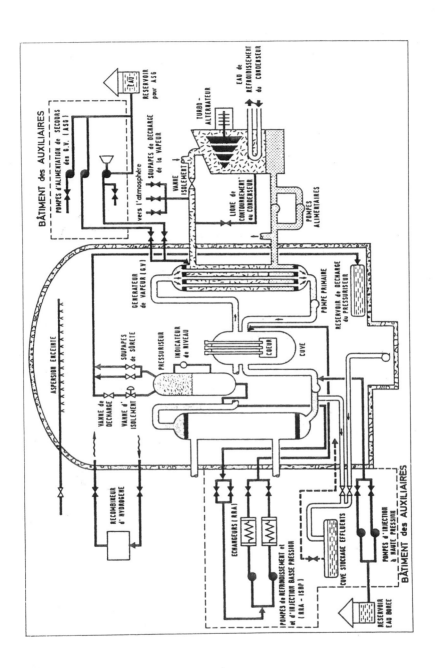

Fig. 16:   Diagram of the centre at TMI
(Source 2, p. 524)

- I wanted to send the alarm panel to hell; it gave us no usable information.

The problem confronting the operators was to understand what was happening. What information should one get? How to sort out the one hundred odd alerts that rang out in a cacophony of hooters and bells? It was not humanly possible to analyse, to 'digest' this avalanche of data. They were trying to grasp the maximum of information, to identify the situation in which they found themselves and perhaps to relate it to one of the situations which they had studied theoretically during their training.

In this vein, one operator noted that the pumps which supply the steam generators with emergency water did start allright. But nobody realised that this water could not get to the generators because two valves were, mistakenly, in the shut position. It took eight minutes until one operator realised this and the valves were opened, permitting this water to cool the primary water. However, according to the experts, this error did not profoundly change the sequence of events. According to P. Tanguy it does not seem, as of today, that this eight minute lapse had a decisive influence on the unfolding of the accident more or less according to the plan of thermohydraulic behaviour of the system. On the contrary, it seems probable that the operators' poor comprehension of the course of events had been influenced by the disturbed situation which they had to face. (3, p. 527)

In fact, one can imagine the surprise of the operators who, while expecting that the residual heat (caused essentially by the products of the fission once the rods had been dropped*, since there had been an emergency stop) had been removed by the emergency water, realised that for eight minutes this had not happened.

This situation could only constitute a handicap for the detection of the accident that was unfolding. For, in fact, the operators were faced with a grave situation defined in the safety reports as being "a loss of primary refrigerant on account of an intermediary break in the steam phase at the pressurizer" (4, p. 14). In fact, if the discharge valves of the pressurizer had opened properly (at operating time: five seconds) (permitting the steam of the pressure unit to discharge into the discharge tank**and thus to limit the increase of pressure in the primary circuit following the initiating incident of loss of secondary water) it would have had to shut off again within twelve seconds when the pressure would have dropped sufficiently. The sequence of shutting again was well established but this valve had remained blocked in an open position while the primary circuit emptied (60t/h) into the discharge tank. The operators who had the information "pressurizer discharge valve closed" took 2 h.20 minutes to understand what had happened. What actually happened was that the discharge tank, which has a limited capacity, opened its valve very quickly (operating time: three minutes), letting the primary steam overflow into the concrete surrounding wall; then there was a rupture of its explosion disc (operating time fifteen minutes). The water then overflowed into the surrounding wall.

The operators knew that they were in a difficult situation since at operating time: two minutes the safety injection started automatically, indicating that there was a break somewhere in the primary circuit. Amidst the

---

*This remaining power is in the order of 7 per cent of the nominal power immediately after the reactor is stopped and then drops rapidly: 4 per cent after 30 seconds, one per cent after two hours.

**See Figure 16, page 110.

multitude of information available the operator picked on the information
"level in the pressurizer" and interpreted the rise of that level erroneously:

> In fact, the rise of the level was quite normal and foreseeable in the case
> of a break in the pressure unit; but it seems that the operators at TMI
> were not trained to deal with this type of situation. (3, p. 526).

This error of interpretation led one operator to reduce the supply of
injected water, by stopping one of the pumps:

> The rapid rise of the level in the pressurizer made me think that the high
> pressure injection was excessive and that we would reach 'solid' state
> (5, p. 2).

One operator later confirmed.

This fear of reaching 'solid' state (this term means a complete filling of
the primary circuit with water) came from what the operators had been taught;
that is to say that in a transitory situation it is important "not to lose the
steam bubble in the pressurizer" because of the danger of losing control of
the pressure in the circuit. The operators entertained this fear for more
than four hours, stopping even the second injection pump and only restarted
it and reduced the supply level at operating time: ten minutes; and even
this despite some doubts:

> The pressurizer was almost full; I couldn't believe it at the time. It was
> too fast ... that was the first damned thing. It had filled up, and we had
> thought it was already full. It should have been full with water but what
> afterwards misled us was that the system did not react as if it was 'solid'.
> We saw no pressure peaks, (5, p. 2) declared one operator two days after
> the accident.

The operators were therefore not aware that the primary circuit was
emptying itself progressively. There had also been the automatic start (at
operating time: 7 minutes 30 seconds) of the draining pump that would send
the contaminated water to the storage tanks in a building that was not water-
tight. The pumping was actually interrupted when an operator realised this
transfer: he did not know where this water came from and whether it was
radioactive.

The operators were also not aware that the water in the primary circuit was
boiling despite the restart of the emergency water supply from the steam
generators. However, the readings of neutron flux were abnormally high (that
warned the operators that there were bubbles in the core); the primary pumps
showed signs of a loss of flow (due to cavitations caused by pumping a
mixture of water and steam). The operators first stopped the two pumps of
loop B (operating time: 1 h. 14 min.) and twenty five minutes later those of
loop A.

b) *The next few hours: the hierarchy of the centre is on the spot.* Two hours
after the start of the accident, the accident situation was still unidentified
The engineer on duty had been on the spot for one hour; other operators,
among them three engineers, arrived in the control room. This was when
telephone contact was established, the first contact with the outside world,
with the engineer who represented the Babcock company, the manufacturer of
the boiler, at the centre, and when the question of the position of the

discharge valve of the pressure unit was broached.* This led to its closure at operating time 2 h. 22 minutes.

The pressure in the primary circuit rose again immediately. The technical experts who analysed the accident in detail confirmed that the operators had information at their disposal which should have led them to the discovery of the failure of that valve:

- The indication of increased temperature downstream from the valve: this indication which is normal at the start of an accident (since the valve is normally open) should have raised the operators' suspicion because it revealed the presence of primary fluid passing through the valve. The operators knew that this valve leaked before the accident (leakage four times higher than normal). This indication was therefore explicable. The idea occurred to them to test this valve**but as they declared a few days after the accident they were reluctant to undertake such a test after the difficulties they had encountered at the reopening of this valve during previous trials (5, p. 2).

- The water level in the discharge tank: this information was not relayed to the control room so that they would have had to check in neighbouring premises.

After the closure of the discharge valve, the operators hoped that natural circulation would establish itself in the primary circuit between the hot point (the core) and the cold points constituted by the steam generators. However, it did not happen; the state of the primary circuit was not conducive (pockets of steam in the high points of the circuit) to natural circulation to be established. The temperatures rose. The operators decided to restart the primary pump.

Until then the different periodical alarms signalled a low level of reactivity. During the third hour of the accident various signals indicated increased levels which led the duty engineer and the watch-keeper of the centre to order a state of emergency on the site a little before 07.00 h.

A few minutes later the chief of the centre arrived in the control room and took over control of the operation by forming a team to deal with the accident and to put the emergency plan into operation. The command post was established in the control room of the other section (TMI 1). General alert was given at 07.24 h by the chief of the centre.

However, the technicians did not succeed in their efforts to get the situation under control. Their problem was the cooling of the core in the knowledge that the primary circuit contained a lot of gas. This was a critical time: at operating time 2h 48 min. ... it was evident that the top part of the core was without cooling fluid, down to two thirds of its height which is 3.66 m (1, p. 100).

---

*To do this it suffices to shut the isolated valve upstream — see diagram.

**The engineer from Babcock who must have known of the incidents which had occurred at the centres of Oconee (June 13, 1975) and David Besse (September 14, 1977) with the discharge valve of the pressurizer should have been forewarned of the problem.

This situation was not foreseen in the original plans and was therefore outside their thinking*. It forced the technicians to improvise and to try and cope as best they could.

At operating time 3 h 13 minutes they had to interrupt the forced circulation in the primary circuit (because of strong vibrations of the pump which had been restarted); they also had to isolate the steam generator of loop B on its secondary site (where minor escapes caused contamination of the secondary water).

The operators decided to release the residual heat into the atmosphere (which constitutes a cold source) and to reduce the pressure in the primary circuit (so as to recreate a liquid condition in the primary circuit). In order to do this they proceeded, height of irony, to reopen the famous discharge valve of the pressurizer and to start the safety injection system, the output of which they were just going to reduce (still in order not to fill up the pressurizer). It seems therefore that in the end the reinstatement of the discharge valve to 'normal' condition after operating time 2 h 22 minutes had done more harm than good.

The radioactivity continued to increase. At operating time 3 h 20 minutes an alarm signal indicated a level of radioactivity in the dome of the enclosing wall of 8 rem/hour which, given the fact that the detector is protected by lead, corresponds to a radioactivity of 800 rem/h. Nearly four hours after the start of the accident the surrounding wall was automatically isolated**. In fact, the surrounding wall was not totally isolated because the pipes continued to transfer the water gathered in the draining trap of the surrounding wall to the storage tanks in the auxiliary building (1, p. 102). The third barrier was therefore by-passed. More than four hours after the start of the accident the operators, after having failed to reduce the pressure in the primary circuit sufficiently (which would have allowed the core to cool through the cooling system normally used when the reactor is not operating) tried to reestablish cooling by means of the steam generators after closing the discharge valves of the pressurizer again and after restoring the safety injection to full out-put. The pressure in the primary circuit was again raised to 140 bar.

c) *The course of events: the operators, the responsible technicians of the centre, the experts (NRC) are all in the control room*. At 10.05 h a group of five experts from the NRC arrived on the site; two of them went to the control room of TMI 2.

The level of radioactivity increased to the extent that the staff had to put on masks with filters, which did not help communications. The technicians battled on to cool the core; but the natural convection in the primary circuit was still not established. At 11.38 h the operators returned to their first idea: to force down the primary pressure in order to cool with the cooling system used when the reactor is not operating.

Whence: stoppage of the safety injection (which had worked at full out-put for nearly three hours) and reopening of the discharge valve.

Eight and a half hours after the start of the accident the pressure in the

---

*We shall see in Chapter 3 what precisely this term means.

**See preceding paragraph on technical characteristics.

primary circuit at last reached 40 bar. The water tanks which belong to the
safety injection system emptied their water into the primary circuit and thus
stopped depressurisation. It would take another seven hours until a situation
was reached which could be called acceptable from the point of view of cooling
the core (this cooling being achieved by the generator of loop A and by
circulation in the primary circuit ascertained by the pumps).

Meanwhile (at operating time 9 h 50 minutes) a dull noise was heard in the
control room. The operators did not realise the significance of this noise
and interpreted it as being due to the snapping of a shock absorber in the
ventilation when in fact it was a hydrogen explosion inside the surrounding
wall (1, p. 107).

During the following night (i.e. from Wednesday to Thursday) the operating
staff divulged the presence of an incondensible gas bubble made up essentially
of hydrogen in the top of the tank. It was now understood that the hydrogen
came from the (exothermic) oxidising reaction of the Zircaloy (of the
casings of the fuel which had reached a high temperature) and that the dull
noise heard before had been an explosion of the hydrogen which had been set
free by the breach of the primary circuit and had combined with the oxygen
of the surrounding wall.

From Thursday morning onwards, a new feeling of disquiet settled on the
technicians concerning the explosive capacity of the bubble of hydrogen
captive in the tank. It proved to be non-existent much later because under
the conditions in the upper part of the tank the oxygen required for an
explosion could not continue to exist. This fear which was not unanimous
among the technicians (the experts from the NRC were convinced of the risk
of an explosion, contrary to those from the centre) was sowing an unbearable
feeling of fear among the population.

As to the hydrogen in the surrounding wall, it could be eliminated by a
system of catalytic recombination, the setting up of which took several days
after the difficulties caused by the level of radioactivity.

2nd: Political confusion and social disquiet

Such an accident obviously concerned, in addition to the operators, two
categories of participants: the various authorities and the public.

a) Wednesday, March 28. From 07.00 to 08.00 h: information given to the
authorities by the operating company.

At 07.02 h, i.e. three hours after the start of the accident (the command
post headed by the chief of the centre had just been set up) the Pennsylvania
Emergency Management Agency was put in the picture. PEMA called the rescue
centres of the three counties concerned. Those responsible at the NRC
received the information at the start of office hours (07.45 h): they could
not be reached earlier.

The authorities learned at 07.35 h that general alert had been raised on
site by the chief of the centre at 07.24 h. The state governor of
Pennsylvania was informed about the situation. Teams were sent out in the
vicinity to measure radioactivity.

*b) 08.25 h: Control of the information slips out of the hands of the
authorities and the operating company.* After a journalist from the local
radio station had alerted his station having noticed police and fire police
in Middletown, the station's director of information telephoned the centre to

talk to their Public Relations service. By mistake he was put through to the control room.

"I cannot discuss (it) just now, we have a problem" (1, p. 103)

was the answer combined with the advice to call the company's headquarters, which gave some details but insisted that there was no danger. Radio Harrisburg announced the accident in its news bulletin on this basis.

At 09.15 h the White House was informed by the NRC.

At 09.26 h Associated Press broadcast the information. This promptness which was partly due to accident gave the officials short notice: a good many among them were put in the picture by the media, the mayor of Harrisburg first learned about the accident from a phone call by a radio station in Boston (!).

c) End of the morning: the first contradictory information. The operating company loses credibility. At the end of the morning the Lieutenant Governor, Mr Scranton, called a press conference and made a brief statement:

> The Metropolitan Edison Company has informed us that there has been an accident at TMI 2. Everything is under control. There is no danger to the public ... There has been a small release of radioactivity into the atmosphere. All safety equipment has functioned correctly. The company keeps constant control of the radioactivity around the site since the start of the accident. No increase in the level of radioactivity has been discovered; the level has remained normal (1, p. 106).

Until midday the Public Relations service of the company continued to deny the existence of radioactive escapes. However, at 11.00 h all non-essential staff left the island.

The Mayor of Middletown, the town where the centre is located, was assured by telephone by the board of Metropolitan Edison that no radioactive leak had occurred. Twenty seconds later he learned from the radio that there had been radioactive escapes. The operating company told him only five hours later.

This confusion in the information given on escape or non-escape of radio-activity originated mainly from a lack of coordination within the operating company, Metropolitan Edison. This caused it to lose its credibility in the eyes of the public and the authorities.

d) Afternoon: The media are on the spot. Information swamps the United States. The information media began to swarm out: numbers of radio and news-paper reporters, photographers, cameramen. Lieutenant Governor Scranton held a second press conference. The style was quite different from the one in the morning:

> The situation is more complex than the operating company had us believe earlier on. We have taken measures. For the time being we think there is still no danger to the public. Metropolitan Edison has given you and has given us contradictory information. We are about to have a meeting with the people in charge of the company and hope to be able to answer most of your questions. There has been an escape of radioactivity outside the centre ... The company has informed us that the steam which contained a measurable quantity of radioactivity escaped between 11.00 and 13.30 h (1, p. 109).

Millions of Americans learned about the accident from the TV news at 19.00 h. The mayor and the municipal council at Goldsboro, 2.5 km from the centre, started talking to the inhabitants of the small community from door to door. They told them what they had learned through the media and through official channels. They brought up the case that the governor might order general evacuation and advised the inhabitants to judge for themselves the opportunity to evacuate without waiting any further while themselves consider- ing this not to be worthwhile.

e) Thursday, March 29: relative calm. The morning papers, regional and national, reported the accident on their front pages on the basis of partial and often contradictory information.

On the spot the level of radioactivity remained high in several parts of the centre. Outside there did not seem to be a serious problem. However, at 14.10 h a helicopter registered a short emission of radiation: 3,000 millirem per hour above the centre. During the afternoon the NRC also became worried about the release of mildly radioactive water from the centre. This was discharge water: the tanks were now full up. This was done without informing the communities or the press, and it did not augur well for public relations. When the information about these releases reached him, the president of NRC insisted that they be stopped immediately. This was at 18.00 h; 180,000 litres had been shed. A solution had to be found for this discharge water of which 1,800,000 litres were already in storage. After hours of discussions the Department of Environmental Resources of the state, issued a press communique in which it acknowledged the, very reluctant, acceptance of the need to release this water. The water releases were stopped shortly after midnight.

At the end of the afternoon the governor held a press conference. A member of the NRC announced to the journalists that there was no longer any danger for people from the island. The Governor did not agree with him; the events were soon to confirm his doubts. At 18.30 h the NRC received an analysis of the cooling water from the reactor; it appeared that the core had been much more seriously damaged than had been thought until then. At 22.00 h this news was telephoned to the Governor's office: There is now a greater possibility of radioactive leakage. Nothing had changed at the centre except this new knowledge which the NRC now had of the situation.

f) Friday, March 30: confusion — 200,000 people leave the area spontaneously.
A 'horrible coincidence', to use a phrase by the commission's president, Kemeny, weighed heavily on this confusion. At 07.10 h, without warning the authorities, the operations supervisor at TMI 2 decided to transfer radio- active gas from a tank (located in an auxiliary building and destined for supply to the primary circuit) to a discharge tank, as the pressure in the former inhibited the circulation of the cooling fluid. The supervisor knew that that this would have damaging consequences on the level of radioactive emission as the whole system involved was not water-tight. In fact, when measures had to be taken, 1,000 millirem/hour were registered at 07.56 h, 1,200 millirem/hour at 08.01 h at 40 metres above the centre. It appeared that there was disquiet at the NRC about the discharge tank; if it did fill up completely there was a risk of rupture of the safety disc and of a pro- longed and substantial escape of radioactivity. Just so: at 09.00 h came a warning that the tank was full. One of the experts made a rapid calculation: the rupture of the disc would cause an emission of 1,200 millirem/hour. Some ten or fifteen seconds later someone announced that 1,200 millirem/hour had been measured at TMI ... This caused consternation. In order to under- stand the situation, the Kemeny report stresses, it must be pointed out that

the communications between the NRC officials and the operating company had
never been good. Later on the NRC found out that the radioactive emission
did not come from the discharge tank, that it had not been full. However at
the time, under the shock and without querying the figure of 1,200 millirem/
hour, action was taken.

The NRC let PEMA know that people high up in the organisation had recommended that the Governor order evacuation. A radius of ten miles was
suggested. However, the director of the Bureau of Radiation Protection
thought that evacuation was not necessary. The director of the emergency
service of Dauphin county had been warned by Metropolitan Edison of the
existence of radioactive emission (08.34 h). Twenty minutes later PEMA
confirmed this but stressed that no evacuation was necessary. At 09.25 h the
director of PEMA informed him that an official evacuation order was expected
within the next five minutes; those responsible at the service in the other
two counties of the region received the same message. The fire police
stations within a radius of ten miles were alerted and an appeal broadcast:
an evacuation order might be given.

Shortly after 10.00 h the Governor talked on the telephone to the president
of the NRC: no evacuation was necessary; nevertheless, a suggestion was made
to the Governor: to request everybody living within a radius of five miles to
stay at home for the next half hour. The Governor agreed and later in the
morning advised everybody living within a radius of ten miles to stay at home.
The Governor also demanded from the president of the NRC that an expert be
sent to him. An hour later the president phoned the Governor and told him
that an expert was being put at his disposal and that special communication
links between TMI, the Governor's office, the White House and the NRC were
being set up.

At 11.40 h the president of the NRC telephoned the Governor and offered
the apologies of his organisation: the NRC had been mistaken and evacuation
was not necessary. Nonetheless, after various discussions, the Governor
decided to recommend evacuation of pregnant women and children of pre-school
age within a radius of five miles; all schools in that area were closed. This
was announced shortly after 12.30 h.

On the part of the operating company, the situation looked hardly brighter
as regards public relations. During a press conference the vice president
of Metroplitan Edison lost credibility again: he knew nothing of the problem
of that 1,200 millirem/hour and acknowledged his refusal to give all available
information. The following day the White House complained to the industrial
group about the large number of contradictory statements which had been
issued by them.

All these things caused a large number of residents to leave. Evacuation
plans were being prepared. A radius of five miles affected 25,000 people.
Within a perimeter of ten miles there were 130,000 people, three hospitals
and old people's homes. 650,000 people, thirteen hospitals and a prison were
within twenty miles.

At 14.00 h H. Denton*arrived on the spot accompanied by a dozen people from
the NRC. The prime preoccupation was the elimination of the hydrogen bubble
that had developed in the reactor. Of course, it was not expected to explode
immediately but one had to get rid of it. Towards 20.30 h H. Denton put the

----

*The expert delegated to the Governor as promised by President Carter.

Governor in the picture: the core was severely damaged; the bubble caused a problem for the cooling of the core; no immediate evacuation was necessary. At 22.00 h the two men called a press conference in the course of which the Governor said again that there was no need for evacuation, retired the council asking them to stay at home but upheld his recommendation that pregnant women and pre-school age children should leave. The Health Department (DHEW) was worried about the possible fall-out of radioactive iodine and undertook the collection of iodide of potassium. An exceptional mobilisation of the industries involved was needed to meet this request; remedy began to arrive the following day. Finally, one controversy remained: Had the White House dissuaded the Governor from requesting the declaration of a state of emergency by President Carter?

During the day 50 per cent of the residents of the area within a five mile radius had spontaneously left as well as a third of the population within the ten mile zone. Altogether about 200,000 people left the area. The event obviously received wide media coverage.

g) Saturday, March 31: the worry about an explosion of the hydrogen bubble.
The big worry about a possible hydrogen explosion came during the weekend. That this had been an unfounded fear, an unfortunate error, the Kemeny report stresses, was never subsequently understood by the public, partly because the NRC never made any effort to admit that it had been mistaken (1, p. 126).

The risk of a hydrogen explosion and possible evacuation of a million people made the front pages of the morning papers. At 09.00 h those respon=sible at Metropolitan Edison indicated at a press conference that the gas bubble had diminished in size by about two thirds since the night before; at the same time H. Denton estimated a reduction of only 10 to 15 per cent. Those responsible at the operating company announced that they would not hold any more press conferences. During the morning a NRC press centre with seven communications people was installed in Middletown; H. Denton was confirmed as official spokesman. At that time 300 journalists were on the spot.

It was known that the danger of an explosion of the bubble would become serious if there was a significant presence of oxygen. Towards midday, after analysis, it appeared that there was indeed a formation of oxygen. The president of the NRC advised H. Denton to warn the Governor of the potential danger. Towards 13.00 h, however, the analyses showed that the oxygen would not cause a dangerous situation for another two or three days. During the night on Saturday other calculations indicated that the percentage of oxygen in the bubble was at the threshold of inflammability. Towards 18.45 h the question of the effect of an explosion had already been studied: the tank would stand up to it.

At 14.45 h the president of the NRC let it be known that a precautionary evacuation of a ten to twenty mile zone might be necessary if the engineers considered it necessary to let the bubble out of the reactor. This could damage the core and also make the bubble explode. In the evening at 20.30 h Associated Press stated the worries of certain NRC experts concerning the possibly imminent explosion of that bubble. During a press conference H. Denton and the Governor confirmed that there was no imminent danger and that there was no difference of opinion among the NRC experts. However there was contradiction, the Kemeny report stresses (1, p. 130), and H. Denton had analyses made ... all the more so as President Carter had announced his arrival for the next day. The Health Department continued to work on a possible evacuation and was determined to proceed with it immediately should the NRC not give assurances about the cooling of the reactor.

h) Sunday, April 1: more confusion. Throughout the night from Saturday to
Sunday the offices in charge of preparing plans for evacuation were inundated
with telephone calls: the citizens worried about the contradictory information
about the hydrogen bubble. The fact that the federal administration had taken
charge of the situation deprived the local authorities rather largely of
access to the information and of their initiative in the action (1, p. 131).
A local member of the House of Representatives who could reach neither the
Governor nor the Lieutenant Governor on the telephone let it be known shortly
before midnight on Saturday that failing information Dauphin county would
issue an evacuation order for the following morning 09.00 h. At 02.00 h a
meeting was suggested; it was to take place at 09.00 h. On this occasion the
director of PEMA too complained about the difficulty he had encountered in
trying to get information. That morning too the Catholic bishop of Harrisburg
authorised the curates of the parishes in the region to give general absolutio
to the faithful during Sunday mass (this canonical dispensation is authorised
only in case of war or serious crisis).

   At 14.00 h President Carter arrived and visited the centre before holding a
press conference. In mid-afternoon new readings showed that the bubble in
the reactor was diminishing. For the rest of the afternoon the NRC knew that
there was no longer any danger of an explosion. Partly because this was not
yet certain, the news was not released.

i) Monday, April 2: caution at the NRC. At 11.00 h at a press conference
H. Denton announced a very substantial diminution of the size of the bubble;
but calculations would still be necessary (the operating company had already
let it be known that the bubble had disappeared). H. Denton acknowledged that
there could have been error or exaggeration on the part of the NRC; he
displayed very cautious optimism so as not to lose credibility. During the
afternoon in a public statement the mayor of Middletown disclosed that he had
given orders to the municipal police to shoot all possible pilferers.

j) Epilogue: end of Act One — Start of Act Two — also set in uncertainty. The
accident at TMI did not end with the disappearance of the bubble; a small
bubble remained, there was still gas in the cooling water and the reactor
itself had been seriously damaged. There were still periodical but small
escapes of radiation and some feared a major escape of radioactive iodine 131.
The schools remained closed. The Governor's recommendation for the evacuation of pregnant women and pre-school age children remained in force. There
was discussion among those responsible as to whether iodide of potassium
should be distributed; in the end and in view of the small doses of dispersed
radioactivity as well as on account of the risk of reviving fear among the
population unnecessarily it was decided not to go ahead with this distribution

   On April 4, the schools were reopened except within a radius of five miles;
the evacuation recommendation remained in force. The NRC wanted a marked
event before lifting this recommendation: a significant reduction of the
temperature of the core for instance.

   The accident did not end with the cooling of the core either. There had
been more than 3.5 million litres of contaminated water inside the confinement wall or in the tanks of the auxiliary building in April; radioactive
gases remained in the surrounding wall; the core was severely damaged; radioactive elements contaminated the walls, the soil and the machines in several
buildings. There was an unprecedented amount of decontamination work to be
done, estimated at a cost of between 80 and 200 million dollars; it would
need several years. In addition the operating company requested permission
from the NRC to release Krypton 85 into the atmosphere in small quantities
(in order to remain within the limits).

Entry inside the surrounding wall will not be possible until the radioactive gases have been extracted. Nobody knows exactly what state the core of the reactor is in. The accident will therefore continue to cause anxiety; workers will continue to be affected by further quantities of radioactivity. Five of them had doses above the norm in August 1979. For the public there remain strong reasons for concern.

3. SOME LESSONS TO BE LEARNED FROM TMI

Without attempting an exhaustive analysis and without substituting for the technicians in their detailed analysis of what happened, in order to draw all possible conclusions a certain number of lessons impose themselves immediately on the neophytes who try to understand as best they can what happened and why it happened in the way it did. The lessons are of two orders: of the technical order i.e. relating to the installation as such and to its way of working; then of the social and political order i.e. relating to the problems of information and relations between the various parties: the operating company, the safety authority and the public.

1st: Technical lessons*

For the layman, the accident at TMI has revealed certain difficulties or insufficiencies likely to exist in the workings of a nuclear centre.

a) The difficulties encountered by the operators in controlling the state of their installation in an accident situation. This accident has shown clearly the profusion of information which the operators found themselves confronted with from the time of the initiating incident: the complexity of the surveillance of the large number of the centre's components. It will be necessary for the technicians to visualise in advance the interfacing of the operators and the total of information in order to draw from them the essential data which are, at the time, 'digestible' by man.

b) The difficulty of conceiving systems of surveillance or information which are simple to interpret and unambiguous. The example of the information 'discharge valve of pressure unit shut' is very revealing since in the event it was not the valve that was shut (it had remained wedged open) but the order to shut which had been given (the information "cut off power supply to the control motor" is read as "valve shut").

c) The difficulty of protecting oneself against failures of a common order. These failures which are difficult to foresee often reveal themselves *a posteriori* as being evident i.e. once the incident has occurred. They require detailed analysis of every system, keeping in mind surrounding systems (geographical proximity can favour such an eventuality) and the functionally linked systems.

In this way the common mode failure of the water supply pumps could have been identified (they had a common auxiliary circuit) or the common mode failure of the closure of the two valves in the emergency water circuit (the valves had probably been closed in order to test the emergency water pumps and it had been forgotten to reopen them).

This research into potential common mode failures is really indispensable

---

*These questions will be taken up again and developed later on (in Chapter 4).

since such failures render all the efforts made by the designers of the system useless (even attempts to improve reliability by duplication of components are made useless).

d) The lack of knowledge of the operators and perhaps also of the designers in the knowledge of the post-accidental situations. The hesitation of the operators in achieving an acceptable situation from the point of view of the cooling of the core, show at which point the systematic study of the cooling of a core by a mixture of water and steam had failed. The operators had to proceed by trial and error skilfully, by the way, to get on top of the situation; starting the safety injection system and then stopping it, closing the discharge sluice of the pressure unit and then opening it. A better knowledge based on reliability analysis would no doubt have permitted faster control and less damage to the core.

The hydrogen explosion in the surrounding wall, wrongly interpreted, was another demonstration of this lack of knowledge.

e) The sloppiness of the operating company in its instructions. It seems that a conscientious attitude on the part of the operating company would have had to lead it to a stoppage of its installation, knowing the existing difficulties with the discharge valve of the pressure unit which in the end turned out to be the source of the accident.

The operators stated to the Commission of Enquiry that they knew that this valve leaked abnormally and that it had some weaknesses (failure to close during previous tests). Keeping in mind the knowledge of this state of dilapidation, the operators were led to gross errors of interpretation and judgement. One can doubtless confirm that if the valve had been in proper order before the accident, the operators would not have needed 2 h 20 minutes, with the same information available, to discover its failure.

This is, incidentally one of the conclusions drawn by P. Tanguy (3, p. 531).

f) Lack of conscientiousness in the design. The designers put three barriers in sequence between the fuel and the public in such a way that in case of failure of the first (casing) and the second (primary circuit) the dangerous products could be confined in the surrounding wall (the third barrier).

The accident at TMI has shown clearly that in fact this third barrier could be by-passed (via the ducts). True, it played a fundamental part since it fulfilled its function well in confining the gaseous products; but a more conscientious design would not have permitted that the water from the primary circuit that had leaked into the draining trap of the surrounding wall, could also be transferred to the exterior of that wall.

The knowledge of the failures which had occurred on the same type of discharge valve of the pressure unit in the centres at Oconee and at David Besse should have led the designers, in this case the Babcock and Wilcox company, to review the design of this valve on its installation.

To conclude our comments on the difficulties and insufficiencies we take up once more the statements made by the Kemeny commission:

A number of factors have contributed to the inappropriate action of the operators such as the insufficiencies in their training, the lack of clarity in their conduct procedures, the failure of organisations to draw the right lessons from previous incidents and the deficiencies in the design

of the control room. These insufficiencies are the work of the operating
company, of the builders and of the Federal Commission which regulates
nuclear energy. This is why the operator error does or does not 'explain'
this particular case: these insufficiencies in the end brought about an
accident like the one at TMI, of this we are convinced.

2nd: Lesson on the capacity for social control of the event

The climate of uncertainty, even fear, which reigned during those few days
is due on the one hand, to a number of difficulties in communications between
the various parties, difficulties leading to erroneous or contradictory
information; on the other hand, to the inefficiency in the preparation of
plans by the various authorities which were to be enacted in case of serious
accident.

More precisely and without wanting to generalise abusively a certain number
of points have to be revealed from the experience gained in Pennsylvania at
the end of March 1979.

a) The difficulty of 'controlling' information. One might think, in such
circumstances, that a good filtering of information would permit a better
management of collective agitation, better anticipation of psychic stress and
also better safeguarding of the chances of the adopted development programme.

The experience at TMI shows how fragile this construction is. One error at
a telephone switchboard, and the Director of a Press Service is connected to
the control room: the information goes on the air. Local and national
personalities are alerted by the radio, at the steering wheel of their cars
while they go to their jobs, or by a phone call from a journalist who works
several hundred kilometres away (Boston). A press conference is organised
but the authorities do not have the information, as we have seen, because of
the reticence of the operating company. A press conference is organised by
the operating company but it does not know that what it wants to silence is
already too well known; it loses its temper: "Nothing obliges me to tell you
what I know". Thus credibility goes out of the window.

Certainly, H. Denton (NRC) appears much better qualified but there is a
great difficulty: How does one 'control' information when it is by its
nature, we are in a disaster situation, incomplete, uncertain, explosive,
sometimes aberrent, including phenomena which threaten to amplify the problem?

b) The difficulty of managing the uncertainty. TMI has confronted people in
responsibility with an accident of a new type. The number of unknowns in the
equation to be solved ruled out assured reasoning. Faced with an explosion,
a fire, an avalanche one 'sees', one 'clears the ground', one narrows down
the unknowns.

Here, the person responsible has to rely on experts who find it difficult
to explain theoretically and empirically what is happening, as H. Lewis notes:

The later development of the accident and in particular the formation of a
hydrogen bubble had not been foreseen by any of the prior analyses partly
because these analyses never went beyond the first phases of the accidental
sequence. The operators therefore soon found themselves faced with a
situation which had not been foreseen and for which they had not been trained
(7, p. 85). Here, the experts had measuring devices which did not give them
reliable readings (apparatus blocked to the maximum, apparatus broken down,
disturbing background noises etc.). The models for comprehension of what was
happening failed: one did not know how to construct new models in real time

(this was the difficulty in understanding the phenomenon of the massive production of hydrogen for instance).

To this background uncertainty must be added that which attached to messages that originated from extremely varied interests. The builders, the operating company, the federal safety agency all had their own objectives; this cannot be without influence on the information given to the decision maker. This is immediately true concerning some of the most competent experts. H. Denton (NRC) did say for instance that he had put the brakes on the production of optimistic messages in the last phases of the critical period in order not to harm the image of the NRC, which doubtless resulted in a blackening of the picture presented; what is true in one sense is true in the other when one considers the behaviour of the operating company. This absence of clarity became still more massive as the number of experts grew. D. Nelkin in particular writes:

Hundreds of specialists from the NRC and other federal and national agencies came to TMI. Other experts came from companies, the nuclear industry, anti-nuclear groups. Their evaluations of the problem, of its dimensions and its causes were just as varied as their interests, their pre-occupation with the future of nuclear policy and the image they wanted to create (6, p. 3).

c) The difficulty of dealing with an unprecedented situation. In the case of TMI the evacuation plans were not ready. The telephone links between the various responsible authorities had not been set up. The crisis headquarters had not been constituted. The authorities (NRC) could not be reached before the opening of offices etc. The confusion could only be massive.

What one had to know how to do was not operational. It was a matter of going beyond the normal limits: dealing with the unprecedented, possible evacuation of a million people based on uncertain information, the uncertainty being impossible to remove until after the critical period.

One ended up by resorting to methods which are incongruous to our age of high technological sophistication: the mayor of Goldsboro ran around the streets with his municipal council and went from door to door with nothing better to offer than: use your own judgement. We dare not tell you to leave your houses (1, p. 111).

4. BALANCE SHEET

The Kemeny report estimated that despite serious damage to the centre the largest part of the radioactivity was contained within the surrounding wall and that the leaks which occurred will have only a negligible effect on the physical health of individuals. The major effect of the accident occurred on the level of psychic stress.

The report acknowledges that even small doses of radiation can have genetic effects. However in the case of TMI they have been really minimal. If there are effects they will be so small in number that they cannot be statistically established.

The financial cost to the operating company is estimated at one or two billion dollars. This cost would become much higher if it were proved to be impossible to put the centre back into operation (1, p. 13).

Finally, the accident has diminished the public's confidence in the nuclear industry. The Commission of Enquiry did not stop at this point. It asked the question: What if .. ? What would have happened if things had not gone the way they did? We shall come back later to this rather uncommon approach. There could have been fusion of the core:

The NRC estimates that if the drying up of the core had occurred earlier in in the course of events fusion would have been quasi certain (10, p. 13).

One may ask oneself: if it had happened, would there have been time to protect the public?

Nevertheless, this particular centre, on account of its exceptional location, the greater than usual thickness of the concrete surrounding wall, would have been able to contain the radioactivity which would have resulted from that fusion, as the Kemeny report stresses before pointing out that one cannot be absolutely certain of the result, that there might also have been faulty handling in the fusion situation ...

In any case, the Kemeny commission concludes, accidents as serious as the one at TMI cannot be permitted to happen in the future (1, p. 15).

Nevertheless, let us dwell on the possible effects of the incident on health. There is no unanimity concerning the amount of radioactivity released: the estimates have varied widely. An independent enquiry carried out by an anti-nuclear engineer gives the following information: on April 11 the level of radioactivity twenty miles from the centre was five times the normal; on April 16 the level of radioactivity thirty miles from the centre was fifty times the normal. The figures are obviously very different from those given by the NRC which estimated that at the end of April the radioactivity five miles from the centre had reverted to an almost normal level.

Let us also remember here the problem of uncertainty. This is an essential dimension of risks and disasters in our time. Various technical reports were able to give evidence of the fact that:

Three hours after the accidental process a large number of indicators of the level of radioactivity went beyond the limit of their scale (8).

The greater part of the detectors was unusable at that time (9, p. 12). This was because the environment contained rare radioactive gases: Xenon and Krypton; the level of background noise was therefore artificially raised. Do we therefore not really know what escaped there*?, asked G. Gibson, one of those responsible at the NRC, in the course of the enquiry. "Exactly", was the answer.

As a French technical report remarks this uncertainty may be prolonged: In the course of a conversation Commissioner Bradford (from the NRC) dwelt on the legal difficulties which may have to be dealt with as a result of the absence of safe data concerning the radioactive leaks during the first days. (10, p. 15).

There will still be the difficulty of being sure about the future. The

---

*This was the chimney of the auxiliary building; the needle of the indicators which were installed there was positioned at the maximum of their scale

removal of the water in the surrounding wall had not yet been started at the end of summer in 1980. One could read these lines in a specialised document:

The officer responsible for the protection against radiation in Pennsylvania (Th. Geruski, Director of the State Bureau for Radiation Protection) estimates that the evacuation of the population living around the accident reactor at TMI "is an imminent eventuality" and will remain so at least during the four years needed for cleaning up and repair (11, p. 2).

5. CONCLUSION

Whatever else might be said about it, the accident at TMI did come as a surprise. The operators were put out of action by phenomena which they did not know. Before control over the situation could be reestablished, scenarios of disaster could have developed. The technical uncertainty was strongly felt for a long time. Effects on health will probably be minimal but assurances in this respect must doubtless remain cautious. The potential danger is actually far from negligible. The economic cost of such an incident is enormous. Finally, as we shall see later on, this event forces a review, to say the least, of certain safety principles.

To conclude this chapter, which we are convinced, requires complementing, spelling out and corrections, we shall quote these essential observations from the report by President Carter:

We are convinced that if the operating companies and the safety authorities do not make changes they will end up by destroying public confidence completely, and they shall be responsible for the elimination of nuclear energy as a viable source of energy (1, p. 25).

One ultimate remark, nevertheless. This judgement by the commission set up by President Carter is especially valid for countries which treat alternative sources of energy lightly or have set things up in such a way as to treat them lightly. In other countries which could not or did not want to equip themselves with sufficient strategic flexibility a different scenario would be more probable in case of a nuclear disaster: recourse to force, the pursuit of nuclear development policy by other means. One must, nonetheless, remain cautious. A large-scale nuclear disaster could also result in, for instance, considerable breakdown at the level of the responsible organisations

REFERENCES

(1) *Report on the President's Commission on the accident at Three Mile Island.* Pergamon Press, New York, octobre 1979 (201 pages).

(2) B. AUGUSTIN et J. M. FAUVE, *L'accident nucléaire de Harrisburg. Analyse d'une crise.* Sofedir, Paris, 1979 (83 pages).

(3) P. TANGUY, L'accident d'Harrisburg. Scénario et bilan. *Revue Générale Nucléaire,* no 5, septembre-octobre 1979, pp. 524-531.

(4) Three Mile Island. *La Gazette Nucléaire,* no 26-27, juillet 1979, publiée par le Groupement de Scientifiques pour l'Information sur l'Energie Nucléaire (G.S.I.E.N.).

(5) *Nucleonics Week*, May 31, 1979.

(6) D. NELKIN, The expert at Three Mile Island. Hastings Center Research
    Group on Ethics and Health Policy Cornell University, October 19, 1979
    (18 pages).

(7) H. LEWIS, La sûreté des réacteurs nucléaires. *Pour la science*, mai 1980,
    pp. 73-89.

(8) Analysis of Three Mile Island-2 Accident; sequence of events Nuclear
    Safety Analysis Center (N.S.A.C.), 1 July 1979.

(9) *Nucleonics Week*, June 28, 1979.

(10) Rapport sur la mission effectuée au Etats-Unis (27-20 juillet 1979) Groupe
    Permanent des Réacteurs Nucléaires, août 1979, (26 pages plus annexes).

(11) *Nucleonics Week*, August 9, 1979.

(12) B. ROCHE et A. CAYOL, Rapport sur la mission effectuée aux Etats-Unis.
    Service Central de Sûreté des Installations Nucléaires Institut de
    Protection et de Sûreté Nucléaire (D.S.N.), 1979 (36 pages).

(13) Rapport de l'Institut de France. Académie des Sciences, 28 septembre 1979.

V. SATURDAY, NOVEMBER 10, 1979: TORONTO

On Saturday, November 10, 1979, a little before midnight, a railway accident involving various dangerous products and mainly wagons of chlorine constrained the Canadian authorities to operate a large-scale evacuation in the suburbs of Toronto (Mississauga). More than 240,000 people were involved and an even larger movement of population was feared. The presence of a still more worrying product than chlorine, polychlorobiphenyl (PCB) was feared. Luckily this was not the case. Under the influence of the very strong heat created by the burning wagons of propane the PCB could have decomposed itself into a series of extremely toxic products.

1. THE ACCIDENT

The railway convoy that approached Toronto on that evening after a trip of 10 h 30 minutes consisted of three locomotives and one hundred and six wagons of which thirty eight carried goods classified as dangerous. Among those wagons twenty eight contained liquid hydrocarbons, five petroleum derivates, five caustic soda, one of them 90 tonnes of chlorine. In the middle of the convoy one therefore found an assortment of twenty four wagons made up of one wagon of chlorine coupled to a wagon of propane on the one side and to a suite of wagons made up of one wagon of styrene, ten wagons of propane-butane, three of caustic soda and two of toluene on the other. During its trip the train had been inspected seven times by a qualified employee and more superficially checked five times.

Nevertheless, lubrication of one bearing (on the axle of one of the wagons) proved to be insufficient. It was an old-fashioned type of bearing: a simple grease box filled with oakum which had to be wetted with oil very frequently. At 23.53 h the axle heated up and cracked. A set of wheels careered off along the track. The train was passing an inhabited area. It stayed on the rails and approached an unoccupied area or rather one occupied by factories and warehouses which separated the first residential zone from the second. It was chance that the accident happened in this particular location.

At 23.56 h the first twenty seven wagons had cleared a level crossing; but, as two witnesses saw, two wagons detached themselves and a series of other wagons were piling on top of each other. There was a first explosion which was heard very far away. The mechanic's mate ran from the head of the train to shut a valve of compressed air which blocked the brakes of the convoy and to detach the wagons at the level of the twenty-seventh, 400 metres from the locomotives. The whole front section of the train which included the largest number of wagons carrying propane and other dangerous products could thus be separated.

At 00.01 h a second explosion occurred. The mechanic's mate said that one wagon was blown vertically into the air. The front section of the train was stopped at 6 km distance and the company, Canadian Pacific (CP) which had been alerted by the mechanic at 00.00 h began to set up a rescue system (notifying the police, the fire brigade, the Federal Rescue Service etc.).

A police superintendent who arrived on the scene a little before midnight had for his first problem the removal of onlookers who had already gathered within a distance of 30 m from the train. The explosion and the clearly visible scene gave, themselves, general alert. The fire police were ready when the first call reached them. At 00.04 h two fire engines were already on the spot but the explosion at 00.01 h hit the rescuers who were setting up

their equipment. The police were alerted by an ambulance which used its radio.
The police network was alerted: serious risk of explosion, unidentified
chemical products. Within the few minutes following the initial explosions
the area was blocked by bottlenecks. The spectators came right up to the
track to get a good view. The police tried to remove them. A third and
fourth explosion carried more power of persuasion.

2. THE ORGANISATION OF RESCUE AND SAFETY FOR THE POPULATION

1st: The uncertainty about the content of the wagons

A command post was set up at a distance of 400 m from the accident. During
the night 500 police and 100 firemen helped by 200 volunteers (who were used
to control the traffic) were at work.

The first attempts by the firemen to approach the fire were hindered by a
series of explosions. After one very strong explosion they received orders
over the radio to turn back. The hosing had no effect on the wagons or on
the nearby warehouse which had caught fire (and the roof of which soon fell
in).

The immediate problem of fire control was complicated by a very serious
uncertainty: it was not known what was inside the piled-up wagons. The train
driver brought a document which gave a list of the wagons and their content.
This document gave no indication of the position of the wagons within the
convoy. Still worse, it seems that the list stated erroneously, the presence
of a wagon of PCB in the train. It took five hours to disprove this information. The firemen could not refer to the panels on the wagons which
indicated the content: they had been burned. The officials from CP arrived
at 01.30 h and it seems that they then confirmed that there was a wagon of
chlorine in the accident train. An official of the region admitted: we needed
quite a lot of time to identify the nature of the goods. The loading note
supplied by CP was a listing for information which nobody could read or
understand.

However, the wagons were piled upon each other and the tangle could only be
sorted out on the afternoon of that Sunday by means of studies made from a
helicopter: the wagon of chlorine rested upon three other wagons of which two
contained propane which meant increased risk of explosion. The wagon with
the chlorine had an 80 cm fissure.

2nd: The response in the face of danger: six successive evacuations

a) Sunday, 01.30 h: first evacuation — 8,000 people. Towards 01.30 h it was
decided to evacuate 8,000 people from an area of a square mile located down-
wind (zone A, see map hereafter). The police and the ambulances patrolled
the roads with sirens and loudspeakers advising evacuation because of "an
immediate danger from chlorine". This was complemented by house-to-house
control. Many people had anticipated this evacuation which was completed by
03.00 h.

Three rescue centres were set up (a shopping centre and two schools) but
they had to be shut down subsequently when the evacuation zone was enlarged.

During this time experts had been alerted, the industry got their chlorine
rescue plan going. In addition, a general appeal to ambulances and their
staff was launched in the Toronto area (2.15 h). At 02.45 h the increasing
seriousness of the situation on the site made a general meeting of all the

officials present necessary: it had to be decided in particular whether the army had to be called in, 150 men were on alert but it would take them four hours to get there on foot. It was decided not to call on federal forces for the time being.

There was, nevertheless, still extreme worry about the chlorine. Because of its quality as an oxidising agent it could inflict serious burns and in gaseous form at a concentration of more than 3 ppm in the air it causes death. It was not known then, it was established only on Monday evening, whether 60-70 per cent of the wagon's content had been dispersed in the air in the first explosion (dispersion up to 1,200 m up and within a radius of 90 km). If the wagon broke up suddenly because of an explosion of one of the propane wagons the wind could blow concentrated chlorine over a vast area. There remained therefore two major problems once the first evacuation had been completed: controlling the fire, anticipating the action in case of an aggravation of the situation (break up of the wagon, change of wind direction).

While the experts were arriving (03.00 h) there was preoccupation about the establishment of relations with the media. An industrial outfit was transformed into a press centre. This outfit was near the accident site. The police actually have a policy of allowing them to stay in such proximity because this is indispensable for journalists, within the measure of safety requirements: the idea is that if this visual proximity is not offered the journalists begin to think that the truth is being hidden from them. Within an hour most of the TV networks were on the spot despite traffic problems. The representatives from the press could approach to within 30 m. They were warned that if the chlorine expanded they would have to take flight with the police, the press turning right and the police left to avoid collisions.

b) Sunday, 04.00 h: Second evacuation — 20,000 people. At 04.00 h a further series of explosions (from the wagons of propane) forced the launching of a further evacuation order (zone B in the diagram).

This time the transport authority had mobilised sixty five buses. The initial explosion had affected the communication facilities of the authority the headquarters of which was in the accident zone. The bus drivers arrived spontaneously; private cars with C.B. radio were used to coordinate the dispersal efforts.

During this second phase 20,000 people were evacuated. The Red Cross took care of provisions and installed information offices. The Salvation Army helped in controlling the traffic and set up a kitchen for the firemen, policemen and others involved in the work. The Scouts and other organisations also helped in organising the evacuation. Volunteer radio operators were at work early on Sunday to deal with supply and information problems.

So, as a second line of defence a lot of organised groups took charge of the problems of organisation.

In the first line, thirty two respirators were procured by 06.00 h and technical assistance requested from the civil defence authorities to deal with the PCB. Higher officials had been alerted at 04.30 h and a first large meeting took place at 06.00 h.

c) Sunday, 08.55 h: Decision to evacuate the hospitals of the area. By 07.00 h the heat spread about by the burning propane wagons was so intense that the command post had to be moved back. It also seemed that the hospital centre of Mississauga would probably have to be evacuated.

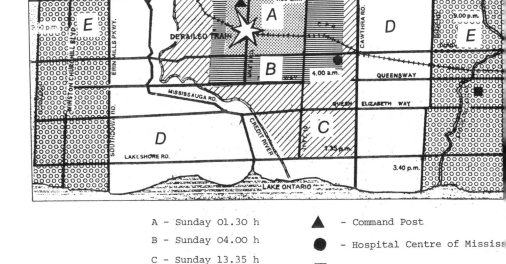

A - Sunday 01.30 h   ▲ - Command Post

B - Sunday 04.00 h   ● - Hospital Centre of Mississ

C - Sunday 13.35 h   ■ - Queensway Hospital

D - Sunday 15.40 h

E - Sunday 21.00 h

Fig. 17: Map of the successive evacuations.

At 08.45 h, considering that this evacuation would possibly take five to
six hours, the decision was taken to proceed with the operation. Twenty
five ambulances and a bus ambulance (with a capacity of twenty places) went
to the scene; a general appeal was launched to obtain further means of transportation; by 09.30 h there were sixty ambulances and seven buses. The people
capable of walking had been evacuated just before 09.40 h, the time when the
evacuation of the one hundred and eighty nine bed-ridden patients began; the
operation was completed by 13.05 h. Within the same period of time two
nursing homes with two hundred and thirty seven and two hundred and two
patients respectively were evacuated. Twenty sick people found themselves
in hospitals which had to be evacuated subsequently and were therefore moved
twice.

At 10.45 h the decision was taken to organise the possible evacuation of a
further hospital in the area (Queensway Hospital).

d) *Sunday, 13.35 h: third evacuation — 42,000 people.* During Sunday morning
the situation on the accident site had further deteriorated: at 09.55 h a
further explosion had caused the spreading of a still larger quantity of gas;
and the wind began to change direction, blowing westwards. The decision to

evacuate one of the rest centres was taken, the operation was carried out
between 12.00 and 14.00 h (there was some confusion: the bus drivers did not
know their exact destinations).

Wind speed increased from 2 to 13 km/h. At 13.55 h general evacuation for
a third zone (zone C) was announced. At that time a total of 70,000 people
had been moved. Special meteorological instruments were installed; the
possible extension of a chlorine cloud was being studied.

e) Sunday, 15.40 h: fourth evacuation. At 15.40 h a further evacuation order
was issued. Despite the extreme reticence of the people responsible at the
Queensway Hospital (230 people seriously ill could only be moved with
difficulty) the evacuation of the hospital centre was being prepared. The
police had been alerted to the fact that in case of a further explosion there
would be only fifteen to twenty minutes to evacuate. The operation was
decided upon and set for completion by 19.00 h.

During the afternoon the evacuation of the residents proceeded calmly. The
police adhered to the rule of issuing frequent bulletins of information. At
14.35 h the fast lanes were closed to traffic. The Red Cross, the Salvation
Army and the other organisations which contributed were advised of the
requirements of food and beds for at least one night. The army supplied
1,200 sleeping bags, 850 air mattresses, 6,300 blankets. Several organisations took part in the organisational effort.

Foreseeing further evacuations the authorities ordered the preparation of
additional refuge centres at 17.05 h.

f) Sunday, 21.00: fifth evacuation; Monday 00.15 h: sixth evacuation. At
19.00 h the wind changed direction and forced a second-time evacuation of
people who had already been evacuated. Between 19.00 and 21.00 h new refuge
centres were set up.

At 21.00 h the evacuation zone was again enlarged; traffic was diverted
more extensively. At 22.45 h it was decided to evacuate a further hospital
and a nursing home near it; this was accomplished between 23.00 and 04.10 h:
300 patients transferred to other hospitals. At 00.15 h, for the 6th time
the limits of the evacuation zone were extended.

3rd: The culmination point: 240,000 people already evacuated, West Toronto
put on alert

At the end of a twenty four hour period 240,000 people had thus been
evacuated from Mississauga and its neighbourhood. A plan for the possible
evacuation of the nearby airport of Malton had been drawn up.

At 01.30 h on Monday morning the police put the western part of Toronto
and Hamilton on alert for a possible evacuation.

The area was closed, the roads blocked, trains rerouted. Police patrolled
the streets and a helicopter kept watch to give warning of any looting.
About 500 police were in action. Plainclothes policemen were dispatched to
check the effectiveness of the roadblocks; weak points were ironed out.
Warning was given that anybody caught in the area would be arrested.

The strategy of the fire police was still to try and control the burning
propane. At dawn on Monday photographs were taken of the chlorine wagon with
special cameras to try and spot the fissures which needed to be sealed.

During the day one gained the impression that it would be necessary to prepare for a real siege: the operation would not be a fast one. That day therefore marked the beginning of a long wait for 240,000 evacuated people. Generally speaking, the information provided by the press was precise and guided by a sense of responsibility. Press conferences and visits to the phantom-town for the media were organised. There were alerts such as the case of a baby found to have scarlet fever in a refuge centre. The waiting was not always without worry. One rescue centre received an order forbidding the occupants to go outside because of the risk of a sudden explosion of chlorine. At 15.45 h on Monday the authorities confirmed that troops would not be called in but additional respirators were requested. They arrived from Halifax in Nova Scotia the following morning (85 apparatus, 65 in reserve).

On Monday evening two more areas were put on alert.

During the day Canadian Pacific Rail announced that it would pay, and that purely as a good-will gesture, only out-of-pocket expenses to evacuees. In Ottawa the federal minister of transport announced that he would introduce legislation on the control of transport of dangerous materials within ten days.

4th: Progressive Control of the situation: return in three phases

a) Tuesday, 13.00 h first phase, 125,000 people authorised to return. At 23.00 h on Monday the first attempt was made to seal the breaches in the chlorine wagon; but the fire delayed the operation until the next morning. The fire was put out at 09.00 h on Tuesday.

At 10.00 h a meeting was organised to study the possibility of a return of part of the population. The evacuated people had thought they were leaving for twenty four hours at the most, and they began to harrass the police at the barriers. Another complication arose: the residents wanted to go home to feed their pets the number of which was estimated at 11,000. Passes were issued after 17.00 h; an animal protection organisation undertook part of the care. As the wind direction changed and the chlorine continued to escape (at the rate of 17 kg/h) it was difficult to reopen the evacuated zones.

At 13.00 h it was possible to decide to let about 125,000 people return to their dwellings. The police thought the return would be orderly but erroneous information led to traffic jams which lasted for seven hours; those who had been authorised to return mixed with those who mistakenly believed they too were authorised. The motorists lost their calm.

As the wagon still leaked the central area remained off limits.

On Wednesday morning the situation on the site had not improved. During the track clearing operations a chlorine cloud escaped. Masks had to be put back on and areas to the south of the accident site had to be reevacuated.

For the central area which had been infiltrated little by little by the evacuees the authorities let it be known that the police had the right to expel the people who had returned to their homes. On the fifth day, Thursday, the pressure exercised by the evacuees grew stronger. An appeal for cooperation had to be launched.

Nevertheless, at dawn on that Thursday the breaches in the wagon which according to the experts still held between 10 and 20 tonnes of chlorine had been completely sealed. An attempt was made to transfer the gas into another tank, and half of this job was accomplished by midnight. In the evening a northerly breeze dispersed the pockets of chlorine which had presented a danger (eight rescuers had actually been hospitalised after having crossed through one of these pockets during the morning). By Friday morning 10 tonnes of chlorine had been tapped off and could be moved away. It was feared that there might again be escapes in the course of the transfer operation.

b) Friday, November 16; 15.00 h: Second phase — 90,000 more people are authorised to return. After six days of absence another 90,000 people were authorised to return to their homes; there remained still 30,000 evacuees.

c) Friday, November 16, 19.42 h: Third phase — the whole area is reopened. Over 16 hours 18 tonnes of chlorine had been taken out of the wagon. At 19.20 h while there were still between 15,000 and 20,000 litres of chlorine in the wagon there was unanimity that all evacuated people could go home. Authorisation was given at 19.42 h.

The authorities warned that the soil around the accident site would have to be removed because it had been contaminated. It was, however, hoped that the ice had prevented a deep penetration by styrene and toluene.

On Monday, November 19, Canadian Pacific Rail opened a claims office for out-of-pocket expenses. The claimants had to sign a document exonerating Canadian Pacific Rail from all later claims.

On November 20, the chlorine wagon, filled with water, was finally removed.

3. BALANCE SHEET

There had been an evacuation of 240,000 people. There had been no panic, no deaths, nobody seriously injured, few acts of vandalism.

The municipality of Mississauga estimated that the cost of the operation amounted to 25 million dollars per day, rescue measures not included i.e.:

    1.5 million dollars of lost sales
    6.0 million dollars loss of manufactured goods
    12.0 million dollars loss of salaries
    2.5 million dollars loss of services and various opportunities.

These figures must be regarded with caution. They represent only an estimate and give only a limited indication of the possible cost of a similar event.

4. CONCLUSION

Compared with what would have happened in case of a rupture of the chlorine wagon or of the wagon supposedly carrying PCB the problem experienced at Toronto from November 10 to 16, 1979 was only a very benign incident.

The event of Toronto was a magnificent alarm: an 'exercise' which the Canadian authorities and all the organisations involved in the affair, in the first and the second line of defence, knew very well how to deal with. One event cannot hide the multiple possible ones which exist in this field. It is important here again not to file the affair away hastily under the usual and all too well known heading;: "Nothing happened at Toronto".

REFERENCES

(1) P. TIMMERMAN, The Mississauga Train Derailment and Evacuation: November 10-17, 1979, Event reconstruction and organizational response. Institute for Environmental Studies, University of Toronto. Mississauga Report no 1, May 1980 (40 pages).

(2) A. WHYTE, D. LIVERMAN and J. WILSON, Preliminary report on survey of households evacuated during the Mississauga chlorine gaz emergency (November 10-16, 1979). Institute for Environmental Studies, University of Toronto. Mississauga Report no 2 (44 pages).

CONCLUSION: FIVE ACCIDENTS, 28 DEAD AND VERY SERIOUS QUESTIONS REMAIN

We have described five recent accidents. On the evidence each of them was very serious. Nevertheless, one might be astonished at the attention given to these events: did they not in the end cause rather a minimal total of victims? Twenty eight deaths, all of which occurred at Flixborough. If one sticks only to statistical considerations this figure appears rather small: what are twenty eight deaths by comparison to the hundreds of victims of railway and mine disasters? Or by comparison to the tens of death registered every week on the roads?

Yet, these accidents have caught the attention of areas, of whole countries, sometimes of the whole industrialised world as in the case of Seveso and Harrisburg. To understand this phenomenon one must not be satisfied with the simple explanation according to which the media were exclusively responsible for this echo: the question has aroused spirits including those of the most guarded among the experts and people in positions of responsibility.

What then underlies these events? Do they have something in common? Are they perhaps a sign of some change in our situation with regard to technological risk, a factor which, however, has never been negligible in history?

This group: Flixborough, Seveso, *Amoco-Cadiz*, TMI, Toronto: does it, in short, constitute a challenge to the industrial societies at the end of the twentieth century?

A broader examination of risk from technological and industrial origins, an analysis across history, permits the formulation of some observations in this respect. This will be the subject of Chapter 2 which will allow us to place the example presented here in a general context and in a structured perspective. They will then appear more clearly for what they really are: a series of grave warnings.

CHAPTER TWO

Radically new threats

*"I see a radical novelty, a 'major risk' without precedent
in the nature of consequences that certain technological
developments can bring about. To the mechanical accident
that can kill or maim the twentieth century has added a
further dimension, even if its proportions are yet unknown:
the disaster that overtakes the integrity of life.*

*In all these cases the risk is run not only by the victims
at the time: it affects life itself by transmission to the
descendants. The statistical scourge of monsters and those
born with an infirmity because of the hazards of nature,
has been extended by the necessities of the human genius
(...). This is where Prometheus is as successful as Nature
... or God by striking his (own) descendants in what is
considered the most precious and the most sacred: life, not
by destroying it but by working its transmission to future
generations".*

J. J. Salomon (1)

I. THE DISASTERS OF THE INDUSTRIAL ERA:
EIGHTEENTH TO TWENTIETH CENTURY

Faced with the risks presented by modern technology an observation has
recently been suggested for consideration: History is strewn with drama such
as mine or railway disasters; the demand by modern society for safety must
not veil the considerable progress that has been made in this field. The
argument deserves closer examination. What were the contemporary dramas of
the rise and development of the industrial age? What has become of them?
These are the very first questions to be asked if one wants to appreciate
subsequently the progress achieved in matters of safety, the shifts and
changes brought about in the field of risk.

We therefore propose hereafter a short account of the typical risks of that
period of industrial upsurge. The purpose of the account is to spell out the
categories and orders of size to be considered. It is not at all an attempt
at an encyclopaedic work. The basic data for this undertaking were derived
from the work of J. R. Nash (2) which constitutes, to our knowledge, the only

available reasonably complete account of the disasters.*

Let us emphasise that, given the perspective of our endeavour, we shall enlarge on some points where reflections on questions of responsibility can be made.

1. THE GENERAL CONTEXT OF THE SAFETY OF POPULATIONS IN COUNTRIES UNDERGOING THE PROCESS OF INDUSTRIALISATION IN THE EIGHTEENTH AND NINETEENTH CENTURY

Without wishing to give a complete picture, one can supply several important elements in the context of 'safety' as it appears at the beginning of the period considered here.

1st. Great scourges which still exist in Europe

Disasters of natural origin are still of great importance even if consider-able progress has been made in this field.

Famine, a constant scourge since the Middle Ages, will soon disappear. Thus, as Jean Fourastie remarks, the year 1709 marks a turning point: from that year onwards "one no longer finds towns and whole regions abandoned to the furies of hunger; one no longer finds men eating children; never again will human flesh be sold in the market at Tournus" (3, p. 77). From 1770 on the crises of agricultural production no longer brought about collective death with the exception of the case of Ireland**which on account of the poor quality of its soil and the implacable yoke imposed on it from outside experienced a hecatomb from 1846 to 1849 (more than a million and a half dead) (4, pp. 91-102; 5, pp. 9-39).

Epidemics are also on the retreat. In France, leprosy had been overcome during the fourteenth and fifteenth centuries, typhus in the eighteenth; the plague raged for the last time in 1722 in Marseille. Smallpox caused 80,000 victims in 1798. Cholera struck still, being brought to Europe in connection with British trade: Paris was struck in 1832 (18,000 dead), in 1849 (10,000 dead), in 1853-54 (11,500 dead), in 1865-66 (11,000 dead), in 1873 (854 dead); in Europe as a whole, cholera claimed millions of victims between 1826 and 1837. Tuberculosis was also rampant (250,000 dead in England between 1851 and 1855). The last great epidemic in European countries was the one of 1917-1919: 'Spanish Influenza' caused more than 20 million deaths (4, pp. 103-120; 2, pp. 732-734; 6, pp. 9-18).

---

*Certainly, it appears insufficient or inexact on certain points, which is inevitable, but the wealth of information which it supplies is such that it constitutes an obligatory stopping point for reflection. It mitigates a deficiency which all those who nowadays work on the subject of major risk deplore.

**The Irish were caught up in a dramatic situation brought about by the following factors: potato disease; a real estate law which made agriculture extremely vulnerable; a blockage of maritime development imposed by England which left them no resources other than those of the soil; the sacred doctrine of laisser-faire which for a long time forbade public assistance; martial law, the cold and typhus; the refusal by the London government to consider the situation. A revolt which only incensed the English still more: to a new demand for help in 1848 the Prime Minister retorted: "Parliament will never accord a loan to Ireland: so great is its fury against seditious, lying and ungrateful Ireland" (5, pp. 9-39).

The other great curtain raiser was war: it was waged many times before the bloodbath of 1914-1918.

Also in the chapter on great natural disasters belongs the case of the destruction of Saint-Pierre in Martinique on May 8, 1902 following the eruption of Mont Pele (30,000 dead, 2, pp. 430-437; 8, pp. 10-15) and also the eruption of Krakatoa (between Java and Sumatra) which in 1883 caused a great flood that killed 40,000 people. In Europe, Italy was struck several times (2, pp. 311-315); the biggest earthquake was the one on December 28, 1908: between Messina, Reggio Calabria and other cities 160,000 dead were counted*(2, pp. 364-370).

2nd: Safety in everyday life

Would everyday life call for more emotion about the phenomenon of accident? For a good part of the population the dramatic nature (which, however, showed improvement during the period) made insecurity a constant feature which a major accident could not accentuate much. The living conditions of the poor at whose expense industrialisation was accomplished are well known. Two illustrations of these are worth mentioning; the first one is brought up by J. Fourastie and deals with the life style of the masses:

The peasant's home, poor but in the fresh air, was exchanged for infected hovels. The working hours of the peasant, bearable in the fields, were adopted in unhealthy, evil-smelling factories. The moral and social framework of the village disappeared and gave way to the anonymous masses confined in inorganic suburbs. It was the hideous era of the proletarisation of man. (3, p. 88).

The second one, taken from a present day British paper, reminds us of the dangers of everyday work:

The machines required frequent greasing but it would cost money to stop them for this purpose. As a result, they were oiled while they were running. It was difficult for an adult to reach the greasing points. Children are smaller. Therefore children would grease the running machines. As a result children were maimed and killed.

The workers nowadays are suffocated by such practices. This is certainly difficult to understand until one realises that the fear of famine is more pressing than the fear of a violent death caused by a machine. The first was a certainty if you did not have work; the second was a wager in which you could take part.

The mouse goes into the trap for its cheese, the fish on the hook for its worm, the worker into the spinning mill for his bread (7, p. 15).

2. DISASTERS OF WHICH THE NATURE IS NOT NEW

1st: The great fires in towns

Very large-scale fires have, sometimes generalised, marked the development of towns throughout history. Thus London suffered badly on a number of

---

*Other cases of massive destruction the prime cause of which had been earthquakes but where the most serious damage was caused by fire following the earth tremor are examined hereafter.

occasions: in the years 700, 982, 1212, 1666 (13,200 houses burned down).
The same happened to Nantes in 1118, Berlin in 1405, Moscow in 1570, Oslo in
1624, Edinburgh in 1700, Lisbon in 1707, Copenhagen in 1728 etc. (2, pp. 654-
662).

This type of disaster struck again in the eighteenth and nineteenth
centuries even though the towns had organised themselved better and better
against the danger. London had another serious fire in 1748 (200 houses
destroyed). Moscow saw 1,800 of its houses on fire in 1752. London suffered
from fire again in 1834: Parliament fell prey to the flames. New York was
afflicted three times (1835, 1845, 1845). Chicago had its big fire in 1871:
18,000 houses burned down, between 250 and 300 dead. Side by side with these
examples one also remembers those of Canton (in 1822, 85 per cent of the city
destroyed) or of Cairo (1824, millions of victims) (2, pp. 654-662).

Together with these immediate fires one must also take into account the
fires following earthquakes. Three cases attract attention: Tokyo in 1857
(107,000 dead); San Francisco in 1906 (700 dead, 75 per cent of the city
devastated, much more by the fire than by the earthquake); Tokyo-Yokohama in
1923: more than 143,000 dead, 300,000 houses destroyed.

Lisbon, 1755. On November 1, 1755 Lisbon was struck by three very strong
earthquakes. After these earth tremors the city was further afflicted by a
big seismic tidal wave and a gigantic fire. The number of deaths was between
50,000 and 100,000. An earth tremor also claimed 10,000 victims in Morocco;
various European cities felt it. The seismic tidal wave was also noticed in
England, in the Antilles and in Northern Europe. The philosophical and
religious agitation was intense as Voltaire's and Rousseau's writings witness;
some went on a search for heretics to burn. The king's secretary was more
pragmatic in his reply to the head of Portugal who questioned himself about
what attitude to take: Sir, we must bury the dead and feed the living (2, pp.
336-339).

San Francisco, 1906. Built between two seismic geological faults, the city
had seen extremely fast development since 1848. Within twelve years the
population grew from 800 inhabitants to 40,000 (in 1860). Its wooden
dwellings brought about a large number of fires (1849, 1850, 1853, 1854); its
location was the reason for several earth tremors (1857, 1865, 1868, 1895).
In 1906 it had 450,000 inhabitants.

The fire chief had warned those in authority on several occasions about the
inefficiency of fire rescue equipment. Six months before the disaster the
National Board of Fire Underwriters had warned: San Francisco has defied all
traditions and all principles by not having been on fire yet. This is to the
credit of the vigilance and efficiency of the fire police but it will not be
possible to escape the inevitable indefintely. The main water supply had
been installed across one of the geological faults.

On April 18, 1906 shortly after 05.00 h a violent tremor was felt. The
earthquake, measuring 8.3 on the Richter scale, was, however, not to be the
most destructive factor. In fact, the gas pipelines were broken; so were the
water supply lines. Ten minutes later the fire raged; an hour later fifty
sites on fire were counted. The fire fighting equipment was insufficient and
a dramatic fact had to be realised: San Francisco had practically no water
left.

On the scene the General in charge of the local garrison took the situation
in hand, knowing the weaknesses of the municipality which was run by a mayor

known to be incapable and corrupt. His military regime soon took on the
shape of (quasi) martial law.

Within a few hours the fire became uncontrollable. Having no water availabl
an attempt was made to stem the spreading of the disaster with dynamite; but
lack of training again caused setbacks and victims. Nevertheless the ferry
was saved and 50,000 people evacuated.

The bill was heavy: between 400 and 1,000 dead, 10 km^2 of the city burned
out, more than 28,000 houses set on fire (2, pp. 490-507; 5, pp. 143-167; 8,
pp. 16-24).

Tokyo-Yokohama, 1923. On September 1, 1923, starting at 11.58 h, the region
of Tokyo-Yokohama was shaken by a series of very strong tremors (8.2 on the
Richter scale). To witness this intensity: the indicators of the local
seismometers just broke. Two hundred and thirty seven tremors were counted
on the first day, 92 the next day, 1,200 within a month. Added to the
tremors were seismic tidal waves that ravaged the coast. Again it was fire
that caused the most damage.

The earthquake had struck at the time lunch was being prepared; hearths
were overturned and fire broke out in some hundreds of places. In Yokohama
the big oil and petrol tanks cracked and more than 100,000 tonnes of hydrocarbon spread through the canals and the river to the port. Yokohama was
80 per cent destroyed and people who fled by boat were caught by the fire.

To make things worse the wind speed increased. Sparks fell on highly
inflammable rubble. Within three quarters of an hour the situation became
dramatic; people trying to escape were turned into torches, bridges caught
fire, the water boiled in the canals. This heat caused "fire storms";
cyclones formed and swept the city, spreading the fire and blowing the flames
in every direction. One of these tornadoes hit an army clothing warehouse:
between 40,000 and 45,000 people had sought shelter there, not many survived.

It was not the best scientific explanations that guided the spirits then.
Using the disaster to his own advantage the emperor put the blame on the
Koreans and the socialists for having displeased the spirits and for being
responsible for the lootings. The terror spread, the Koreans fled where they
could. Martial law was imposed on September 2.

The lack of water, the lack of food, communication problems (carrier
pigeons were used for messages) made the fight against the disaster and even
just survival very difficult. Diseases appeared with epidemics of dysentry
and typhoid: more than 3,700 people were killed by them.

In the end, 200,000 injured, more than 140,000 dead, 500,000 homeless were
counted in the area. More than 300,000 houses had been destroyed (2, pp. 285
288; 6, pp. 19-45; 8, pp. 41-47).

2nd: The great fires in buildings

House fires are not typical risks of the industrial era: the great historic
event is the destruction of the Alexandria library in 500 AD. These disaster
were still not sufficiently prepared for in the eighteenth, nineteenth and
twentieth centuries as the following account shows (2, pp. 654-662).

        1772: Theatre at Zaragoza (Spain) 77 dead
        1781: Palais Royal, Paris, 20 dead
        1836: Theatre at St. Petersburg (Russia) 700 dead

1845: Theatre at Canton (China) 1,670 dead
1863: Jesuit church, Santiago (Chile) 2,000 dead
1876: Brooklyn Theatre (USA) 295 dead
1881: Ring Theatre, Vienna (Austria) 850 dead
1887: Opera Comique, Paris, 115 dead
1897: Charity Bazaar, Paris, 150 dead
1899: Hotel Windsor, New York, 92 dead
1903: Iroquoi Theatre, Chicago, 602 dead
1938: Nouvelles Galeries and Hotel Norilles, Marseille, 100 dead
1940: Night Club, Natchez (Miss., USA) 198 dead
1942: Night Club, Boston (USA) 491 dead

The Ring Theatre, Vienna, 1881. Following an incident the fire caught the
stage curtains and immediately spread over the stalls which held a capacity
crowd on that evening of December 8, 1881. This caused panic, all the more
so as the lights suddenly went out and, against the regulations, there was
no emergency lighting.

In addition, the emergency exit signs were either non-existent or insufficient in the corridors of the dress circles. At the time when the flames
penetrated the stage curtain and leapt out into the auditorium the spectators
in the dress circles who tried to escape rushed off in the wrong directions
which led into dead-ends. Tens of them who had discovered a real emergency
exit found themselves in front of padlocked doors.

That was not yet all of the negligence. From the start of the fire the
fire brigade on duty and the stage hands had fled without starting the alarm
(the key to it was in the pocket of one of the firemen) and without lowering
the safety curtain.

They did not dream of activating the five fire hydrants installed on the
stage and intended for such instances (5, pp. 56-57).

Outside one police officer declared very hastily: there is nobody left in
there. The action by the fire brigade was directed towards the protection of
adjacent buildings: any chance of extinguishing the fire had been excluded
from the start. Some people nevertheless broke into the theatre to save the
disaster victims, but to do this the doors had to be smashed with axes; right
at the start of the fire a policeman had thought it wise to lock them up to
prevent the crowd from trying to rescue family or friends caught in the fire.

The Opera-Comique, Paris, 1887. This again was a rapidly spreading fire;
there was an auditorium caught up in panic on account of the fumes and gases,
the darkness (the emergency lights were insufficient to penetrate the smoke
screen), the padlocked doors, the malfunction of the fire fighting equipment
inside. But in contrast to the Vienna disaster the equipment outside was
efficient: the fire police used the 24 metre ladders for the first time (5,
pp. 71-86).

However, this drama had been foreseen. The unsafe state of the theatre or
at least of the part reserved for the staff had been recognised. A member of
parliament had even raised a question on the subject in the chamber thirteen
days previously. The minister himself had emphasised the seriousness of the
situation, showing a touching perspicacity in the use of statistics:

The Minister of Education and the Arts:

I repeat that this situation is quite dangerous, and it is certain that if

fire broke out in the Opera-Comique, and this eventuality is unfortunately
nearly certain within given time ... (Shouts of diverse opinions). With your
permission: there is no theatre that has not been on fire, even several times,
within a century. This is a statistical fact; it follows that we can consider
it probable that the Opera Comique will have a fire ... (Laughter). I hope,
however, that this will not be soon.

In the present situation if the fire occurred during a performance there
would be disaster. It is certain, as has just been mentioned, that we would
see several hundred people perish. This is a very serious responsibility, an
eventuality which deserves the highest degree of attention from government
and parliament.

Now, the question is how to provide for this, and here we are faced with
difficulty ... All I can do is to submit the question to my colleague in
charge of Finance (Laughter).

If the Minister of Finance feels he can accept these propositions we would
together draft this law and submit it to the budget commission.

Here, then, gentlemen, is the state of affairs; this is what I propose to
do (Very good! Very good!).

The Speaker: The incident is closed. (9, p. 988).

That was on May 12, 1887. On May 25, one hundred and fifteen people died
in the ruins of the Opera-Comique. Without the efficient intervention of the
fire police the toll would have been higher still. This drama did not change
attitudes: in 1923 the Opera Comique was again on fire, claiming 103 victims.

The Charity Bazaar, Paris, 1897. The charity sale organised by the principal
French and European aristocratic families for the benefit of works of welfare
took place, in that year, in a sort of hangar, hastily constructed and
decorated with highly inflammable material. The light bulb of a cinematograph
proved to be defective. An unfortunate use of ether, a match and the fire
was spreading about within seconds; the ceiling collapsed on 1,200 people
crowded together inside. Scenes of panic and terror developed in the hangar,
good manners often gave way to savage brutality. More than 120 victims were
counted (5, pp. 87-106).

Michel Winock has shown (10) the repercussions of this drama clearly. It
unleashed a number of forces that found in it a means or a pretext for
expression: the gruesome, the perverse, the war of the sexes, the class
struggle, irrationality, antisemitism, the collective fear of death. We
retain two reflections that are pertinent to our subject. The first one is
by G. Clemenceau who was astonished that there could be two standards of
emotion depending on whether it was elegant ladies of the aristocracy or
miners who became victims of a fire-damp explosion (by the way: nine years
later G. Clemenceau had to face the drama of Corrieres). The second one is
by the Dominican padre who preached at the commemorative service in Notre
Dame at which the President of the Republic and members of the government were
present. His message centred on two points: the sin of pride in this
scientific century*; the divine wrath at the departure from the straight and
narrow path**by the oldest daughter of the church:

---

*An allusion to the supposed prime cause of the disaster: the light bulb of
 the cinematograph.

**Taken to be the deeper cause of the tragedy.

- He (God) wanted to teach a terrible lesson at the pride of this century in which man incessantly talks about his triumph against God.

From the flame which it (the scientific century) claims to have snatched from Thine hands like the classical Prometheus Thou hast made the instrument of Thy revenge.

France has deserved this chastisement by a further departure from her traditions. Instead of marching at the head of Christian civilisation she has consented to follow like a servant or a slave doctrines which are as alien to her spirit as they are to her baptism.

France, having chosen the evil road of apostasy, has been visited by the "angel of destruction" (10, p. 38).

3rd: The great maritime disasters

The risks of maritime navigation have been experienced through the ages. Industrial development was to make safer ships available; but the growth in size of the vessels made the tolls heavier when the drama occurred. The major defeat in this field was the tragedy of the Titanic in 1912. Before dealing with some of its key factors we submit the following recapitulation which gives an order of magnitude (2, pp. 675-709):

			Registered peaks (fleets excluded)			
Century	No. Victims	No. events	Civilian		Military	
			Year	Deaths	Year	Deaths
16th	100	1	1586	450		
	500	2				
	1,000	1 (fleet)				
	Thousands	1 (Armada 1588)				
17th	100	15	1656	644 (collision)		
	500	3				
	1,000	2 (fleet)				
	Thousands	2 (fleet)				
18th	100	20	1770	700	1772	900
	500	4				
	1,000	2 (fleet)				
	Thousands	4 (fleet)				
19th	250	66	1866	738	1811	2,000 (coll.)
	500	10				
	1,000	1				
	Thousands	1			1865	1,517 (fire)
20th (1900-1949)	250	41	1912	1,547 (Titanic)		
	500	12				
	1,000	11	1948	2,750	1949	6,000 (evac.)
	Thousands	2	(China, mine explosion)			

Table 6: The great maritime disasters

The Titanic, 1912. The *Titanic* "the biggest vessel of all time" (53 metres high, more than 250 metres long), the most powerful (55,000 HP), the fastest (24-25 knots), the most luxurious was the pride of the British White Star Line.

On April 10, 1912 it left Southampton for its first Atlantic crossing.

Confidence in this giant was limitless. The *Titanic* had been equipped with a partitioning system with automatic shutters which divided the hull into 16 compartments. Lloyds of London had issued it with a certificate of unsinkability even though the partitions, strangely, were not high enough to shut every compartment hermetically. It was thought that in case of trouble there would always be time to intervene before the water reached the height of the partition and spilled over into the adjacent compartment. Such was the blind confidence that one of the ship's officers thought he could assure a female passenger at embarkation by declaring: Not even God himself could sink this vessel.

Aboard this liner were 2,207 people of which 1,316 passengers: the flower of the international financial aristocracy in the first class, 706 emigrants in third class (plus goods worth half a billion dollars). The first two days of this first crossing of the *Titanic* passed without problems. In the morning of the third day, April 14, the temperature was even mild for the season. However, the weather conditions were changing, and the *Titanic* was warned of this several times.

At 09.00 h a telegram warned of the presence of drifting ice in the area it was approaching. Early in the afternoon, two further telegrams confirmed this information: the *Titanic* was speeding straight into a dangerous area.

At 19.30 h while the temperature was dropping further a fourth message confirmed to the liner that it was inside the danger area.

Nobody worried. There was actually no question of rerouting or accepting delay, or of slowing down. Only the lookouts were instructed to pay special attention. The *Titanic* continued to speed on like a fireball through the night. Nevertheless, a fifth signal was sent to the liner.

In the radio room the first operator, Philips, was in touch with the station at Cape Race. Suddenly his frequency was picked up by a transmission from a nearby freighter, the Californian, which transmitted:

Listen, old chap, we are blocked in here with ice all around us ...

Shut up, Philips replied savagely. I am talking to Cape Race, and you are messing up my communication!

It was 23.40 h. At this time, the lookout, incredulous and before long paralysed, before raising the alarm saw an iceberg shaping up quite near. "An iceberg straight ahead of us! We'll hit it!" On the bridge the officer made the boat turn. There was only a small impact; few people noticed it. The ship stopped, the captain was put in the picture; he ordered an inspection.

Some passengers began to worry: There is nothing serious, set your minds at rest! Stay in your cabins. The crew is taking care of your safety.

Some of them were not duped. In the smoking room for instance the tremor had been felt and a sailor was heard shouting: We have hit an iceberg.

In fact, the iceberg had ripped the hull open over a length of 90 metres. The six forward compartments were flooded.

At 00.50 h the captain had to plan the evacuation of the ship. The distress signal was sent out but was not picked up by the only vessel in the immediate vicinity, the *Californian* which was stopped ten miles away (having stopped in view of the danger from the ice). At the time radio watch was not obligatory.

The evacuation turned into disorder and terror: the classical evacuation exercises had not been carried out, the sailors did not know their assignments. There were lifeboats for only about half of the embarked people. At least 1,000 people were condemned to drown. There were no life belts for the third class passengers. In first class there were sour faces about these life-vests being passed on and soon the life-boats were boarded to which the emigrants from third class had no access: force had to be used to safeguard the privileges of the first class passengers; a good number of them hardly condescended to embark on these frail small boats which they were offered. Why quit the unsinkable vessel? The noise of the escaping steam did not facilitate operations.

The 'water-tight' partitions showed their limits, and the vessel sank faster and faster. While the lifeboats were lowered, some of them half empty, the emigrants went into the attack. Shots were fired and stopped their attempt. Six hundred and sixty people had embarked on lifeboats. One thousand five hundred people were left on the wreck, the stern of which soon rose to a vertical position. From his lifeboat the chairman of the Star Line saw his jewel sink into the sea. The survivors were rescued: there were 705 out of the 2,207 who had embarked on the *Titanic*.

Among those who perished the largest number were emigrants, people who had been unable to afford a place in first class (11, pp. 45-46).

3. THE NEW GREAT RISKS OF THE INDUSTRIAL ERA

1st. Mining disasters

Mine disasters have struck English workers since the seventeenth century; the first half of the twentieth century was the most murderous period. In Europe the main reference point remains the disaster at Corrieres (1906, 1,099 dead).

The table on the following page gives an indication of the scale of the major mining accidents (2, pp. 710-720).

Courrieres, 1906. On the morning of March 10, 1906 about 1,780 workers had gone down into the three pits of the mine at Courrieres which was considered to be one of the safest in the basin of the Pas-de-Calais. Towards 06.30 h, however, a very large explosion occurred; gas invaded the tunnels which were transformed into furnaces. 1,099 dead were counted. It was not a fire-damp explosion but a phenomenon that was little known in France at the time, a 'dust explosion' i.e. a rapid inflammation of large quantities of dust in the air.

These were the only statements made unanimously. Emotions, some families had lost up to seven members; social antagonisms, problems between the staff of the mine and the government engineers who were legally in charge of the rescue operations and the enquiry, rivalries between young and old unionists, an electoral aspect, the sometimes unsatisfactory information given by the

Period	No. accidents	No. victims	Average per accident	Maximum re-corded dead	Distrib./ Category
1700-1749	5	198	40	69 (GB 1708) 60 (GB 1710)	50 -: 2
1750-1799	6	180	30	39 (GB 1767) 39 (GB 1799)	50 -: 0
1800-1849	32	1,513	47	102 (GB 1835	50 -:10 100 -: 1
1850-1899	194	11,614	60	550 (Upper Si-lesia 1895) 361 (GB 1866)	100 -:25 200 -: 4 300 -: 1 400 -: 0 500 -: 1
1900-1949	255	30,000-33,000	118-130	3,700 (GDR 1949*) 3,000 (China 1931) 1,549 (China 1942) 1,099 (China 1906)	50 -:59 100 -:35 200 -:14 300 -: 8 400 -: 3 1,000 -: 4

---

*Accident in a uranium mine in the GDR. According to sources there were:
2,300 dead (*Berlin Telegraph*), 1 dead (Soviet source), 3,700 dead (chief
of Leipzig fire police).

Table 7: Mining disasters

government (which had actually resigned at the time of the drama; soon afterwards reformed with G. Clemenceau at the Interior Ministry), there were so
many factors that promised confusion, controversy, violence and finally
repression.

Courrieres was more than a mining disaster. Not only did it leave behind
several hundred invalids for life, 562 widows, 1,133 orphans as well as hunger
cold, misery and bitterness but it was also going to be the cause of rage and
the response to it: martial law declared by Clemenceau who sent in troops to
patrol the miners' dwellings. There were arrests of responsible unionists, an
unfounded theory of a 'plot' before the resumption of 'normal' work and the
holding of the programmed elections.

About the antecedents of the accident and the responsibility of the operating
company

The following two series of observations can be offered. The first one
reflects the action and the feeling of the miner's delegate; the second one
gives the analysis by the General Mining Council which debated the accident
at its sessions of May 10 and 17, 1907.

a) The point of view of the miners. The pronouncements and reports by the miners' delegate Simon are at the root of the rumours about the murderous negligence of the mining management at Courrieres. Since November 28, 1905 i.e. nearly four months before the disaster, Simon had pointed out in his inspection report the lack of air in the tunnels, the large quantity of coal dust in the atmosphere and the need to moisten it. On February 16 of the following year he recommended not to let the workers go down into pit No. 3 and to supply more air. On the 17th he made the same remarks. On March 3 i.e. a week before the explosion he repeated, even more strongly, the same recommendations; as the company paid no attention to his observations, little by little very heavy noxious gases accumulated in the abandoned seams and a fire broke out which the engineers were unable to control. Trying to contain it they decided to wall it in with fire-proof cement bricks. Delegate Simon stood up against this solution which he considered extremely dangerous and advocated flooding of the seam in which the fire raged.

The workers employed on the construction of the wall also were aware of the danger: "Several of us did not go down into the mine on Saturday because they foresaw the misfortune. We noticed definite signs of unrest with the horses which scared us." (12).

b) The analysis of the General Mining Council. Considering that if, as shall be shown, it has not been possible, despite the most persevering and attentive investigations, to establish the exact cause that started the fire which resulted in the disaster of March 10, 1906 it cannot be denied that its extension appears to have been due to the spreading, subsequent to various circumstances, of the ignition of dust through the whole expanse of the working area of pits Numbers 2, 3, and 4-11 over a length of about 3 km and a width equal to 1,500 metres surface.

Considering, as concerns the start of the fire, that everything points to its starting in the Lecoeuvre tunnel, without this actually being possible to establish with absolute certainty, it remains then impossible to establish whether this ignition must be attributed to an unforeseen eruption of firedamp or to the explosion of a shot or again to that of a packet of explosives and one can in this respect only establish hypotheses.

That in these circumstances neither the use of open-flame lamps in the Lecoeuvre tunnel instead of safety lamps, the use of which was obligatory on this site in the terms of Article 74 of the regulation of February 8, 1905, nor the use of explosives of the type Favier No. 1 instead of safety explosives which, it seems, was obligatory because of the prefectorial order of March 28, 1898 could not be established as having had a definite cause and effect relationship with the accident and as being susceptible to attribute on account of this a responsibility to the operating company.

Considering that it emerges, on the other hand, from all statements on observations made, that the cause of the accident cannot be found with the fire in the Cecile seam:

That this fire, the importance of which has been considerably exaggerated and against the dangers of which all necessary measures had already been taken, did not in itself constitute a serious cause of danger of the sort that would forbid access to the mine for the workers.

That there is therefore no responsibility established on this ground;

Considering that one can nevertheless establish as having contributed to a

considerable degree to the seriousness of the disaster certain general
arrangements consisting mainly in the freely established communications
between the tunnels Numbers 2, 3, and 4-11 and in the imperfection of the
ventilation which resulted at the same time from a rather irregular mode of
distribution and the absence of embanking in the large strata;

That the spreading of the explosion over such a vast expanse was in effect
the consequence of the fact that the working areas of the three tunnels in
question intercommunicated to a large extent;

But that these arrangements really were proved to be so defective only by
the fact of the disaster itself;

That as the mine at Courrieres was not plagued by fire-damp its division
into independent areas of limited size did not seem called for, no more than
the ventilation, the communication between the tunnels appeared on the
contrary justified by safety considerations, particularly in order to ensure
escape for the staff in case of an accident in one of them, especially in
case of an incursion of water;

That as regards danger from dust neither experience nor lessons derived
from practice permitted the suspicion, in a mine with no fire-damp, of the
possibility of a fire of such magnitude, explosions of dust alone, in the
absence of fire-damp, previously recorded in France having never spread over
more than 50-80 metres from their point of origin, exceptionally to 180
metres at the mine of Decize (accident on February 18, 1890).

That therefore these arrangements, no matter how open to criticism they may
appear today on account of the consequences they had, could not before the
accident be criticised;

It is (our) opinion that the local service engineers were correct in concluding that the incident cannot have legal consequences (13, pp. 484-486).

About the rescue operation

The rescue operations were conducted in 3 phases:

March 10-11:

Attempts were made by all possible means to go to the rescue of the victims
and to assess the size of the disaster.

March 11-30:

The government engineers who took charge of organising the rescue and the
enquiry held the conviction (supported by various witnesses from among the
miners and from their own inspections in the mine) that there were no
survivors in the mine and that there was too much danger to the rescuers.*

---

*In its support for this evaluation which it considers well founded the
  General Mining Council in its Notice of 1907 brings up the following two facts:

On December 12, 1866 an outbreak of fire claimed 334 victims at Oaks
Colliery in Yorkshire, England. The following day, December 13, the rescuers
were taken by surprise by a further explosion. Twenty eight of them perished
and the mine had to be closed without recovering the bodies or any survivors
who might have been left.

On June 14, 1894 at Karwin (Austria) a first explosion had claimed one
hundred and sixty five victims. The following day, June 15, a second explosion
killed seventy rescuers and again the mine had to be closed.

The reversal of the ventilation was ordered, the mouth of pit No. 3 closed and the battle against the fire launched.

After March 30:

   Things took a dramatic turn on March 30: thirteen workers emerged from the mine alive. From the time when, contrary to expectations, there appeared some chance of rescuing survivors one could for a moment forget the caution that the fire had demanded in the second phase of the rescue. Without worrying about the danger to which one might be exposed explorations in all parts of the mine, to which one could at all penetrate, were quickly organised.

   On the rescue opinions differed even more strongly than on the previous point.

a) The viewpoint of the delegates for the safety of the mine workers. We take up the observations of the two delegates, members of the commission set up by the Ministry of Public Works, who submitted a minority report:

First Mistake: The refusal to free the access to pit No. 3.

   We have noted with regret that Mr Bar, the Chief Engineer of the Courrieres company has rejected the suggestion by Mr Reumaux, Engineer and General Agent of the mines at Lens who soon after the disaster and as soon as he arrived on the accident site, demanded that the tangle of beams and planks be broken up, which had formed a blockage at a depth of 170 metres after the explosion and completely obstructed pit No. 3. The mistake is in our opinion even more serious since Mr Bar only objected that he feared that the lining of the mineshaft would be damaged which lining at that depth no longer exists. The chief mining engineer Mr Petitjean, the miners' delegate Simon, the mine worker Vincent, Mr Thiery, the Director of the mining company at Douchy, have for their part deplored that the opening of pit No. 3 had to be proceeded with, be it with dynamite, or by employing some heavy weight; this approach had to be the most favourable for the rescue explorations undertaken in search of survivors. The events that followed actually confirmed these opinions.

   ... We are led to conclude that on account of the refusal by Mr Bar to permit the crushing of the obstacle floor which obstructed pit No. 3 the Courrieres company has incurred the most serious responsibility and committed inexcusable mistake (14, pp p. 164).

Second Mistake: The reversal of the ventilation on March 12.

   In these circumstances departing from the mistaken idea that there could be no more survivors, and while memorable precedents made it obligatory to keep up hopes regardless, it was decided to reverse the ventilation; by definitely condemning pit No. 3 as an approach for penetration and rescue, it was changed from air supply pit to air exit pit and abandoned pits No. 3 and 4-11 in the same act, even though on the day of the disaster at 6 o'clock in the evening workers had come out alive through these tunnels (15, pp. 465-466).

Third Mistake: No consultation with the safety delegates.

   We must put on record above all that at the council of engineers held at Courrieres-Operations on March 11, the day after the disaster, a council in which twenty five engineers took part, no reference was made to the lamps, the experience and practical knowledge of any of the miners' delegates.

... On March 16 a further council of engineers from the mining corps and engineers from companies in the vicinity met. Despite orders from the minister the miners were not consulted etc.* (14, p. 469).

On March 15 a telegram from the Minister of Public Works asked the Inspector General, Mr Delafond, to "pursue the enquiry, associating to this enquiry the direct cooperation of the miners' delegates from the districts concerned". At no time did the minister give Mr Delafond the order to consult the miners' delegates on the organisation of the rescue operations (14, p. 469).

Fourth Mistake: The setting up of an isolation barrier which walled in for good all those who might have survived after the explosion.

Fifth Accusation: The most serious and the one that "the majority of the commission thought almost useless to record" (14, p. 472), the lack of consultation as expressing a rather more commercial than humanitarian preoccupation:

This fact (the lack of communication) gives us all the impression that the issue in this consultation was only to save the mine and that there was no longer any preoccupation among the engineers with the rescue of the surviving miners (14, p. 472).

b) *The viewpoint of the majority of the commission.* Having presented and rejected the remarks by the minority the other members of the commission concluded:

1. The operations have been performed, from the start, by the government engineers in accordance with the legal arrangements governing the mines in such cases. The responsibility of representatives of the company in this respect cannot be called into question.

    The safety delegates of the mine workers had not been legally heard; they could present all their observations by entering them in their registers; they have made no use of this facility.

2. There is no indication that would permit the assumption that miners who had survived the asphyxiation of the first few days perished later on in the mine for lack of effort which could have been made. The autopsy has in fact shown that the miners which have been claimed to have died long after the disaster were burned and asphyxiated at the beginning.

    Attempts at self-preservation, of which traces have been found in the course of the excavation work, were made by surviving or dead miners as early as the first day.

    The eight miners who originally were together with the thirteen who escaped on March 30 died from asphyxiation, five of them on the first day three on the second or third day while trying to get to the mine shaft.

    The survivors did not find anyone alive in the mine.

_____

*There is a certain inexactitude which the majority challenges (but the correction perhaps further strengthens the interest in the question raised).

3. The rescue work was particularly difficult on account of the exceptional size of the workings and the tangle of the tunnels struck by the accident.

The programme and the means adopted for the execution of this work conformed to the standards of the profession and were dictated by the very circumstances of the accident.

The removal of obstacles from pit No. 3 by violent means could not be undertaken because of the dangers involved and because of the particularly serious consequences which could result for subsequent rescue work.

Pit No. 3 being inaccessible to traffic the reversal of the air stream was justified by the apprehension that one had reason to have, especially after what the delegates had said, concerning the intensity and the dangers of a spread of the fire in Cecile; it permitted to have, in case this happened, the most practical ventilation; it contributed to the removal of the noxious gases which during the first few days had prevented a breakthrough to the thirteen survivors.

The setting up of barriers in the access ways at Josephine and Julie had been made necessary by the new fire that broke out in Josephine after the accident; this fire created a particularly dangerous situation for the workers employed in the rescue operation and demanded an especially cautious approach in order to avoid a further disaster which one could not sufficiently control.

These barriers, while their presence was considered necessary, did not actually harm any of the survivors.

4. Summing up; the commission feels that no reproach can be made to anybody for the organisation and implementation of the rescue operation after the disaster.

The above report having been read to the fully assembled commission on Tuesday, May 8, at Douai, Messrs Cordier and Evrard*confirm that their personal conclusions have been faithfully represented; they declare to be unable to change them.

The President of the Commission

signed: Carnot

(14, pp. 481-483)

As concerns the most serious accusation raised by the minority against the company and those in charge the majority without considering such accusations worth mentioning ... uses, nevertheless, the opportunity to confirm again, as the Minister of Public Works has already done in Parliament on April 3, that Mr Delafond has always concerned himself exclusively with the mission incumbent upon him concerning the victims without thinking even for a moment of saving a few tonnes or a few thousand tonnes of coal.

c) *The opinion of the General Mining Council.* Concerning the rescue operations and the respective statements after the completion of the enquiry by the special commission under the presidency of the Inspector General, Mr Carnot:

---

*The two minority members.

Considering that the final statements with regard to the fire that occurred
in the Cecile seam have established that this fire consisted only in the
burning of limited timberwork (some 2,000 kg of wood), infinitely less
important and less worrying subsequently than the witnesses given at the start
of the rescue operations led one to believe, particularly those of the miners'
delegate from pit No. 3;

That these witnesses have weighed heavily on all decisions taken, making
one believe that there was an imminent danger which was far from being as
serious as it was said to be;

Considering the excavation of pit No. 3, for which the engineers in charge
of the rescue have been reproached for not having used violent means as
required, has in the end taken less than thirty seven working days, and this
by using stronger means than were originally available and in circumstances
which were much more favourable.

That it follows from this that the continuation of the work in conditions
which were dangerous for those who had to carry it out could not within
reasonable time have brought any useful result;

Considering, on the other hand, that the reversal of the air stream which
at a time was so strongly criticised and the start of the ventilator of pit
No. 4 brought about, according to the statements made later, the gradual
improvement of the air in the tunnels south of pit No. 3 thanks to which the
four "survivors" were able to leave their point of refuge and proceed without
being asphyxiated to the exit of pit No. 2;

The opinion is that it emerges from the statements made after the completion
of the commission's work quoted above that abandoning the excavation work on
pit No. 3 and reversing the air stream were justified by the event and that
it is specifically this reversal of the air stream which made it possible for
the survivors to escape death (13, pp. 490-491).

2nd. Railway disasters

Railways were developed since 1830. The first big disaster occurred between
Versailles and Paris on May 8, 1842; more than 60 dead were counted. This
means of transport was responsible for an increasing number of victims up to
the middle of the twentieth century. The table on the following page gives
a brief summary of all the events that occurred and their seriousness (2,
pp. 736-743).

These facts expressed in figures, established on the basis of a series of
disasters, have to be adjusted upwards if one takes into account all accidents
that occurred. The *Revue Scientifique* published more complete statistics in
1882 as P. Legrand reports (4, pp. 165-166). We have selected some of these
figures:

Germany 1879   541 derailments and collisions in motion

                      2,727 accidents altogether

                        411 killed, 1,322 injured

England 1881   42 killed, 1,161 injured

France 1866-1877   773 accidents

                        218 killed, 2,158 injured.

Period	No. accidents	No. Victims	Average per accident	Maximum recorded	Distribution/ Category	
1833-1849	11	90-136	8-12	54-100 (Versailles 1842)	10 -:	1
1850-1900	76	3,500	46	216 (Mexico 1881)	50 -:	14
				200 (Turkey 1882)	100 -:	5
				178 (Russia 1882)	200 -:	2
1900-1949	191	12,000	60	600 (Mexico 1915)	50 -:	28
				543 (France 1917)*	100 -:	11
				500 (Rumania 1927)	200 -:	7
				500 (Spain 1944)	300 -:	0
				426 (Italy 1944)	400 -:	1
					500 -:	3
					600 -:	1

*Troop transport.

Table 8: Railway Disasters

Breaking of axles (as on May 8, 1842) on the line Versailles-Paris, loaded with passengers (4, pp. 152-154), switching point errors as at Quintinshill (GB) in 1922 (chain collisions, 227 dead, 246 injured, mainly military personnel, 8, pp. 34-37), highly inflammable materials, inadequate arrangements (such as the shutting of doors in a way that they could not be opened as on the Versailles-Paris train in 1842) were among the factors that explain the size of the tribute paid by the railways. The *Revue Scientifique* offered the following analysis in 1882:

To get an exact picture of how well or how badly a network is run as regards accidents one would have to know all the conditions under which it works and particularly all the problems that occur. It is for instance certain that a track that has only minor gradients and bends with a large radius presents less dangers to the passengers and to the operators than one that has not been constructed with the same advantages ... There will always remain, despite the greatest precautions, the cases of force majeure such as the breaking of axles, the cracking of rails or their displacement by evildoers, the fortuitous obstacles in the way of the trains caused by flooding, by accumulations of snow, level crossings for people and animals, derailments caused by obstacles put on the track deliberately etc. (5, pp. 165-166).

Lagny-Pomponne, 1933. On December 23, 1933, at about 20.05 h, 25 km from Paris, the express train from Paris to Strassburg ran into the Paris to Nancy train which had stopped and was just about to start moving again. Going at 105 km/h the express train pulverised four coaches of the other train which was packed like all other trains just before Christmas Eve. Excluding the

disaster of Saint-Jean-de-Maurienne (a train packed with military personnel
which derailed in 1971), this was the worst disaster in French railway history.
There were two hundred and thirty dead, one hundred and forty injured, and a
further collision with the train coming from Reims was avoided only thanks to
the presence of mind of the conductor of the Nancy train who had run along
the track and set up lamps and signals. The night, the fog, the distance
from Paris made the rescue difficult.

At first, the mechanic and the driver of the express train were arrested
but this double arrest did not have the expected effect: it caused intense
emotions; the two railwaymen were set free two days later. The legal enquiry
concluded that the "mistake" was due to the equipment and the weather
conditions.

The railwaymen's federation proceeded with its own enquiry from which the
following observations are taken:

The rolling stock was in such a defective state that it could only be called
"an apalling mess" as attempts were made to arrange the large number of
additional trains that were wanted for the Christmas season. In addition,
shortage of staff caused considerable delays.

Trains departed at too short intervals and "without instructions for
reduced speed having been issued to the mechanics".

The signal system had been known to be unsatisfactory since 1926/27 as can
be seen from the studies published in the *Revue Generale des Chemins de Fer*.
The visibility of the signals was insufficient and the fog on December 23
constituted an aggravating factor; even the functioning of the signals was
faulty.

Following the disaster a new regulation was worked out which specifically
excluded the use of coaches not made of metal for the transport of passengers
(6, pp. 85-104).

Couronnes, 1903. Even below the earth's surface the railway can cause
disaster, as witness the event on August 10, 1903 on the Paris Metro, Porte
Dauphin — Nation line.

18.53 h

The train No. 43 runs along the platform of Barbes station towards Nation.
Thick smoke emerges from it; a short circuit has caused a fire. The
passengers are evacuated, and the back engine pushes the train out of the
station. Passing Jaures, the conductor requests an extinguisher. At the
next station (Combat, later Colonel Fabien) the fire increases in intensity.
The train cannot continue.

19.23 h

The following train has to leave its passengers on the platform of Jaures
station; it (then) catches up with the accident train. Flames lash out again;
the convoy stops at 25 metres from Menilmontant station which has been
evacuated.

Thick smoke spreads toward the station which the train has just passed.
In there is a stationary train, overloaded because it has taken on the
passengers evacuated from train No. 43 at Barbes station and those of the
next train who had been evacuated at Jaures station. Many among the 250
passengers protest: they want to be reimbursed.

The air became too thick to breathe. People wanted to get out but this was no longer easy, given the darkness and the smoke. There was trampling, asphyxiation, error of direction; the temperature rose to 80°C at the top of the stairs. It was the next morning before the fire police managed to get down to the Couronnes and Menilmontant stations. Eighty four dead were counted. Some observations must be put on record:

Apart from the recriminations from passengers who were determined to claim their 'threepence' before making sure of their escape, the following must be considered: on the site of the accident there were in the end three trainloads of passengers; electrical installations were rudimentary; the coaches were made of wood; the Metro on this line was only three years old; finally: the power supply circuit for the trains was the same as the one for the stations; an accident on a train caused a power failure, the sequence of events from the accident to disaster was therefore quasi-automatic (15).

3rd. Explosion of gunpowder and ammunitions

The risk presented by gunpowder was felt in a precocious fashion: in 1645 a third of the city of Boston was destroyed by an explosion of this origin (2, p. 654). But the danger became much more acute from the eighteenth century onwards (2, pp. 654-662):

1769:	A quarter of the town San Nazzarro (Lombardy, Italy) destroyed by a gunpowder explosion: 3,000 dead.
1856:	Lightning strikes a warehouse on the island of Rhodes: 4,000 dead.
1905-1914:	3 big accidents in the USA, CHINA, USA — 19, 20, 30 dead.
1911:	2 accidents (USA, Belgium 31,110 dead)
1916:	7 accidents (of which one in Russia — 1,000 dead; one in Austria — 195 dead; one in France — at Double Coronne — 30 dead).
1917:	5 accidents (of which one in Archangel, Russia — 1,500 dead; one in Bohemia — 1,000 dead; one at Halifax, Nova Scotia, Canada on December 6 — 1,600 dead).
1918:	4 accidents (of which one at Hamont, Belgium — 1,750 dead (train carrying explosives) one in Austria — 382 dead; one in USA — 210 dead).
1919-1929:	10 accidents (of which in 1919 at Longwy, France — 64 dead (train); the Peking arsenal in 1925 — 300 dead).
1930-1940:	3 accidents (of which one at Lanchow, China — 2,000 dead in 1935; one in Madrid in 1938 — hundreds dead).
1940-1944:	8 accidents (of which one in Yugoslavia — 1,500 dead in 1941; one in Port Chicago in California — 321 dead in 1944; one in Bombay — 1,376 dead in 1944).

4th. Explosions of factories and installations

Industrial development led to the use of explosive products in connection with pressure apparatus. New types of accidents appeared.

1858:	Explosion in the London docks — 2,000 (?) dead.
1869:	Boiler explosion at Indianapolis — 27 dead amongst a crowd of 15,000 people gathered for a state fair.

1901:   Explosion in a clothing factory at Manchester — 14 dead.

1907:   Explosion in a steelworks at Pittsburg, USA — 59 dead, many
             vanished.

1912:   Explosion of a locomotive's boiler at San Antonio, Texas —
             800 kg fragment projected over a distance of 400 metres,
             another of 450 kg over 700 metres; 26 dead, 32 injured.

1915:   Explosion of a car petrol storage tank in Oklahoma — 44 dead,
             2 blocks of flats destroyed.

1915:   Dust explosion in a Swiss factory — 30 dead.

1917:   Explosion in a factory at Montreal — 25 dead.

1917:   Explosion of 3 factories at St Petersburg, Russia — 100 dead.

1921:   Explosion at Oppau in a factory of Badische Anilin (BASF),
             Germany — 565 dead, 4,000 injured, town destroyed.

1926:   Explosion in an electrometallurgical factory at St Auban,
             France — 19 dead.

1927:   Explosion of a cistern of hydrocarbon at Pittsburg, USA —
             28 dead.

1928:   Explosion in a factory in Massachusetts, USA — 23 dead.

1933:   Explosion of a hydrocarbon storage at Neuenkirchen, Germany
             — 100 dead.

1933:   Explosion in a rubber factory in Shanghai — 8 dead.

1939:   Explosion of a cellulose factory (with release of chlorine)
             at Brachto in Transylvania — 62 dead.

1942:   Explosion of a chemical factory in the province of Limbourg,
             Belgium — 200 dead, 1,000 injured.

Halifax, Nova Scotia, Canada, 1917. The French freighter *Mont Blanc* came
from New York where it had loaded 5,000 tonnes of explosives and inflammable
goods; at Halifax it was to join a British cruiser which should escort it on
its way to Europe. At the time it arrived in Halifax on December 7, 1917 the
Mont Blanc was struck by another ship. Scuttling was not possible: the fire
had already started and soon the explosion was felt up to 100 km away. Half
the town was in ruins (3,000 houses, 6 km² destroyed. Out of 550 children in
the Halifax area there were seven survivors; at least 1,200 (4,000) victims
were counted, more than 8,000 injured. The rescue equipment was destroyed.
Snow fell soon and helped fighting the fires which had broken out but made
rescue operations difficult and hit the 25,000 homeless hard. Martial law
was declared (2, pp. 227-228; 8, pp. 38-40).

Bombay, India, 1944. This is a replica of the explosion at Halifax. On
April 12, 1944 a freighter of 7,200 tonnes carrying nearly 1,400 tonnes of
highly explosive equipment (torpedoes, mines, incendiary bombs) exploded in
the port of Bombay. Nearly 1,400 dead and 3,000 injured were counted. The
port was wiped out (2).

5th. Great dam breaks

Dams have existed since the times of antiquity. However, the industrial
era has given them a new function in addition to that of irrigation: the
generation of electricity. Technology was also further developed. There was
a move from earth-filled to masonry-type dams (out of 500 dams on record in
1830 only sixteen were of this type). As prototypes they were in the

beginning disappointing; in the USA where people worried less about failures because of the very distant locations of many of the dams, in that country where one had to reckon with insufficient knowledge of raw materials and with the audacity of the pioneer spirit, people went ahead without employing any engineers (out of fifty five great American dams before 1900 there were 19 dam failures).

At the opposite end of the scale, the Japanese dams, built by a society with a very ancient, rich civilisation and thousand-year-old practices, have defied the centuries.

Let us add here that the dams built by the English in India have encountered very many failures (the rate is even higher than in the United States). Actually, seven out of thirty eight dam failures which occurred before 1900 happened on constructions made by the French in Algeria, a country the hydrology of which was very little known at the time of construction (16, p. 10-11).

Since the nineteenth century the viability of constructions was substantially improved. The number of dam failures recorded during construction and for the first twenty years after being put into service has developed as follows: (17)

Dams built between: 1850 and 1899: 4.0% (out of 600 constructions)
1900 and 1909: 3.5% (out of 400 constructions)
1910 and 1919: 2.6% (out of 600 constructions)
1920 and 1929: 1.9% (out of 1,000 constructions)
1930 and 1949: 0.7% (out of 1,900 constructions)

P. Goublet has prepared the following list of disastrous accidents (we restrict ourselves here to the prewar period):

1802: Puentes, Spain — 608 dead
1864: Dale Dyke, Great Britain — 250 dead
1968: Iruka, Japan — 1,200 dead
1889: South Fork River, USA — 2,000 — 4,000 dead
1895: Bougey, France — between 86 and more than 100 dead
1911: Austin, USA — between 80 and more than 700 dead
1923: Gleno, Italy — 100 — 600 dead
1928: San Francisco, USA — 400 — 2,000 dead.

6th Air Disasters

Ikaros' dream came true in the eighteenth century with the launching of balloons carrying men in their gondolas. The Montgolfier brothers designed the vehicle; some heroes had to be found but the king refused permission and allowed the first test only with prisoners under death sentence. However, Pilatre de Rozier succeeded in getting permission and went up into the air on November 21, 1783. From that day on the balloon starts multiplied. Accidents occurred. Pilatre was the first victim: pressed for time and money and the challenge he crashed with his collaborator on June 13, 1785 while trying to cross the Channel. Between 1802 and 1885 at least twenty five fatal accidents were recorded (disappearance, fire, asphyxiation, crash) (4, pp. 193-228).

The early twentieth century saw the launching of airships, vessels of quite a different dimension, designed for the transport of a large number of passengers over long distances. These machines were built by the German count Zeppelin who founded the Zeppelin company with Dr Eckener. Between 1910 and 1944 it arranged more than 2,000 flights and carried more than 10,000

passengers in its first four vessels. The British launched themselves on the
same course with the idea of building up a transcontinental fleet for use in
its relations with the Commonwealth. A series of grave accidents crushed the
hopes placed in this technical venture (2, p. 661).

1913:   Destruction by fire of the German airship LZ-18 — 28 dead.
1919:   Explosion of Zeppelin L-59 — 23 dead.
1921:   Break-up of the British airship R-38 — 44 dead.
1922:   Crash of the Italian airship Roma — 34 dead.
1923:   Disappearance of the airship Dixmude, operated by France — 52 dead
1929:   Break-up of the American airship Shenandoah — 14 dead.
1930:   Crash of the English airship R-101 — 48 dead.
1931:   Crash of the American airship Aken (?Akron) — 73 dead.
1937:   Explosion of the German Zeppelin Hindenburg — 36 dead.

The accident of the Dixmude marked the demise of the airship in France.
The R-101 accident put an end to the English attempt. The fire of the
Hindenburg under the eyes of the American press and at a time when the Reich
rather needed combat aircraft dealt a fatal blow to the German programme.
The disappearance of the largest, most elegant of those 'lighter than air'
ships signalled the death of the airship (at least for a long time) as a
means of transport.

As for the Hindenburg, the question may be raised whether recent
changes and a maintenance fault, which had not escaped Dr Eckener, but were
brushed aside by the military chiefs for propaganda reasons, did not play a
part in its tragic end. With the R-101 there is no doubt: the causes of the
accident were not to be found just on the technical side. The same applies
to the case of the Dixmude.

The disappearance of the Dixmude, 1923. The most beautiful flower of French
aeronautics came out of German workshops and was delivered under the terms of
the Versailles treaty. After various disappointments (due to sabotage at the
point of origin) the Dixmude could take to the air.

The use that was made of it was much too strenuous: some of the operating
parts of the airship were modified; test flights were made without maintenance and without sufficient examination. The (intended) record run overruled
these safety considerations. Even incidents that occurred during a flight in
1923 did not induce more caution. Political demands led to the Dixmude
taking to the air again; its commander had argued in vain: as it is, the
Dixmude is certainly fragile and incapable of intensive service. It was
designed for war raids and not for cruising.

However, on December 18, 1923, the airship had to go on a flight over the
Sahara. There were only thirty nine life-vests and forty parachutes on board
for fifty one passengers. There was no landing base other than the one from
which it took off. Never had an airship been so intensely employed before a
cruise of such importance. No checks had been made. On December 21 it
disappeared in the sea during its return journey.

The Commission of Enquiry had some difficulty in establishing its credibility when it concluded: "No responsibility can be established". Its
explanation, that the accident was caused by lightning, was disputed by most
of the experts.

Fifty one people, among them the elite of French aeronautics, disappeared
in this accident which established a new record in the accounts of those in

charge of the programme: the number of people killed in one and the same air accident (6, pp. 47-60).

The disaster of the R-101, 1930

To set up its intercontinental fleet the British built two airships, the R-100 and the R-101. The two competed with each other the second one being (financially) backed by the government; the air ministry was squeezed, the more so as the trials of the R-100 had yielded good results: the competitor had made a return trip to Canada.

After a trial on June 28, 1930 during which the R-101 dropped dangerously it was decided to extend its length in order to make it less 'heavy'; one was in a hurry; the secretary of state wanted a voyage to India within three months. The departure actually took place on October 4. To the haste were added certain arrangements intended to have a publicity effect: one wanted to make a luxury hotel of it, and in order to compensate for the surplus weight of the silver-plate the crew had to do without parachutes. On October 5 the R-101 crashed near Beauvais in France.

The Commission of Enquiry was explicit:

The R-101 left for India while one could see that it had not (yet) completed the trial periods of the experimental stage (...). The conclusion cannot be avoided that the R-101 would not have left for India in the evening of October 4 had there not been reasons of a public nature according to which it was highly desirable that this was done if at all possible.

This can be added to the comment made by the only surviving officer: the R-101 has proved one thing, namely that politics and experimental work do not mix.

The shock caused by the disaster was so great that all further development of airships in the UK was stopped. The R-100 was grounded for a year before being scrapped and sold by weight for £400. Worse still: so many men of talent perished in the R-101, the elite of English aviation engineers took part in the historic voyage, that the development of civil aviation in England was set back by several years. (8, pp. 48-54).

7th. Collapse of large structures

a) Bridges. In the lists of accidents drawn up by R. Nash (2, pp. 721-723) one finds twelve cases of bridge collapses of which four in the United States cost many lives. In Europe, the major disaster occurred at Yarmouth in England: a bridge crumbled under the weight of spectators of an aquatic event; there were 250 dead. In France the latest accident of the period occurred at Libourne: fifteen victims were mourned. There was also the collapse of a bridge in Scotland in 1879 as a train passed over it which gave additional force to the wind which blew at gale force; there were no survivors, one hundred people were cast into the water.

b) Collapse of buildings. There are fifteen accident cases recorded in Nash's list (2, pp. 712-723). In France there were: the collapse of the Palais de Justice at Thiers in 1885 (30 dead), of a building in Vincennes in 1929 (nineteen dead) and of the Palais de Justice in Bastia in 1932 (fifteen dead).

8th. Intoxication and Poisoning

Accidents of this kind were reported (2, pp. 721-723):

1923: Poisoned rice in China - 22 dead.

1930: Poisoned soup in Bombay - 30 dead.

1936: Chemical poisoning of rice in Japan, 15 dead.

1938: Poisoned rice in Japan - 15 dead.

REFERENCES

(1) J. J. SALOMON, De Lisbonne (1755) à Harrisburg (1979). Le risque technologique majeur: un formidable défi. *Futuribles 2000*, no 28, novembre
1979, pp. 5-10.

(2) J. R. NASH, *Darkest hours — A narrative encyclopedia of worldwide
disasters from ancient times to the present*. Nelson Hall, Chicago 1976
(812 pages).

(3) J. FOURASTIE, *Machinisme et bien-être. Niveau de vie et genre de vie en
France de 1700 à nos jours*. Les Editions de Minuit, Paris, 1962 (251
pages).

(4) P. LEGRAND, Fléaux et catastrophes jusqu'au XIXe siècle. *Les grandes
catastrophes*. Tome 1, Editions Famot, Genève, 1977 (246 pages).

(5) Ch. A. TIXIE, Grandes catastrophes de la fin du XIXe siècle et au début
du XXe. *Les grandes catastrophes*. Tome 3, Editions Famot, Genève, 1977
(247 pages).

(6) Ch. A. TIXIE, Grandes catastrophes de 1781 à nos jours, *Les grandes
catastrophes*. Tome 4, Editions Famot, Genève, 1977 (253 pages).

(7) P. KINNERSLY, *The hazards of work: how to fight them*. Pluto Press,
London, 1973.

(8) F. KENNETH, *Les grandes catastrophes du XXe siècle*. Editions Princesse,
Paris, 1976 (152 pages).

(9) Chambre des députés, quatrième législature, Session ordinaire de 1887,
comptes rendus in extenso, 52 séances, *Séance du jeudi* 12 mai 1887,
pp. 987-988).

(10) M. WINNOCK, L'incendie du Bazar de la Charité. *L'Histoire*, no 2, mai-
juin 1978, pp. 32-41.

(11) A. AMZIEV, La nuit infernale du Titanic. *Les grandes catastrophes
maritimes*. Les dossiers "histoire de la mer". Ed. Arnaud de Vesgre,
Neuilly, 1979.

(12) D. COOPER-RICHET, Drame à la mine. *Le Monde*, dimanche 25 novembre 1979.

(13) Avis du Conseil Général des Mines sur l'accident des mines de Courrières
du 10 mars 1906. *Annales des Mines*, Tome XII, 1907, pp. 484-492, En
annexe au texte de Ch. E. Heurteau: La catastrophe de Courrières.

(14) Rapport de la commission chargée par M. le Ministre des Travaux Publics,
des Postes et des Télégraphes, de procéder à une enquête sur les
conditions dans lesquelles ont été effectués par les ingénieurs de l'Etat

les travaux de sauvetage à la suite de la catastrophe survenue aux mines de Courrières le 10 mars 1906. *Annales de Mines*, Tome XII, 1907, pp. 445-483. En annexe au texte de Ch. E. Heurteau: La catastrophe de Courrières.

(15) J. PERRIN, La R.A.T.P. multiple les précautions pour prévenir les risques de courts circuits et d'incendies dans le métro. La nuit, si tragique de Couronnes. *Le Monde*, 8 août 1978.

(16) A. GOUBET, Risques associés aux barrages. Note du Comité technique permanent des barrages, juin 1979, (26 pages).

(17) Note interne. Service Technique de l'Energie Electrique et des Grands Barrages Direction du Gaz, de l'Electricité et du Charbon. Ministère de l'Industrie, du Commerce et de l'Artisanat (11 pages).

II. DISASTERS IN LARGE-SCALE INDUSTRY.
THE POST-WAR PERIOD

Still concerned with making available a better appreciation of actual risks we shall now examine how dangers linked with the general development of technology and industry in the post-war era were capable of causing disasters. These have been numerous and grave. New risks have appeared, increasing a feeling of insecurity while everyday life appeared to be safer.

Here again synthesizing studies fail. However, we shall find good reference material amongst them, mainly that supplied by J. R. Nash, already quoted (1), and by R. Audurand (2). The list of events established by R. Audurand will be put forward in an annex. Within our text we shall abide, as before, by categorisation, illustration and order of magnitude without missing out, where appropriate, on questions of responsibility.

1. THE SAFETY CONTEXT IN INDUSTRIALISED COUNTRIES SINCE THE WAR.

1st. Great risks of natural origin

Famine has disappeared in industrialised countries; the same is true of epidemics which puts the developed regions of the world in quite a different situation from that of the remaining three quarters of the planet (for instance: at least 800,000 died in Nigeria in 1968 from famine; 200,000 in all of Africa and India in 1973/74*.

Great dangers such as earthquakes, typhoons and seismic tidal waves can no longer cause a feeling of insecurity to the people in industrialised countries, apart from Japan and, to a certain extent, the United States on account of tornadoes (for instance cyclone Camille in 1969: two hundred and fifty eight dead). One figure clearly shows the disparity between the rich and the poor of this earth: about 95 per cent of the deaths caused by such disasters occur in the Third World**(3, p. 3). The progress made in the field of communication and information sometimes enhances the awareness of these disasters of the poor.

Some disasters that occurred from time to time in Europe must, nevertheless, be mentioned. Heavy storms have been experienced in northern Europe, for instance in 1949 (thirty nine dead), 1951 (twenty two dead), 1951 (sixty three dead), 1954 (fifty eight dead) etc. Recently, on August 14, 1979, a storm caused fifteen deaths in the Irish Sea among the contestants in the Fastnet race. Hurricane Capella, in 1976, caused damage estimated at more than three billion Deutschmark (4, p. 38).

These storms can cause floods. Those in Holland in 1953 (1,853 dead), the worst floods since 1521, and in North Germany in 1962 (three hundred and forty three dead) were the most serious ones. (1, pp. 670-672). Sometimes one must also face cold waves (1954: one hundred dead; 1956: nine hundred and seven dead) or heat waves (1957: three hundred and forty dead).

---

*These figures actually give a wrong picture of the general state of malnutrition in poor countries.

**It must, however, be mentioned that three quarters of the losses of wealth caused by these disasters have occurred in the rich countries which reflect the economic disparity between the 'Ones' and the 'Others'.

The earthquake risk also remains with us: the one in Friaul in Italy (May 6, 1976, more than a thousand deaths) is one example.*

2nd. Risks connected with the occupancy of land

Inasmuch as space is invested, built up, occupied — or, sometimes, by contrast deserted — new problems are met with that can give rise to disasters. Some events have heightened attention in this field even if it is not proven that there has really been, in these cases, risk creation or thoughtless exposure to dangers from nature. The outstanding events in France were mainly: the avalanche at Val d'Isere (February 10, 1970: thirty nine dead), the landslide on the Plateau d'Assy (Haute Savoie, April 16, 1970: seventy one dead of which fifty six were children), the floods in south-western France (July 1977, subsequent to the fall of 154 mm of rain in the night from July 7 to 8) or the one of Morlaix in 1974 (7, 8). The possibility of major rises of the Loire's water level where the flood-threatened area is heavily inhabited is also a worrying factor.

In case of exceptional rises there could be up to 2-3 billion FF of damage and 300,000 people affected (9, 10). For this reason an important protection programme has been set up (which will also be useful for the maintenance of minimum water levels required by the nuclear-electric centres).

Madam,

This morning's paper publishes an advertisement informing hunters that hunting is forbidden on your land.

I have noticed that this news has been unfavourably received by the population in general and by the hunters in particular. I believe you have taken a decision which runs counter to your interests and that on further thought you will change it.

Please keep in mind that fire kills more birds in a day than hunters kill in a year.

I should like to add that if your woodland is intact this is not due to chance but to the spirit of solidarity of the inhabitants of X ... who always immediately went to the scene and fought the fire.

I am afraid that if you uphold the ban this will no longer be the same in the future.

As you will remember, it is now some years since a fire was started accidentally on the edge of the road which runs along the border of your property; you arrived there when we were already on the spot.

If Miss Y ... never banned hunting it was for the reason just mentioned.

I am convinced that you will consider my request and remain, dear madam,

Yours faithfully,

Signed: The Mayor

------

*In the Third World earthquakes are more disastrous (Guatemala 1976: 22,000 dead).

A recent letter from a mayor in the south of France to an owner of woodland (Source: Quoted in a study by the Institute for Woodland Development: "For a contractual policy of opening green land to the public", report for the Ministry of the Environment, December 1974, vol. 1, p. 175).

In the same order of things is the progressive disappearance of mediter-ranean woodland because of large fires. These fires signal among other things dangerous use of land (spread of secondary homes, departure of the inhabitants, diverse refuse dumps, lack of maintenance of the environment etc.). Disaster here is in large measure the result of a dilapidated situation which makes the environment very vulnerable (11), also the result of societal problems (see p.   ).

3rd. Safety in everyday life

Among the risk factors typical of today's industrial society the following are the main:

a) Risk from cars. In addition to some outstanding occurrences such as the disaster of Le Mans in 1955 (eighty three dead) or bus accidents, there is the daily tribute being paid to this technological innovation: 15,000 killed in Germany, 13,000 in France in 1977. Every year, world-wide, there are about 250,000 deaths and 7.5 million injured on the roads (5, p. 4).

b) Risk at work. Even if working conditions nowadays are less dangerous than in the past, even if the number of victims is on the decline, work accidents in France account for nearly 2,000 deaths a year. In 1978 there were 1,606 deaths on the records of the social security system (6, p. 32) to which must be added the 238 deaths of people not covered by the system. For the three years 1976/77/78 the total number of deaths stands at 5,228 (this is the figure for wage earners covered by the general system).

For work-related diseases which are recognised as such and for wage earners covered by the general social security system the total number of deaths is two hundred and nine (6).

c) Pollution risk. Two big events may serve as examples:

London 1952: The fog in December of that year impeded the exchange of air in the capital; this caused general poisoning of the atmosphere. 4,000 immediate deaths*were counted, and the figure of 8,000 was mentioned for not immediate deaths (1, p. 344).

Minimata 1956-1973: The Chisso company set up at Minimata in 1907; it under-took the manufacture of acetaldehydes in 1932. Soon worrying phenomena appeared; the first human victim was examined in 1956. In 1959 the Chisso company was no longer unaware of its responsibility for the situation. In 1962, one hundred and twenty one cases of disease were acknowledged, 44 people died after prolonged agony. The fishermen had only the choice between polluted fish and famine. Subsequent to measures taken elsewhere (Canada, Japan) the victims rejected compensation as a solution. The court case opened in 1972. In 1973 sentence was pronounced on Chisso. The 'strange disease' of Minamata had caused two hundred and forty three deaths; 1,300 people have officially been pronounced affected; but the real figure was estimated at 10,000 (12, 13).

---

*4,700 according to Ian Burton (3, p. 73).

In order to draw up a telling balance sheet today one would need to have in-depth epidemiological studies made. Too often these go wrong (even in those cases where specialists and people in authority suspect the existence of increased risk) so that one does not get very far with the examination.

d) Life style. Pollution, stress and types of food can be the origin of certain diseases (cardio-vascular ailments, cancer ...) that can be attributed to the general living situation of the individual (rather than a specific cause).

4th. The very great risks surrounding the safety problem

The post-war period has seen growing awareness of the precarious conditions of man's life and survival. The military question is the most acute; since the creation of thermonuclear weapons and their manufacture in large numbers in some countries, the safety of the world has rested on the equilibrium of terror.

Other than that, worries have been created on account of the large problems of ecological equilibrium: change of climate, carbon dioxide concentrations in the atmosphere etc.

Such is, in brief outlines, the picture in which the specific fact of technological disaster in the post-war era features.

2. DISASTERS KNOWN FROM THE PAST

1st. Fire

(Whole) towns no longer burn down. However, buildings still fall prey to the flames from time to time (1, pp. 660-663):

1946: Hotel La Salle, Chicago — 61 dead
1946: Hotel Winecoff, Atlanta, USA — 119 dead
1947: Danang, West Berlin — 86 dead
1947: Theatre Select, Rueil, France — 87 dead
1949: Hospital, Effingham/Illinios, USA — 77 dead
1967: L'Innovation Department Store, Brussels — 322 dead.

2nd. Navigation

Since 1950 the contribution made by navigation is much less heavy on passengers. Still, according to Nash's list (1, pp. 701-709) four accidents in the bracket of two hundred and fifty to five hundred victims were counted plus a more serious one (1954: seven hundred and ninety four dead on a ferry in Japan).

3rd. Mining

Between 1950 and 1975, ninety three mining disasters were counted worldwide (1, pp. 710-720), causing a total of 6,700 to 7,000 victims with peaks like the accident at Umata in Japan in 1963 (four hundred and fifty two dead) or the one at Wankie in Rhodesia on June 6, 1972 (four hundred and twenty seven dead). In addition to these major incidents there were two accidents claiming between three hundred and four hundred victims, three in the next lower numbers bracket, seven accidents killing between one hundred and two hundred people, twenty causing the death of between fifty and one hundred workers. A clear reduction in the frequency of accidents in Europe is noticeable. In France, the disaster at Lievin on December 27, 1974 caused

forty two deaths. The major mining incident in Europe occurred on the
surface at Aberfan in Wales.

Aberfan 1966:

On October 21, 1966 at 09.15 h, 140, 000 tonnes of material from slag heap
Number 7 towering above Aberfan, soaked with water, slid down the slope and
destroyed a school as well as eighteen houses. The collapse caused the death
of one hundred and forty four people of which one hundred and sixteen children
(15, 16, pp. 87-93). (We shall come back later to this case which is a
completely clear one where helplessness of disaster prevention is concerned.)

4th. Railways

From 1950 to 1975 about one hundred and seventy serious accidents were
counted world-wide which claimed more than 9,000 victims (1, pp. 740-743).
In the bracket of 100 to 150 victims there were twelve accidents, six in the
one hundred and fifty to two hundred bracket, three claimed more than two
hundred victims each. (Mexico 1972: two hundred and four deaths; Mexico
1955: three hundred deaths; Pakistan 1957: three hundred deaths).

Western Europe appears to be less and less afflicted, the Third World
countries more and more (eleven accidents out of the twenty one that claimed
more than one hundred deaths during the period occurred in non-developed
countries).

The latest big disaster in France dates from 1972: it was the one at
Vierzy (June 16, 1972) which was caused by the collapse of a tunnel just
before the arrival of two passenger trains. One hundred and seven victims
were counted. England had her most recent large accident (with more than
one hundred deaths) in 1952. These are the major incidents in Europe between
1950 and 1975.

5th. Explosions

Accidents caused by ammunitions and fireworks were less serious than in
earlier times and less frequent even though they have not been completely
eliminated. In France for instance accidents occurred at Pont de Buis
(Finistere) and again at Saint Marcel d'Ardeche.

Saint Marcel d'Ardeche 1962. The explosion of the gunpowder factory of Banc
Rouge at Saint Marcel d'Ardeche on April 9, 1962 claimed eighteen dead and
and fifty one injured. At the summary court at Privas the experts held
against the two people in charge:

- Faulty operating methods which did not conform to the rules of the
  profession;

- Installation of workshops without authorisation and without approved
  instructions by the directorate and the inspectorate of works and the
  board of the national gunpowder factory at Sorgues;

- Fault in the constant supervision of the mixer;

- Lack of qualification among the staff (the superintendent of the mixing
  operation was a delivery van driver);

- Too large a concentration of staff in the vicinity of the most dangerous
  workshop of the factory.

On these charges, one of the defence lawyers commented:

> Controls were carried out; the Director has never been advised or ordered to take specific safety measures which he is now accused of not having applied (17).

During the period from 1950 to 1975 twelve major accidents were counted world-wide (according to Nash, 1, pp. 661-662) of which one in Colombia claimed 1,200 dead in 1956 (explosion of an ammunition convoy), one in Havana in 1960 (One hundred dead). There was also an explosion and a fire in a Titan 2 missile silo in the USA in 1965 (fifty three dead).

Another type of explosion occurred twice during the post-war period; these were major accidents caused by transport of ammonia nitrate, a very explosive material. Twenty dead, five hundred injured and heavy material damage were reported at Brest when the Norwegian ship *Ocean Liberty* exploded on July 28, 1947. Three months earlier the town of Texas City had been partly destroyed by an explosion.

Texas City 1947

On April 16, 1947 at 09.12 h the French liberty ship *Grandcamp* exploded in the port of Texas City with its 2,300 tonnes of fertilizer which consisted of ammonia nitrate. A fire had broken out and the captain had given the order not to use water: Stop! Don't pour water on the cargo or you will lose it. Steam was used instead from which a gigantic explosion resulted.

There were five hundred and fifty two dead, two hundred disappeared, 3,000 injured. All windows in Texas City were broken and half of those in Galveston, 16 km away. A one tonne object was hurled over a distance of 400 metres, the ship was reduced to fragments which were blown nearly 5 km high and over an area of 9 km. Two aircraft flying overhead were destroyed. The big petrochemical factory of Monsanto was severely damaged when the hydrocarbon tanks caught fire. More than 3,000 dwellings were destroyed. Water and electricity were cut. Fires developed: by noon hundreds of them raged.

The following day disaster came about from a second ship, also loaded with explosive products. The explosion on the *Grandcamp* had caused a fire on it. The tugs called in to remove this second 'bomb' were not strong enough: the ship had been imbedded into another vessel. The second explosion revived fires and caused hundreds of deaths. Shortly afterwards the chief of police told the press: All of Texas City will be destroyed if the wind veers south. Luckily, this did not happen. But the situation was sufficiently dramatic to cause panic. A rumour circulated: the chlorine tanks leaked. Some demanded masks and started running through the streets carrying protective items that were useless — and this did not help to calm the spirits (1, pp. 545-550).

On a quite different scale, which, however, one must not forget, there are the accidents caused by the escape of gas in households such as in Argenteuil in 1970 and at La Courneuve in 1978.

6th. Dam bursts

The reliability of dams has improved since 1950. According to the census taken by A. Goubet (18, pp. 21-22) the following cases were recorded after 1950:

1959:   Vega de Terra, Spain 144 to nearly 400 dead (?)
1959:   Malpasset, France — 421 dead
1960:   Aros, Brasil — 1,000 dead
1961:   Babi Yar, USSR — 145 dead
1961:   Hyokiri, Korea — 250 dead
1963:   Quebrada La Chapa, Colombia — 250 dead
1967:   Semper, Indonesia — 200 dead
1967:   Nanksagar, India — 100 dead
1976:   Del Monte, Colombia — 80 dead
1976:   Santo Thomas, Philippines — 80 dead.

It seems that dam breaks affect the industrialised countries less and less*.

The Third World countries are not in the same situation. We shall come
back to this later.

3. BIG ACCIDENTS CONNECTED WITH NEWLY ADOPTED TECHNOLOGIES

1st. Fires in buildings incorporating highly inflammable materials

In addition to the case of the very rapidly spreading fire at CES Pailleron
which on February 6, 1973 caused twenty deaths there are the cases of the
dance hall at Saint Laurent du Pont, the "Cinq-Sept" and of the entertainment
complex "Summerland" in the Isle of Man (GB).

The Cinq-Sept 1970. On Saturday, November 1, 1970 this dance hall in the
Isere was crammed. The decorations were made of highly inflammable materials.
The main entrance featured a turnstile; the other exits had been bolted to
prevent frauds. A match that was dropped on a cushion caused the fire, a
mass of whirling fumes and flames. The plastic material collapsed. The turnstile was quickly blocked. Within a minute the blazing mass left no more
chance to its prisoners; 146 young people met their deaths in it. (16, p. 118)

Safety measures in the club were practically non-existant: insufficient
emergency exits (even if they had not been bolted); lack of luminous signs
for these exits, lack of fire fighting equipment, no telephone. For the
court the Cinq-Sept was the materialisation of everything that is forbidden
by the regulations (16, pp. 113-118).

Summerland, the entertainment complex in the Isle of Man, 1973. The Summerland project was an original idea: to create a tourist attraction with great
capacity on the Isle of Man that would offer the visitors the charms of
mediterranean climate. For this purpose a large complex under a plastic bell
was built in which a riviera-like temperature was maintained. Opened in 1971,
the centre had seen great success in 1972. The following year it was to be th
the theatre of a gigantic fire (15, 19).

The architects were chosen in 1965. A bureau from the Isle of Man (a small
company of two architects which had never worked with more than six
technicians) joined forces with a larger group that specialised in leisure
industry constructions. In this way the great Summerland complex was designed
consisting of seven levels, the access being on the fourth, the outstanding
features being the west and south faces and a roof made of Oroglas (acrylic
building material).

----

*The threat has not been reduced to zero, however, as we shall see later.

The necessary building permits were obtained in 1967, 1968 and 1971 respectively. Work was carried out rapidly because one wanted to open for the 1971 season. In a pamphlet the centre was presented as perfectly safe; the question of fire prevention had been given special attention during the construction; the structures were non-inflammable; any fire would be confined to the hall in which it broke out. The public welcomed this complex. By contrast, it received a mitigated welcome in the world of construction: true this was a first attempt at constituting an artificial microclimate but there had been a systematic reduction of standards in the whole undertaking. However, success was assured in 1972 and the centre contributed 13 per cent to the tourist income of the island.

On August 2, 1973 Summerland was to be the theatre of a gigantic fire. A kiosk which had been used as an entrance gate and had been damaged by a thunderstorm two months earlier had been dismantled and partly removed. There remained some elements including electric wires and various bits of waste that were left at the end of an outside terrace against the wall of Summerland, a galbestos wall. Three young boys set fire to this rubble at 19.40 h. Within a few minutes the kiosk was on fire. The flames ran along the partitioning wall, inflammable material caught fire. The fire invaded the fourth level; fumes spread about; the flames quickly attacked levels 5, 6 and 7. When the wall and the Oroglas ceiling caught fire an enormous conflagration developed through the whole elevation of the building, the fire spreading from east to west, destroying everything inflammable on the fourth level.

Attempts at extinguishing the fire were fruitless. When the fire service — who were called late — arrived they had to declare their impotence: they could perhaps save the adjacent buildings but for Summerland it was too late. Attempts were made to avoid panic but this was impossible. The flames, the gas, the darkness — the automatic emergency generator did not start — the scuffles in the (poorly designed) passages, the emergency doors blocked from the outside or marked 'Private' or not leading to any safe place, parents swimming against the stream in search of their children: there was no longer much chance for action. Fifty three dead were counted.

Authorisation for departure from standards given without sufficient strictness. The construction, of course, required permits. In particular observation of rule No. 39 of the Isle of Man law was mandatory which prescribes that all buildings must have non-inflammable outside walls which resist fire for two hours. Rule No. 50 requires that the roof must offer appropriate protection against fire from a neighbouring building. According to rule No. 47 fire prevention devices must be set up for all inflammable walls.

A first request for a permit was filed in October 1967. The authorities commented that the Oroglas did not meet the requirements of rule No. 39. The local architects pleaded that even if it was not fire-resistant — which is what the Chief Fire Officer had stressed — it was not inflammable. The Chief Fire Officer concluded his examination of the application by stressing: since the complex does not present exposure to a risk coming from another building and as it is unlikely that there would be trouble at the level of the emergency exits I raise no objection against the project. The official permit stated however, that there was deviation from rule No. 39.

The second permit was requested in July 1968: the local engineer in charge of the application remarked that one of the materials used (the Galbestos) instead of the concrete foreseen in the initial project did not conform to the requirements of rule No. 39; it was inflammable and not resistant to fire. The permit was issued nevertheless, on the same terms as in 1967.

Finally in 1971, the external construction having been completed a third request, concerning the interior arrangements, was filed. The rules for theatres (of 1923) had to be conformed to which foresee the presentation of a plan and the supply of an estimate of the number of people the building would accommodate. These two requirements were not respected. The Chief Fire Officer was mainly preoccupied with the fact that he did not know how to apply the regulation for theatres to such an unusual complex. Finally he recommended (on June 8, 1971) to issue the permit specifying, however, that all appropriate safety arrangements had to be adopted without delay. The permit was in fact issued.

The materials used. The concrete which had originally been foreseen as the building material for the face of the building that was attacked by the kiosk fire (east face) had been replaced by an inflammable building material, the Galbestos, which did not resist fire for two hours but only for a few minutes. This is why the fire could spread to the inside of the building. The literature on the subject of this building material was so inexplicit that the architect had thought it was non-inflammable.

The decorator for his part had also substituted an inflammable material, Decalin, for a safer material that had been foreseen at the start. He did not know the properties of this Decalin nor that it was inflammable.

As for the Oroglas, it had never been subjected to significant tests for the scale of its use in Summerland.

The evacuation: an accumulation of design and management faults. The commission was to take up quite a number of serious faults:

insufficient signposting;

difficult access for fire fighters;

most emergency exits on the same side of the building;

possible chimney effect above the main exit;

insufficient space at the bottom of the main escalator and at the main exits (congestion); two other exits defective;

doors not functioning as fire screens, did not shut;

emergency doors not equipped with panic bars but with keys (a key had to be fetched which meant running to an office);

emergency doors marked "Private";

emergency doors obstructed by a parking lot;

emergency doors not leading to a safe place;

darkness all over on account of poor maintenance of the emergency generator.

2nd. Risks presented by highrise buildings

Highrise buildings give cause for concern especially when the recommendations concerning the building materials to be used inside are not adhered to.

Sao Paulo 1974. The Joelma tower was twenty five stories high; the first ten stories were reserved for parking. Above these there were usually 1,000 people working; at the time the fire broke out only five hundred people were in the building. The fire spread at great speed on account of the ventilation system and of inflammable building materials that had been used despite open criticism from the mayor which had been repeated for weeks.

In addition, safety measures were dropped during construction. A single
unprotected staircase (in a central position) served the whole building while
at least two enclosed staircases would have been necessary

Helicopters, impeded by currents of heat from the fire, by fumes reducing
visibility for the pilots and under certain conditions stopping the engines
and by the environment (buildings, TV antennae etc.) could not be considered
normal means of rescue.

Appeals for calm by the fire service were fruitless: many jumped in order
to escape the flames and met a quick death under the eyes of 10,000 veritable
'spectators' who blocked all approaches and impeded the work of the fire
service who by acts of sheer heroism managed to save some people.

It soon became clear to the authorities that only a few people could be
saved, and they authorised TV to say so. This heightened the impulsive
sensations of fascination and horror (...) Three hundred thousand cars*(?)
soon squeezed each other into the nearby streets while the fire grew worse
(16, p. 121).

In the end a heavy-duty helicopter could make several shuttles between the
roof and a place of safety; it managed to save about one hundred people
before the roof collapsed. One hundred and seventy seven victims were
counted (1, p. 292; two hundred and twenty according to 16, p. 119) and two
hundred and ninety three suffered severe burns.

It was ensured that new regulations were enforced. The director of the
Police technical department indicated that there were not enough laboratories
to set up a prevention plan to test all building materials (16, p. 123).

3rd. Aircraft accidents

The scale of air disasters follows the growth in the capacity of aircraft
and the increasing density of air traffic (1, pp. 632-641).

Until 1949 the figure of fifty five deaths**was not surpassed (in-flight
collission over Washington, November 1, 1949). In 1950 the peak reached
eighty dead (Wales, March 12); it rose to one hundred and twenty eight deaths
in 1956***(collision between a Super Constellation and a DC 7, June 30,
Arizona), and to one hundred and thirty six deaths in 1960 (collision between
a DC 8 and a Super Constellation, December 16, New York).

While the average number of victims per accident increased, the peaks (also)
continued to rise: one hundred and fifty five deaths in 1969 (crash of a DC
9 in Venezuela), one hundred and sixty two deaths in 1971 (Boeing 727 hit in
flight by an airforce plane — in that year there had been two hundred near-
misses in Japan, six hundred in the United States; one hundred and seventy
six deaths in 1972 (crash of an Ilyushin 62 in the USSR, unconfirmed); one
hundred and seventy six deaths in 1973 (in Nigeria, crash of a Boeing 707).
Recently new orders of magnitude have been recorded.

---

*Cars that were soon abandoned by their drivers who wanted to get a better
view and cared little about the problems this caused for ambulances and for
the fire engines.

**If one leaves aside the crash of an American bomber on an English school
in 1944 (Freckleton, August 23, seventy six dead).

***In civil aviation (an American military aircraft crashed in 1953 claiming
one hundred and twenty nine deaths).

Ermenonville 1974. On March 3, 1974, soon after take-off from Orly, a Turkish Airlines DC 10 with three hundred and thirty four passengers and twelve crew members on board crashed at Ermenonville. The enquiry showed that these three hundred and forty six people went to their deaths because of the faulty lock on a luggage door that was torn off at an altitude of 3,600 metres; this in turn damaged the floor of the passenger cabin and the three control systems which passed through there (44).

An incident with a similar cause had already occurred two years earlier on a DC 10 (June 12, 1972); the cables had not been damaged on that occasion and the pilot succeeded in making a safe landing at Windsor, Ontario. The necessary modifications demanded by the American federal administration were not made. The administration thought that its correspondence with Mc Donnel Douglas, the manufacturer, was sufficient and that it was not necessary to proceed with the issue of a directive. They were wrong.

Two years after the incident at Windsor some aircraft had not yet undergone the necessary modifications; two DC 10s were even built and sold without these adjustments. Three days after the disaster of Ermenonville an imperative instruction was sent to all DC 10 users. The confidence had cost the lives of three hundred and forty six people (1, pp. 579-584; 16, pp. 125-129).

Santa Cruz 1977. On March 27, 1977 on the airport of Santa Cruz de Tenerife, Canary Islands, two Boeing 747s collided. There were nearly six hundred immediate deaths (21) and a final toll of six hundred and twelve (21).

Chicago 1979. On March 25, 1979 an American Airlines DC 10 crashed at Chicago causing the deaths of two hundred and seventy five people. All DC 10s were grounded for a prolonged period by the American administration (22).

Generally speaking, the average annual number of victims of air accidents is in the order of 1,000 (USSR and China not included). In 1974 it was 1,155 (45).

4th. Accidents on oil platforms

Alexander Kielland 1980. The platform served as a floating hotel for people working on the North Sea oil field Edder (a satellite of Ekofisk) in the Norwegian zone. It sank on March 27, 1980. One hundred and twenty three dead were counted (23).

Gulf of Bohai 1980. A Chinese platform subsided in a thunderstorm in the Gulf of Bohai in June 1980. The accident caused seventy deaths. (24).

4. DISASTERS LINKED WITH LARGE SCALE INDUSTRY

1st. Fires and gas explosions in fixed installations

The manufacturing, stocking and utilisation of highly inflammable and explosive products on the one hand, the work with temperatures and pressures that are much higher than in the past on the other have led to large scale industrial accidents and this despite the precautions taken, the latter themselves being more developed than hitherto.

Below we shall give some measures of magnitude for the phenomenon of conflagration and explosion in large industrial enterprises as well as some cases for illustration, especially the one at Feyzin, the French case of reference.

A recent study by American consultants in the petrochemical industry has produced a list of large scale accidents (the criterion being financial: losses above 10 million dollars) that have occurred since 1950. The table below summarises their study (25):

	No. cases	Total loss in million $	Average loss in million $
1950–59	7	173	24.7
1960–69	16	404	25.2
1970–79	46	1,318	28.6
Total	69	1,895	27.5

Table 9. Estimate of losses incurred by the petrochemical industry during the last three decades because of big accidents: nearly 2 billion dollars.
(Source: 25)

Cleveland 1944. A small tank filled with liquified natural gas — 4,200 cubic metres compared with the tanks of hundreds of thousands of cubic metres which nowadays hold gases that are in some cases more dangerous, but which are luckily of a different design — cracked and burst into flames on October 20, 1944; another small tank cracked twenty minutes later. The heat released was so intense that it set buildings 500 metres away on fire. In addition, the gas expanded through sewers and subterranean pipelines and exploded all over the town. Streets were ripped open, pavements blown up, sewer covers hurled across buildings. There were one hundred and thirty six dead and two hundred injured; seventy nine houses, two factories and two hundred cars destroyed; the bill came to 6.8 million 1944 dollars. Yet, there had been two favourable factors: the time (the accident occurred in the early afternoon when there were few people in the streets) and the wind which blew away from the inhabited area, thereby reducing the impact of the accident (26, pp. 11-12 and pp. 33-36).

Rotterdam refinery 1968. A cloud of explosive gas formed accidentally on January 20, 1968 and found an ignition source. This was a detonation comparable to that of an explosive charge of 10-20 tonnes of TNT. All tanks, installations and buildings within a radius of 200 metres were destroyed. The detonation killed two men and injured another two; 3,500 injured were counted among the inhabitants of neighbouring areas; the shock wave destroyed windows within a radius of 3-5 km. Damages amounted to 125 million 1968 FF.

CDF — Chimie 1972. On January 30, 1972 an explosion occurred in an ammonia synthesis unit at Mazingarbe in the Pas de Calais. The explosion of the reactor — height 25 metres, outer diameter 1.10 metres, volume 160 tonnes, operating pressure 400 bar — caused considerable damage. Two fragments of 2.5 tonnes and 1 tonne were hurled through the control room 25 metres away; three compressors were destroyed by a one tonne piece which bounced on them and crashed 170 metres away. Another one tonne piece crashed toluene pipes 90 metres from the place of the explosion; yet a further piece of the same weight smashed through a villa 300 or 400 metres away after having bounced in the garden. Damage from the blast was incurred up to 2 km away (windows broken, doors twisted, walls cracked, roofs damaged).

Feyzin 1966. The refinery at Feyzin was put on stream in 1964; it was to process 1.7 million tonnes a year; it had had auxiliary installations which included mainly overhead tanks of liquid hydrocarbons of 300,000 m^3 capacity. These stores were in zone B of the refinery area. There were, among others, two spherical tanks which could hold 2,000 m^3 of butane and 1,200 m^3 of propane (the four spheres of propane were numbered from T.61,440 to T.61,443). The following extracts are from the verdict handed down by the Court of Appeal at Grenoble in 1971 (27, 28).

Preliminary technical explanations. In order to avoid excessive internal pressure in the sphere in case of an accident or a conflagration each sphere is equipped with a safety device consisting of two valves installed in the upper part of the sphere, yielding 73 tonnes/hour of gaseous products.

Each also contains a cooling device consisting of two rings of vaporisers installed at the top, the median part and the lower part of the tank respectively. This device is directly connected to the fire extinguishing network by a valve. Its average yield is 2,200 litres/minute for the propane spheres.

The nature of liquified hydrocarbons requires frequent draining during storage to eliminate the water and soda mixed with the product which after pouring off accumulate in the lower part of the tank.

The draining of the spheres is effected by means of two valves located at 5 centimetres distance from each other and are operated by a square key-lock, the lower valve serving as an evacuation pipe that 'drips into a square draining trap of 50 centimetres side-length and 1 metre depth, linked to the network for used water from the refinery.

On the other hand, gas samples are taken from time to time for analysis of the manufactured products and checking of their standards (28, pp. 38-39).

Antecedents of the accident of January 4, 1966. The draining of manufactured products practised from the start of storage on site (June 12, 1964) had brought to light some problems arising from the device:

- The valves were too close to each other, as the passage of propane from liquid to gaseous state which took place at a temperature of minus 44°C caused an almost simultaneous icing up of both valves;

- Their control by removable keys rather than wheel key-lock presented risks of gas escape in case one of these keys being dropped;

- Their diameter (2 inches) was too large;

- With the draining trap located at the feet of the operator, which meant that he was frequently splashed and sometimes suffered burns to his face and hands by the gushing of liquid into this opening.

- The valves were often difficult to operate.

- Finally, the access to the valves was made difficult by the presence of pipes which the operators had to step over in order to carry out the draining.

Employees had told the management about these problems; things remained practically as they were. Two serious incidents had occurred that gave substance to the apprehension and fears expressed:

a) On August 6, 1964 at about 23.00 h one Robert Tinjod, an operator's mate, had opened — before massively draining the butane sphere 462 — completely the two valves of the tank, letting the liquid flow normally into the drainage tank, and he had climbed on top of the sphere in order to check the gauge there, thinking he had enough time before finishing the draining operation. It was then that the gas shot out in force.

Tinjod who wanted to shut the valves which were iced up by the passing gas froze his right hand slightly and had to be treated in hospital.

The draining taps were shut by a manufacturing engineer and one of the firemen on duty who were helped by a favourable wind.

b) On February 26, 1965 at 11.05 h one Isaac Bittoun a chemist, had been assigned with his colleague Godde to carry out the draining of the propane sphere 440 to take a gas sample.

In these ill-defined circumstances, after the usual emission of water and soda the propane shot out and burned the two men. The safety workers Leseurre and Rossit, after being alerted, intervened. The first one was also burned but the second one managed to shut the valve. The alert had been serious.

This last incident which if the wind had not again been favourable could have developed into disaster even though the motorway had not yet been opened to traffic had subsequently caused the issue of a service bulletin on the method of draining the spheres (March 4, 1965) by Mr Ory, the Chief of Technical Services. It said in particular that after the keys had been attached to the two draining valves the valve on the sphere side was to be opened completely, then the valve on the atmosphere side partly opened, without ever opening it completely in order to be sure that it could be closed, as soon as gas appeared, the closure of the draining valve or, in normal circumstances, of the valve on the sphere side, and then shut the second valve.

Additionally, this instruction indicated, for the control draining on the bottom of the sphere, the facility of using the piping between the two taps as a lock-chamber i.e. by opening the valve on the sphere side, shutting it again immediately, then opening the second valve to the atmosphere in order to empty out the content of the line.

It finally made it obligatory that the taking of laboratory samples had to be done in the presence of a safety officer and that draining was to be carried out by two people.

This bulletin which was entered into the service manual and posted in the pump rooms was generally known to the staff but had never been backed up by practice exercises. Also, some operators kept to their own ideas about the question and to the procedures previously practised (28, pp. 39-41).

The conflagration on January 4, 1966. On January 4, 1966 it had been decided to clean propane sphere 433 at the end of sample taking. Taking part were: Robert Dechaumet, operator's mate, Raymond Fossey, safety officer and Bernard Duval, laboratory helper.

In contravention of the instructions in the service bulletin from Ory this operation was carried out at 06.40 h, i.e. in complete darkness; the lower part of the sphere was lit by the diffused light of a candalabrum and

horizontal projectors placed at a certain distance. The temperature was
between 4 and 5°C, and there was virtually no wind.

Contrary to instructions Dechaumet first half-opened the lower valve, then
fully opened the upper valve, as it emerges from the experts' statements on
the pieces recovered as well as from those made by Fossey. The latter whose
function it was to watch the work and to intervene if need be did not budge
but looked on from a distance. Some dirt ran into the drainage tank, then
suddenly the gas shot out in force and struck the operator in the face and
on the body.

Dechaument, caught in the cloud, lost his safety goggles and involuntarily
unhooked the operating key of the upper valve the fixing nut of which had
actually not previously been tightened on the operating square.

Fossey shouted: "You have opened it too wide." Dechaumet who had recovered
slightly tried to shut the upper valve but did not succeed in putting the
key back on because of the icing caused by the escape of gas. He forgot to
try and close the lower valve on which the key was still fixed and refused
to keep trying.

Meanwhile Fossey and Duval had raised the alert over the telephone and the
"généphone". The three safety officers, Rossit, Roy and Fossey, tried in
turn to stop the escape, without success.

Gas escaped from the sphere which at 05.00 h in the morning had held
693 m^3 of propane at the rate of about 3.3 m^3 per second according to the
calculations made by the experts. The gas mixture, being heavier than the
air and there being hardly any wind blowing, the propane expanded by gravity
in the direction of the motorway. Nobody thought of alerting the fire
service, the gendarmerie and the CRS.

The cloud, approximately 1.50 metres high, reached the motorway on which
there were a number of vehicles between 06.55 h and 07.05 h. Employees from
the refinery and from the guard of the factory then intervened on the motorway and on the CD 4 road to stop the traffic. At 07.15 h Robert Amouroux,
driving his CV4 Renault, arrived on the scene; he was going from Serezin du
Rhone (Isere) to Feyzin to take up his duties in a company working for the
refinery. When he arrived at the cross-over linking the CD 4 with the motorway and crossed the gas cloud the latter, no doubt as a result of a spark
produced by the vehicle, caught fire,

Panic-stricken Amouroux stopped his car and got out; his clothes caught
fire; he ran and threw himself into a ditch a few metres away. He was found,
a quarter of an hour later, severely burned, and taken to hospital in Lyons
where he died on January 8, 1966.

The scene had been observed by the neighbouring customs post who telephoned
the gendarmerie at Saint Symphorien d'Ozon which immediately sent their
available staff to the scene. The CRS for their part acting on their own had
obtained information on what was happening and shared the work required with
the gendarmerie: stopping vehicles on the exposed roads, isolating the danger
zone, evacuating the houses and the school of the Razes area of Feyzin which
was in serious danger.

Sphere 443 had caught fire: it was a drinks retailer who telephoned the
fire brigade in Lyons at 07.12 h. Two other phone calls were received from
the refinery a bit later. The direct telephone line had not been used.

At the factory general alert was raised by a siren while the three professional firemen on duty who had been unable to plug the escape tried in vain to extinguish the fire of the sphere by attacking it with powder extinguisher and activating the fixed cooling system of the eight spheres and of the two liquified hydrocarbon towers.

The stock of powder (1,500 kg) being quickly exhausted, Rossit, the chief of the group, tried unsuccessfully to use the foam extinguisher which he had available. This piece of equipment could not function due to lack of water suction; a foam launcher could not be used for lack of pressure.

In fact, while the fire fighting network of the refinery was designed to deliver a maximum of 800 m^3/hour of water the simultaneous opening of the cooling systems for the propane and butane tanks by the safety officers required the use of 1,128 m^3/hour. Therefore, from the beginning of the fight against the fire, water was in dangerously short supply. The situation was aggravated by the fact that the neighbouring Rhone Gas Company which also used the water supply network of the refinery had, as a precaution, also started the cooling system for its two propane spheres and was hosing them with a fire hose.

The fire brigade from Lyons arrived on the spot from 07.33 h onward in successive pickets led in turn by the Adjutant Prevost, Commander Legras (from 07.43 h) and Commander Pierret (from 07.46 h). They joined their efforts with those of the professional and auxiliary firemen from the refinery and were in turn joined by members of the fire fighting team of the nearby Rhodiaceta factory at Saint Fons (Rhone) who arrived at 08.20 h and the fire pioneers of Vienne who after being alerted by the Commander from Lyons arrived at 08.28 h.

As chief of the first intervention picket from Lyons adjutant Prevost occupied himself immediately with sphere 443 which he tried to extinguish with the help of the foam launchers. Being unable to succeed he abandoned the burning tank and concentrated his efforts on the neighbouring propane tank 442.

The rescuers giving up the attempt to extinguish the fire devoted themselves exclusively to the cooling of the other tanks to prevent them from catching fire and hoping that sphere 443 would empty its content which burned as soon as it entered the atmosphere.

However, faced with the drop in pressure already mentioned, Adjutant Prevost and subsequently Commanders Legras and Pierret decided to put a special highpowered fire engine for hydrocarbon fires on suction in the Rhone canal, but for lack of adequate fittings this was sucked in and could only be recovered after some twenty minutes.

On the other hand, the rescuers were handicapped by the customs enclosure the doors of which were padlocked. Employees of the refinery forced the padlocks and then demolished the enclosure with an excavator.

Meanwhile, reinforcements had continued to arrive and authority was passed first to Commander Legras, then to Commander Pierret.

At 07.45 h the important event mentioned above occurred: <u>the release of the safety valve of sphere 443</u>; the gas which escaped through it caught fire immediately causing a fire column of some ten metres in height. <u>This incident</u>

was interpreted as reassuring by some of the people in charge at the refinery*:
it indicated according to them that the sphere would empty itself completely.
They told Commander Pierret and some of his co-workers so.

However, some of the rescuers were gripped by a mute apprehension born of
the considerable increase of flames enveloping sphere 443 and the growing
turmoil caused by the conflagration.

As to the manner in which the accident was attacked, Commanders Legras and
later Pierret had confirmed the measures taken by Adjutant Prevost, restricting themselves to a role of preventing the spread of the accident by hosing
the tanks that were likely to catch fire.

The lowering of pressure constrained the rescuers to a dangerously close
approach to the tanks as the water from their launchers reached the top only
with difficulty. This dangerous situation determined Commander Legras to pull
his men back after they had fixed their launchers in firm hosing positions.

Nearly one hundred and seventy people were then in area B.7/1 and in the
other areas of zone B. They were firemen from Lyons and Vienne, professional
and auxiliary firemen from the refinery and from neighbouring companies or
companies working for the refinery, the director, department heads, employees
of the factory, supervisors and staff from neighbouring factories and
spectators.

The explosion of sphere 443 which occurred at 08.45 h struck most of these
people. Added to the waves of burning gas caused by the deflagration were
pieces of steel, some of them of considerable weight, that were hurled in
some instances over several hundred metres.

Seventeen rescuers succumbed to the explosion or later on to their severe
burns. Among the eighty four injured (...) forty two suffered complete
disablement for work for more than three months.

However, the explosion had extinguished the fire in the whole of areas
B.7/1 and B.7/2 and the southern part of area B.11. The rescuers whose
courage had been above praise and some of whom had saved the lives of
colleagues in danger while risking their own lives then fell back, taking the
injured with them.

On account of this the explosion of sphere 442 at 09.45 h did not cause
further victims but, like the preceding one, did cause much material damage
as far as 16 km away at Vienne.

Between the two blown-up spheres a crater, 35 metres long, 15.40 metres
wide and 2.10 metres deep had opened up (28, pp. 41-45).

2nd. Dispersion of toxic and highly toxic products

A first sub-family of gases includes products with a toxicity similar to
that of chlorine and ammonia.

Baton Rouge 1976. On December 10, 1976, 90 tonnes of chlorine escaped from
a tank at Baton Rouge (Louisiana) and caused the evacuation of 10,000 people
and the blocking of the Mississipi river over a length of 80 km (29, p. 11).

---

*Our underlining.

Blair 1970. On November 16, 1970 in Nebraska there was an escape from a 32,000 tonne ammonia tank for two and a half hours, causing a release of 140-160 tonnes. The cloud that developed covered 365 hectares up to 2,500 metres from the tank forming a layer of between 2.5 and 9 metres height but claiming no victims (rural area) (22, p. 12).

Putchffstroom 1973. On July 13, 1973, 18 tonnes of ammonia escaped at Putchffstroom, South Africa, leaving eighteen dead of which six were outside. The cloud spread over the town (2, p. 126).

Les Grandes Armoises 1969. Release of 4 tonnes of ammonia during transfer from a fixed to a mobile cistern on May 12, 1969 in the Ardennes. Vegetation was burned over an area of 2 km by 450 metres; various animals were killed. Inhabitants were warned in time and evacuated (2, p. 124).

A second sub-family comprising even more dangerous products can be distinguished: arsenic, hexafluor of uranium, hydrofluoric acid*, acrolèine**, phosgene***etc.

Manfredonia 1976. Between 10 and 30 tonnes of arsenical salt spread following the rupture of an ammonia production tower in Italy in September 1976. Theoretically, 100 milligrams are sufficient to kill a man (31).

Pierrelatte 1965, 1977, 1977; Cadarache 1977. Multiple accidents, the products in question being hexafluor of uranium and fluorhydric acid. In the majority of cases, however, there was no external pollution; no intoxication, no injuries and no deaths (2, pp. 123, 128-129).

Pierre Benite (Lyons) 1976, 1976, 1978, 1978. Multiple incidents at an outfit manufacturing acrolèine, a product of high toxicity (concentration threshold: 0.1 ppm). On July 10, 1976 a wagon of acrolèine tumbled into the Rhone river: the 21 tonnes killed all the fauna down to Vienne (320 tonnes of fish). On December 19, 1976 a container holding 5 tonnes of acrolèine did not stand up to an accidental polymerisation of the product; by chance some electricity cables snapped and the acrolèine caught fire: there was no formation of a toxic cloud that would have been capable of intoxicating the housing areas of this suburb of Lyons. On July 12, 1978 a new escape of acrolèine. People living in the neighbourhood were put to considerable discomfort (33). On October 12, 1978 some hundred kilograms of acrolèine were released into the atmosphere. Discomfort for several thousand people in the area: twelve people admitted to hospital for observation (1, pp. 127-129).

3rd. Transport accidents

a) Land transport. The case of Missisauga-Toronto, one of the most commented-on, is not at all an isolated one. As far as rail transport is concerned the North American continent appears particularly affected. On the day after the accident of the Canadian Pacific train three wagons of liquified propane

---

*Threshold: 3 ppm equals 2 milligrams/m^3. Lesions, intoxications (toxicological file INRS No. 6) (32).

**Threshold: 0.1 ppm equals 0.25 grams/m^3. As dangerous as gas used in warfare (file INRS No. 57).

***Threshold: 0.1 ppm. Suffocating, peracute intoxication, fast killer (file INRS No. 72).

exploded in Florida following a derailment (34). The day after a thousand families in Michigan were evacuated following the derailment of a wagon containing fluorhydric acid (35). R. Audurand reports two important events in Florida: in Youngstown in 1978 the derailment of a train caused an escape of chlorine; there were eight dead, one hundred injured, 3,500 evacuated in an area of 100 km². At Crestiew in April 1979, 5,000 were evacuated following the derailment of a convoy of twenty eight wagons of ammonia and chlorine (2, p. 129).

In Europe there were two big cases, the first one occurred some time ago, the second one very much in everyone's memory.

Ludwigshafen 1948. The large BASF complex at Ludwigshafen (FRG) employing 22,000 people was partially destroyed on July 28, 1948 by a big explosion. There were two hundred and forty five dead and thousands injured inside as well as outside. It would appear that a wagon containing dangerous products had been left standing instead of being quickly unloaded. It caused an explosion which was soon followed by three additional deflagrations. In addition to the destruction of part of the complex windows were broken within a radius of 8 km (1, pp. 257-258; 28, p. 129).

Los Alfaques 1978. On July 11, 1978 a road tanker carrying 18 tonnes of liquified propylene under pressure exploded near a camping site at San Carlos, Spain. The radiation of heat from the fireball was extremely intense; there were two hundred and sixteen dead and several hundred people had burns (2, p. 128).

Lievin 1968. Explosion of a tank wagon at the chemical (nitric fertilizer) factory of Grande Paroisse (Pas de Calais). Release of 19 tonnes of ammonia. Six dead, twenty people living in the neighbourhood had to be hospitalised with intoxication (2, p. 124).

Saint Amand les Eaux 1973. On February 1, 1973 at 17.30 h an 18 tonne truck of liquified propane under pressure (5-7 bar) overturned in the middle of the town of Saint Amand les Eaux, northern France, on a road bend. The propane escaped, evaporated and formed a gas cloud which spread along the street over about 120 metres. It caught fire when it came in contact with a heat source. At 17.36 h when the rescuers arrived the fire raged. A few minutes later a violent explosion occurred, killing four people instantly and injuring forty others. The tank had broken up into three main parts after the explosion: the front half, practically intact, was found vertically implanted in the nearest building; the bottom had been hurled over a distance of about 450 metres from the site of the accident; the rear half had been ripped open and hurled against a house that was completely destroyed. An Ami 6 car which had been behind the truck when the explosion occurred was hurled over the wall running along the street and over a distance of about 70 metres. In the end there were nine dead, thirty seven injured, some twenty houses damaged and the town was declared a disaster zone (2, p. 125).

To these rail and road accidents must be added those involving gas and oil pipe (line)s.

Port Hudson 1970. This is the very rare case of a detonation (very fast explosion). 112 m³ of liquid propane had leaked within the twenty four minutes between the escape and the explosion. A propane and air mixture developed which spread over a surface of about 4 hectares and had a volume of at least 30,000 to 60,000 m³. The blast caused by this explosion was equivalent to that of about 50 tonnes of TNT: the detonation added to a

combustion of residual volume that was richer in propane after a whirl of
flames. By chance all this happened in a non-inhabited area (27, p. 30).
The event could have resulted in a very large scale disaster had it happened
in another place. Someone walking 800 metres away was thrown to the ground.
A policeman driving 25 km away from the site of the explosion saw his car
make a swerve. Three hundred and fifty kilometres away at Kansas City the sky
was seen turning red. Up to a distance of 800 metres buildings were seriously
damaged; up to 3 km away 60 per cent of the windows were broken; up to 10 km
away 30 per cent were broken. In addition, because high pressure can build
up according to the surface structure of the terrain there was a pocket of
destruction at 13 km from the place of the deflagration and another at 20 km
distance. An enquiry report estimated that in an urban area everything would
have been destroyed over 4 hectares and people would have been in grave danger
in an area of 120 hectares (37).

Huimanguille 1978. The rupture of a gas pipeline on November 1, 1978 caused
the deaths of fifty eight people, 800 km southeast of Mexico City (38).

Pavia 1980. The oil pipe line from Genoa to Milan fractured on April 21,
1980 and polluted the river Po over a stretch of 100 km. The several
thousand tonnes of petrol spilled caused fears of conflagrations (navigation
was suspended, bridges closed), release of toxic vapours, a severe change in
the ecological equilibrium, pollution of the groundwater. The river Po flows
through the richest part of Italy (39, 40).

b) Maritime transport. Petrol tanker accidents have attracted most attention
since the wreck of the *Torrey Canyon* on March 18, 1967. Up to the *Amoco
Cadiz* disaster, which does not mark the end of such accidents, there have been
a series of events the repetition of which causes concern. Considering only
the French and British coasts here there were (41, pp. 133-134):

- On August 19, 1969 the collision between the *Gironde* and another ship —
  2,000 tonnes were spilled.

- On October 23, 1970 the collision between the *Pacific Glory* and the *Allegro*:
  10,000 tonnes were spilled.

- On May 15, 1971 the collision between the *Herculo* and another ship — 300
  tonnes were spilled.

- On November 26, 1974 the collision between the *Chaumont* and the *Peter
  Maersk* in the access channel of the port of Le Havre. 1,700 tonnes were
  spilled, 20 km of coastline polluted.

- On January 24, 1976 the *Olympic Bravery*, a completely new petrol tanker
  of 275,000 tonnes, left Brest for a Norwegian fjord where it was to be
  anchored on account of commercial problems which inhibited its utilisation.
  It found a nearer resting place: after leaving Brest and subsequent to an
  engine breakdown which did not lead to a call for assistance it foundered
  on the coast of Ouessant. After rupture of the hull the 1,250 tonnes of
  engine fuel (the tanker was empty) spread and polluted the coast of the
  island which made implementation of the Polmar plan necessary.

- On October 17, 1976 the East German tanker *Boehlen* sank northeast of the
  island of Sein; the pumping of the petrol that had not been spilled cost
  155 million FF and three human lives.

- On May 8, 1978 the *Eleny V* was rammed in the North Sea, and 24 km of
  British coastline were polluted.

- On March 7, 1980 the Madagascan tanker *Tanio* carrying 26,000 tonnes of heavy fuel oil and 900 tonnes of engine fuel broke apart. The front part sank during the day. There were four dead. The rear part was tugged in an emergency (2 miles off dangerous shallows) and later towed towards the port of Le Havre. On March 9 the rose-coloured granite coast was struck. The damage to the avifauna seems to have been on the scale of the one caused by the *Amoco-Cadiz* (20,000 birds killed) because of the type of fuel and the location of the oil slicks near the (bird) reservation of Sept Isles (42). The pollution of the coast turned out to be serious as the days went by. The Polmar plan was considered under its administrative and financial aspects and implemented after a few days' delay, immediate measures having been taken, however, from the time of the advice of the accident (supply of barriers and pollution fighting units, setting up of command posts etc.). Cleaning up proved to be difficult. The rose-coloured rocks were affected in depth. A decision had also to be taken on the problem of the wreck which presented a "Damocles sword" for the coast. Assurances of the past had at last to be given up; Aymar Achille Fould declared:

Those who imagine that if all precautions were taken there would never again be any accident have never seen the sea (43).

REFERENCES

(1) J. R. NASH, *Darkest Hours*. Nelson Hall, Chicago, 1976 (812 pages).

(2) R. ANDURAND, Le rapport de sûreté et son application dans l'industrie. *Annales des Mines*, no 7-8, juillet-août 1979, pp. 115-138.

(3) I. BURTON, R. W. KATES and G. F. WHITE, *The environment as hazard*. Oxford University Press, New-York, 1978 (240 pages).

(4) Centenaire de la Munich Rück: 1880-1980. *Münchener Rückversicherungs Gesellschaft*, Munich, 1980 (102 pages).

(5) T. BENJAMIN, Une étude dans quinze pays de quelques facteurs déterminant le nombre et la gravité des accidents de la route. Première partie: données statistiques de base International Drivers' Behaviour Research Association, Courbevoie, France, 1977 (45 pages).

(6) Statistiques nationales d'accidents du travail. Années 1976-1977-1978. Caisse Nationale de l'Assurance Maladie des Travailleurs Salariés, juillet 1980.

(7) *Le Monde*, 13 février 1974; 14 février 1974

(8) *Le Monde*, 14 juillet 1977; 24-25 juillet 1977.

(9) L'eau en Loire-Bretagne: spécial Villerest. Bulletin trimestriel du Comité et de l'Agence de Bassin Loire-Bretagne, no 11, mars-avril 1976.

(10) P. LAGADEC, Protection et mise en valeur de la coulée verte ligérienne Mécanismes de solidarité pour la prévention et la mise en valeur des grands espaces verts péri-urbains et coupures vertes d'intérêt régional, Vol. 2. Laboratoire d'Econométrie, Ecole Polytechnique et Mission de l'Environnement Rural et Urbain, 1976 (94 pages).

(11) P. LAGADEC, Eléments pour une prospective de l'espace naturel forestier méditerranéen. Centre International de Recherche sur l'Environnement et le Développement, juillet 1977.

(12) F. NEGRIER, L'assurance face aux risques catastrophiques. *L'Argus International*, no 13, juillet-aout 1979, pp. 254-274

(13) F. GIGON, *Le 400è chat ou les pollués de Minamata*. R. Laffont, Paris 1975.

(14) *Le Monde*, 29-30 décembre 1974 (Jugement: *Le Monde*, 22 juin 1978).

(15) V. BIGNELL, Ch. PYM, G. PETERS, *Catastrophic failures*. The Open University Press, Faculty of Technology, 1977.

(16) F. KENNET, *Les grandes catastrophes du XXè siècle*. Editions Princesse, Paris, 1976 (152 pages).

(17) *Le Monde*, 28 février — 1er mars 1965. (Jugements: *Le Monde*, 27 mars 1965; appel: *Le Monde*, 13 novembre 1965).

(18) A. GOUBET, Risques associés aux barrages. Note du Comité technique permanent des barrages, juin 1979 (26 pages).

(19) P. LAGADEC, Développement, environnement et politique vis-à-vis du risque: le cas britannique, tome 3: cinq catastrophes. Laboratoire d'Econométrie, Ecole Polytechnique, avril 1978 (113 pages).

(20) *Le Monde*, 29 mars 1977.

(21) *Le Monde*, 27-28 avril 1980.

(22) *Le Monde*, 27 mars 1979.

(23) *Le Monde*, 30-31 mars 1980; *Le Monde*, 1er avril 1980.

(24) *Le Monde*, 8 juillet 1980.

(25) A review of catastrophic property damage losses. M. and M. Protection Consultants — A technical Service of Marsh and Lennan, March 1980 (10 pages).

(26) L. DAVIS, *Frozen fire. Where will it happen next?* Friends of the Earth, San Francisco, 1979 (298 pages).

(27) G. MARLIER, Les explosions de vapeurs de gaz liquéfiés (BLEVE). ELF-France — Groupe d'Etudes de Sécurité de l'Industrie Pétrolière. Note du 7 mai 1980 (8 pages).

(28) Extraits des Minutes du Secrétariat-Greffe de la Cour d'Appel de Grenoble, no 220/71, jeudi 25 mars 1971 (68 pages).

(29) Advisory Committee on Major Hazards: Second Report. Health and Safety Commission, Her Majesty's Stationery Office (H.M.S.O.), London 1980.

(30) R. A. STREHLOW and W. E. BAKER, *The Characterisation and Evaluation of Accidental Explosions*. N.A.S.A., National Technical Information Service, Springfield, U.S.A., 1975.

(31) Rapport du Ministère de la Santé et Rapport de l'Inspection Provinciale du Travail. Annexe au Rapport Parlementaire d'Enquête sur la catastrophe de Seveso: Camera dei Deputati VII Legislatura. Commissione Parlamentare di inchiesta sulla fuga di sostanze tossi che avvenuta il 10 Luglio 1976 rello stabilimento I.C.M.E.S.A. e sui rischi potenziali per la salute et per l'embiente derivanti da attivita'industriali (Legge 16 Giugno 1977, n. 357). Juillet 1978 (470 pages).

(32) Fiches toxicologiques. Recueil des fiches parues dans les cahiers de notes documentaires Institut National de Recherche et de Sécurité (I.N.R.S.).

(33) *Le Monde*, 16-17 juillet 1978.

(34) *Le Monde*, 13 novembre 1979.

(35) *Le Monde*, 14 novembre 1979.

(36) *Le Monde*, 17 janvier 1979.

(37) D. BURGESS, Detonation of a flammable cloud following a propane pipeline break: December 9, 1970, explosion in Port Hudson. United States Department of the Interior, U.S. Buro of Mines, 1973 (26 pages).

(38) *Le Monde*, 4 novembre 1978.

(39) *Le Monde*, 24 avril 1980.

(40) *Le Monde*, 25 avril 1980.

(41) Rapport de la Commission d'Enquête du Sénat, présénte par A. Colin Seconde session ordinaire 1977-1978, juin 1978, no 486 (289 pages).

(42) L'accident du Tanio. Centre de Documentation de Recherche et d'Expérimentation sur les Pollutions accidentelles des Eaux (C.E.D.R.E.), juin 1980 (29 pages).

(43) *Le Quotidien de Paris*, 14 mars 1980.

(44) P. EDDY, E. POTTER and B. PAGE, *Destination désastre*. Grasset, Paris 1976.

(45) *Flight International*, 23 janvier 1975.

III. THE THREATS PRESENTED BY COMPLEX TECHNOLOGICAL
SYSTEMS AND THE LARGE SCALE INDUSTRIAL
CONCENTRATION. ON THE THRESHOLD OF THE TWENTY-
FIRST CENTURY

We shall not dwell here on the heritage to be saved in matters of risk and particularly industrial risk. Many tendencies of the past persist often countered by more and more adapted measures or sometimes, on the contrary, reinforced by other factors such as the ageing of installations. The following remark by a British expert may suffice to stress the still very real fact of major technological risk:

Based on past events I suggest the following as being a typical disaster in 1979:

- Frequency: One a month somewhere in the world.

- Circumstances: Loss of retention in a storage tank of a hydrocarbon processing plant.

- Consequences: explosions followed by conflagration.

- Implicated products: Hydrocarbons in C3 or C4 (1 case out of 3).

- Quantity released in the form of a cloud: 18 tonnes.

- Radius of cloud: 80 metres.

- TNT equivalent: 7 tonnes.

- Deaths: 11

- Damage: 14 million US dollars.

Naturally, many future incidents will have less serious consequences but on the other hand, from time to time, perhaps once every five years, there will be a Cleveland, a Ludwigshafen, a Feyzin, a Flixborough or a San Carlos (Los Alfaques) (1, p. 13).

These multiple accidents which regularly attract attention must not impede our sight of other changes that are happening. Beyond the big industrial accidents, of which we have already had multiple examples, there are much more serious menaces shaping up. True, they have not yet really shown themselves but as we shall see the cases presented in the first chapter are already signs of their reality. It is a matter of better understanding the warning which the incidents at Seveso, Harrisburg and the *Amoco Cadiz* constituted ... . Previous developments have permitted the demarcation of the general universe of technological and industrial risk. Now we must bring the essential into focus: the menaces typical of the technico-scientific developments of the late twentieth century.

1. BEYOND PAST EVENTS: MENACES TO BE STUDIED

"More fear than hurt", "pollution of the minds" ... The commonly encountered remarks in connection with disasters could no doubt be taken up strongly and apparently convincingly if one were satisfied with the hasty accounts drawn up after the events recounted in chapter one. Nuclear risk? There were no deaths, not even at Harrisburg. Chemical accidents? The worst accidental spillage of chlorine — at Baton Rouge: 90 tonnes — did not claim a single victim; the most serious release of toxic substance, the dioxin at Seveso, has caused hardly any victims if one heeds the official figures.

With courage and lucidity the British government and its administration
have not followed this tempting propensity which is so facile and so
dangerous in the long term. Some short extracts from official texts deserve
to be shown respect for their rarity as regards form and depth as well as for
their pertinence.

In the course of this century there has been a major acceleration in the
growth of new industries in this country and in the rest of the world. New
and complex technologies are constantly being introduced and dangerous substances are being handled at an unprecedented scale. The new technologies
have brought great economic advantages but also new risks for the workers
and for the public in general (...). We therefore have to take the responsibility of evaluating where and in what measure these risks exist. Past
experience of industrial accidents is no guide on which one could rely in
order to know what might happen in the future. The fact that major accidents
have not occurred in the past does not mean that they could not occur in the
future (2, p. 25).

A study of the past cannot on its own serve as a means of predicting what
can happen in different circumstances (3, p. 10).

It is very difficult to draw conclusions from history in times of rapid
change. Normally mankind learns from disasters, and this learning process
reduces the repetition of disasters (...). However, the introduction of new
processes and the scale on which the existing processes have been developed
create new problems to which history can have only limited and inadequate
answers (3, p. 17).

By the same token, the Americans have permitted the asking of outright
questions concerning possible accident scenarios. This is one of the issues
of the enquiry commission under J. Kemeny which had to analyse the TMI
accident "What if?" (4, p. 15). What else could have happened, the enquirers
asked who ordered a special technical study of this question (5).

In this particular case reflection has led to the discovery of seventeen
accident scenarios other than the one which actually occurred in March 1979
(5).

In two cases the final situation was less serious than the one experienced;
in eleven other configurations the results were comparable to the ones
experienced; in four scenarios the consequences would have been heavier.
The commission does not stop here. Since there could have been partial
fusion of the core a study of a core fusion and its consequences was carried
out. One came to the conclusion that because of the excellent quality of the
rock basis on which the centre is built, on account of the special thickness
of the walls of the main building in this centre a fusion of the core would
not have had very serious consequences for the environment. The commission
pushed research still further with remarkable honesty: it stressed the
limited nature of its investigation. There could have been in addition: an
operating error subsequent to a fusion; a particular vulnerability in the
electric wiring or the plumbing that passes through the confinement wall; a
less robust centre than the one where the accident occurred could have been
involved. The commission stresses: only some scenarios have been examined;
the situations of core fusions are not well known; more scenarios than the
ones examined could have occurred (5; 4, p. 15).

We must emphasise again the role of the reflex reaction that what did not
happen must not cause worry "lest one falls victim to catastrophism". This

reaction is rejected as we shall see by the British and American authorities.
Councils of safety experts, already quoted, hold the same opinion. In their
study of chemical accidents during the last thirty years these authors add a
series of stresses as warnings to their statistical tables; their very first
observation is this: the history of accidents over the past thirty years is
no absolute indicator of the type and the scale of accidents that may occur
in the future (6, p. 1).

It must be understood, as Harold Lewis stresses, not without some provo-
cation, that this is not a question of succumbing to an apocalyptic vertigo.
It is particularly important to consider, as we shall do later on the
precautions taken, the probability of events considered, in approaching risk
as a product of scale and probability, as Harold Lewis also stresses, not
without reason, after the case of TMI:

> Those among us who are considered to be knowledgeable on this subject
> frequently find their friends asking them the following question: so we
> have been lucky that all went well but, tell me, what is the worst that
> could have happened? I usually answer by saying that the worst could have
> been a chain reaction of other accidents such as a general breakdown of
> the electricity supply, an earthquake, a conflagration or any other
> phenomenon that might have caused the fusion of the core of the reactor.
> The latter could have reacted on the water in the tank or in the confine-
> ment wall causing a steam explosion that could have blown the roof off the
> confinement wall. The radioactivity could then have escaped into the
> atmosphere just at the time when a typhoon swept the area. In it's mad
> course across the country the deadly cloud would then have given off above
> each big city the exact quantity of product necessary to annihilate the
> population. Nothing is impossible.

> My friends then get furious but I keep telling them that my response is not
> a subterfuge. It shows that the question of knowing what could have
> happened makes sense only when it is accompanied by an estimation of the
> probability of the event. In other words, it is essential that these
> safety problems as well as others, be treated by choosing as one's
> objective a quantitative evaluation of the risks. One can ask about the
> consequences of an event that is likely to happen once in a hundred years,
> once in a thousand years or in a million years but it does not make sense
> simply to ask what could have happened (7, p. 89).

The remarks made by H. Lewis must be taken seriously which we shall not
fail to do. They do not forbid stopping for a while at those 'possibilities'
which in the past were often hardly known, sometimes claimed to be
impossibilities. There are essentially two terms of reference: the consequences of the accident and it's probability. The end result has been a
bit neglected in the past. Today it is important to consider it as a whole,
without limiting oneself to the second term of reference.

The scale of accidents that are likely to happen is such that one needs to
have a better knowledge of the phenomena that come into play in order to be
able, among other things, to define policies and in case of a favourable
option to emphasise the requirements in terms of probability of occurrence.*

---

*As F. R. Farmer suggested in 1967: one can draw a curve in a plan expressing
the functions of consequences and probability which separates the acceptable
from the unacceptable.

Fig. 18: Drawing by Konk.
(Source: *Le Monde*, March 30/31, 1980)

Generally speaking, it will be difficult to establish policies for industrial development without this knowledge of potential accident phenomena:
their severity, their overall impact on the community affected in its
organisation, its reproduction, its living and working conditions, its
resources etc.

Before going into this examination it is worth emphasising the difficulty
of the attempt because of the great uncertainty in this field. A specialist
in the chemical field points out:

The attainable precision when one tries to circumscribe the risk through
theoretical analysis remains always limited: the most experienced authors
themselves estimate that for a spread of reliability of 90 per cent the
results vary with:

- a factor of ten in each direction for frequency;

- a factor of between two and fifty per cent for consequences (66, p. 6/449).

2. VAST FIELDS OF STUDY TO BE COVERED

Research is still little developed in matters of great technological and
industrial risks, we shall only bring up a certain number of points here which
deserve reflection. It seems that three large areas can be distinguished:

1st. Threats related to energy

a) Nuclear energy. If one undertakes to go beyond the limits within which most people often restrain their thinking ("a centre isn't a bomb"; "it is a statistically established fact that the nuclear industry is the safest of all industries" (9, pp. 7-9)) one is quickly faced with non-homogeneous studies* as concerns theoretically possible accidents.

From among these studies we quote the one by F. R. Farmer**which envisages the case of a severe accident in a centre located in an average populated area (a town of several thousand inhabitants a few kilometres from the reactor; a million people within a radius of 30-50 km; a rural population with a density of 150-200 per 100 km^2) and occurring in 'average' weather conditions (this could be more or less serious with a factor of 10).

The release envisaged would be about 10 per cent of the gaseous and volatile fission products contained in the core of which 5 million curies is iodine 131 and 0.6 million curies is caesium 137. Within a radius of 90 km one may estimate the number of thyroid cancers appearing within twenty years at 1,000. The irradiation of the population due to the radioactive cloud and to the inhalation of caesium 137 would cause fifty cases of leukaemia and a similar number of lung cancers would be induced by inhalation of ruthenium. The level of irradiation due to contamination of the soil (mainly on account of the caesium) would lead to the evacuation of the area at the time of the accident. The sector around the reactor within a radius of 15 km could not be reoccupied for at least a year. The contamination of grazing land due to iodine 131 would require control of milk consumption up to a distance of 150 km in the direction in which the wind had been blowing (10, pp. 233-234).

Other studies give different figures: between some ten and several hundred thousand dead if one believes what the GSIEN (Groupe Scientifique pour l'information sur l'Energie Nucleaire) reports:

The Americans, eighteen years before working out the WASH 1.400 report (called the "Rasmussen Report") had published an equally official but much more pessimistic report, the WASH 740; for a 1,000 MW PWR reactor this document estimated the number of victims in case of release of half of the gaseous fission products from the core of the reactor at 40,000, this on the assumption of a maximal accident such as the total breakdown of the reactor's cooling system.

Nowadays the British give estimates which are hardly more reassuring for the release of 10 per cent of fission products from a 1,000 MW reactor: between 11 and 3,500 deaths, 'depending on conditions' (location of the centre, wind etc.). The margin of uncertainty is considerable. The same source (the 6th report by the Royal Commission on Environmental Pollution, published in September 1976) indicates that for a fast breeder of 1,000 MW there would be between ten and one hundred times more victims, i.e. between 110 and 350,000 (9, p. 10).

---

*These studies are also relatively few in number because they deal with accidents 'beyond assessment' (the term will be explained in the chapter on 'Utilisation of Science'). They come more often from safety authorities than from industry, the latter not taking into account major accidents of this type of theoretical possibility in their designs and installations.

**F. R. Farmer was in charge of nuclear safety at the U.K. Atomic Energy Authority.

Distance band (km) $x_1 - x_2$	Population		2.5% release				5% release				7.5% release				10% release			
			Mortalities		Morbidities		Mortalities		Morbidities		Mortalities		Morbidities		Mortalities		Morbidities	
	No. in band	Total to x	No. in band	Total to x	No. in band	Total to x	No. in band	Total to x	No. in band	Total to x	No. in band	Total to x	No. in band	Total to x	No. in band	Total to x	No. in band	Total to x
Remote site																		
0 – 2.5	2.1×10^2	2.1×10^2	3.9×10^{-1}	3.9×10^{-1}	2.0×10^1	2.0×10^1	2.4×10^0	2.4×10^0	1.8×10^2	1.8×10^2	6.6×10^0	6.6×10^0	2.0×10^2	2.0×10^2	2.6×10^1	2.6×10^1	1.8×10^2	1.8×10^2
2.5 – 5.5	7.4×10^2	9.5×10^2	0	3.9×10^{-1}	0	2.0×10^1	1.6×10^0	4.0×10^0	1.1×10^2	2.9×10^2	5.2×10^0	1.2×10^1	4.1×10^2	6.1×10^2	1.1×10^1	3.6×10^1	6.3×10^2	8.1×10^2
5.5 – 6.5	2.5×10^2	1.2×10^3	0	3.9×10^{-1}	0	2.0×10^1	0	4.0×10^0	0	2.9×10^2	4.5×10^{-1}	1.2×10^1	1.0×10^1	6.2×10^2	9.1×10^{-1}	3.7×10^1	7.3×10^1	8.8×10^2
6.5 – 10.5	1.4×10^3	2.6×10^3	0	3.9×10^{-1}	0	2.0×10^1	0	4.0×10^0	0	2.9×10^2	0	1.2×10^1	0	6.2×10^2	4.2×10^{-1}	3.8×10^1	1.8×10^2	9.0×10^2
Total		2.6×10^3		3.9×10^{-1}		2.0×10^1		4.0×10^0		2.9×10^2		1.2×10^1		6.2×10^2		3.8×10^1		9.0×10^2
Semi-urban site																		
0 – 2.5	1.0×10^3	1.0×10^3	2.3×10^0	2.3×10^0	1.1×10^2	1.1×10^2	1.3×10^1	1.3×10^1	9.3×10^2	9.3×10^2	3.4×10^1	3.4×10^1	9.7×10^2	9.7×10^2	1.3×10^2	1.3×10^2	8.7×10^2	8.7×10^2
2.5 – 5.5	1.9×10^4	2.0×10^4	0	2.3×10^0	0	1.1×10^2	3.4×10^1	4.7×10^1	2.4×10^3	3.4×10^3	1.2×10^2	1.6×10^2	9.6×10^3	1.1×10^4	2.6×10^2	3.9×10^2	1.5×10^4	1.6×10^4
5.5 – 6.5	1.6×10^4	3.6×10^4	0	2.3×10^0	0	1.1×10^2	0	4.7×10^1	0	3.4×10^3	2.9×10^1	1.9×10^2	6.6×10^2	1.1×10^4	5.8×10^1	4.4×10^2	4.7×10^3	2.0×10^4
6.5 – 10.5	3.2×10^4	6.8×10^4	0	2.3×10^0	0	1.1×10^2	0	4.7×10^1	0	3.4×10^3	0	1.9×10^2	0	1.1×10^4	4.6×10^1	4.9×10^2	2.0×10^3	2.2×10^4
Total		6.0×10^4		2.3×10^0		1.1×10^2		4.7×10^1		3.4×10^3		1.9×10^2		1.1×10^4		4.9×10^2		2.2×10^4

Table 10: Number of immediate deaths and deferred deaths following the dispersal of different fractions of radioactive products from the core. (Source: 11, p. 66)

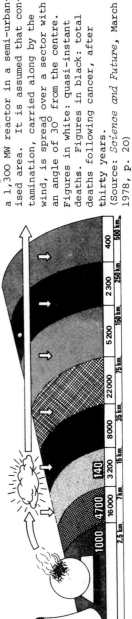

Fig. 19: Accident involving a sudden escape of 10 per cent of the fuel of a 1,300 MW reactor in a semi-urbanised area. It is assumed that contamination, carried along by the wind, is spread over a sector with an angle of 30° from the centre. Figures in white: quasi-instant deaths. Figures in black: total deaths following cancer, after thirty years. (Source: *Science and Future*, March 1978, p. 20)

On the fast breeder type of reactor the National Protection Board, the official radiation protection advisory organisation in GB, has published (in August 1977) an estimate of the radiological consequences of accidental escapes from a fast breeder reactor of 1,000 MW, averaging assumptions on the quantity of fuel released into the atmosphere, on population dispersion and on climatic conditions. The results are given above (11).

This type of study must look not only at production centres. It is necessary to know the consequences of maximum accidents likely to occur in research centres (which are not always located in the countryside), in transport*, not forgetting space flights for instance (propelling engines employing radioactive systems; three crashes are already known: *Cosmos* 954 in northern Canada in January 1978, an American satellite in the Indian ocean in 1974, the lunar module *Apollo 13* in the Pacific in 1970) or military activities (accident at Palomares in January 1966).

In France there are no such studies, at least not officially. It does not look as if there were at present precise documents on the maximum possible risk presented by centres such as Le Pellerin, Nogent sur Seine, Bugey, Creys-Malville or by a treatment centre like the one at La Hague. The answer given by M. de Torquat, the chief of SCSIN (Service Central de la Surete des Installations Nucleaires) to the question on the maximum possible accident, a question asked after the TMI accident in the context of the France-Inter programme "The telephone rings", shows clearly the insufficiencies in this field:

Let us take quite simply the case of a meteorite falling on and pulverising a centre. What is the maximum that can happen? Even if this type of exercise is sometimes a bit exaggerated, as I have just said, one comes to conclusions which show that hazard ranges of 100 km are absolutely out of proportion to the problem, i.e. the radiological consequences of the most severe accidents that can occur, and the case in Pennsylvania that we are talking about is actually not a major case of this type, but the most severe radiological consequences in the most severe cases would certainly not reach beyond five or at the most ten kilometres and that consequently the centre of Nogent 100 km from Paris, it is about 100 km from the centre of Paris, can in no way, whatever the circumstances, have consequences for the Parisians (France-Inter, March 30, 1970, 19.20 h).

The conclusions seem therefore difficult to establish. Against these declarations by the safety authorities one may compare these lines from B. Laponche who, before examining the means provided for the avoidance of such disasters, points out the contours of the potential danger:

The ordinary water reactor of a 1,000 MW nuclear centre (the chain of PWRs built in France includes reactors of 900 MW and 1,300 MW) contains 90 tonnes of fuel. When the reactor is in operation the total radioactivity of this fuel, especially on account of its fission products, is in the order of 15 billion curies. This figure is enormous, considering that the quantities of radiation which are fatal to humans are figured in thousandths

---

*As noted for instance in the report by the French National Assembly on the *Amoco Cadiz* accident: The Commission has never lost sight of (...) the particularly serious dangers which are caused by maritime transport of dangerous substances other than hydrocarbons, notably not only chemical products but certainly also radioactive substances. The delay that has occurred on these two fronts requires, on the contrary, increased watchfulness.

of curies. A serious accident causing several thousand cases of cancer
and genetic anomalies within the area where the centre is located (an area
with average population density, without a big city) would correspond to
the release of several million curies into the environment, i.e. some 10
per cent only of the gaseous fission products contained in the reactor.
If the total of gaseous fission products and a part of the plutonium and
other fission products in the reactor were released, this would mean the
poisoning of an entire region, particularly by means of watercourses,
agricultural products and ground water. The number of victims would
obviously depend on the location of the centre (near to or far from a large
city), the rapidity of the accident, evacuation possibilities etc. These
different factors play an essential part in the evaluation of the possible
consequences of a serious accident. There exists therefore a considerable
potential danger (...) (12).

b) Liquified gas. The storage and the transport of liquified gas (particularly if done in tanks under pressure, but the menace is not negligible even
when done in cryogenic tanks) present potential risks (13-21).

In France there are already more than forty refrigerated hydrocarbon
storages the unit capacity of which can reach tens of thousands of cubic
metres and even more*.

The tendency is towards the increase of these storage capacities because of
the still increasing demand for energy and the new resources of butane and
propane which will arise from the recovery of gas in oil fields. The increase in the consumption of imported liquified hydrocarbons must be about
4-5 per cent per year. For the next ten years the total storage capacity
(already above 630,00 m^3) is expected to double. Nearly all this storage
will be located on already existing sites: Berre, Carling, Dunkirk, Fos sur
Mer, Gonfreville-l'Orcher, Lavera, Montoire de Bretagne, Port Jerome.

These developments will increase the traffic of methane ships which also
present considerable risks.

What are the risks to be considered? Restricting ourselves to the essen-
tials the following elements can be set forth (knowing that there are still
many unknown factors).

Subsequent to the rupture of a tank one discovers the spillage of liquid
on the soil (or on the sea in the case of a ship), its partial vapourisation
(more pronounced on water than on land), the formation of a layer of gas and
gas derivates. The danger areas can rapidly (within a quarter of an hour)
reach considerable dimensions (at least 1-2 km) if the conditions for
dispersal are bad.

Three types of serious events may occur if there is no dispersal and if the
circumstances permit the reaction of the air-gas mixture to a heat source or
under the effect of an explosion.

Deflagration - The ignition of a slick or cloud upon contact with a heat

---

*Examples: Natural gas storage of 80,000 m^3 at Fos sur Mer; ethylene storage
of 13,250 m^3 at Carling, of 17,500 m^3 at Dunkirk; propylene storage of
16,500 m^3 at Dunkirk; propane storage at 35,000 m^3 at Gonfreville-l'Orcher;
two tanks of 120,000 m^3 unit capacity are under construction at Montoire de
Bretagne etc.

source may spread in the form of a deflagration: the fire front may move at a speed of some tens of metres per second; a pressure wave of several hundred millibars may accompany this phenomenon. This was the case at Cleveland (1944), the example already quoted in which small tanks were involved (about 4,000 m^3).

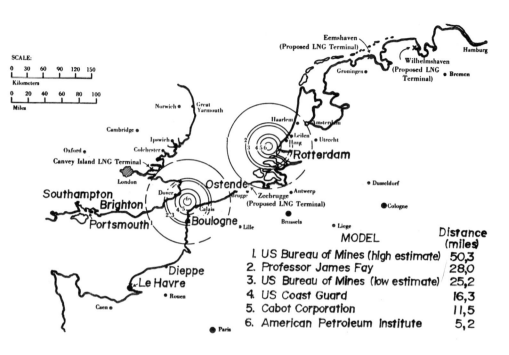

MODEL	Distance (miles)
1. US Bureau of Mines (high estimate)	50,3
2. Professor James Fay	28,0
3. US Bureau of Mines (low estimate)	25,2
4. US Coast Guard	16,3
5. Cabot Corporation	11,5
6. American Petroleum Institute	5,2

Fig. 20:   Uncertainties about the risks presented by slicks
of gas derivatives: the North Sea.
(Source: 13; p. 181)

<u>Detonation</u> - This event is held to be much more unlikely to occur and even impossible with certain gases. In this case there is a spread of a wave of several bars of pressure at supersonic speed (Mach 6-8); the temperature of the flame is well above the one in the preceding case (here, temperatures approaching 2,900°C could be reached). With this phenomenon the TNT equivalents used in some instances are of some importance: in some studies, as in the case of a detonation, one kilogram of hydrocarbon is considered to correspond to ten kilograms of TNT. The major reference for detonation is the case of Port Hudson (1970) which was already mentioned; there are few other cases on record.

<u>Ignition at great speed</u> - If ignition occurs immediately and large quantities of gas are involved, a 'fireball' can develop which spreads considerably and with extremely strong heat radiation. If one follows certain British experts, ignition of materials may occur up to a distance of several kilometres (5 to 8). A small accident of this kind is the one at Los Alfaques (there were

only 18 tonnes of propylene involved). The intensity of the phenomenon produced can finally cause 'fire storms' (there has been one such case, though from other causes, in the example at Tokyo in 1923).

The large number of uncertainties which exist about these phenomena are still stressed. Many factors can be reassuring: negative test results, theoretical calculations, precautionary measures taken, encouraging statistics. However they cannot exclude the possibility that there may have been too optimistic an interpretation in this study, that as yet unknown phenomena may show up etc. It seems that there are not many studies of the theoretically possible maximum risks. A study by the US Federal Power Commission on the eventuality of a methane accident in the port of New York-Staten Island showed that 800,000 people were in the danger zone, that 42,000 people might be killed or severely burned (13, p. 18).

As concerns shipping accidents there is one grave precedent case:

The conflagration of the gas tanker Yugo Maru in the bay of Tokyo 1974 - On November 9, 1974 the *Yugo Maru* with a capacity of 80,000 m^3 of liquified gas was in collision with a Liberian freighter *Pacific Ares*. This happened in the bay of Tokyo while the Japanese ship had an escort vessel that was equipped for fighting fires of chemical origin. The *Yugo Maru* carried propane, butane and naphtha. The collision caused an escape of naphtha which resulted in flames 60 m high and extreme heat. The chemical foam proved ineffective. Attempts were made to avoid at least the ignition of the butane and propane tanks. However, a petrol tank exploded; the flames reached a height of 600 m. This set the gas which escaped from the valves of the other tanks on fire. The two ships drifted towards the towns of Yokosuka and Yokohama: one had to wait for a reduction of the heat radiation before tugging them to safer places.

Five dead were counted on the *Yugo Maru*, twenty eight on the *Pacific Ares*: there was only one survivor on this ship, the mechanic who held out for fifteen hours in the machine room with his gas mask.

The *Yugo Maru* was towed to the centre of the bay of Tokyo and was left there to burn out. Ten days later it was still burning and the danger of an explosion was still great. The authorities decided to have it towed out to the high sea. The wind and the waves aggravated the accident; there were explosions. One could not tow it out 90 km as was necessary; it had to be abandoned at half the distance as the ship became too dangerous.

Still on fire, the *Yugo Maru* drifted into the navigation lanes that lead to the bay of Tokyo so that on November 27/28 the Japanese Defence Ministry decided to sink it. Attacked with a bomb, a torpedo and with rockets the ship went down, still on fire. Left to itself it would have burned for another five months (13, pp. 79-81).

c) Offshore oil operations. The risks presented by fossil energy operations off the coasts were demonstrated in three notorious cases (which are not the only ones):

Santa Barbara 1969. On June 28, 1969 the well at Santa Barbara, California, erupted. The quantity of oil spilled was estimated at 15,000 tonnes, and 65 km of coastline were polluted (22, p. 30).

Ekofisk 1977. Between April 22 and 30, 1977 an incident on the Ekofisk platform let 12,700 tonnes of petrol escape into the sea. The sealing of the accident well could be rapidly accomplished (23).

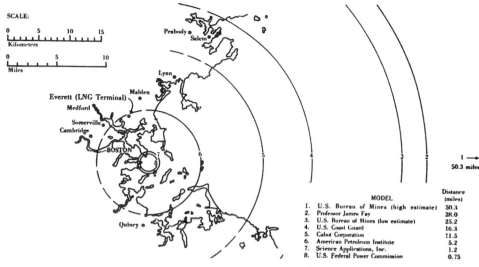

Fig. 21:   Uncertainties presented by slicks of gas
derivatives: the port of Boston.
(Source: 13, p. 178)

The distances indicated correspond to the supposed distances at which inflammable
slicks are still found (concentration in the air: 5 per cent subsequent to the
rapid vapourisation of 25,000 m^3 LNG on the water.

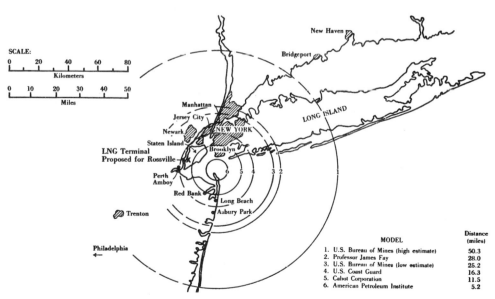

Fig. 22:   Uncertainties about the risk presented by slicks of gas
derivatives: the port of New York.
(Source: 13, p. 179)

Ixtoc 1, 1979-1980. On June 3, 1979 the Ixtoc 1 well in the Gulf of Mexico
saw an accident that caused one of the worst ecological disasters produced
by man in the maritime environment.

An official Mexican report gives the following explanatory details:

1st: The loss of control of the well is due to the fact that while drilling
at a depth of 3,624 m a 'cavern' was found unexpectedly and by chance, a
cavern of undetermined size, which in the beginning caused a loss of re-
circulation of the liquids used to create a hydrostatic column and for transporting rock elements cut by the drilling tool, liquids known under the name
of 'drilling mud'; this cavity was the cause of the fact that neither the
company contracted by PEMEX*nor the geologist on board got any indication at
the surface that would have pointed to the presence of hydrocarbons at depths
between 3,620 m and 3,627, otherwise they would have made the necessary
arrangements.

2nd: The other situation that bore directly on the accident was that when
the drill entered the cavern it had an approximate weight of 105.4 tonnes
and its penetration into the cavern was one metre. This resulted in the
potential energy of the shaft converting itself into kinetic energy which was
capable of and sufficient to fracture the stratum of paleocene where the oil-
carrying layer was. Subsequently, the great internal pressure in this layer
began to act in such a way as to push back, after a short while, the areas
of lower pressure and to overcome the weak hydrostatic column of mud.

As a result of this, when the drill was withdrawn and replaced by more mud
and the drill had still to be raised another 200 metres the internal pressure
of the well was higher and created an implosion of the mud and the ejection
of the drill and finally the release of oil and gas.

3rd: It is highly probable that upon the exit of a flux of petrol, gas,
mud and stones under strong pressure in the centre of the derrick, fire
breaks out because of a spark created in the upper part of the derrick. As
the gas and the oil emerged the conflagration was fast and general (24, p. 5).

The eruption was at 03.00 h; at 03.15 h an explosion followed the conflagration; the platform had to be abandoned. One could not get to the bottom
of the accident until nine and a half months later, on March 22, 1980.
Between these two dates more than half a million tonnes of hydrocarbon were
spilled into the sea.

At the height of the accident about 4,000 tonnes of petrol leaked daily
from the accident well. This yield was subsequently reduced: in February
1980 PEMEX confirmed that the flow of oil had been reduced considerably to
one sixtieth of a tonne per day (from the American side, however, a number of
experts estimated the daily flow at 450 tonnes). Legal proceedings have now
started (a large number of victims are claiming more than 400 million dollars
in damages plus interest) as well as scientific studies: the researchers from
the environment protection agencies and from the American agency for the
oceans and the atmosphere are asking the federal government for seven million
dollars so as to be able to follow up the impact of the eruption on the
environment. This ecological follow-up and the general environmental studies
seem very desirable as the knowledge of these phenomena appears to be weak if
one believes the scientific coordinator of the American agency for the ocean

---

*PEMEX: Petroleos Mexicanos, Mexican national well operations company.

and the atmosphere (NOAA): We are at present unable to answer most of the questions and particularly the essential question: what has really happened to this mass of hydrocarbon which has been moving about the Gulf of Mexico for approximately nine months? (25, 26, 27).

These events indicate the types of threats that may be raised by a petroleum extracting industry even if it is very conscientious about safety. The previous prospecting permits granted in the Mediterranean for instance (Decree by the Ministry of Industry of December 6, 1979, *Journal officiel* of January 12, 1980) seen in this light can raise passions*.

2nd. Threats linked with the chemical industry

On the level of production and storage the two most serious problems are the possible escape of massive quantities of toxic gas (such as chlorine) and, still more serious, the dispersion of extremely toxic substances (even in minimal quantities, a few kilograms).

A large storage may have a major accident and the immediate consequences could be as serious as those of a nuclear accident (immediate effects). A Seveso on a larger scale may occur and this may mean the condemnation of an area, its inhabitants and their descendants for a very long time.

To the accidents involving production and storage of which Seveso is so far the most serious warning must be added those which may happen on the level of research, transport, trade, waste storage. Here too, some warnings have been reported.

a) Menaces connected with transport. They did not escape the attention of, for instance, the parliamentary enquiry commission which studied the *Amoco Cadiz* affair. The parliamentarians stressed: this type of pollution would be of quite a different nature from the one of the black tides even though the latter were already very serious.

---

*Dr Bombard may have reproached himself for his outcries but others like R. Latarjet have got excited about the same subject.

What is one to think of the search for and the exploitation of petrol in the deep sea? Already at minor depths in the order of 100 m where the whole operation up to the entry into the soil is accessible to inspection and to repairs the risks are large. The accident at Santa Barbara on the Californian coast and at Ekofisk in the North Sea have given serious warnings. But imagine drillings at 2,000 m depth in an area with non-negligible seismic risks like the ones foreseen in the Mediterranean. Suppose they unlock (as is hoped) even in a moderate stratum, say, 100 million tonnes. What precautions will be taken? Even a minor movement of the seabed would be sufficient to shear the pipe where it emerges and to open the stratum into the sea, at too great a depth for intervention. The stratum would empty itself. The hydrocarbon, being lighter than water, would rise to the surface and there spread a thin layer over an immensely large expanse, so vast that the bacteria which digest these hydrocarbons when the latter are not too abundant (as in the case of cleaning of ships and 'small' known accidents) would no doubt lose the battle and be annihilated. Here is not the place to describe what could happen. May it suffice to say: no isolated nuclear explosion could cause a comparable biological cataclysm (28, p. 19).

Le Cavtat 1974; Le Tagnari 1971. While oil pollution remains, quantitatively, the major risk to which coastal countries are exposed other transports, less known to the general public and characterised by the appearance of a special- ised fleet for the transport of dangerous substances, often in bulk, expose coastal regions to risks which we consider to be underestimated by the authorities.

Many chemical products are poisonous to living organisms and their trans- port in large quantities can present substantial risks for the marine environment (or for coasts, given gaseous emissions).

It is for this reason that the US Government has considered the transport of chlorine and hydrofluoric acid in bulk to present unacceptable risks.

On the other hand, certain products such as tetraethylic lead or compounds of mercury are not only extremely dangerous even in minimal does but are absorbed by marine organisms. They climb up the food chain and reach man as a consumer of fish, shellfish or mussels. Two recent accidents have caused ecological disaster on a large scale.

There was the sinking of the *Cavtat* near Fiume in the Adriatic in 1976 which carried 50 tonnes of tetraethylic lead in water-tight metal barrels. After a prolonged stay at the bottom of the sea these barrels would have deteriorated and their contents ... would have made the Adriatic a biologi-- cally dead sea.

It is true, and we congratulate ourselves on it, that the coming together of two circumstances has prevented such a disaster: the wreck was at little depth and the governments concerned, aware of the danger, carried out, at great expense, a recovery operation with divers. All barrels were recovered.

More serious were the consequences of the sinking of the ship *Tagnari* a few hundred metres off the Uruguayan coast in 1971. This freighter carried barrels of isothiocyanate of methyl, imino-ethane and 25 tonnes of mercury compounds as well as other toxic products. The metal containers were fissured following storms in April 1978*.

On account of the emanations, villages had to be moved inland. At the same time thousands of marine animals were swept on to the beaches following the absorption of contaminated organisms. The Uruguayan and Brasilian authorities did not draw up a definite balance sheet of this pollution. It is known that mercury acts long-term but is generally fatal as could be seen in the bay of Minamata where it caused the deaths of two hundred and thirty four people.

Even though today this fleet is numerically small it will develop rapidly because of the intensification of trade in energy and chemicals. Despite

---

*These particularly dangerous products caused the anarchical, rapid and massive multiplication of algae and polluted microorganisms (red tide) the ingestion of which caused the death of mussels, fish and later of human beings after sneezing and profuse nose bleeding. From there they exited into the interior of the earth (...). After the sinking of the ship which belonged to Lloyd Brasileiro the authorities did not consider it worthwhile refloating because the cost of the operation would have been higher than its value, cargo included.

precautions taken on the level of design, construction, equipment of ships as well as with regard to the packing of cargoes the transports of methane, fuels or radioactive waste and the most noxious chemical products compared to which the one for which our commission was set up would appear minimal (11, pp. 52-53).

The Channel 1971. A serious alert was raised in the Channel in December 1971. If one follows the report of *Le Monde* on the case one finds that a maritime accident led to the dispersal of the contents of various containers of dangerous products, among them (29)

- some tens of 200 litre barrels of di-ethylamine in watery solution*;

- 286 drums of sodium cyanide**;

- 4 barrels of equally toxic di-isocyanate of toluene***;

- containers of nitric acid;

- containers of ammonia.

The Gino 1980. The sinking of the *Gino* with her 40,000 tonnes of carbon black which contained more than 10 tonnes of benzopyrene (a carcinogenic compound) is also something quite different from a black tide. As a notice from the Institut Scientifique et Technique des Peches Maritimes of January 22, 1980 indicates there remains a latent and diffuse risk of contamination for a large area. A large mud hole, a critical ecological zone, could be affected (30, 31).

b) Measures connected with trade.

Some serious warnings must be remembered here:

Thalidomide 1959-1961. The thalidomide affair is one of the first disasters of this type since in 1961 the baleful genetic effects of this otherwise inoffensive drug were definitely established. At that time the product had already been widely used and the birth of 6,000 malformed children in twenty countries are attributed to it.

In England four hundred victims have received compensation and the distributors, the Distillers company, have paid compensation amounting to about 38 million dollars. In Germany the licence holder has paid a contribution of 1 million dollars for the compensation of victims.

---

*Di-ethylamine is an extremely inflammable liquid which can form explosive mixtures in the air. The di-ethylamine vapours irritate the mucosa of the eyes and the respiratory organs. The concentration threshold of the vapours of di-ethylamine in the air was set at 25 ppm = 75 mg/m^3 (toxicological file of INRS No. 114).

**Sodium cyanide can cause cyanohydric poisoning by inhalation, skin contact, accidental ingestion or in other ways. Peracute poisoning causes death within a few minutes. The concentration threshold in the air is 5 mg/m^3 (Toxicological file of INRS No. 11).

***The vapours of 2-4 di-isocyanate of toluene (which develop at room temperature) are very irritating to the eyes and the respiratory system. Threshold of concentration in the air: 0.02 ppm = 0.14 mg/m^3 (Toxicological file of INRS No. 46).

This affair has particularly alerted the insurance companies to the dangers of third party liability covers because it was the first time that such amounts had been awarded and, particularly in Great Britain the distributor of the product took recourse to certain Lloyds syndicates on policies going back to 1958, a time when nobody expected that damage awards could reach such proportions (32, p. 262).

Hexachlorophene (Talc Morhange) 1972. This case of poisoning by a talcum powder containing a high percentage of hexachlorophene has already been reported. It caused the deaths of thirty six babies and left one hundred and forty five handicapped (32, p. 262).

Enterovioform. Enterovioform, an inoffensive product in all countries where it has been used, proved to have disastrous effects in Japan where it caused 1,000 deaths and 30,000 disabled (32, p. 262).

Iraq 1971-72. In September 1971 a large number of people were poisoned by the consumption of goods containing mercury; 6,500 people were hospitalised and four hundred and fifty nine deaths were counted; another poisoning, again from mercury, occurred in the following year, claiming hundreds of victims (33, p. 735).

c) Menaces connected with waste. The question of waste has caused much attention following some recent events.

Herfa Neurode 1978. Public opinion in Germany was stirred up in 1978 by the 'chemical waste bin' found in the old mine of Herfa Neurode (FRG). Two hundred thousand tonnes of arsenic, cyanide and highly noxious insecticides which the industry did no know what to do with after the ban of these products in the United States had already been buried there. Even today the FRG sells (at a very high price, by the way) the opportunity to get rid of dangerous waste to some of its neighbours (34).

Lekkerberk 1980. On May 20, 1980, 900 people from this village, 15 km from Rotterdam had to be avacuated to permit the changing of extremely polluted soil on which their houses stood. These people were inconvenienced by chemical waste which had been deposited, ten years earlier, on land earmarked for construction. The waste had been buried secretly by a transport company, but the barrels which held the products began to crack. In spring 1980 it was thought that 100,000 m^3 of earth would have to be removed. The Dutch Interior Ministry had to foot the bill for the operation (the cleaning work took several months) and to compensate those who refused to go back to Lekkerberk (35).

Love Canal 1978-1980. The case of 'Love Canal' blew up with a bang in May 1980 when the spectre of Seveso rose behind some analysis results on people living in that part of Niagara Falls, USA. In fact the affair had been brewing for a long time as the following account shows which attracts attention to a quite new phenomenon:*the risk of a major accident by deferred effect (35, 36, 37, 38):

1947: The chemical company Hooker Chemicals and Plastics Corporation whose headquarters is in Houston bought the land in question for the purpose of dumping waste.

---

*Even if in that particular case the alarm was raised rather precipitately.

1952: The company had already buried 20,800 t of products.

1953: Hooker relinquished the land for one dollar to a school office with
       a clause renouncing all responsibility in case the buried products
       should prove a nuisance. The office built a school and sold the
       remainder of the land to a promoter who built individual houses for
       families with modest incomes.

1977: Torrential rains caused the canal to overflow periodically, bringing
       to the surface what experts finally identified as eighty two chemical
       compounds of industrial origin, of which eleven are known to be
       carcinogenic. Among these eleven products was dioxin.

1978: In August, after enquiries had established a standard rate of cancer
       among the inhabitants of Love Canal, the Health Commissioner of the
       state of New York recommended evacuation of all pregnant women and
       children under two years of age. In December, after the state of
       New York had refused to rehouse the most endangered families, the
       inhabitants organised protest demonstrations. There were sixteen
       arrests.

1979: A subcommission of the House of Representatives revealed that the
       Hooker company had known since June 1958 that there were escapes of
       toxic waste at Love Canal. In November a federal report indicated
       that cancer cases among the inhabitants of Love Canal had risen to
       one in ten. In December the Justice Department started a lawsuit
       for damages plus interest against the Hooker company (125 million
       dollars).

1980: In February the case took on national proportions. The Federal
       Agency for the Protection of the Environment (EPA) announced that
       four products suspected to be carcinogenic had been discovered in
       the air samples taken near the contaminated area.

In May (the 17th) the federal government informed the inhabitants of the
results of two enquiries carried out respectively by a Houston Laboratory
(Biogenics Corporation, this was an exploratory study destined to collect
preliminary information for the lawsuit intended by the Justice Department
against Hooker Chemicals) and by two physicians from Buffalo (B. Paigen from
the Roswell Park Memorial Institute and S. Barron from the State University
of New York). These results caused panic. The first report indicated that
chromosomal defects normally affecting one person in a hundred were found in
eleven out of thirty six people examined; the second one established nervous
damages among the thirty six people examined.

In May (the 21st) President Carter declared a state of emergency for Love
Canal and ordered the evacuation of seven hundred and ten families (2,500
people) who still lived there. The evacuations were carried out at the
expense of the government. The quasi-certainty of no longer being able to
sell one's house, the worry about health, the uncertainty about possible
effects on procreation raised fear among the inhabitants. The fact that
this affair was not an isolated one also conferred on it national significance:

The affair of Love Canal is felt so acutely because several other cases of
waste from the chemical industry have come up in recent months. The most
serious one was the conflagration of 34,000 barrels of chemical waste last
April which burned for 48 hours at Elizabeth, New Jersey. They had been
dumped illegally years earlier and their presence was made known on several
occasions by residents (36).

In addition a federal report was published at the time indicating that a very large number of dumps of dangerous chemicals existed in the United States. In 1978 the EPA had inspected fifty such dumps: in forty eight cases leakages were discovered which contaminated nearby watercourses.

In parallel with the development of 'the affair' the two studies were, however, revaluated with some difficulty (the author of the Houston study was unwilling to submit his basic data). On May 27 a study group set up by the EPA declared that the study on chromosomal damages was quite invalid, mainly because of the lack of a control group. A previous group, set up to check the same study had come to an identical conclusion on May 21: even for an exploratory study the Houston study was lacking the required scientific exactitude. The Buffalo study also suffered from this lack of a control group. It seems that, pressed by the Justice Department (in the course of the lawsuit) the EPA had not allowed its contracting parties the necessary time for a serious study.

Valid analyses of the health situation had to wait several months. However, harm had already been done: psychological shock. Cancers and malformations were now images imprinted on the minds, rightly or wrongly. Uncertainty had arisen. The reactions observed in Vietnam veterans who were affected by the dispersal of 'orange agent' (with strong dioxin concentrations): depression, irritation, fatigue, nausea, vertigo ... The people resented that no provision had been made for this kind of situation in the political and institutional system. The men, more than the women, seemed affected: they could no longer protect their families and their possessions*: whilst they were otherwise scrupulously law-abiding: it seemed that it was the women who took the steps.

This had repercussions on marriages: from two hundred and thirty seven families who were evacuated in 1978 40 per cent separations and divorces were reported. Often the woman wanted to leave and the husband wanted to stay put where his identity was assured. The uncertainty and the indecision of the people in charge had an adverse psychological effect on these people; by contrast, however, the community strengthened its bonds.

Whatever the effective reality or non-reality of the chemical danger in this case, one has to recognise that a new problem had arisen:

The situation is uncertain, there are strong fears, the public authorities are embarassed. The best reflection is perhaps the one made by a member of the Red Cross: "Give us a conflagration, a typhoon, a flood". (38, p. 244).

A true natural disaster against which one cannot fight ...

3rd. Menaces linked with the biological sciences and genetic engineering

This third field could, in the future, present extremely serious dangers (39). Short of a sufficient analysis on our part we shall not elaborate this point as we have the previous ones. We shall only declare that here again the problem of risk remains sometimes outside the field of reference of the best known literature (like the report "Science de la Vie et Societe, Biology and Society, which was sent to the French president (40)); that problems arise, however, from the scientific circles themselves (41); that communities have already been stirred up like the municipality of Cambridge in the USA which dreaded the experiments carried on at Harvard University (42).

---

*The responsibility devolving on the 'head of the family'.

As in other fields, one notices sometimes an extreme sensitivity to questions touching on risk as if the raising of questions constituted in itself a danger, almost an injury to the dignity of the researcher.

The subject is, however, doubtless not totally devoid of substance. Let us follow, as an example, J. Deutsch for a moment:

Relaxing of safety conditions - After the American standards the French ones concerning the safety of genetic manipulations have been lowered by the DGRST (October 1979). It became, for instance, possible to transplant human DNA into bacteria in very low safety conditions ("confinement level" Pl or P2). Since the conference of Asilomar which in February 1975 permitted the resumption of genetic manipulations the safety standards were founded upon the idea that the experiments were all the more dangerous the closer DNA donor was in evolution to the human species. One may still fear in fact the 'awakening' by genetic manipulation of latent viruses in the DNA that one transplants. One must know that the Pl and P2 levels correspond to ordinary laboratory installations and that there is only one single laboratory in Europe with the very high safety level P4, the European laboratory at Heidelberg. This lowering of standards is sufficiently important to envisage questioning of the construction project for a manipulation room at safety level P3 at the new agronomic centre of Toulouse.

Accelerated development in this sector - The relaxation of safety conditions occurs at a time when more and more teams and more widely distributed ones undertake genetic manipulations in research laboratories and when industry envisages, in France, to launch itself on this course, as Ch. Merieux, P-DG from the Institut Merieux, a subsidiary of Rhone-Poulenc, declared on Channel 2 2 on September 24, 1979.

Little known or wrongly known risks - Are there recent data that permit the justification of such lowering? The scientific experiments which permit an evaluation of the risks of genetic manipulations are difficult and rare but recent data do not invite optimism, on the contrary. Thus the American microbiologist R. Curtiss has studied in particular the properties of one of the main hosts used in genetic manipulations, the coli bacillus E. coli K12, and even set up pedigrees of so-called 'weakened' coli bacillae or EK2 of which it was hoped that they could not survive outside laboratory conditions. The data from Curtiss have been widely used for the establishment of the American standards of 1977*.

The recent data show that the probability figures established by Curtiss have sometimes been underestimated to considerable degrees.

Thus the probability of transmission of recombinant DNA in a coli bacillus to other bacillae by conjugation (sexual transmission) calculated initially by Curtiss at 10^{-19} which is very weak could be calculated at 10^{-6} on the basis of new data**.

This is, by the way, Curtiss' present position who when also reexamining the survival of the 'weakened' pedigree 1776 showed that it could survive for up to four days in human intestines***.

---

*R. Curtiss, letter to D. Frederickson, director of the NIH, April 12, 1977, distributed in France by the DGRST.

**A. Mendel, ref. No. 39

***R. Curtiss, quoted in *'Nature'*, 279, 360, 1979.

A second example is hardly more encouraging. One of the risks frequently quoted concerns voluntary or accidental manipulations of pathogenic viruses. If a manipulated bacteria that contained such a virus retained a pathogenic potential the epidemiology of the disease would thereby be completely changed in an unpredictable manner. Cancer could become an epidemic disease.

Experiments have been made to test this risk: the DNA of the polyome virus which is carcinogenic in mice and hamsters has been cloned into the weakened coli bacillus 1776. The manipulated bacteria and the DNA from these bacteria were injected into mice for testing on the one hand their capacity for producing viruses (infectious potential) and on the other their capacity for producing tumours (oncogenous potential).

The results achieved*show that the bacteria themselves are not tumouro-genous but that the DNA issuing from these bacteria present an oncogenous potential similar to that of the viral DNA itself. This is in itself worry-ing, and more so is the experiment open to criticism because the hamsters were infected with manipulated coli bacillae by subcutaneous injection which is not the most probable mode of infection**. Strangely, these results were interpreted optimistically (...) (42, pp. 484-485).

The risks connected with biotechnology extend even further than those of the nineteenth century. A boiler exploded (then) and killed workers around it. Here, the consequences could be very long-term and difficult to log (one might think, for instance, of an endemic form of cancer); the knowledge of the effects could only be very limited even though these effects would be extremely serious.

It may reasonably be objected that precautions are taken. They do not rule out the questions which certain events have legitimised:

Birmingham 1978. On September 11, 1978 in Birmingham a university employee died of smallpox; she worked just above a laboratory for research into the virus of this disease. The enquiry report was crushing for the management of the laboratory (there was a second victim: the research director committed suicide) (43)***.

San Diego University, USA, 1980. Professor S. I. Kennedy has attracted the attention of the authorities to the way he has carried out his experiments at the University of San Diego, California. The members of the Bio Safe Committee in charge of the control of the research work of the University's biologists concluded after the enquiry that Professor Kennedy had violated, perhaps deliberately, the federal standards to which the researchers working on the synthesis of the DNA must conform (44).

As these first slip-ups already suggest (to which the problems connected with malevolence must be added, of which there has been an example in the south of France recently) one cannot be satisfied, in this field as well as in others, with ritual formulae: "There has been no disaster so far". "All precautions are taken but naturally an accident is always possible ...".

----

*Israel *et al.*, *'Science'*, 203, 883-892, 1979.

**B. Rosenberg and L. D. Simon, *'Nature'*, 282, 773, 1979.

***This was not a case of genetic manipulation. Nevertheless, the problem, confinement not observed, is a general issue.

3. SOME REASONS FOR THE PRESENT DANGER

1st. The scale of operations

Giantism is a prime element that changes the nature of present-day risks. To the fierce growth on a quantitative scale corresponds a qualitative jump in the potential consequences of an accident.

This applies to the case of the *Amoco Cadiz* and its 230,000 t of crude oil. Some data should be recalled here: 1951, 1957, 1968, 1977 have, respectively, seen the launching or the entering into operation of tankers of 40,000, 85,000, 132,000, 200,000 and 550,000 t capacity. Between 1970 and 1977 the number of oil tankers in the category from 200,000 t to 300,000 t rose from thirty five to five hundred and thirty eight and from six to one hundred and eighteen in the category above 300,000 t (45, p. 55).

South Africa 1977. On December 16, 1977 two oil tankers of 330,000 t each collided off the South African coast; luckily one of them was empty.

Nowadays the same tendency exists with methane tankers. In 1969 there were only three methane tankers; ten years later the figure had risen to seventy nine. Up to the 1970s their capacity was about 30,000 m^3; this figure was quadrupled by the end of the decade: 120,000 to 130,000 m^3. In design circles there is already thought of still bigger ships (200,000 m^3), one company even suggested a figure of 330,000 m^3.

The storage of liquified gas has seen the same trend towards giantism as we have seen (despite the reticence of a number of experts).

In the nuclear field too the installations are getting bigger and bigger.

2nd. The nature of products stored

In three dimension the products in use prove to be more dangerous than in the past: the sharp toxicity, the possible general impact on the environment, the long term risks which they can involve. The number of these products is also increasing.

The preceding developments have shown, particularly in the case of Seveso, the menaces that exist on account of these products. Let us add an 'odd item' which is seldom mentioned.

Stockholm 1980. In March 1980, 250,000 people narrowly escaped deadly poisoning according to Swedish experts. A barrel containing 50 kg of potassium cyanide was found, for unknown reasons, in a container that was going to the compacting press of Stockholm's household refuse collection service. One of the employees noticed at the last moment the skull and crossbones painted on the barrel and stopped the machine. This barrel came from the L. M. Ericson factories who use it for the surface treatment of metals.

Given the expanse of towns, the location of industries, nowadays sometimes particularly dangerous in the heart of urban areas, the menace can be more than serious (48).

Genoa 1978. In September 1978 there was an escape of a toxic product at a factory located in the agglomeration of Genoa (800,000 inhabitants). There were three deaths and some fifty people had to be hospitalised (49).

3rd. Concentration of activities

In certain cases the concentration of activities increases accident probability.

a) Increased density of maritime traffic. Estimates of daily traffic in the Channel are as follows (in volume and in both directions): oil tankers: 10,000 t; oil/methane tankers: 10,000 t; chemical cargo vessels: 10,000 t; oil/minerals carriers: 10,000 t. These figures will increase considerably with the intensified oil activities in the North Sea (50, p. 138).

b) Increased density of road transport. This question was brought into the limelight with the accident at Saint Amand les Eaux: two years ago we counted an average of forty five trucks per hour in both directions, i.e. ninety vehicles on roads like the one where the disaster occurred: narrow, not adapted to this kind of traffic, the mayor of this community declared (51).

c) Increased density of industrial areas. The concentration of industrial installations often at the gates of substantial agglomerations causes fears of various accidents. The chain reaction which has already occurred in certain cases is particularly worrying. Here are some examples.

Texas City 1978. On May 30, 1978 a group of four storage tanks part of an alkylation process installation began to leak for unknown reasons (rupture of a propane tank was suspected). The explosion amplified the disaster: the four storage tanks, three others nearby, two spheres of propylene disintegrated when they were hit by flying fragments. The alkylation unit and the gas storage installation were destroyed. Flying fragments damaged the fire fighting equipment. Damages were estimated at more than 30 million dollars (6, p. 9).

Ponce, USA, 1979. This petrochemical complex had an accident on December 11, 1979 when a pressurised gas tank broke up. The fire which followed caused 10-15 million dollars worth of damage. What is more important, fragments were hurled over 600 m to a neighbouring factory one of the three units of which caught fire. Ten million dollars worth of damage occurred there (6, p. 10).

Bantry Bay 1979. On July 1, 1979 the French petrol tanker *Betelgeuse* exploded in Bantry Bay, Ireland. Fifty dead were mourned. There also remained a worrying factor: the nearby installations of Gulf (Oil) contained several million tonnes of petrol; damages could not have been restricted to the terminal area (52, 53).

The increased density of the industrial network is no secondary problem. In various regions substantial menaces must now be faced. Two cases can be brought up as illustrations, not because they are the most difficult but because they have been the subject of studies on risk identification (or, in the second case, of more or less preliminary work)*.

Canvey Island. The Canvey Island area (the island itself and the adjacent area of Thurrock) in the Thames estuary offers a particularly clear example of accumulation of major risk in one place. Being near London and offering space and access to the sea (a canal in the immediate proximity) the area soon attracted industry which imposed itself at the expense of recreation. The latter is, however, still alive: Londoners have flocked to the isle of

---

*Two general maps of French chemical industry are presented on 212-213 and give the first idea of the situation.

Canvey which now counts 33,000 permanent residents, and 2,000 caravans at the feet of enormous storage tanks for liquified gas from the island terminal await the city dweller who at the weekend looks for 'rest' at 300 m from one of the largest industrial complexes in the U.K.

Installations connected with petrol, liquified gas, the chemical industry and storage of all kinds nowadays clutter the countryside and particularly the living conditions of the place. This is one of the most dangerous sites in the country. The agglomeration of Canvey is particularly exposed as a study made by the British government shows (2, 54).

We give here two views of the area. However, let us spell out the risk level created by this accumulation of industry (2, p. 32).

Canvey: Industrial and residential area.
(Source: *Evening Echo*)

Probability of a disaster causing

- more than 1,500 deaths 17×10^{-4}*
- more than 18,000 deaths 1×10^{-4}

Two more refineries are planned; this would raise the probability figures to:

- more than 1,500 deaths 21.1×10^{-4}
- more than 18,000 deaths 1.8×10^{-4}

Several improvements have been suggested and undertaken which reduce the risk:

	Refinery projects not incl.	Projects incl.
- more than 1,500 deaths	4.2×10^{-4}	5.7×10^{-4}
- more than 18,000 deaths	0.3×10^{-4}	0.5×10^{-4}

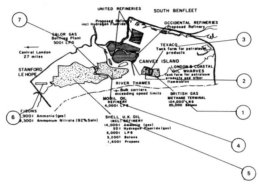

1 (C) Methane Terminal (British Gas): storage of LNG, 100,000 t (about to be closed; storage of liquified butane, 20,000 t (recently closed); arrival of methane tankers.
2 (A) Storage unit for inflammable products (London and Coastal Oil Wharves): storage 300,000 t, arrival of ships, oil pipeline.
3 (B) Storage unit for hydrocarbons (Texaco): storage unit of 80,000 t.
4   Mobil refinery: processing unit; storage of hydrocarbons: 1.5 million t; storage of propane-butane: 4,000 t.
5   Shell refinery: processing unit; storage of hydrocarbons: 3.5 million t; storage of liquified ammonia: 14,000 t; storage of hydrofluoric acid (gas): 20 t; storage of propane: 3,200 t; storage of butane: 1,600 t.
6   Ammonia treatment unit (Fisons): storage of ammonia: 1,900 t; storage of ammonia nitrate (solution): 9,300 t.
7   Propane-Butane treatment unit (Calor Gas): storage of 500 t.
–   Unloading of chemical products (at Pitsea)
–   Thames traffic canal very near the piers for the methane and oil tankers.

Fig. 23: Canvey-Thurrock: the general complex on the Thames
          waterfront. (Source: 67, p. 2-176)

_____

*All figures are given for a probability within 10,000 and per annum: $(xx10^{-4})$

Lyons, the 'corridor of the chemical industry'

A very high density of dangerous installations exists nowadays in the south of Lyons' agglomeration zone. This zone which constitutes what is called the 'corridor of the chemical industry' is crossed by the A7 motorway and the Paris-Marseille railway line which are channels of heavy traffic; transport of dangerous products aggravate the safety problems of the 'corridor' and every serious accident on the level of the fixed installations of the zone is a potential danger for the traffic which runs on these essential national and international communication lines (connection between northern Europe and the Mediterranean).

1 - Storage of petrol	- Storage of butane (one 1500m³ sphere) and of propane (two spheres, one 1000m and one 500m³
2 - Sulphuric acid plant (with storage of sulphur, oleum and acid) - Acrolein plant (with storage of acrolein and propylene) - Hydrofluoric acid plant	- Boron trifluoride plant - Plant for the monomer of Soreflon (P.T.F.E.) - Sodium chlorite plant - Liquid chlorine plant

3 Storage and use of:
  - Hydrochloric acid
  - Sulphuric acid
  - Chlorine
	- Hydrogen cyanide
	- 65% Perchloric acid
	- Acetic anhydride
	- Chlorosulphuric acid

4 Storage and use of: - Dichloroethane, isopropylamine (anhydrous), monomethylamine, ethylene oxide	- Phosgene - Various toxic products

5 Storage and use of:
  - Liquid ammonia
  - Ammonia solution
  - Diketene
  - Phosphorus oxychloride
	- Phosgene
	- Acetic arhydoide
	- Thionyl chloride
	- Chlorine
	- Hydrogen chloride in sulphuric acid

6 - Storage of vinyl chloride monomer (one 5200m³ sphere) - Nitric acid area (with stocks) - Sulphuric acid area (with storage of acid, sulphur and oleum	- Polymerisation plant for vinyl chloride and production plant for organic peroxides - Chlorine area (storage of chlorine, production of PVC and perchloric acid)
7 Storage and use of:	- Chlorosilanes - Silicon tetrachloride
8 - Storage of liquid ammonia	- Sodium cyanide plant
9 - Production and compression of acetylene	

10 Storage and use of:
  - Chlorine
  - Chlorosilanes
	- Liquid ammonia
	- Hydrogen chloride (gas)
	- Silicon tetrachloride

11 Methane cracking plant - Storage of ammonia (four 160m³ spheres)	- Adiponitrile hydrogenation plant
12 - Storage of oxygen	- Storage of hydrogen
13 - Storage close to the road - Risks from gas	- General storage of liquidfied hydrocarbons - Complete production of hydrogen sulphide
14 - Storage of butane (one 1000m³ sphere) and propane (one 600m³ sphere)	- Packaging plant for butane and propane - Storage in mobile containers (consisting of railway trucks, and road tankers)
15 - Dangerous units situated on the edge of the autoroute	

Table 11: The most specifically sensitive installations
            of the 'corridor of the chemical industry'
            June '79

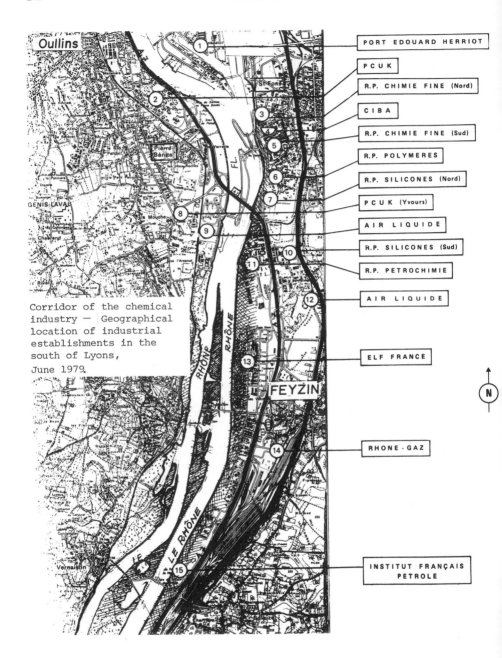

PORT EDOUARD HERRIOT

PCUK

R.P. CHIMIE FINE (Nord)

CIBA

R.P. CHIMIE FINE (Sud)

R.P. POLYMERES

R.P. SILICONES (Nord)

PCUK (Yvours)

AIR LIQUIDE

R.P. SILICONES (Sud)

R.P. PETROCHIMIE

AIR LIQUIDE

ELF FRANCE

RHONE-GAZ

INSTITUT FRANÇAIS PETROLE

Corridor of the chemical industry — Geographical location of industrial establishments in the south of Lyons, June 1979.

Fig. 24:   The corridor of the chemical industry.
(Source: Inventory of the Mining Administration
(1979, based on an IGN map))

These are the exclusive documents presented by Le Point: two maps which give the picture of industrial risk in France. Two official work tools on which the experts in charge of the control of "establishments classified as dangerous and unhealthy" base themselves to enforce their surveillance.

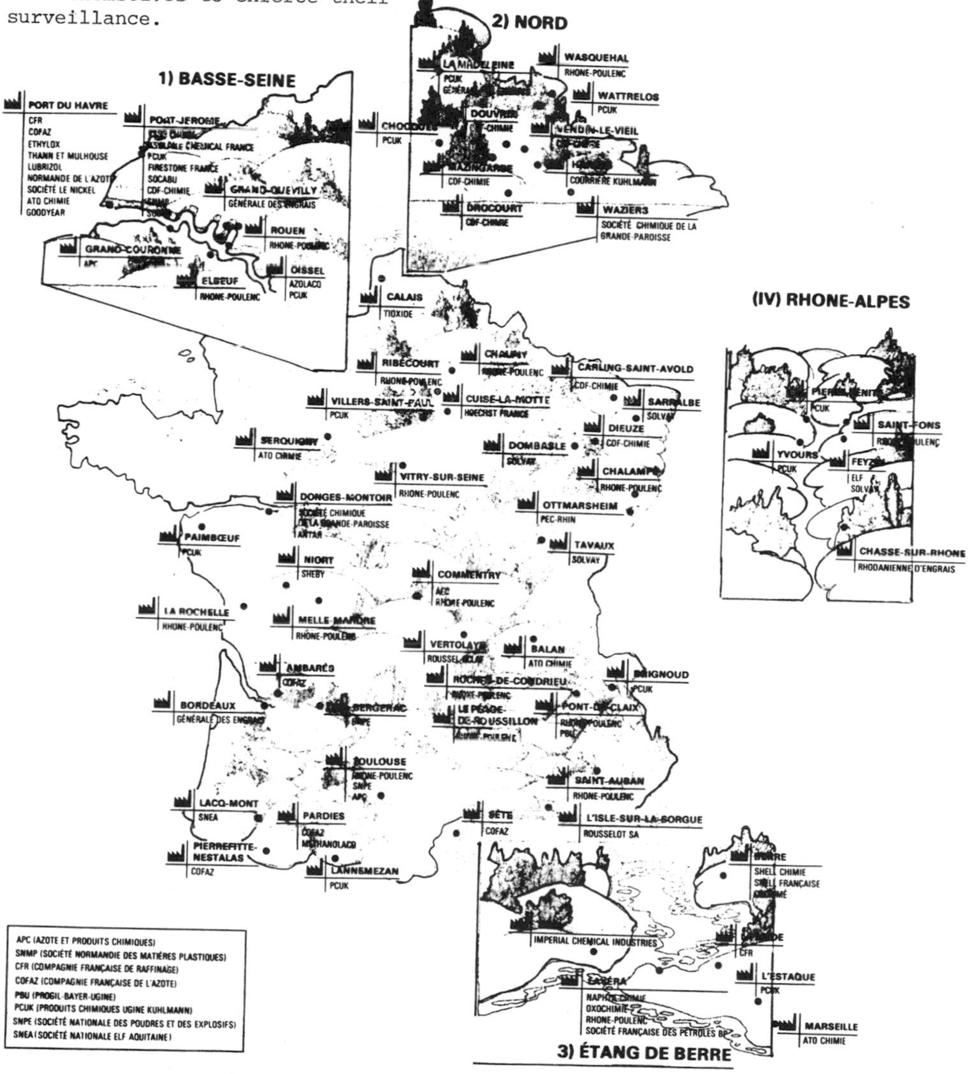

Fig. 25: The 100 chemical and petrochemical factories the danger potential of which requires close control. (Source: Le Point, 292, April 24, 1978)

Fig. 26: Map of large storages of dangerous chemicals.
Today one should add to these particularly the installations
of Stocknord at Dunkirk and of Gaz de France at Montoire de
Bretagne.

(Source: Le Point, 292, April 24, 1978)

4th. Risks caused by malevolence, sabotage and organised attack

Over the years this menace has lost its theoretical nature. From the many known, if not recorded, examples we quote:

Trieste 1972. On August 4, 1972 dynamite charges were placed on the discharge pipes of the large storage tanks at Trieste in Italy. Two of them were completely destroyed and a third one seriously damaged; three more were affected. One of the tank tops collapsed, and the burning liquid overflowed the 3 metre high barrier that surrounds the tank. Burning petrol flowed as far as the terminal's control room 600 m away. There were more than 10 million dollars worth of damage (6, p. 6).

Alfortville May 1978. Explosion of gas ducts of Gaz de France (55).

Lemoniz 1978. In March 1978 a criminal attack was made on a nuclear centre near Bilbao in Spain (responsibility claimed by the military branch of ETA) (56).

Salisbury 1978. Rockets and bullets were fired at the storage area of the terminal in Salisbury, Rhodesia. Twenty two out of twenty five storage tanks for diesel fuel, kerosene and petrol were destroyed in the fire that developed and burned for three days. The terminal which covers 16 hectares was largely destroyed (6, p. 9).

Narita 1978. A commando raid threw air traffic over the whole of Japan into disarray for four hours by the severance of a cable near Tokyo which links all control towers of the country.

Canvey island 1979. The IRA exploded a bomb on a kerosene tank at Canvey island on January 17, 1979. By chance the latter was almost empty and a disaster was avoided (58).

Soleure 1979. The weekly *'Le Point'* gives some additional information concerning the nuclear industry:

The attacks against nuclear installations have been expensive. In May 1975 two bombs delayed the completion of the nuclear centre at Fessenheim/Alsace for several months. The following month the main computer of the Framatome company — the builders of French centres — was destroyed at Courbevoie. At Argenteuil the test laboratories for valves of the same company were damaged. In August 1975 two bombs caused the temporary closure of the centre at Brennilis in Finistere. In November 1976 it was the turn of the Paris offices of a company manufacturing nuclear fuels, then the electronic control room of the uranium mine of Magnac/Haute Vienne. In May 1979 the electric circuits of the centre at Bugey/Ain were sabotaged. In October 1979 the cables of a control room of the European uranium enrichment factory, Eurodif at Le Tricastin, were cut. There is no point in carrying on with this enumeration: one would never finish counting up the damages to which the nuclear factories in France were subjected. In many cases the experts noticed considerable technical competence on the part of raiders (60, pp. 92-93).

4. TECHNOLOGY IN ITS SOCIAL AND NATURAL CONTEXTS

1st. A universe in crisis

The general universe in which the industrial countries move shows many traits that are not new. Famine and misery for the poor. War in these poor

countries as well.   Natural disasters are always more numerous and more
deadly in the Third World.

However, perhaps this universe is about to change.   The facts speak against
the hopes for development: the poorest, deprived of energy, founder in underdevelopment.   The indebtedness of the 'countries in the process of development' (if one can still use this expression) has doubled between 1976 and
1980.   Every year hunger kills 15 million people, of which 5 million are
children.   The 4.5 billion human beings of 1980 will have become 6 billion by
the end of the century.

This fantastic discrepancy between the wealth of some and the misery of the
others — as Andre Fontaine writes (61) — brings war; it will not be difficult
to find ammunitions for these conflicts when worldwide military expenses
reach "1 billion dollars a minute", i.e. four times what they were at the
time of the Korean war, and the arms trade has increased sixfold since the
1950-59 period according to the international peace research institute in
Stockholm.   "Four tonnes of explosives *per capita*" are enough to supply
confrontations.

These confrontations will no longer necessarily follow the well-known
rules of the post-war period which were sanctioned at Yalta.   The first
revealing false note was sounded in 1980 with the violation of a sacred
principle: the disrespect for an embassy.   The Iranian revolution is perhaps
a first sign of a change that is happening in the relations between nations
as well as in the relation between a society and industry, technical progress
and rationality.

In this universe, pregnant with upheavals, the new techniques contribute
their co-efficient of insecurity.   One can report some of their traits.

2nd. High risk technology and industry in a world pregnant with menaces and disruptions

a) Risks created by man and natural risks.   The interactions between man's
creations and the forces of nature is nowadays a more real problem than it
has been in the past.

There is on the one hand the risk of unleashing natural forces by means of
giant constructions.   This was the case with certain large dams (Grandval in
France 1961, Monteynard in France 1963, intensity 4.9; Cariba Bassa in
Rhodesia 1961, intensity 6; Kayna in India 1963, intensity 6.5, 177 deaths;
Kremasta in Greece 1965, intensity 6.2 etc). (62, p. 30).

On the other hand and in particular there is the risk of a large number of
installations (and not only small heaters and heating gas as in the cases of
San Francisco, 1906, and Tokyo, 1923) playing a trigger role for natural
forces.

Three cases are recalled by way of warning:

- The destruction and fire of the hydrocarbon tanks at Kodiak and Crescent
  City in California subsequent to a seismic tidal wave in 1967 (62, p. 139)

- The chain reaction of damages that occurred in Japan following an earth
  tremor in 1978 on the Izu peninsula.   Following a dam break subsequent to
  an earthquake a chemical factory was damaged and toxic salts released into
  the river Kanogawa and into the bay of Suruga which are now polluted for
  an indefinite period of time (32, p. 258).

- The retrospective worry in connection with the eruption of Mount St. Helen in the state of Washington in April 1980: there was a nuclear centre not far away. Luckily it was not operative at the time (63).

b) Technology transfers to the Third World and the transfers of risk. The population density and sometimes the insufficiency of controls and follow-up, the differences in sensitivity to aggressions, the safe knowledge of natural conditions that are more subject to disturbance than in countries with more temperate climates etc. result in possibly increased technological risks for Third World countries which import technology. We recall the most urgent warning of this kind: on August 11, 1979 in the region of Morvi, 500 km west of Bombay, a dam broke; there were at least 25,000 deaths (60; 65).

c) Risks created by man and political and social instability. The acts of malevolence and sabotage referred to before must here be related to the general context that brings them about. Here, situations determine at the same time extremely serious actions and provide for those who want to get rid of them a first class "raw material" (in terms of effect if not in political effectiveness). Societies which create desperadoes or madmen and at the same time give them the means for disproportionate actions cannot be sure of their safety.

If one goes on to the classical question of armed conflicts between countries one must recognise there as well a disruption at the level of necessary means for battle. The party attacked offers the aggressor magnificent targets for a considerable destructive potential, targets which are highly vulnerable into the bargain. On this bias the civilian catches up with the military and the defence of territory proves, on account of this as well, more complex than in the past.

REFERENCES

(1) V. C. MARSHALL, Bulk Storage and handling of flammable gases and liquids Operational methods of Hazards Evaluation. OYEZ International Business Communication, London, ler octobre 1979 (Traduction par R. ANDURAND, CEA-IPSN).

(2) Health and Safety Executive, Canvey, an investigation of potential hazards from operations in Canvey Island. Thurrock area, Health and Safety Commission, London, HMSO, 1978.

(3) Advisory Committee on Major Hazards: second report. Health and Safety Commission, HMSO, London, 1978.

(4) *Report of the President's Commission on the accident at Three Mile Island.* Pergamon Press, New York, October 1979 (201 pages).

(5) President's commission on the accident at Three Mile Island. Technical staff analysis report — Alternative event sequences, 1979 (Unpublished).

(6) M. and M. Protection Consultants, A review of catastrophic property damage losses (30 years-worldwide) Marsh and Mc Lennan Plaza, Chicago, March 1980 (10 pages).

(7) H. LEWIS, La sûreté des réacteurs nucléaires. *Pour la Science,* mai 1980, pp. 73-89.

(8) F. R. FARMER, Siting Criteria, a new approach. Proceedings of AEA Con-
     ference on containment and siting of nuclear power plants. Vienne,
     avril 1967.

(9) Nucléaire Face à Face. Que Choisir? Numéro spécial, 1978.

(10) F. R. FARMER, Principales and standards of reactor safety. Agence
     Internationale de l'énergie atomique. Symposium, Jülich, 5-9 février
     1973.

     Le dossier électronucléaire. Syndicat CFDT de l'énergie atomique,
     Seuil, 1980, pp. 233-234.

(11) G. N. KELLY, J. A. JONES and B. W. HUNT, An estimate of the radiological
     consequences of notional accidental releases of radioactivity from a
     fast breeder reactor. National Radiological Protection Board, Harwell,
     Didcot, Oxon, August 1977.

(12) B. LAPONCHE, Pour une autre politique de l'énergie. *Faire*, no 57-58,
     août-septembre 1980, pp. 23-60.

(13) L. NIEDRINGHAUS DAVIS, Frozen Fire, Where will it happen next? Friends
     of the earth, San Francisco, 1978 (298 pages).

(14) Transportation of liquefied natural gas. Congress of the United States,
     Office of technology assessment, Library of congress catalog, no 77
     600048, September 1977.

(15) T. A. KLETZ, Consider vapor cloud dangers when you build a plant.
     Hydrocarbon Processing, October 1979, pp. 205-212.

(16) G. MARLIER, Les explosions de vapeurs de gaz liquéfiés (B.L.E.V.E.) ELF
     France, Service Sécurité Raffinage — Groupe d'études de sécurité de
     l'industrie pétrolière (GESIP), 7 mai 1980 (8 pages).

(17) B. MAURER, R. HESS, H. GIESBRECHT and W. LEUCKEL, Modelling of vapour
     cloud dispersion and deflagration after bursting of tanks filled with
     liquefied gas. Loss prevention and safety promotion in the process
     industries, preprints. 2nd International Symposium, Heidelberg, 6-9
     September 1977.

(18) J. RIGOURD, C. MICHOT et J. DANGREAUX, Détermination de l'explosibilité
     des produits solides ou liquides Laboratoire des substances explosives
     du Centre d'Etudes et de Recherches des Charbonnages de France.
     Colloque sur la sécurité dans l'industrie chimique, Mulhouse, 27-29
     septembre 1978 (9 pages).

(19) D. H. SLATER, Vapour clouds. *Chemistry and Industry*, 6 May 1978, pp.
     295-302.

(20) Lettre du Dr RASBASH (Department of fire safety engineering, University
     of Edinburgh) à Sir Bernard Braine (Member of Parliament), 5 mars 1979.

(21) G. LE RICOUSSE, La sécurité industrielle et les grands stockages de gaz
     naturel liquefié. *Les Travaux Publics*, 3e trimestre 1978, pp. 47-54.

(22) Pour une politique de lutte contre la pollution des mers. Rapport du
     groupe interministériel des problèmes de pollution de la mer Documen-
     tation française, Paris 1973 (271 pages).

(23) D. FISHER, A decision analysis of the oil blowout at Bravo platform. IIASA, Luxemburg, RM 78.6., January 1978 (35 pages).

(24) R. VALERO CHAVEZ, J. SALAS ALBA et J. AYMES CONCKE, Rapport sur les causes du sinistre touchant un puits de pétrole et ses conséquences. Présenté par C. Lic. Oscar Flores Sanchez, Procureur Général de République. Mexico, 24 juillet 1979 (6 pages). (Documentation du C.E.D.R.E. Brest le 7 août 1979).

(25) *Le Monde*, 27 mars 1980.

(26) *Le Monde*, 2 avril 1980.

(27) *Le Monde*, 10 avril 1980.

(28) E. et A. PARKER et R. DESCOURS, *La vérité sur l'énergie nucléaire*. Menges, Paris, 1978 (178 pages).

(29) *Le Monde*, 26 janvier 1972.

(30) J. DENIS, La marée noire blanchie par le silence. *Science et Vie*, avril 1980, no 751, pp. 88-91.

(31) Etude des gisements coquilliers du littoral breton. Campagne Gino I, Etude de la pollution autour de l'épave du Gino (du 5 au 25 février 1980). Institut Scientifique et Technique des Pêches Maritimes, Nantes, 22 janvier 1980.

(32) F. NEGRIER, L'assurance face aux risques catastrophiques. *L'Argus International*, no 13, juillet-août 1979, pp. 254-273.

(33) J. R. NASH, *Darkest hours — A narrative encyclopedia of worldwide disasters from ancient times to the present*. Nelson Hall, Chicago, 1976 (812 pages).

(34) *Le Monde*, 13-14 août 1978.

(35) *Le Monde*, 25-26 mai 1980.

(36) *Le Monde*, 24 mai 1980.

(37) G. B. KOLATA, False alarm caused by botched study. *Science*, Vol. 208, 13 June 1980, pp. 1239-1242.

(38) C. HOLDEN, Love canal residents under stress. *Science*, Vol. 208, 13 June 1980, pp. 1242-1244.

(39) A. MENDEL, (ouvrage collectif) Les manipulations génétiques. *Le Seuil*, Paris, 1980 (332 pages).

(40) F. GROS, F. JACOB et P. ROYER, Sciences de la vie et Société. *Le Seuil*, Paris, 1979.

(41) J. DEUTSCH, Manipulations génétiques: guérir le "vertige des biologistes". *La Recherche*, no 110, avril 1980, pp. 484-485.

(42) F. MENDELSOHN, "Frankenstein at Harvard" The public politics of recombinant DNA research. The Social Assessment of Science — Proceedings. Universität Bielefeld, 1978, pp. 57-78.

(43) M. BLANC, Quand les laboratoires manipulent des microbes. *La Recherche*, Vol. 10, no 99, avril 1979, pp. 388-390.

(44) *Le Monde*, 12 septembre 1980.

(45) Rapport de la Commission d'Enquête de l'Assemblée Nationale présenté par H. Baudoin. Première session ordinaire 1978-1979, novembre 1978, no 665. Tome 1 (333 pages).

(46) *Le Monde*, 18-19 décembre 1977.

(47) Mitre Corporation. A study of dangerous substances in French industry — identification of high priority organic chemicals. Ministère de l'Environnement et du Cadre de Vie (non daté).

(48) *Le Quotidien de Paris*, 10 mars 1980.

(49) *Le Monde*, 21 septembre 1978.

(50) Rapport de la Commission d'Enquête du Sénat présenté par A. Colin. Seconde session ordinaire 1977-1978, juin 1978, no 486 (289 pages).

(51) *Le Monde*, 3 février 1973.

(52) *Le Monde*, 10 janvier 1979.

(53) *Le Monde*, 26 juillet 1980.

(54) P. LAGADEC, Le problème de la sûreté d'un grand complexe industriel — Le cas de Canvey Island. Ministère de l'Environnement — Fondation Royaumont — Ecole Polytechnique, 1979 (98 pages).

(55) *Le Matin*, 1er juin 1978.

(56) *Le Quotidien de Paris*, 20 mars 1978.

(57) *Le Matin*, 22 mai 1978.

(58) *Le Devoir*, 5 novembre 1979.

(59) House of Commons. Parliamentary debates. H.M.S.O. Londres. Hansard, 18 January 1978. Columns 1988-1994.

(60) R. MIHAIL, La délinquance nucléaire. *Le Point*, no 395, 14 avril 1980, pp. 91-93.

(61) A. FONTAINE, Quatre tonnes d'explosifs par tête. *Le Monde*, 12 juin 1980.

(62) B. A. BOLT, W. L. HORN, G. A. MACDONALD and R. F. SCOTT, *Geological Hazards*. Springer Verlag, Berlin, 1975 (328 pages).

(63) *Le Monde*, 27 mai 1980; 28 mai 1980.

(64) *Le Monde*, 15 août 1979.

(65) Communications avec l'Ambassade de l'Inde à Paris (1979 et 1980).

(66) V. PILZ, What is wrong with risk analysis? Loss Prevention and Safety
     Promotion in the Process Industries. Preprints, pp. 6/448-6/454. 3rd
     International Symposium Basle, Switzerland, Sept. 15-19, 1980.

(67) A. SPIEGELMAN, Hazard control in the chemical and allied industries.
     Preprints, pp. 1/129-1/137. 3rd International Symposium Basle,
     Switzerland, Sept. 15-19, 1980.

CONCLUSION: DESPITE AN APPEARANCE OF GREATER SAFETY —
MENACES OF QUITE DIFFERENT SERIOUSNESS

It is by their very nature extremely delicate to attempt an analysis of
the disruption and continuities which a situation presents when one does not
have the perspective of a certain distance in time. However, the issue of
major technological risk today does not permit us to wait quietly for a time
when one could establish a diagnosis without fear of error. We have therefore tried to set up as from now and in present circumstances of data some
range-poles in technological development viewed from the angle of risk.

This examination leads us to do justice to the often spread impression
according to which safety is much better assured than it used to be and that
the industrial risk is small compared to other dangers such as war.

It appears in fact that the risks of the past (mining: Courrieres 1906;
navigation: the *Titanic* 1912; railways: Lagny 1933; dams: South Fork River
1889 etc.) have often been conjured up these days, at least in developed
countries. Whence a first — founded — impression of safety. The evidence
according to which a thermonuclear bomb is something different from a simple
factory needed no demonstration. Nevertheless, we have seen the appearance,
particularly after World War II, of the phenomenon of very large industrial
accidents which went far beyond the confines of a factory or installation
(Cleveland 1944; Ludwigshafen 1948; Los Alfaques 1977 etc.).

This risk still exists today even if, like the precedent, it can be taken
to be better and better known and doubtless controlled (this will be examined
later on). It seems that we are now entering a third phase. One recognises
only some traits through a series of cases of which some have been studied
in the first chapter. The preceding developments have permitted to spell out
the menace. It has at least two dimensions: space and time.

In terms of space the seriousness of potential destructions is today much
greater than ever. In such and such a town these destructions would be
massive if a giant storage tank with liquified gas broke up, if a methane
tanker had a serious accident and if very unfavourable conditions prevailed
at the time*.

In terms of time certain products involved can have very long-term effects
as the pharmacologists Sullivan and Barlow remind us who consider the possible
effects of these substances on a foetus:

- immediate effects such as death and abortion;
- effects recognisable at birth: fatal, very serious or cureable;
- effects not measurable before several years have passed: mental
  retardation, cardiac problems;
- effects not measurable until after very many years have passed if the
  foetus has been affected by carcinogenic substances;
- effects which appear only in future generations as animal experiments
  have shown**.

---

*The chain of conditions (in this statement) does not necessarily
give cause for a smile: the most unfavourable conditions in question
here are the most favourable for a ship's accident for instance.

**F. M. Sullivan and S. M. Barlow, Congenital malformation and other
reproductive hazards from environmental chemicals. Long term hazards
from environmental chemicals, The Royal Society, London 1979, pp. 91-110.

The change in the nature of industrial risk is thus recognisable in two aspects: much larger areas than before (leaving out some exceptions like Halifax 1917, Bombay 1944, Texas City 1947) can be hit by a disaster.

An incomparably longer period of time can become the measure of these disasters (infinite on the scale of a generation). In other words, one can no longer say as for instance the health authorities at Seveso wrongly did: the area affected is limited, isolated and evacuated. Fire, which is what is implied in this type of reasoning, is no longer the reference type of accident.

These new data find their place in a context which seems much more complex and open than before in terms of rules of behaviour as well as with regard to the confrontation between groups, countries and ideologies.

This causes us to put into the centre of our analytical table the question of social responsibility in the face of major risk, because the control of these menaces cannot be effected mechanically. This is all the more true since, as we have shown in a multiplicity of cases, societies have often proved greatly irresponsible and ignorant in matters of technological risk.

What has been done to control these risks? What have we got in hand today to throttle the existing menaces? Are the means adequate to the end?

The second part of this study will be devoted precisely to these questions.

PART TWO

The management of major technological risk

CHAPTER THREE

Means and tools of management

I   A panoply of means

II Utilisation of Science

Foresight and preventive measures against damages are only too often caught up with and surpassed by even greater dangers (...)

The institution of insurance results from human reason. In large measure it permits the material repair of the consequences of human failures.

But it would logically find its limits at the moment when man no longer had the capacity to order the problems of his existence reasonably.

Muenchener Rueck

I. A PANOPLY OF MEANS

The general improvement of living conditions, the demands in matters of the environment, the more serious menaces weighing down on the framework of life and work must lead the industrial societies to giving themselves new means to manage the inconveniences of their development*. The question of risk created by man was approached and treated like the more general one of the protection of the environment.

In addition to the setting up of techniques and scientific tools worked out by specialists (and which we shall examine in the following subchapter II) more general frameworks have been defined in response to identified risks and reported disasters. In this way means of prevention have little by little been spelled out and set up side by side with means to combat disasters

---

*Between 1970 and 1979 we have seen a flurry of laws adopted for this purpose in the United States (1, p. 1/129):
 - Environmental Protection Act;
 - Occupational Safety and Health Act;
 - Toxic Substances Control Act;
 - Consumer Product Safety Act;
 - Clean Air Act;
 - Water Pollution Control Act;
 - Hazardous Materials Transportation Act;
 - Mine Safety and Health Act;
 - Resource Conservation and Recovery Act.

and arrangements to help victims. We shall study these three questions in
succession.

For the examination of legal and institutional arrangements concerning
prevention, a subject which is naturally deserving of considerable attention,
three countries have been chosen as examples: in addition to France we have
chosen Great Britain (because of very interesting innovations which this
country has introduced in its legislation and administration in the last six
years) and Italy (because this country offers an instructive counterpoint
image). In each case we have limited ourselves to the central elements of
the arrangements, the more specifically industrial legislation, even if some
references may be made to supportive legislation. Let us stress again that
it has seemed essential to us to locate these legislations and administrative
means historically. This dynamic approach permits a better comprehension of
the origins of the available means and their strength but also their limits,
a topic that will be dealt with later on.

1. MEANS FOR THE PREVENTION OF INDUSTRIAL RISK

1st. The case of France

The central text is the law of July 19, 1976 concerning installations
classified for the protection of the environment, completed by the decree of
September 21, 1977 (2). It regulates the major risks of technological and
industrial origin, the case of nuclear risks being left separate, which we
shall examine subsequent to the general legislation of classified installations. In order better to understand the nature and the scope of this law,
it seems useful to spell out its origins. The history of this matter shows
three landmarks, the dates of the laws for the control of industrial activity:
1810, 1917 and finally 1976. Let us examine these successive phases one
after the other (3, 4, 5, 6,).

*a) Before the French Revolution, a profusion of regulations mainly concerning
public health in towns*. The regulation of work activity is not of recent
date nor even from the era of the industrial revolution. Since antiquity,
noise, fumes, release of dirty water have been controlled, sometimes forbidden inside city walls. The preoccupation with public health in towns,
expressions of which one sees clearly under Philippe Auguste for instance
(road sweeping, drainage of used water, paving of streets in Paris, 1148) was
not without effect on certain trades: in the fourteenth century the ordi-
nances by the Provost of Paris forbade the raising of animals; in 1363 a
letter forbade the dumping of waste from slaughterhouses on public roads; a
royal ordinance prescribed the use of special pits for the disposal of
polluted water from similar establishments ... With this royal ordinance of
April 30, 1363 a second preoccupation appeared: the risk of fire. All
industries using fire found themselves regulated.

Little by little every trade was subjected to arrangements expressing the
two preoccupations with public health and the prevention of fire. On the
eve of the revolution, the whole pattern of existing regulations appeared
insufficient and incomplete, at the same time inauspicious for industrial
development and worrying for safety.

In addition to a general overhaul there was need for a new approach to the
problem because of the expected spread of certain industries which could be
considered as dangerous, this after the works by Lavoisier and Berthollet.
However the legislator did not know how to innovate. On the contrary, with
a law of September 21, 1791 he locked the system in the state in which it was.

The confusion of Paris spread to the provinces, each region wanted to have its own regulations.

b) The imperial decree of October 15, 1810 concerning unhealthy factories and workshops. Concerned about the development of the nation's industry which was subjected to incoherence and arbitrariness that discouraged initiative and even the simple pursuit of existing activities and in order to put an end to the afflux of complaints raised by manufacturers Bonaparte consulted the Institut de France in 1805. From this consultation resulted the report of the 26th Frimaire of the year XII in the system of dates of the calendar of the French Revolution edited by G. de Morveau and Chaptal and signed by de Cuvier. The basic ideas of that text are still at the bottom of present legislation ... . In short, it puts forward the problem of co-existence between industry and its neighbourhood. The fifth paragraph of the report deserves to be remembered for the clarity with which it expresses the fundamental principles.

It is therefore a necessity of the first order for the prosperity of the crafts, that limits be imposed which leave nothing any more to the arbi-trariness of magistrates and which outlines for the manufacturer the sphere within which he can exercise his industry freely and safely and which guarantees to the owner of neighbouring land that there will be no danger for his health or for the products of his land (3, p. 14).

This report needed to be completed because industrial expansion demanded that the necessary modalities for the desired coexistence be more precisely defined. The need for a nomenclature made itself felt. The institute was consulted a second time, still in 1805; from this resulted a further report dated from 1809. It was then that the principle of industry classification was set up, the activities causing the most nuisance (1st category) having to be moved away from residential areas; added to this was the principle of an authorisation having to be applied for from the competent authorities upon the creation of an enterprise of this type. These principles too under-lie present legislation. They are found in the imperial decree of 1810 of which we quote the essential articles (3, pp. 9-11):

Article 1: From the date of publication of the present decree factories and workshops which issue unhealthy or unpleasant odours may not be set up without permission by administrative authority; these establishments shall be divided into three categories:

The first category shall comprise those which must be moved away from private residential areas;

The second, factories and workshops the removal of which from residential areas is not absolutely necessary but the setting up of which nevertheless must not be permitted unless it has been made sure that their operations are carried out in such a way as to not incommodate neighbouring property owners or to cause them damage.

In the third category shall be included establishments which may stay with-out inconvenience to residents but must remain subject to police surveillance.

Article 3: Permits for factories of the first category shall only be granted if the following requirements have been met:

The application for a permit shall be presented to the police commissioner and shall be posted on his orders in all communities within a radius of 5 km.

Within this period every property owner shall have the right to present his reasons for objection.

The mayors of the communities shall have the same right.

Article 4: If there are objections the Prefectorial Council shall give its decision unless there is a decision by the Council of State.

Article 5: If there is no opposition the permit shall be granted, if there is reason for it, upon the decision of the police commissioner and the report from our Interior Minister.

Article 9: The local authority shall indicate the location where factories and workshops coming in the first category may be established and shall spell out the distance from private residences. Any individual building in the neighbourhood of these factories or workshops after their establishment has been permitted shall no longer be admitted to request their removal.

Article 10: The division into three categories of establishments that issue unhealthy or unpleasant odours shall take place in accordance with the table appended to this decree. It shall serve as a rule in each case when a request for the setting up of such an establishment has to be decided on.

Article 11: The arrangements of this decree have no retroactive effect; it follows that all establishments which are at present operating shall continue to do so freely except for damages which may be payable by those who prejudice the properties of their neighbours; damages shall be arbitrated by the courts.

Article 12: Nevertheless, in case of serious impediments of public health, of agriculture or general interest the factories and workshops of the first category which cause them can be eliminated by means of a decree from our Council of State after having heard the local police, taken the advice of the police commissioners and having rejected the defences by the manufacturers or the inhabitants.

This law was to remain in force for more than a century, each government finding its simplified and centralising character a first-class tool.

c) *The law of December 19, 1917 on dangerous, inconveniencing or unhealthy establishments.* The decree of 1810 (amended somewhat in points of detail in 1815) was, however, found to be too restrictive. Reports (1864, 1899) recommended a softening of the arrangements in force, particularly the definition of a category of establishment for which a simple declaration would be sufficient. Various draft laws were proposed (1889, 1903) but a new law had to wait until 1917 to be adopted.

The decree of 1810 was liable to criticism on one count: it did not pronounce on the necessary means for the control of these factories and workshops which it categorised and thus mitigated a lot of the restrictions which were so painfully felt by the operators.

Since 1866 a well documented study by M. de Freycinet (Dunod, Paris, p. 60) regretted the situation and demanded the creation of a government inspection service:

Every police commissioner, advised by his Hygiene Council, issues for his Department the factory permits and inserts the clause which he considers important for public health. Surveillance is exercised by local authorities

and in most cases by ordinary police officers. There does not exist, as in
Belgium or in Prussia or even in England, a government inspector with central
authority. The execution of orders also leaves much to be desired ... (3,
p. 480).

In 1907, in a report to the Senate, Dr Chautemps gave a description of the
situation which was no more complementary, except in the case of the Department of the Seine:

At present, the Department of the Seine finds itself in a privileged
situation; it has in fact a special corps of inspectors, appointed in competition and presenting from a scientific point of view all desirable
guarantees. In almost all the rest of France the inspection of classified
establisments is done by mayors, police officers and rural policemen, that
is to say by people whose impartiality and care for the public interest, no
matter how devoted they are, most often cannot make up for incompetence when
it is a matter of evaluating manipulations and recognising the presence of
chemical products (3, p. 480).

One notices that when the position of inspector had been created the power
of these agents remained rather theoretical: they had neither the right to
enter the establishments without the assent of the owners nor the right to
draw up an official report on infringements found (3, p. 480).

Various projects and propositions for laws were prepared (1883, 1887, 1889)
to correct these defficiencies but one had to wait until 1917 before one
could see improvements of the system put into operation. What are the main
contributions of this law of 1917? It enlarged the field within which the
arrangements made in matters of safety would be applicable; it stressed the
problem of dangers which until then had taken second rank after that of
nuisances:

Article 1: The mills, workshops, factories, shops, work sites and all industrial and commercial establishments which present causes of danger or
inconvenience be it for safety, or for agriculture, are subject to surveillance by the administrative authority under the conditions determined by this
law (3, p. 12).

Otherwise the law takes up the principle of differentiation between the
three categories, the third category being released from the regime of
authorisation and now only subject to declaration. All this is defined in
the legislation. It is also made clear (Article 21) that the police
commissioners can entrust inspection work to "any official or any other
member of the Department's Hygiene Council or to a Health Commission which
appears to them competent on account of its function or its competence".
In the more industrialised Departments inspectors for the classified
establishments can be instituted, appointed by the police commissioner after
a competition and remunerated on the Department's budget, their allowances
to be determined by the General Council upon proposal by the police
commissioner ... . These inspectors have the right of entry into classified
establishments (Article 21).

d) *The law of July 19, 1976 concerning the installations classified for the
protection of the environment.* The law of 1917 had remained in force for
nearly sixty years. During this period, industry and its environment had of
course developed. Increased urbanisation had been a prime cause for the
modification of the law of 1917: in 1932 a law on urbanisation foresaw a new
principle: reserved residential areas within which no establishments of the
first and second category could be authorised. This principle of exclusion
was further strengthened in 1943.

Secondly, it appeared again that the public authorities did not have the necessary means to carry out their task of controlling adherence to this legislation. P. Gousset the author of fundamental works on the law governing classified establishments, does not hesitate to write in terms which recall the papers by Dr Chautemps (1907) and M. de Freycinet (1866).

(...) One can make one more serious reproach still to the legislation on classified establishments: i.e. that it is often poorly applied. As we have shown, the inspection services, particularly in the province, do not have by far the specialised personnel in sufficient numbers and equipped with the necessary means to ascertain at the same time the examination of new businesses and the regular control of existing establishments. It is absolutely necessary that the present inspectors, most of whom work part time, be replaced by a specialised and structured group of which the best element should work in Paris with the central administration in order to work out a sustained policy combatting industrial nuisances (3, p. 546).

Given the new scope of industrial risks the problem could not be viewed with indifference. The accident at Feyzin was to stir up public conscience. On June 28, 1966 the Minister of Industry entrusted the inspection of "factories for the treatment of petrol and its derivatives and residues" to the chiefs of the mineralogical departments of the Mining Administration. This was the beginning of the establishment of a structured service for the control of major risk. The engineers from the Mining Administration were the only ones with very broad experience in matters of industrial safety. They had a double qualification. On account of their very longstanding responsibilities in the Mining Administration. On the one hand they had direct experience of safety because they were charged with investigating accidents (as we have seen in the case of Courrieres) and on the other hand they had general knowledge of safety in industry covering: refineries, production, transport and distribution of gas; transport of hydrocarbons and dangerous products by pipeline, steam and gas pressure apparatus, technical control of automotive vehicles (7, p. 42). There was one sensitive point: these services still had to be given the means to take effective control of dangerous establishments. This was a problem the size of which largely surpassed the means at the disposal of the Mining Administration (leaving out Paris and du Nord where for a long time the inspection of classified establishments had been partly entrusted to the engineers from the mineralogical departments).

These directions were confirmed and generalised in a circular letter from the Minister of Industry dated June 28, 1968: a structure was to be progressively established for the control of nearly all dangerous, inconveniencing or unhealthy establishments under the responsibility of the engineer from the Mining Administration in charge of the mineralogical department. At central level coordination was to be ensured by a Division for the Prevention of Industrial Nuisances, created for this purpose at the Ministry of Industry. A Directorate of the Industrial Environment was created within the framework of the ministry in 1969. When the Ministry of the Environment was created in 1971 these coordination duties were transferred to the new ministerial department which still exercises this responsibility by means of the Service of the Industrial Environment (SEI) of the Directorate for the Prevention of Pollutions and Nuisances. The Industry and Mining Services, as technical units, are put under the same responsibility even if their management remains at the Ministry of Industry.

The take-over by the Mining Service with its structure and pertinent competence of high risk installations was confirmed and reinforced by a circular letter on March 23, 1973 (2). These services, renamed "Interdepartmental Services of Industry and Mines" took the place of the previous regime the representatives of which were not generally characterised by technical competence but rather by legal or administrative training or experience (7, p. 42). It was decided in addition to pursue a programme over several years, of reinforcing the Mining Service for the inspection of classified establishments.

These new means, developed and coordinated on the national level by the SEI were to find a more adequate basis in the law of 1976 which was adopted to replace the law of 1917.

What are the contributions of this new law which are detailed in the decree of September 21, 1977? Let us mention some essential points here.

The field of application of the law of 1917 is considerably enlarged. a) On the one hand the "judicial" approach to the problem of risk is abandoned: the law of 1976 concerns all activities mentioned in the nomenclature (established by the Council of State with reference to the volume of activity and the nature of products), whether this activity is exercised by a natural person or a legal body, whether public or private. The law of 1917 referred only to activities of an industrial or commercial nature and excluded government establishments; petroleum installations also escaped the regulation for classified establishments: they were reintegrated into the general regime by the law of 1976. The same applies (in part at least) to the railways which have a separate regime. However, basic nuclear installations have been kept outside the field of application of the law of 1976*.

Let us mention also that transport remains outside the 1976 legislation.

b) On the other hand the interests protected by the law are also enlarged. The preoccupation with 'nature and the environment' in general singularly broadens the concern for the protection of the neighbourhood. In addition, and this is essential for our purpose, the notion of 'safety' makes its appearance in the text of the law: there is no longer reference just to the 'neighbourhood' which with present day risks would be insufficient.

These are the provision of the first article of the law:

Article 1: Subject to the provisions of this law are the factories, workshops, warehouses, work sites, quarries and generally installations operated or maintained by any person or legal entity, public or private, that can present dangers or inconveniences be it to the comfort of the neighbourhood, to health, safety, public hygiene, agriculture, the protection of nature and the environment or the conservation of sites and monuments.

The control of installations is clarified, the powers of the administration are strengthened. a) The previous classification is simplified by a regrouping of the first two categories of the law of 1917. From now on two types of installations are considered: those subject to authorisation (the most dangerous ones) and those subject to declaration (the previous third category) (Article 3).

b) The police commissioner has his functions as arbiter in the issue of permits strengthened. On the central level the administrations may draw up,

---

*One notices that at all times key elements of economic life seem to have been set outside the general regime and dealt with in special laws.

in addition to the circular letters as in the previous regime, orders giving
technical regulations by branch of activity (Article 7). The central
administration steps into the role of the police commissioner where permits
involving installations that affect several counties or regions are concerned.
(Article 5).

c) The police commissioner can in the event, after warning has not been
heeded, order a suspension of activity without having to refer to the minister
as was the rule previously, and this suspension may not relieve the operator
of his obligation to pay the salaries of his employees (while previously the
ordering of this requirement was left to the discretion of the police
commissioner) (Article 25). By decree of the Council of State, upon advice
from the competent consultative committee, the higher council for classified
installations, the discontinuation of an installation which presents "dangers
or inconveniences which cannot be eliminated by the measures foreseen in this
law" can be ordered. (Article 25).

The responsibility of the operator is clarified and strengthened. The decree
of September 21, 1977 foresees a certain number of obligations for industry.
A general obligation and three more specific rules concerning the problem of
safety shall be mentioned here.

a) The operator has to submit with his permit application a study of the
impact as foreseen in the law on the protection of nature of July 10, 1976
and this regardless of the cost of the project (the law of July 10, 1976
foresees a minimum of 600,000 FF and exonerates projects of lower cost from
the study). This obligation is spelled out in Article 2 of the decree,
paragraph 4; the industrialist has to attach to his application:

Article 2, 4: The impact study foreseen in the article of the law of July
10, 1976. This study shall indicate the elements which characterise the
existing situation with regard to the interests covered by Article 1 of the
law of July 19, 1976 and shall bring out the foreseeable effects of the
installation on its environment as concerns these interests.

b) The operator shall also attach to his application a study of the risks
presented by his project. This is spelled out in Article 3, 5 of the decree;
there shall be submitted:

Article 3, 5: A study setting out the dangers which the installation may
present in case of an accident and justifying the measures to reduce the
probability of its occurrence and its effects as determined under the
responsibility of the applicant. This study shall spell out in particular,
taking into consideration the public means of rescue known to him, the
consistency and the organisation of the private means of rescue at the disposal of the applicant or such as he has ascertained to be available for
combatting the effects of an eventual accident.

c) The operator shall advise the administration of any significant modification of the intended installation:

Article 20: Any modification made by the applicant of the installation, of
its manner of utilisation or its environment of such nature as to cause a
notable change of the elements of his application for a permit shall be
brought to the attention of the police commissioner together with all information required for evaluation before it is carried out.

d) Finally an essential article: all incidents must be reported to the
administrative authority:

Article 38: The operator of an installation subject to authorisation or declaration has a duty to declare without delay to the inspectorate for classified installations all accidents or incidents that have occurred on account of the functioning of this installation which are of such nature as to have a bearing on the interests mentioned in Article 1 of the law of July 19, 1976.

Information of the public. The permit for the opening of a classified installation may only be granted subsequent to an enquiry carried out by the police commissioner's office and destined to inform the police commissioner of the opinions of persons and organisations of the area concerned in the project (Article 5 of the law and the decree). A prefectorial order prescribes the public enquiry which shall take one month; it specifies the hours and the place where the public can inform itself about the application, the name of the enquiry officer, the perimeter within which the notice to the public shall be posted. The file submitted for enquiry includes all documents submitted to the administration by the applicant with the exception, as the case may be, of such parts as are considered to be industrial secrets.

The enquiry will result in a report by the enquiry officer put to the public in the open register and to consultative advice or to the municipal councils concerned. The operator may reply in writing (within twenty two days). The officer then has eight days to formulate a considered opinion on the subject. Everybody can peruse the reply by the applicant and the reasoned conclusions by the enquiry officer at the police commissioner's office. Subsequently, if there has been a positive decision by the police commissioner, a copy of the order of authorisation may be viewed at the mayor's office (4, p. 96).

Various consultations. The police commissioner is held to consult a certain number of services for advice: the inspectorate for classified installations, the county's civil defence services, the health and social services, the equipment services and the agricultural services in every case; other services may also be consulted. In addition, the county's Hygiene council (CDH) shall also be consulted; this is a proceeding in which representatives of a number of social groups get together (administration, employers, electorate representatives, associations etc). However, the advice of the CDH is in no way binding on the police commissioner.

In the case of large projects the advice of the General Council may also be obligatory (Article 15 of the decree). For projects involving several departments the General Council may be consulted, and the Ministry of the Environment rather than the police commissioner is responsible for the issueance of the permit. (Article 16 of the decree). This change of competence is also generally applicable for installations belonging to government services or organisations (Article 27 of the law). Where the minister is responsible for the issueance of the permit the Higher Council for classified installations, an open organisation such as the CDH is on county level, is consulted.

These are the essential elements of the central law on matters of industrial risk in France.

Without being exhaustive we may add two complements to this exhibit. On the one hand there exists general basic legislation which further strengthens the means available for the control of risks. On the other hand there is specific legislation which regulates certain activities that are not covered by the law on classified installations.

From among the general laws which supply a basis for the law of July 19, 1976 the following three may be mentioned:

Law of July 10, 1976 concerning the protection of nature (decree of October 12, 1977 (8). This law stipulates a general obligation for the respect of the natural and human environment; its decree spells out the content of the impact study which every applicant must submit for a project covered by the law. We mention here Article 1 of this law:

Article 1: The protection of natural spaces and landscapes, the protection of animal and vegetable species, the maintenance of the biological equilibrium to which they contribute and the protection of natural resources from all causes of degradation which threaten them, are in the public interest.

and Article 2.2 of the decree:

Article 2.2 of the decree: The study of the impact on the environment must specifically include: an analysis of the effect on the environment and in particular on sites and countryside, on fauna and flora, on the natural environment and the biological equilibrium and, as the case may be, on the comfort of the neighbourhood (noise, vibration, odours, light emission) and on hygiene and public health.

The code on town planning. The inclusion of risk in the document on town planning (POS) is foreseen in the code on town planning, Article R. 123-18.2. This article establishes the principle that specific measures which can go as far as a ban, may be opposed by third parties where the possible occurrence of a risk justifies it.

The circular of February 24, 1976 by the Ministry of Equipment and the Interior Ministry completes the preceding law and specifically takes into account industrial risks. This circular spells out the participation of those in charge of civil safety in the preparation of the documents on town planning and the contents of these documents (SDAU and POS), in particular the demarcation of areas exposed to risks, whether natural or other, the measures of protection against these risks, the provision of a rescue organisation system (9, p. 98).

Beyond the POSs various developments relative to a chosen perimeter may be subject to special rules that become necessary because of the existence of classified installations (Article 2.421-8 of the law of December 31, 1976 on the reform of town planning (5, pp. 15-20).

The law of July 15, 1975 concerning the dumping of waste and the recovery of materials. We mention the following articles from this law (10):

Article 2: Any person who produces or is in possession of wastes in conditions which may cause noxious effects on the soil, on flora or fauna (...) or pollute air or water (...) or generally affect human health and the environment is held to arrange their disposal in accordance with the provisions of this law in such a way as to avoid these effects.

Article 6 (11): The manufacture, the storage for the purpose of sale, the offer for sale, the sale and the making available to a user in any form whatsoever of products which generate waste may be regulated with a view to the disposal of such waste or, if needed, banned.

Article 8: Companies manufacturing, importing, transporting or disposing

of wastes which fall into the categories defined by decree to be capable, as
they are or when they are disposed of, of causing such harmful effects as
are mentioned in Article 2 are held to submit to the administration all
information concerning their origin, their nature, their characteristics,
their quantities, their destination and the way the wastes which they cause
are remitted to third parties or taken in charge or disposed of.

The decrees on the application of section VI of the law have been published.
They refer to the creation of a National Committee for the disposal and the
recovery of wastes (decree No. 76-472 of May 25, 1976) and of the National
Agency for the disposal and recovery of wastes (decree No. 76-473) (9, p. 104).

As concerns the legal and regulatory arrangements for specific sectors we
mention two examples:

*Law of July 12, 1977 concerning the control of chemical products (decree of
January 15, 1979)*. From this law (11) which excludes radioactive substances
from its field of application the following provisions must be stressed:

Article 1: The provisions of this law aim at protecting man and his environment against the risks that may result from chemical substances, i.e.
elements and their combinations as they are found in their natural state or
are manufactured by industry, be it in a pure state or in preparations.

Article 3: Before starting the manufacture for commercial purpose or the
import of a chemical substance which has not been on the French market any
manufacturer or importer must submit a declaration to the competent administrative authority. If the substance presents dangers to man and his
environment he shall indicate the precautions to be taken to counter them
(...).

This declaration is part of a technical file which gives the elements for
evaluation of such unacceptable dangers and risks as the substance may
present to man and his environment.

The content of this file is spelled out in Article 3 of the decree:

Article 3 of the decree: If the conditions of marketing the substance, in
particular the quantities or the type of use or distribution, or if the
result of a test as provided for in the preceding article justify it the
file must provide complementary data, particularly on the possibilities of
accumulation of the substance in the receiving environment and the people
and animals living there, its possible diffusion from one environment to
another, its effect on chemical processes in nature, its various forms of
toxicity at medium and long term for man and other living beings and
especially its mutagenic, carcinogenic, teratogenic properties and finally
all data which are required for the evaluation of the dangers to man and his
environment caused by the substance.

Article 6: The administrative authority shall keep secret all information
concerning the exploitation and manufacture of the substances and preparations while making sure in appropriate form the publication of toxicological information obtained from the examination of the files on said substances
and preparations (...).

Article 9 of the decree: A commission for the evaluation of the ecotoxicity
of chemical substance is to be set up and attached to the Ministry of the
Environment. It was created by ministerial order of June 8, 1979.

This commission deals with the technical files and gives its opinion on all questions submitted to it by the minister and formulates the recommendations concerning the pollution of the environment by chemical products.

By order of April 10, 1980 (10) the Minister for Health and Social Security set up an organisation of pharmacovigilance with the aim of collecting information on all serious accidents apparently connected with the use of pharmaceutical products and of all incidents and accidents of which he has reason to suspect that they may be connected with the use of pharmaceutical products and to set up the enquiries that have been decided on the national level (Article 2).

By the same order an organisation for toxicovigilance was established with the aim of systematically collecting, from all organisations called upon to treat acute and chemical intoxications, such chemical information on accidents apparently linked with the use of non-medicinal toxic products (Article 10).

The two activities are coordinated by a commission for toxico-pharmaco-vigilance (Article 13).

Regulations for nuclear installations

The laws. During the debate on the adoption of the 1976 law on classified installations it had to be stressed several times by the minister that nuclear centres are outside the field of application of the new law and would be covered by special laws.

We mention here only some aspects of this legislation on nuclear installations which legislation has been growing incessantly since the 1960s.

In the first place it must be emphasised that in France (an exceptional situation among OECD countries) there is no general law on nuclear installations. All the laws governing this activity are therefore drawn up by the single regulatory authority, and parliament has only to pronounce on them. Thus one finds a large number of provisions given in the form of decrees which apply various laws like those of 1917 on classified establishments, of 1961 on atmospheric pollutions, odours etc.

The essential law among these multiple provisions is the decree of December 11, 1963 (13) which establishes the list of "basic nuclear installations", stipulates that they can only be created after a permit has been issued by decree after a public enquiry and advice from an interministerial commission and an approving opinion from the Minister for Health. They are subject to control by inspectors especially assigned to this task and control by inspectors from the Central Service for the Protection against Ionising Radiation (SCPRI). This law was completed and modified by a second decree of March 27, 1973.

The organisations established (13, 14, 15, 16). The Minister of Industry is the minister in charge (directly or in the final instance) of:

The organisation for study and research: the Atomic Energy Commission (CEA).

The operator: Electricité de France.

The national safety service: the Central Safety Service for Nuclear installations (SCSIN). Created by decree on March 13, 1973 this service is in

charge of preparing and implementing all technical activity concerning
nuclear safety and its regulations; it coordinates the safety studies
carried out by the various organisations involved. It is in charge of
guiding all authorisation procedures.

The safety organisation set up inside the CEA (order of November 2, 1976):
the Institute for Nuclear Protection and Safety (IPSN). The IPSN is in
charge of carrying out the nuclear protection and safety studies which are
requested by the administrations involved.

The organisation in charge of preparing the safety evaluation reports which
are submitted to the permanent groups (see below): the Department for Nuclear
Safety (DSN). As a part of IPSN it provides technical support for SCSIN.
It was created by a decision from the Minister of Industry on March 27,
1973 (modified on December 17, 1976).

The Minister for Health is in charge of: the service in charge of the
protection of persons: the Central Service for the Protection against Ionising
Radiation (created on November 13, 1956).

In addition there are:

The Interministerial Committee for Nuclear Safety, created by decree of
August 4, 1975. It is chaired by the Prime Minister and has a general
secretariat. It has a coordinating function.

"Two permanent groups" of experts created by the ministerial decision and
instruction of March 27, 1973 from the Ministry of Industry (one for the
nuclear centres, the other for all other nuclear installations) are in charge
of examining the safety of the installations and work out a certain number of
'recommendations'. The group in charge of nuclear centres consists mainly of
experts from the CEA and the EDF who have to give their personal opinion on
safety.

The Interministerial Commission for basic Nuclear Installations (created by
decree of December 11, 1963) provides the opportunity for various ministries
to express their opinion to the Minister of Industry on requests for the
creation or modification of basic nuclear installations.

The authorisation procedure for the creation and control of installations.
In short: the regulatory activity of the public authorities in this field is
exercised along three principal and mutually complementary lines:

- a system of individual authorisations for each installation;
- the drawing up and application of general technical rules;
- surveillance.

The request for authorisation is accompanied by a "preliminary safety
report". This report is analysed by the Department for Nuclear Safety at
the IPSN which reports this analysis before the "Permanent Group" involved.
This group subsequently transmits the file to the SCSIN which gives its
opinion on the safety of the installation and spells out the conditions that
must accompany the authorisation. The SCSIN establishes a draft decree of
authorisation and consults the Interministerial Commission for basic Nuclear
Installations; the approval from the Minister of Health is then requested
before the authorisation is signed. In the case of reactors a "provisional
safety report" is established six months before fuel is loaded; it goes
through the channels up to the SCSIN which proposes the authorisation of the

loading to the minister. Finally, before normal operation starts, the operator has to submit a "final safety report" which leads, under the conditions provided, to the authorisation for the start of operations.

In parallel with this there is a public enquiry at local level to obtain the declaration of public usefulness. This consultation is of the same type as the one provided for classified installations. The conclusions by the enquiry officer are sent to the Minister of Industry by the police commissioner. They then enter into the national procedure which will take its course and may finally lead to the necessary authorisations.

2nd. The case of Great Britain

Confronted with the shock of Flixborough the British Government, as we have seen, went beyond the simple setting up of a commission of enquiry. On June 27, 1974 it announced its intention to set up a committee charged with the examination in its entirety of the problem of industrial risk. In this way Flixborough was to determine, or at least ease, changes in the administrative organisation of industrial safety management; on the level of legislation the explosion of June 1, 1974 was to bring about an impulse or at least a more determined conviction among parliamentarians to adopt a new law to replace the Factories Act of 1961 which was founded on largely obsolete principles. This new law, the Health and Safety at Work Act, was promulgated on July 31, 1974. It was to make the long wanted administrative reorganisation possible, that had been proposed by Lord Robens' Committee on Safety and Health in their report in 1972*.

Flixborough, like Feyzin in France in its way, permitted the strengthening of convictions and made attempts at realisation more urgent. One must not, of course, see this event as 'the' cause of the rigidities observed (the same is also true for Feyzin). It is therefore indicated to present the evolution of the British system of legal and administrative regulations of major risk in general fashion at least in its general features. The case is very interesting because of the dynamism observable in this respect on the other side of the Channel during the last decade (17).

a) A very old tradition, somewhat upset after the second World War.

Eighteenth and Nineteenth centuries. The British industry of the period was certainly not prepared to tolerate a control of its activities (18). A quick glance is sufficient here (19; 20). When in 1848 the daily working hours were reduced to ten hours per day for women and children the Manchester manufacturers responded with the setting up of an association aimed at the abandonment of such "undue restrictions" and "inadmissible interferences". In 1856 they won their case in parliament. Against this background we may mention a few dates which mark a certain, at least theoretical, concern** about regulation:

    1833:   The Factory Act regulates the use of children and women in the textile industry. Powerless, the judges saw themselves prevented from appointing the first four inspectors to enforce the law.

---

*The background to the Robens' report is described in *"Hazard Control Policy in Britain"* by John C. Chicken, Pergamon Oxford 1975.

**A concern in which one must also recognise the signs of a rivalry between landowners and industrialists, the former viewing the introduction of regulations as a means of limiting the powers of their rivals.

1842:   The Mines Regulations Act forbade the employment of women and children underground.

1860:   Industrial legislation expands beyond the textile and mining industries.

1875:   Explosives Act.

1878:   Law on the protection of dangerous machine parts.

1901:   The Factories and Workshops Consolidation Act introduces a first overall codification of hygiene and safety matters.

Twentieth century. The laws multiply so as to face the various aspects of industrial development. For instance: laws on petroleum (1928), new law on factories (1937), on work accidents (1946), on radioactive substances (1960), on nuclear installations (1965) etc.

In the late 1960s. The focal point is the Factories Act (revised a third time in 1961). Once again preparations are being made to amend this law on factories because it has appeared less and less adequate. Consultations for this purpose started in 1967. The results leave one perplexed: the law becomes monstrously big without, however, covering all workers (several millions of them remain without legal protection).

1970. A new orientation takes shape: the 'hold-all' approach, called the 'telephone directory method', will be abandoned in favour of a basic law defining mainly principles to be completed as and when required by the evolution of techniques by regulations or specific codes of conduct. Among these main principles is the preoccupation with major risk. It had already been clearly intended in 1966.

b) 1967-1972: time of reflection on industrial legislation and particularly on questions of major risk*.

1967: The report by the Chief Inspector of Factories and its consequences. It is in fact in 1967 that the problem of the protection of people who are exposed to the risks of large scale industrial accidents appears in an official document. This was the report by the Chief Inspector of Factories (annual report for 1967) from which we quote the following lines:

Since the start of the industrial revolution disasters causing a consider-able number of deaths have periodically occurred in industry. At the end of the nineteenth century boiler explosions were recurring events (...). Legislation, inspection and a better appreciation of risks have substantially reduced the dangers from these occurrences; but the enlarged scale of modern industry, the infinitely larger size of factories; the higher speed of most machines, have created the possibility of explosions (...) at least comparable in scale and violence to the explosions of the past (...). In addition the size of modern industry has led to the storage and use of very large quantities (often thousands of tonnes) of potentially dangerous materials such as acrylonitrile, liquified gas and liquid oxygen (...) (21, p. 328).

---

*For a general description of the system of British industrial laws cf. 17. We restrict ourselves here to the problem of major risk, an important but not the only dimension in the recasting of the British system after 1967.

Along these lines the Department of Employment (21) created a working group in May 1969 (Major Hazards Working Group): as an interministerial group it had to arouse reflection about the problem of risk in every administration. It appeared that the ministries were very poorly informed about the new industrial risks in the country (except for special cases such as the nuclear industry). Consequently, a technical working group was set up under the Chief Inspector of Factories to identify highly dangerous substances and the volume determining major risk on the level of storage. The various administrations agreed on a list and contacts were established with employers' organisations and the unions in November 1970. During the same months the Department of the Environment issued a circular recommending that the Factory Inspectorate be consulted before permits for dangerous installations were issued (these were issued by local authorities). In addition, a special working group was set up within the Chemical Branch of the Factory Inspectorate to deal with the special dangers of the chemical industry.

1970–1972: the committee chaired by Lord Robens. These various impulses were soon relayed through the great efforts put forward by the committee under the chairmanship of Lord Robens (1970–1972). On May 29, 1970 Mrs Barbara Castle, then Secretary of State for Employment and Productivity, set up a special committee to conduct an enquiry to examine the whole existing and desirable legislation in matters of hygiene and safety. Lord Robens was entrusted with the chairmanship of the group. The mission with which he was entrusted was not only to carry out an analysis and to formulate suggestions on the administration of industrial risks but also: to question and make recommendations on supplementary arrangements to be provided for the protection of the public against risks other than those of the more general pollution of the environment and connected with industrial activities, commercial establishments and public works sites (22, p. 5).

We shall not dwell on the general recommendations of the report published in 1972. Inspired often by a liberalism of another era — "there are too many laws", "self-control would do the job" ... — they are hardly remembered (even if as the always displayed British tradition of consensus wants it, they were very well received). Certain pertinent criticisms received more attention: laws overlapped each other while large areas were left unregulated; split administration is not a very apt tool; nine different laws, five ministries directly involved, seven inspection groups but millions of people unprotected. In normal circumstances the stick is too big, and in case of danger the system is outright disastrous as the case of the fire at Dudgeon Wharf shows which was reported by the committee.

Dudgeon Wharf 1970. The fire brigade had been questioned about the risks involved in the demolition of some empty tanks which had held a particular hydrocarbon. After a visit to the site they referred the issue to the Factory Inspectorate. The inspector assigned to the task asked himself, after a visit, whether the affair came under the Factory or under the Construction Regulation Act of 1961 and referred it to his superior. Nobody can be blamed. The system required that clear answers were given. The real question was nevertheless: who is best equipped to deal with the problem? Two weeks later the proper authority had still not been ascertained ... but the explosion occurred. We do not want to say that the organisational 'buck-passing' was the cause of the disaster but this illustrates a situation to which the present laws and administrative organisation can lead: uncertainty, delays, communication problems (...) due to questions of demarcation. It tends towards an inability to respond to the problems. It has much to do with the complexity of the organisation.

The case of Dudgeon Wharf is a special one: generally speaking, the effect of fragmentation is too diffuse to venture an expert opinion. However the damaging effect is no less irritating (22, pp. 10-11).

These statements led the Robens committee to recommend the setting up of a unified authority regrouping all administrations involved and based on a law comprising the problem in its totality. But we shall dwell here*mainly on the very interesting analyses by this committee on matters of major risk.

a) Fundamental rethinking of the safety issues. Prudence commands to ask ourselves: have we reached a plateau in matters of safety? Have we not, on account of our traditionalist approach to safety and control, come to the point of diminishing returns? Despite strengthened measures on the level of safety, is there nothing to be feared on account of the scale and the increasing complexity of modern industry that creates new risks and is capable of modifying present tendencies in the long term?

We have doubtless reached a point in time when the problems of safety must be radically rethought. It is true, where for instance toxic substances are concerned, there is nothing very new: asbestos was used in the past and caused problems then. What is new is the rapid increase in the number of chemical substances and compounds brought out and marketed by industry; it is the scale of their use. Without giving way to scaremongering one must at least ask whether our traditional approach to these questions is still pertinent. A second illustration: inflammable and explosive materials. Despite the progress made with regard to control it would be foolish to put forward the hypothesis that the improbable could not happen (22, pp. 3-4).

b) Always to come in after the event. It is a disastrous fact that preventive legislation has seen the light of day only after unforeseen disasters. There is the tragedy of thalidomide which led to a new system of control for medicines; there is the Aberfan disaster which was followed by new legislation on the control of mine pits; there are the deaths of eight people in a fire at Eastwood Mills (Keighley) which resulted in a change of the Factories Act of 1959 ... The safety system should present an answer to the future and not as it is now only to the past (22, p. 4).

c) An evolved society. Again reviewing our approach to safety issues there are reasons of another nature. The present body of laws on hygiene and safety has its roots in previous legislation which was introduced in a totally different social context. The attitudes, the expectations towards authority and decision making have changed. More and more, employers and employees expect to take an active part in the elaboration and implementation of such laws. How does the traditional approach to regulation meet and use these expectations?

We, therefore, have no reason whatsoever to be sure that what we used to do with regard to risk is adequate to a situation which we know to be in a process of evolution (22, p. 4).

These preoccupations of the Robens committee were not to remain dead letters. In 1974 the move was made from reflection to action.

d) 1974-1976: The recasting of a system that was one and a half centuries old.
The Health and Safety at Work Act of 1974 provided the global framework which

---

*We follow the text of the report very closely without offering a word-by-word translation.

had been sought for a more adequate approach to industrial risk (23). A unified authority was to regroup all inspection bodies and be given rather large autonomy. Being a technical organisation, this 'Health and Safety Executive' was to be covered by a political body, the 'Health and Safety Commission', a tripartite body made up of representatives*from the employers' side, the workers and the local communities, the whole thing coming under the ministerial authority of the Department of Employment. This law is, at the same time, more comprehensive, more flexible and more effective, appealing to the responsibility of the agents involved while also providing for very strict constraints in case of manifest infringements.

These general principles are applied to the problem of major risk which the law does not ignore. The employer is also responsible for the safety of people whom he does not employ (Article 3). In the same way the employee must pay attention to the health and safety of others (Article 7). All this, however, "as far as is reasonably possible" (23).

To exercise its new responsibilities, especially in matters of major risk, the Health and Safety Executive (HSE) had to set up a number of technical and consultative bodies.

The Major Hazard Branch. This is the realisation of the desire that was expressed after Flixborough to see the creation within the Factory Inspectorate of a group of experts who are capable of assisting the administration in defining a policy to control major risk.

The Advisory Committee on Major Hazards. This is one of the consultative committees at the HSE. It consists of five independent groups: identification of major risks, safety, interaction of risks, planning, research.

This committee whose members come from various sources (industry, works councils, universities ...) was set up at the end of 1974 with the following mandate (24):

To identify the installations with the exception of nuclear installations** which may present potential risks for its employees, for the public and for the environment and to formulate advice on the measures to be taken for their control (appropriate to the nature and importance of the danger); the establishment, the siting, the design, the functioning, the maintenance and the development of these high risk installations; the industrial and non-industrial development in the neighbourhood of such installations.

The 'Risk Appraisal Group'. The group was set up to make special studies on given problems concerning major risk. When a file is sent to the HSE for advice to a regional authority it is examined by this evaluation group which consists of high-level specialists and inspectors and represents a strong expert capacity. All important files are reviewed by this group. Since its establishment in 1974 the 'Risk Appraisal Group' has dealt with more than three hundred cases (especially the rebuilding of the factory at Flixborough and the oil operations in Scotland). After examination it returns the files with its advice to the HSE which returns them to the regional authorities accompanied by its judgement.

---

*'Representatives', not 'delegates' as British custom requires (the delegate is bound to his basis which makes the attainment of consensus too difficult)

**Dealt with in another context.

Later on we shall come back to the limitations of this system. Here we shall stress one of the most outstanding achievements under the responsibility of the HSE: the safety study of the industrial area of Canvey in the Thames estuary. This was an achievement which is "unique in the U.K. and probably in the world" (25, p. V).

3rd. The case of Italy

Feyzin and Flixborough occurred while France and Great Britain had already initiated reflections on the changes to be made in the existing systems. Seveso occurred in a context which was much less prepared to respond to the event and, more generally, to face the difficulties of major risk.

a) The burden of history. Italy (26) is doubly handicapped by her history in facing up to today's major risks (27). The late unification of the country in the first place laid a responsibility on the mayors of the 10,000 communities of the peninsula which they obviously could not shoulder.

The fascism of the 1930s was to accentuate the difficulties. It confirmed the power given to the chiefs of local communities, the 'podesta', representative of the police commissioner, appointed to replace the traditionally elected mayor, which removed all local control of risks caused by industry. Some central bodies were created (such as the national association for the control of combustions, ANCC, for pressure apparatus) but the reluctance to control industry did not make them effective services. The ban of unions which were replaced by organisations controlled by the employers removed a last assurance. A basic law, the Testo Unico, on public health issues was passed in 1934 but, still unchanged forty five years later, it appears most insufficient, prescribing for instance the removal of certain installations from the centre of communities (a rather derisory provision nowadays when the neighbourhood of communities is no longer rural land but an urban network) or regulating still only dangerous substances "used, stored or transported" (dioxin is neither used, nor stored nor transported).

After the war, with the opening of the borders, the reinstatement of unions, the shock was very hard. There was only one easy approach: turning to the exploitation of the only factor of production that was still freely accessible: the environment. This was then the policy of the accepted risk. At this price and without any iron or oil of its own Italy was to become one of the most industrialised countries of the world. L. Conti emphasises:

This is the effect of the outdated laws which to this day have left us with ossified organisations, a lack of technical groups in the public administration, the decadence of education and scientific research (27, p. 7).

b) A certain legal and administrative void. What is the situation at the time of the Seveso disaster? A basic law is available to the mayor: the Testo Unico which, as we have seen, is no longer an adequate tool. Some central bodies (of the ANCC type) should in principle help the mayor in exercising controls but for lack of means this remains very theoretical. To remove from the system what could still be effective each organisational body has developed its own rules and jealously emphasises its prerogatives which often created an illusion of safety when the existing rules were applied at the start.

Some lines written by Italian observers illustrate the sad reality of 1976:

The impotence of the prime responsibility: the mayor. A small community does

not have the benefit of a competent legal service and of solid technicians,
therefore no support for its mayor and consequently the overall competence
accorded to the mayor by the Testo Unico ends up by dissolving into pure
formality and the Damocles sword of legal sanctions which dangles above the
heads of all mayors in Italy and which for this very fact does not come down
on anybody's head (28, p. 55).

<u>The administration: not a tool but a lot of divisions</u>. One notices that in
a system of fragmented standards to which corresponds a multitude of organisational bodies everybody wants to be in charge of particular functions. The
practical effect of such a situation is that every organisational body tends
to have a narrow vision of its competence without consideration for the safety
problem as a whole. In addition, such an approach gets narrowed down even
more as the technical administration understands its function in a formal
fashion and restricts itself to acts of communication between other services;
these acts hardly accomplish the protection of public interests but only the
fulfillment of prescribed duties and relieving the service in question
cheaply of its responsibility (29, p. 101).

<u>Rules born under the sign of partitioning</u>. The rules in force concerning
safety valves seemed to be made so that the gases should be prevented from
reaching airtight chambers which would, however, be necessary in order to
avoid the release of toxic substances in case of a rupture. This amounts to
separating the concept of release under pressure (a phenomenon belonging to
the thermodynamic and mechanical field) from the properties of the substance
that is released (a phenomenon belonging by contrast to the field of
chemistry) (28, p. 55).

c) Some deficient — in some cases: worrying — laws. Various laws have been
introduced to complement the one of 1934:

<u>The law on atmospheric pollution of July 13, 1966</u>. It covers only part of the
field and controls only some ten parameters, regulating the emissions without
paying attention to the load capacities of the areas.

<u>The Merli law of May 10, 1976 on water pollution</u>. It started by giving the
freedom to pollute (on the level previously attained by everybody) for three
years by abrogating all previous regulations. For the future it set a
restraint only on the concentration of pollutants (heavy use of water therefore satisfies the regulation which is rather worrying where the Italian
piped water system is concerned) and provides a massive transfer of responsibility on the administration if one follows Laura Conti:

> The recent Merli law on the discharge of liquids is an example of how laws
> must not be written. For instance, it does not set any fine on the empty-
> ing of a demijohn of corrosive sublimate into a river but only for emptying
> it without administrative permission, so that instead of making the one who
> pollutes an offender it becomes an offence of the official who should have
> refused permission and who by not having refused it explicitly, has
> implicitly permitted it (incredibly, the law provides in fact that permis-
> sion is tacitly given if it has not been explicitly refused (28, p. 53).

d) A set of rules that must be understood in its entirety. Seveso occurs in
this setting which presents two opposite images.

<u>The Theory</u>. As expressed in Articles 32 and 41 of the constitution of the
Italian republic (perhaps the only one in Europe to mention the question of
industrial safety):

The republic protects public health as a fundamental right of the individual
and as a collective interest.

Private economic initiative is free; it must not be exercised in contradiction to social usefulness or in such a way as to cause harm to the safety,
freedom and dignity of man (Articles 32 and 41).

The Practice. As expressed in this observation which was made on the
occasion of the Seveso drama: public control presupposes a coherent and up-
to-date legislation with a service network equipped with tools, men and
financial means adequate to the requirements of a technically advanced
society. Our country lacks sufficient legislation and does not have the
public structures to ensure respect for it (30, pp. 38-39), or in that other
observation which was also made after the accident of July 10, 1976.

The Truth. Is nowadays that there exists no public structure in a position
to give preventive advice and no regulation which obliges an industrial
employer to declare the risks of his activity and the preventive measures
he expects to take.

Today the permit for an installation is like a blank cheque given to the
industrial employer because no administration is in charge of control. One
may simply refer to some archaic and obsolete regulation such as the one
giving a list of unhealthy activities which in theory may not be installed
in urban agglomerations. The application of this law would lead to the
closure of a great many of the factories in Lombardy (30, p. 39).

One must, nevertheless, understand that a society cannot, not even under
the shock of a drama like the one at Seveso, fundamentally change a
legislative and regulatory entity which represents more than itself, i.e.
the image of an entire social organism. A year and a half after Seveso,
parliament in Rome adopted a draft law mentioning (in Article 24) that safety
issues would have to be analysed in the context of the "requirements of
production" (31, p. 11).

This notwithstanding, consultations are underway between various ministries
to work out new proposals in matters of industrial safety. These discussions
might reach the legislative level within the next few years. It would remain
to give the administration the necessary means and then to aim at proper
functioning of the system that would be established.

2. THE BATTLE AGAINST DISASTERS

We shall restrict ourselves here to the case of France and to some
essential reminders on means provided for the battle against disaster.

1st. History

1810: Up to the Revolution rescue organisation was left to charitable
organisations, usually religious orders. In Paris, however, professional
fire fighters appeared in 1716. A law of August 16-20, 1790 entrusted the
mayors with providing suitable precautions against accidents and calamitous
plagues. (32, p. 12). An order of Messidor, year IX, put the Paris professionals under the authority of the police commissioner of the department
Seine and the prefect. The fire at the Austrian embassy in 1810 led
Napoleon to dissolve the civilian body and to entrust the fire fighting
service for the city of Paris to an infantry battalion (33, p. 44).

<u>Twentieth century</u>: The city of Paris was therefore better protected but this was not the case for the remainder of the country. It is true that the principles of the law of 1790 were confirmed in the law of April 5, 1889 but they were hardly put into practice. The municipalities, mainly for lack of means, remained poorly organised where the battle against accidents was concerned. While in 1898 for the first time the Finance Act had provided subsidies to the communities for their fire fighting services the notion of safety remained still very vague and on the eve of the second World War two thirds of French communities were without fire fighters (32, p. 3).

<u>1938</u>: The fire at the "Nouvelles Galeries" in Marseille in 1938 was decisive. It caused the necessary psychological shock, bringing it home that the rescue bodies could no longer be left to the exclusive care of local communities and that they had to be adapted to the size of risks ensuing. The modern means of transport, the communication networks gave rise to a re-thinking of the organisation which until then had been conceived strictly on a local level. Thus the idea was born that the government should guide, sustain and co-ordinate the efforts of the municipalities with a view to intervening directly with powerful means in extreme cases. (32, p. 3).

<u>The second World War and the cold war</u>: The laws of 1790 and 1884 dealt with civilian protection in peace times. The law of July 11, 1938 provided the organisation for wartime. The service created for this purpose on the national level, Directorate of civil defence, was to be placed under the authority of the Ministry of Defence. It was later transferred (law of March 16, 1942) to the Interior Ministry. In 1943 (law No. 597 of September 20) the Directorate of Civil Defence of the fire police was reorganised under a Directorate General for Civilian Protection.

The post-war period, across various events (creation of the French nuclear force, the Korean war, the large forest fires in Les Landes in December 1949), saw the confirmation of this integrated organisation of the defence against accidents seriously affecting the population. The decree of November 17, 1951 (No. 51-1314) upheld the option of 1943 and created the National Civil Protection Service which is in charge of ensuring, within the Interior Ministry, the survival of the nation in case of war and of protecting the life and the assets of the citizens in peacetime.

The service still had to be given a statute and financial means. An effort was agreed upon in 1952 (decree of January 17: creation of a certain number of jobs). An important step was taken in 1975: the service was ranked among the directorates general and the other directorates of the Central Admin-istration of the Interior Ministry under the name of Civil Safety Directorate (32, p. 3).

<u>2nd. The Civil Safety Directorate and the battle plans</u>

The principle of a 'rescue organisation' in the form of plans established in advance and in a uniform manner for the whole country was established by the interministerial order of February 5, 1952. The ORSEC plan is at the same time a repertory of means in terms of men and equipment and a document that outlines the tasks and the command organisation of rescue. Its frame-work is the department. The police commissioner as representative of the government is responsible for its implementation, the training of staff, the start, conduct and halt of operations. For the execution of his command duties he has at his disposal a general staff the head of which is the so-called department director of Civil Safety and an *ad hoc* group ("conference of heads of service" consisting mainly of technicians chosen according to the type of accident).

The operating principle of this ORSEC organisation is the coordinated utilisation of existing means. It does not provide a doubling of the basic means already existing in the departments whose day-to-day function is the fighting of current accidents and incidents; it is mainly the corps of fire fighters: 190,000 volunteers, 10,000 professionals and 7,000 military (in Paris and Marseille) which are also organised at department level (Department Fire Protection Service), on the communal level (Rescue Centre) and around more important points (principal rescue centres)*.

The ORSEC plan permits the addition of other means which the police commissioner considers useful (ambulances, public works machinery). If needed, reinforcements may be supplied by neighbouring departments and even by the region or on the national scale. If the situation requires it, the Civil Safety Directorate which has a general staff and operations room ensures coordination on ministerial level.

The ORSEC plans were completed by a series of additional plans like the one concerning hydrocarbons (worked out after the Feyzin disaster in 1967), or marine pollution (Polmar 1971); we also mention the ORSEC Tox plan (1972) for accidents involving dangerous substances and the ORSEC Rad plan (1974) for cases of danger from radioactivity.

Specific Intervention Plans. Complementing these structural plans "Specific Intervention Plans" (PPI) are increasingly established which are destined to guide the rescue operations in case of a severe accident more precisely. These plans deal closely with operations on the levels of internal emergencies as established by companies (such as refineries, petrol storages, hydrocarbon storages, nuclear centres etc.). These internal plans aim at internal accidents and are implemented under the responsibility of the factory director or the chief of the establishment until the ORSEC plan comes into operation which marks the transfer of command to the police commissioner (34, 35, 36).

On December 29, 1978 the Interior Ministry (Civil Safety Directorate) published a diagram of PPI types and its "recommendations for the presentation to and the activation of the population in the vicinity of electronuclear centres in case of incident or accidents which might eventually have radiological consequences" (34, pp. 91-109).

According to this text each specific plan must include:

A common body gathering the following elements:

- the authorities concerned,
- the methods for spreading the alert,
- the record of rescue means,
- the command structure,
- maps and plans.

An operational document stating according to the level of alert:

- the authorities concerned,
- a diagram for the diffusion of the alert,

---

*The strength of the fire fighting force, their equipment, their officering are determined by the order of February, 24, 1969. The density of the population concerned is the central criterion for the evaluation of necessary means.

- means of transmission,
- identification procedure for messages (preparation of standard type telegrams),
- methods of transmission,
- organisation charts for the command structures,
- description of the functions of participating parties,
- nomenclature of means,
- a statement of possible intervention scenarios.

3. THE COMPENSATION OF VICTIMS

1st. The commercial approach: insurance

The gathering of a rather large number of individuals who are exposed to the same risks, prior contribution by everybody, compensation in case of misfortune are the fundamental data of the insurance mechanism such as it has functioned since the most ancient times (37, p. 4). In order to function the mechanism requires, however, that one obvious condition is met: that the risk is insurable. This is the case in which a risk is recognised by a substantial mass of individuals who are sufficient in number to split between themselves in minimal fractions the cost of eventual damages. Not insurable are risks which are too multiple or too severe and affect those insured no longer in isolation but as a mass and attain the proportions of cataclysms (38, p. 4).

These conditions are met where automobile risks are concerned, in the insurance of 'water damage', theft etc. This is no longer the case when one gets into the field of disasters the typical features of which are their singularity and the exceptional weight of their effects. This is how the International Assembly of Insurance Technicians codified the notion of disaster in 1947:

A risk is defined as of a disaster nature if it is due to extraordinary causes such as natural events or human conflicts which strike people or things of which the immediate effects are on a scale and have considerable economic impact which does not show a characteristic of foreseeable periodicity and which therefore does not respond to statistical regularity according to contemporary scientific understanding (39, p. 17).

These few statements give a measure of the audacity of those who were the first to launch themselves into 'insurance' which in the beginning was based on a wager rather than on probability calculation. In the fourteenth century it was the banker's loan which permitted the fitting out of a ship and the attempt of a 'big venture', a rather risky undertaking. If the ship sank one owed nothing to the banker; if one arrived safely one paid the loan back plus half the cargo. This formula changed with the appearance of an inter-mediary, the insurer, who reduced the risk to the banker and the cost to the ship-owner. The first contract of this kind which has survived to the present was made in Genoa in 1347. Audacity was also needed for the launching of fire insurance: the Hand in Hand, the oldest insurance company in the world, was founded in 1696, only thirty years after the great fire of London.

One could not rely exclusively on audacity and chance. Insurance therefore very soon led to the introduction of techniques for dividing the risks which on account of the amounts involved, the exposure to disastrous events or the dangers of accumulation occurring surpassed the capacity of a single company or a single market.

a) Co-Insurance. Historically this was the first of the techniques for dividing risks. The same risk was divided horizontally between several companies, each one being responsible only for that part of the total cover which corresponded to its own capacity (40, p. 741).

Industrial development has spread this practice. One establishment may thus be covered by some 60 companies. To facilitate the working of the system one of the companies, called "apéritrice" took charge of the file and served as intermediary between the co-insurers and the insured.

b) Reinsurance. This is the vertical complement of co-insurance; it can be effected several times over. The insurer uses this mechanism to unload the risks which might exceed his own capacity on account of:

- their size which might hurt the homogeneity of his portfolio of risks;
- their statistical probability which may be little known;
- their small number which may cause unacceptable deviations.

The reinsurer receives therefore a very large variety of business coming from a large number of countries. The size of reinsurance companies, the diversity of their business, the technical means at their disposal allow the functioning of this mechanism.

Some figures*are of interest:

Number of contracts in the portfolio of a large reinsurer. The number of businesses insured for between 50 and 100 per cent of the full maximum of retention**is of the following order (41, p. 273):

- 150 zones of seismic accumulations
- 1,000 petrochemical or chemical risks
- 100 jumbo jets
- 50 climatic zones — hailstorms
- 100 ships
- 50 drilling platforms
- 50 land surface storm zones
- 150 nuclear reactors
- 1,000 construction or public works sites
- 300 large guarantee or credit companies.

Sums available for various events

Two to three billion dollars for the impact of an earthquake on a big city
300-700 million dollars on a drilling platform or a large building site
400 million dollars for third party liability in passenger aviation (80 for the body of a Boeing 747 for comparison)
100 million dollars for pharmaceutical third party liability (41, p. 271).

Overall capacities of the reinsurance markets

The American market collects for its own risks	4.5 billion $
The London market collects for its own risks	3.5 billion $
The rest of the world collects	15.0 billion $

---

*Supplied by F. Negrier, director general of Societe Commerciale de Reassurance (SCOR). The SCOR together with the Swiss reinsurance company ranks after the Muenchener Rueckversicherungs-Gesellschaft who was the first reinsurance company in the world, founded in 1880.

**Maximum limit of cover for which the insurer can commit himself.

World-wide reinsurance collections are in the vicinity of 25 billion dollars (41, p. 272).

Amounts paid out by the market (insurance and reinsurance) on the occasion of various disasters:

Hurricane Betsy, USA 1965   700 million $ (41, p. 271)
Hurricane Capella, Europe 1976   520 million $ (42, p. 38)

Ford, Cologne, FRG 1977   230 million $ (41, p. 26)

On October 20, 1977 the central spare parts store of Ford was almost completely destroyed by fire: 75,000 m^2 of warehouse space out of 100,000 m^2. The insurance covered material damages plus lost profit caused by the accident which was one of the costliest of the post-war period.

Flixborough, GB 1974.   60 million $.

The losses are made up of 48 million for material damages, 10 million for interruption of operations and 2 million for third party liability (41, p. 961).

Tenerife, Canary Islands 1977.   160 million S.

The losses are made up of 63 million for aircraft and 97 million for damages to passengers and third parties (41, p. 261).

Finally we shall emphasise, with F. Negrier, a further feature which the reinsurers are trying to take on: a certain attraction for adventure or the more unconventional risk.

The cover for disasters and new risks would be defaulting if reinsurance did not have an original feature which is doubtless its motor (motive force).

If like direct insurance it bases itself on the law of chance and large numbers by untiringly creating homogeneous risk communities, it sometimes shows very generous hospitality by welcoming if not in its own house then at least in its ante-room to heterogeneous communities of unknown, dangerous, unbalanced risks which it accepts without passport in order to redistribute them more tamely in the large worldwide circulation of risk (41, p. 273).

c) *Insurance pools.* For very special and very substantial risks the companies are sometimes very reluctant to commit themselves. But "aware of the practical problems raised by a refusal to insure" (40, p. 170) they create mechanisms called 'Pools' for these sectoral risks. They are groups of insurers and reinsurers who thanks to the gathering together of their means allow the mounting of a specific enterprise.

Nuclear pools. These are the oldest ones. They exist in the United States, Sweden and Great Britain. In France a pool was created in 1959. It gathers together almost all French insurance and reinsurance companies as well as foreign companies, more than one hundred companies. These pools strengthen each other through the links which they establish between themselves by means of reinsurance. In France cover is extended in this way to the operator of nuclear centres to the extent of 50 million FF per accident (13, pp. 376-378). The overall network of these jointly responsible pools provides a capacity in the order of $600 million (41, p. 270).

Pharmaceutical pools. Various precedents (thalidomide, enterovioform etc.) have been the cause for setting up this mechanism which offers third party liability cover to manufacturers. The pool has a capacity of about $100 million (41, p. 970).

GARPOL: French guarantee pool for pollution risk. The object of this body
(44) is to broaden the cover available in France against harmful effects.
Until 1974 damages caused by air and water pollution were usually excluded
from third party covers except for pollution of river, lake and underground
water if due to accident i.e. if they were unforeseeable and sudden. In May
1974 these restrictions were lifted by the French Federation of Insurance
Companies; a circular from the Technical Accident group of this body suggested
the abandonment of the requirement of a sudden causative effect and the
extension of cover to other harmful effects (with the proviso that standards
and satisfactory maintenance of antipollution equipment were observed);
however, the amounts available for such cover did not exceed 500,000 or 1
million FF per year.

In July 1977 the insurers became aware that the cover expected by industrial
customers could no longer be made subject to the requirement of a 'sudden
cause'; they thought they could provide a service by attaching to the third
party cover a cover for expenses involved in isolating and cleaning up
polluted premises; the demand was for covers largely exceeding 1 million FF
per year and contract. To respond to these standard requirements, virtually
the whole French insurance and reinsurance market gathered together in GARPOL
which as a whole permits the mobilisation of a capacity of about 22 million
FF. The cover does not apply to single sudden events but more generally to
cases of 'isolated, repeated or continued' damage. The mechanism can permit
the extension of commitment limits to 5-10 million FF for third party cover
and to 500,000 to 1 million FF for 'expense reimbursements', cleaning up.
For some particularly substantial risks these ceilings can be raised to the
limit of the overall capacity of GARPOL (22 million FF).

As a balancing off, there are contractual provisions concerning prevention:

- it is only considered to be an accident if the damage is fortuitous i.e.
  unforeseeable by the insured; if damages are not due to lack or insufficiency of maintenance known to the insured;

- the insured must himself, in every accident, carry a margin of 10 per cent
  of the limit amount of the cover (for third party and expense recovery
  cases). He risks to lose the benefits of his contract if he tries to
  cover himself elsewhere for this margin.

- The insurer is entitled to visit the insured installations at any time
  without prior notice. In case of disagreement on measures to be taken the
  contract can be terminated.

- Contracts of this type are issued only for periods not exceeding one year;
  their renewal therefore requires the insurer's formal agreement (43, p. 5).

However, since recourse to insurance can no longer suffice for certain
risks (insufficient cover and exonerations which do not permit all desirable
covers) other means have been developed: compensation funds and international
conventions between governments.

2nd. New compensation mechanisms: the compensation funds

The weight of precedents connected with the actual or potential accidents
of modern industry have led the creators of risk to organise themselves.
Two fields of activity in particular have attracted attention: maritime
transport of hydrocarbons after the accident of the *Torrey Canyon* which
showed up the insufficiency of the classical means of restitution (only a
quarter of the bill could have been settled if classical maritime law had

been applied) and the difficulty of their implementation (the ship-owner had
to be proved to be at fault); the field of nuclear energy where very soon the
question of rules to be provided for the responsibility in case of damages
arose.

Two requirements had to be considered: the compensation of victims (for
humanitarian reasons but also in order not to risk a ban on the activity
or more or less severe impediments to its development); the limitation of the
responsibility of the operators who could not expect to be able to carry the
full burden of the events they might start.

A compromise solution was found and given blessing by the signing of
international conventions. Operators accepted to pay compensation for
damages based on the principle of 'objective liability' i.e. they accepted
to pay without there being a need for the claimant to prove the operator at
fault. In return they obtained the limitation of their liability. The
victim could have recourse to a compensation fund; the operator could
concentrate on his business because one of the prime features of a fund is
the imposition of a ceiling (45; 46).

These principles have been applied, as we have said, in various fields of
which we shall quote a few examples:

a) Compensation mechanism for damages due to pollution of the sea by hydro-carbons (47, pp. 304-307).

Conventions. The Brussels convention of 1969 on third party liability for
damages due to hydrocarbon pollution has substantially modified the law which
was in force until then (Brussels convention of 1957 which was restricted to
the implementation of the principle of limitation of the ship-owner's
liability) by providing the implementation of the following arrangement: for
ships carrying hydrocarbons in bulk liability is laid on the owner (however
not exclusively in case third parties are at fault); this liability is of the
'objective' type but is in turn limited (to about 77 million FF); the owner
has to create a fund of this size to prove his solvency.

This arrangement was criticised on two counts: the insufficiency of cover
provided for potential victims and the financial burden placed on individual
ship-owners.

The Brussels convention of 1971*attempted therefore corrections of the
1969 text. A further fund, complementing the first one, was provided within
limits of 166 million FF; the burden for the ship-owners was reduced. This
fund operates under the control of governments and is financed by the
petroleum industry as a whole (47, pp. 302-306).

Private professional accords. Since before the implementation of these
conventions the professions had established accords the framework of which
was a form of mutuality; they too are founded on the principle of objective
liability (with the same reservations as those in the 1969 convention).
Under the terms of the TOVALOP accord**which came into force in 1969 the
signing ship-owners undertook to reimburse the governments for expense the

---

*This convention was still not in force at the time of the *Amoco Cadiz*
  accident.

**Tanker Owners Voluntary Agreement concerning Liability for Oil Pollution.

latter have incurred in order to avoid a polluting slick reaching the coast and to repair damages actually inflicted. The liability of the ship-owner was originally limited to $100 per gross barrel and had a ceiling of $10 million; these limits were raised in June 1978 to $160 per gross barrel and $16.8 million*.

The CRISTAL**accord came into force in 1971 and provides a complementary compensation financed by contributions from the petroleum industry in proportion to imports. The expense for cleaning up and the compensation of victims can be covered up to an overall limit of $36 million since June 1, 1977***.

These accords are valuable particularly if the ship involved in a pollution does not come from a member state of the Brussels convention.

b) _Compensation mechanisms for damages in the nuclear field_. Between 1960 and 1971 five international conventions dealing with nuclear third party liability have come to light. The 1960 Paris and the 1963 Vienna conventions established the basic principles of amends in cases of nuclear accidents in installations:

- objective liability which avoids the search for a party at fault;
- placing the liability on the operator which avoids the search for the liable party (and anticipates the search for anyone other than the operator);
- liability limited in amount and in time so that it remains bearable and insurable;
- obligation for financial cover in order to ensure the certainty of compensation for the victims;
- special rules on the competence of the courts and for the execution of verdicts to facilitate such compensation to the maximum (13, p. 133).

A third convention (Brussels 1963) aimed at the increase of available cover through participation of public funds that were reserved for this purpose, a national solidarity. The convention foresaw a compensation mechanism in three stages:

- a grouping of private insurance or other financial cover at the expense of the operator; the amount could vary between 5 and 15 million units of account of the European Monetary Agreement (AME****).

- intervention by the government concerned above the previously mentioned ceiling up to 70 million units of account (about 350 million FF);

---

*This was the approximate equivalent of the 77 million FF from the 1969 convention before the fall of the dollar's exchange rate.

**Contract regarding an interim supplement to tanker liability for oil pollution.

***Which is the approximate equivalent of the 16 million FF ceiling of the 1971 convention.

****The unit is equivalent to about 0.89 g of fine gold. However, since June 29, 1976 a directive from the Council of the EEC has modified the definition of the European unit of account which no longer refers to gold but to a basket representing an average value of the currencies of the member states.

- intervention by the whole of the contracting governments up to 120 million units of account (about 600 million FF).

In the United States the Price Anderson Act of 1957 foresees the possibility of compensation up to $560 million (being 140 million to the debit of insurers, 335 million to the debit of sixty seven American nuclear centres, the remaining 85 million to the debit of government funds) (48, p. 302).

The repair of damages due to transports of radioactive substances is the object of a special convention (Brussels 1971) which institutes, as in the case of hydrocarbons, a rule of objective liability, limited and assigned (47, p. 308).

3rd. Public aid, the liability of government

Nowadays it has become normal to turn to the government to obtain amends in case of an accident; it is equally habitual in our day and age to see that the public authorities ensure that this demand is met favourably. On the level of principle a right seems to have been acquired for measures to be taken (even if those struck by an accident, as in the case of the black tides, do not fail to complain about slowness and deficiencies).

What are the roots of this national assumption of responsibility? We may, with J. F. Pontier, the author of a recent study on "public calamities" (49) put forward a certain number of factors that are important for the reflection on present industrial risks and menaces of the future which have been identified. The examination is carried out on the case of France.

a) State aid in the case of "public calamities": a general framework providing certain compensation for those struck by accident.

The precedent of war damages. The attention of the public authorities was first drawn to accident victims in the case of war damages. Under the Ancien Regime (i.e. before the French Revolution) this preoccupation was mostly absent. In the nineteenth century the state was reluctant to recognise its responsibility. Thus, for instance, the law of September 11, 1871 rejects the project established by a commission to take stock of the damages caused by the 1870-71 war and to recommend compensation for damages; the law, in the end, speaks only of 'indemnification'. As the basis for this law Thiers had just invoked "the interest of the state which itself has suffered much and deserves consideration". The idea of national solidarity was not to the taste of the parliamentarians of the time either.

We notice, however, the interlude of the Revolution which for a while brought some innovations. The decree of August 14-16, 1793 affirmed the right of victims to compensation in the wake of war:

The Convention declares in the name of the nation that it will indemnify all citizens for losses which they have incurred from the enemy's invasion of French territory by demolitions or strikes which our defence required ... (Article 1).

One had to wait until the twentieth century to see these principles revived which had been abandoned since 1795. The new criterion of war (affecting the whole of the country); the fact that the wars had been won ("Germany will pay"), the moral recognition of the idea of national solidarity led to the laws of 1919 and 1946; the first one took up a law introduced in 1914 and established the "right to compensation"; the second one (October 28, 1946, Article 12) opens up a right to "integral compensation".

The size of the disaster actually made this law (and no doubt the first one as well) more than a simple compensation law: it was in the words of M. Waline, as quoted by J. M. Pontier, "a law of national solidarity and a law of encouragement for the rebuilding of public interest"*(49, pp. 8-17).

The precedent of agricultural calamities. The agricultural profession in France drew attention to the disasters suffered in the agricultural sector on account of bad weather, drought and diseases. A first measure was encouraged (1930-1950): access to favourable loans was facilitated. In the discussion of the law of July 10, 1964 which sets up a system of coverage against agricultural disasters the operators were encouraged to take out insurance: "It is not possible, declared the Minister of Agriculture, to tell the farmer that he will be covered against abnormal risks if he has not tried in the first place to organise himself against normal risks"**. This was taken up in Article 2 of the law***which defines agricultural disasters:

Agricultural disasters in the sense of this law are uninsurable damages of exceptional importance due to abnormal variations of a natural agent when the technical means of preventive or curative action that are normally employed in agriculture could not be used or have proved to be insufficient or inoperative (48, pp. 18-26).

'Calamitous events'. What responsibility does the state have to assume for other calamities? Since there is no law on the subject we have to consider several factors in order to arrive at an answer to the question (49, pp. 32-61).

The Practice. Despite the absence of a law defining conduct in case of a disaster we notice that the public authorities do not stay on the sidelines in such situations as the following examples show.

Toulon-La Seyne 1899. On March 5, 1899 a gunpowder factory located between the two towns exploded, killing fifty four people and leaving seventy two injured within a one kilometre radius; many houses collapsed. The government considered then that "the state has a duty to mitigate and to amend as much as possible the effects of this disaster" and it submitted a draft law on extraordinary aid for adoption (49, p. 48).

La Courneuve 1918. Subsequent to this explosion which was connected with the war effort the government demanded the adoption of a law to counter "the immediate consequences of the disaster and (...) to permit the families to return to their homes and (...) to resume a normal life"****(49, p. 39).

Is it the uninsurable nature of an accident which institutes public compensation? One might think so if one refers to various statements.

---

*M. Waline, *Treatise on Administrative Law*, 8th edition, para. 1, 305, p. 743.

**E. Pisani, *Journal Officiel*, parliam. debates, Nat. Assembly, 2nd session on June 16, 1964, p. 648.

***Law No. 64-706 of July 10, establishing a system of coverage against agricultural disasters, *Journal Officiel*, July 12, 1964, p. 6 202.

****Draft law instructing the Interior Minister to open supplementary credits for aid to the victims of various explosions by J. Pams and L. L. Klotz, *Journal Officiel*, Doc. Parl. Chambre, 1918, annex No. 4 462, p. 415.

Private persons "must above all count on their personal precautionary effort"*quoted by J. M. Pontier (49, p. 42); "the public authority does not have to concern itself with damages which inasmuch as they represent normal risks can be covered by insurance bodies"**quoted by J. M. Pontier (1, pp. 42-43); and yet:

Celliers/Savoie 1919. This community was destroyed by fire; the accident which could have been insurable (at least theoretically) was considered a public calamity and became the object of aid from the public authorities (49, p. 45).

Is it then the exceptional nature of the damage which opens up the possibility of compensation? One might think so; but there are examples to the contrary of which the extreme case is the following:

Toulouse 1926. On April 11, 1926 the bell-tower of the Dalbade in Toulon collapsed. This monument, 83 metres high, destroyed or damaged several buildings in its fall and caused the death of several people. Parliament considered it a public calamity (49, p. 46).

These observations lead us to conclude that the notion of public calamity has in practice rather fluid outlines, leaving much space for political judgement. J. M. Pontier writes:

It is from the moment when an event is felt to be a public calamity that it is effectively considered such by the public authorities. It is always difficult to tell why an event has been considered in this way. It may be for the reasons analysed before; the particularly unexpected nature of the disaster, the severity of the damages caused, the absurdity of the phenomenon etc.

One can describe the result better than the psychological mechanism that explains this feeling: one may speak in this respect of a 'consensus populi' which is the expression of an emotion shared in the face of damages suffered by a part of the population. The information and above all the mass communication media have sensed an essential role in the creation of this feeling in our time, in the process of arousing opinion (49, p. 47).

Statements in parliament. Broad definitions of public calamity were laid down by Parliament in 1923 and 1946; here are the two texts (49, pp. 16-17):

Since the law proclaimed the equality and solidarity of all Frenchmen in the face of the burdens of war it may, by analogy, seem equitable that the community (...) should contribute (...) to the exceptional burdens i.e. their relief and restitution of assets destroyed by disasters***.

---

*Report by P. Marraud in the name of the Finance Commission, *Journal Officiel*, Doc. Parl., S. 1927, annex No. 295, p. 475.

**Report on the reasons for the draft law presented by H. Cheron and Ch. de Lasteyrie, *Journal Officiel*, Doc. Parl., chambre, 1922, annex No. 4 438, p. 689.

***Milau report in the name of the Finance Commission, *Journal Officiel*, Doc. Parl., chambre 1923, annex No. 512, p. 848.

We have accepted the necessary principle of solidarity of all Frenchmen in the face of the totality of all risks of which war is the main one*.

Orders by the Council of State. We recall the order issued in the affair of La Courneuve following a consecutive recourse to the explosion. J. M. Pontier writes:

> The Council of State, going further than its Governmental Commissioner who proposed recognition of the state's responsibility based on fault, decided that the storage and the maintenance of dangerous machinery in the proximity of an important agglomeration in conditions of improvised organisation "carried risks in excess of the limits of the ones which normally result from the neighbourhood" and such risks are "of a nature, in case of an accident occurring outside all events of war, to invoke, regardless of any fault, the liability of the state". With this decision the Council of State changed what until then had only been aid into a right to compensation (49, p. 39).

This order became law and was applied in the case of an explosion of an ammunitions wagon**.

Constitutional provisions. The principles of equality and solidarity in the face of tasks resulting from national calamities has a constitutional value as the preamble to the Constitution of October 27, 1946***affirms:

> The nation proclaims the solidarity and equality of all Frenchmen in the face of tasks resulting from national calamities (49, p. 49).

The administrative definition of public calamity. In the absence of a specific law we have to refer to a circular from the Interior Minister (No. 76-72 of February 6, 1976) to find a precise and official definition of the notion:

> The generic term 'public calamities' describes all natural cataclysms or calamitous events such as cyclones, tornadoes, tempests, floods, landslides and avalanches, earthquakes and explosions which have caused, on a communal level or exceptionally on account of their severity on an individual level, the destruction or a substantial deterioration of movable or immovable assets.

We see therefore a solemn affirmation of the principles of solidarity and equality which, however, are not translated into any law. One might think of some factors that limit this state aid (scale of the phenomenon, uninsurable nature) but practice enlarged the field of intervention.

There is still a further question. 'Force majeure' could be most often involved in the natural cataclysms of which the legislator and the administrator thought when setting out laws. What becomes of the discussion when 'the calamitous event' makes it apparent, as in our time, that man is more

---

*J. *Off. Constit. Nat. Assembly*, August 30, 1946, p. 3 410.

**C. E., October 21, 1966, Minister of the Army vs. SNSF. Rec. 557, D. 1967, 164, Concl. Baudouin, JCP 1967, 11 15 198, note Blaevoet, AJDA 1967, 37, Chr. Lecet and Massot.

***The Constitution of 1958 in its preamble proclaims its attachment to this law as the Constitutional Council of July 16, 1971 was to confirm (49, p. 49).

involved in bringing about accidents? What if the government that has
tolerated an activity is also more involved in these events?

We have seen various responses being offered: private aid organised under
the control of the government (or governments) for marine pollution by
hydrocarbons; parapublic funds for damages of nuclear origin. There is also
the argument for public aid in case of accidents resulting from space
flights*. If there is still no general framework providing compensation to
be paid for disasters of technological and industrial origin one principle
nevertheless shapes up: the principle of search for a certain balance between
the need for compensation felt by the victims and the need of the operators
and the governments to develop certain technologies and industries.

Compensation becomes a better organised relief but not a right which would
open up the possibility of integral compensation with a legal obligation to
compensate; the victim finds himself less deprived on account of a trend
towards objectivation of liability as we have noticed previously (i.e. by
dispensation from the burden of proof). On the other hand the operators and
the governments are guaranteed a limit to their involvement by the non-
adoption of the principle of integral compensation. P. Girod writes:

Such a system represents only a distant relationship with the issues of
liability. It seems in fact that these mechanisms are inspired by the
empirical search for equitable compensation rather than by formal
principles of liability (50, p. 269).

The implementation of relief. A fundamental principle in this matter (in
the case of France) is that the relief initiative belongs to the executive
powers and not to the legislature. Taking up the options of the Third
Republic**in this respect, Article 40 of the Constitution of the Fifth
Republic decrees:

The propositions and amendments formulated by the members of parliament
are not admissible when their adoption would result either in a reduction of
public resources or in the creation or aggravation of public burden.

This is certainly the case with all proposals for the allocation of relief:
it is therefore legally impossible for parliamentarians to propose laws for
the relief of accident victims (49, p. 138). However, practice shows that
parliament does obtain measures of this nature from the Executive (branch of
government) without opposition by the government.

In the same manner the assessment of the amount of damages comes under the
executive branch, its representatives and the experts it appoints for this
purpose (even if the members of parliament are not kept out of this task).

---

*Vote by the UN General Assembly on November 30, 1971. Besides nuclear
 risks presented by spacecraft one must also take into account more
 'classical' accidents (49, pp. 45-46):
   - September 1960: wreckage from an American craft damaging a farm in South
     Africa; November 1960: people injured in Cuba; September 5, 1962: A
     10 kg iron piece from the soviet Sputnik IV falls on a street in
     Manitowock, Wisconsin, USA; September 5, 1969: five people injured on a
     Japanese ship by a fall of non-disintegrated wreckage.

**Journal Officiel, January 12, 1877, p. 285.

Two primary types of aid must be considered. In the first place: emergency credits. According to the circular No. 76-72 of February 6, 1972 from the Interior Minister their purpose is to help families in a difficult situation the day after the accident to cope with their immediate essential needs such as food, housing and clothes. These means are credits which are included in the budget of the Interior Ministry every year under the title "Emergency relief for victims of public calamities". This title is managed by the Directorate of Civil Safety. The credits are allocated according to a simplified procedure upon request from the police commissioner before any precise assessment of damages.

Further, there is a "Relief Fund for victims of accidents and calamities" created by Article 75 of the law of August 4, 1956. This is a special appropriation account defined by the order of January 2, 1959 for tracing "operations which, following the provisions of a finance law enacted by government initiative, are financed by means of special resources". The account is destined for the centralisation of resources of all kinds for aid to accident victims; it serves to finance expenses related to the allocation of aid in cash or in kind (49, p. 161). The recorder of the Finance Commission of the Council of the Republic commented on this Article 75 as follows:

This in fact is a pure accounting measure which permits in case of a calamity to centralise in a single open Treasury account the total of collections*.

The fund comprises, as receipts, the subsidies allocated by the state, the communities, public establishments, possibly the amount of donations and the total of collections of a national nature operated among the public on government initiative, destined for aid to victims of accidents and calami- ties (49, p. 163).

There is therefore no 'budget.' but simply an accounting measure which permits action in case of need. The outlines for the use of this fund are set forth by an interministerial body: the "Coordination Committee for Relief to Accident Victims", instituted by a decree of November 27, 1953**.

Loans are provided for assistance which can thus be paid for emergencies or for partial compensation the final purpose of which is to help revive industry, commerce and agriculture.***

They are granted for a long term (forty five years) in variable amounts according to their importance to the state. Finally, certain fiscal arrangements provide tax relief (but no exemption which has always been refused on the grounds that public finances must not be thrown off balance).

4th. Private bodies of general interest

Among these bodies we mention the Red Cross, Catholic Relief, Popular

---

*M. Pellenc, report in the name of the Finance Commission on the draft law adopted by the National Assembly, *J. Off.*, *Doc. Parl.*, Council of the Republic, 1956, annex No. 587, p. 791.

**Decree No. 53. 11 75, *J. Off.*, Dec. 2, 1953, p. 10 734.

***Denoix report in the name of the Finance Commission in charge of examining the draft law concerning loans to accident victims, *J. Off. Doc. Parl.*, S. 1910, annex No. 96, p. 428.

Relief etc. These are joined *ad hoc* by 'support committees'. These private bodies can collect substantial funds; the latter are most frequently channelled through the Treasury.

REFERENCES

1. PREVENTION

(1) A. SPIEGELMAN, Hazard control in the chemical and allied industries. 3rd International Symposium on loss prevention and safety promotion in the process industries. Basle/Switzerland. September 15-19, 1980. Preprints. Vol. 2, pp. 1/129 à 1/137.

(2) Installations classées pour la protection de l'environnement. *Journal Officiel de la République française*. No 10001 - 1. 1977

(3) P. GOUSSET, *Le droit des établissements classés*. Dunod, Paris, 1968.

(4) C. GABOLDE, *Les installations classées pour la protection de l'environnement*. Sirey, Paris, 1978.

(5) A BOQUET, *Guide des installations classées*. Ed. du Moniteur, Paris, 1979.

(6) C. LE PAGE-JESSUA et C. HUGLO, La législation sur les nuisances industrielles. *Annales des Mines*, juillet-août 1979, pp. 29-40.

(7) J. COLLIOT et B. de FONT-REAULX, La prise en charge de l'inspection des installations classées par les services de l'Industrie et des Mines. *Annales des Mines*, juillet-août 1979, pp. 41-46.

(8) - Loi no 76-629 du 10 juillet 1976 relative à la protection de la nature. *Journal Officiel*, 12-13 juillet 1976, pp. 4203-4206.

    - Décret no 77-1141 du 12 octobre 1977.

(9) N. NOWICKI, La loi, la prévention et l'information. *Futuribles*, no 28, novembre 1979, pp. 98-106.

(10) Loi no 75-633 du 15 juillet 1975 relative à l'élimination des déchets et à la récupération des matériaux.

(11) Loi no 77-771 du 12 juillet 1977 sur le contrôle des produits chimiques. *Journal Officiel*, 13 juillet 1977, pp. 3701-3703.

(12) Organisation de la pharmaco-vigilance et de la toxico-vigilance. *Journal Officiel*, 10 mai 1980.

(13) Commissariat à l'Energie Atomique. *Droit nucléaire*. Eyrolles, Paris, 1979.

(14) C. de TORQUAT et J. BOURGEOIS, La sûreté des installations nucléaires en France: organisation, technique et législation. Electricité de France, Guide International de l'Energie Nucléaire 1978 (51 p.).

(15) M. BURTHERET et Mme De CORMIS, Le régime administratif des centrales nucléaires. *Revue Générale Nucléaire*, no 3, mai-juin 1980, pp. 249-258.

(16) Syndicat CFDT de l'Energie Atomique.   Le dossier électronucléaire.
     Le Seuil, Paris, 1980.

(17) P. LAGADEC, *Développement, environnement et politique vis-à-vis du
     risque: le cas britannique.*   Tome 1.   Laboratoire d'Econométrie de
     l'Ecole Polytechnique, mars 1978.

(18) F. ·ENGELS, *La situation de la classe laborieuse en Angleterre.*   Ed.
     Sociales, Paris, 1973, réédition.

(19) P. KINNERSLY, *The hazards of work: how to fight them.*   Pluto Press,
     London, 1973, pp. 235-238.

(20) C. AMBROSI et M. TACEL, *Histoire économique des grandes puissances.*
     Delagrave, Paris, 1963, p. 154.

(21) H. M. Chief Inspector of Factories — Report for 1967.   Department of
     Employment.   *Safety and Health at Work. Report of the Committee* (1970-
     1972).   Selected Written Evidence, London, HMSO, 1972, pp. 170 332.

(22) *Safety and Health at Work. Report of the Committee,* HMSO, London, 1972.

(23) Health and Safety at Work Act.   London, HMSO, 1974.

(24) Advisory Committee on Major Hazards — First Report.   Health and Safety
     Executive, London, HMSO, 1976.

(25) Health and Safety Executive.   Canvey, an investigation of potential
     hazards from operations in the Canvey Island/Thurrock area.   Health and
     Safety Commission, London, HMSO, 1978.

(26) P. LAGADEC, Développement, environnement et politique vis-à-vis du
     risque: le cas de l'Italie — Seveso.   Laboratoire d'Econométrie de
     l'Ecole Polytechnique, mars 1979 (280 pages).

(27) L. CONTI, Les lois italiennes à l'égard des risques majeurs.   Texte
     pour une réunion de travail organisée par le Laboratoire d'Econométrie
     de l'Ecole Polytechnique, mai 1978 (15 pages).

(28) L. CONTI, Trop d'échéances manquées. *Survivre à Seveso?*   Presses
     Universitaires de Grenoble/Maspero, 1976, pp. 45-58.

(29) Camera dei Deputati VII legislatura.   Commissione Parlamentare di
     inchiesta sulla fuga di sostanze tossi che avvenuta il 10 luglio 1976
     rello stabilimento Icmesa e sui rischi potenziali per la salute e per
     l'ambiante derivanti da attivita' industriali (Legge 16 giugno 1977,
     n. 357).   Juillet 1978 (470 pages) (Notre traduction).

(30) S. ZEDDA, La leçon de chloracné, *Survivre à Seveso?*   Presses
     Universitaires de Grenoble/Maspero, 1976, pp. 21-44.

(31) Disegno di legge, approvato dalla Camera dei deputati nella seduta del
     22 giugno 1978, risultante dall'unificazione.   Senato della Repubblica,
     VII legislatura.

2. BATTLE

(32) P. PISA, La sécurité civile.   Direction de la Sécurité Civile, (non
     daté) (16 pages).

(33) A. P. BROC, La protection civile.   P. U. F., Coll. Que Sais-Je?, Paris, 1977, no 1682 (126 pages).

(34) Recommandations pour la préparation et la mise en oeuvre des mesures de protection des populations au voisinage des centrales électronucléaires en cas d'incidents ou d'accidents pouvant comporter éventuellement des conséquences radiologiques. Schéma-type de plan particulier d'intervention des secours. Ministère de l'Intérieur, Direction de la Sécurité Civile, Services Opérationnels, 29 décembre 1978.

(35) C. GERONDEAU et M. BERTHIER, Les plans particuliers d'intervention relatifs aux centrales électronucléaires. *Annales des Mines*, juin 1980, pp. 181-186.

(36) X. HAMELIN, A. CHENARD, R. GOUHIER et P. PERNIN, Rapport d'information déposé par la Commission de la Production et des Echanges sur l'organisation de la protection de la population en cas d'accident pouvant entraîner des émissions radioactives. Assemblée Nationale, seconde session ordinaire de 1978-1979, no 1200, Annexe au procès-verbal de la séance du 25 juin 1979.

3. INDEMNIFICATION

(37) M. SAINT-YVES, L'assurance -- Pourquoi? Comment? *Le Centurion*, Paris, 1978.

(38) C. STRULOVICI, L'assurance incendie — introduction. Centre National de Prévention et de Protection, Octobre 1978.

(39) E. COPPOLA Di CANZANO, Couverture des risques catastrophiques provoqués par des calamités naturelles et de risques politiques. Rendez-Vous de septembre, 1979, Monte-Carlo, pp. 17-24.

(40) P. BLANC, La réassurance. M. B. M. Risk Management. Monte-Carlo, 21-22 février 1979.

(41) F. NEGRIER, L'A. B. C. de la réassurance. A. F. Réassurance, 1977, no 1321.

(42) Centenaire de la Münich Rück: 1880-1980 Münchener Rückversicherungs Gesellschaft. Münich, 1980 (102 pages).

(43) Incendie du magasin central des pièces détachées de la société Ford-Werke AG à Cologne Merkenich, 20 octobre 1977. Gerling Institut für Schadenforschung und Schadenverhütung, 1978.

(44) Assurance des risques de pollution en France: le G.A.R.P.O.L. Groupement Technique Accidents — Association Générale des Sociétés d'Assurances contre les accidents. Novembre 1978 (7 pages).

(45) P. M. DUPUY, *La responsabilité internationale des Etats pour les dommages d'origine technologique et industrielle.* Ed. A. Pedone, Paris, 1976.

(46) M. REMOND GOUILLOUD, De la responsabilité de l'apprenti sorcier. *Futuribles*, no 28, novembre 1979, pp. 65-74.

(47) Rapport de la Commission d'Enquête de l'Assemblée Nationale présenté
     par H. Baudoin. Première session ordinaire 1978-1979, novembre 1978,
     no 665. Tome 1 (333 pages).

(48) A. MELLY, Three Mile Island. *L'Argus International*, no 13, juillet-
     août 1979, pp. 297-303.

(49) J. M. PONTIER, *Les calamités publiques*. Berger Levrault, Paris, 1980.

(50) P. GIROD, La réparation du dommage écologique. L.G.D.J., Paris, 1974.

II. THE UTILISATION OF SCIENCE AND SPOTTING
TECHNIQUES FOR SAFETY

The last few decades have seen an important change in the traditional
relations between industry and science. In short, there has been a move from
large scale patchwork based on certain basic scientific principles to the
implementation, on an ever vaster scale, of technological systems based
almost entirely on models. Because of this the notion of concrete experience,
naturally without being dropped altogether*, tends to give way to theoretical
comprehension which anticipates the event and largely dispenses with the old
trial and error learning process. If the law and the institutions have had
difficulties in perceiving what this change induced on the level of risk and
responsibility — and this still more so as the technologies introduced are
intrinsically more dangerous — science for its part has not remained inactive.

Scientific tools have been called upon for safety as well. Two important
innovations have come to light: the definition of new principles for the
approach to the issue of safety; the systematic trial application for these
principles in structured practice. Faced with discoveries "as serious and
as important as the discovery of fire", as Frederic Joliot said with regard
to the atom, it was normal and highly desirable not to wait as long as had
been the case with conflagrations.

We shall choose here some specific fields for examination: the safety
analyses for nuclear installations: the safety studies for Concorde and for
weapons systems; the first implementations in the classical industrial field,
especially in the chemical industry and finally the innovations brought about
by the rare safety studies applied to industrial areas with a high concentration of risk.

1. RISK STUDIES ORGANISED IN AN OVERALL APPROACH

The publication by Ch. Starr**in the magazine *Science* in 1969 (1) is often
the first stone in the edifice which is now being completed, at least in its
essential structures, is thus in the field of technological risk studies.

We shall not examine here the social effect of the approach or the tools
used. We can already see that the experts wanted to know the risks and to
measure them while the politicians wanted to pass over them once again:
getting the elements for judgement which would permit them to base their
decisions and to justify them. These expectations, latent if not expressed,
have given the overall approach its comprehensive form. It consists of
three important phases which we shall present successively (1 to 10).

---

*One would, of course, not switch to actual experimentation: simulations
on mathematical models can only be based on a panoply of experimental
bases: study of a phenomenon, mock-up of an apparatus or a system,
instrumentation of service apparatus etc.

**Even if this kind of paternity always raises discussions. In this case,
H. Otway emphasises for instance that a number of people in Anglo-Saxon
countries have dealt with this matter earlier on. They did so in spare
time and doubtless without the official support from which this work
benefited in the 1970s.

1st. Identification

It is important, first and foremost, to be able to get together the elements of the answer to the question: what constitutes a menace?

With D. W. Rowe we can spell out the field of questioning. It is a matter of identifying or spelling out better in particular:

- the new risks;
- the risks of which the scale may have changed, for instance because of a context that has developed; there may have been an important increase of danger of a cumulative nature or on account of an exceeded threshold or synergic phenomena;
- long known risks that have not been studied in depth;
- risks of which social awareness has grown; it may be that a risk that has been dominant in the past has been eliminated or reduced which creates new priorities in the battle against dangers; there may have been a change of power in society which calls attention to other risks; it may also be that certain social groups become more particularly concerned about a risk which until then had been diffuse and was the object of only lesser concern.

What does 'spelling out better' mean? The distinction made in the English language between 'hazard' (the physical possibility of the occurrence of an event) and 'risk' (the effective realisation of this possibility, this realisation being approached as a probability) is very interesting here. It permits an understanding of how the work of 'risk' identification could develop in two complementary directions as V. C. Marshal remarks (9, p. 6-397).

- the analysis of possibilities ('hazard analysis') — the study of menaces which we have carried out earlier on is of this kind;
- the analysis of sequences of events ('risk analysis') — the first step of a qualitative nature in the probability examination of the phenomena. We shall hereafter examine the methods and tools used to carry out these complementary analyses.

a) The identification of dangers. It is therefore a matter of spelling out the types of existing dangers, of menaces for the future: their nature, their possible consequences. Various lines of analysis may be pursued, among them, as we have seen in the preceding chapter:

- historical and statistical study;
- the study of acute dangers;
- study of menaces for the immediate or a more remote future.

Among these studies we recall those of the advisory committee on major hazards (10-11) which tend to spell out the broad categories of risks (explosion, disastrous conflagration, escape of gas or toxic substances). In each case the committee examines the experience of the past and what is new in the respective situations (explosion of non-confined gas, possible formation of fire storms, dispersion of highly dangerous substances in quantities of less than a kilogram while up till then the unit of reference was the tonne).

The work often requires very complex research on the theoretical plane (behaviour of a gas cloud for instance) as well as on the practical (experiences to be carried to a significant scale).

All these 'possibles' will be, if they occur, only the end result of a
process.   It is therefore important to study the chains of events that can
lead to these phenomena.

b) The identification of dangerous sequences.   With R. W. Kates (4, 5) we
can distinguish three related methods of general marking of the structure of
these processes; the precise analysis is based on an essential tool: the
construction of 'trees' of which we distinguish two complementary categories.

Three exploratory methods

Systematic study*

It consists in the systematic examination of everything that could contribute to the recurrence of one or the other risk.   The identification process
uses a standard procedure for the classification of products, processes,
phenomena relative to the risks which are connected with them.   If the
examination procedure is clear, this is not the case with the very vast
field examined which makes the work into a sort of 'angling' which never
exhausts the questions asked.

R. W. Kates quotes a certain number of examples of this type of study; we
mention them for illustration even if they are serious problems rather than
immediate accidents.   Thus a programme by the Environment Mutagen Society in
the United States has for its object the examination of "tagébéicité"(?).
One starts from the preoccupation: what are the potential mutagenic effects
of chemical products? and develops researches in a large number of directions,
all starting from this question.   There are also the works of the National
Academy of Science on ocean pollutants which aim at determining which
pollutants present a potentially important risk, which others present less
serious dangers and finally which others should be the object of supplementary studies (12).   Other examples are the works on atmospheric pollution
and the multiple attacks on the earth's ozone layer (13).

Observation programmes.   Observation programmes entering into specific
questions and into dangers suspected to result from them do not explore
risks; they concentrate on the examination of some key indicators which
provide information on the analysed system.   The Global Environmental
Monitoring Systems of the United Nations which operate by following up on
some eighteen pollutants, grouped in eight product categories (14), illustrate this type of work.

Specific diagnoses.   The works starts from the observation of the abnormal;
it continues with the consideration of symptoms, development and possibly
treatment in progress concerning the danger observed.   The general examination of development possibilities of an abnormal event may start from
scenarios.   We have seen this for example in the case of the TMI accident:
the commission set up by President Carter has asked itself about other
accident sequences that could have occurred (15).   Generally speaking, these
scenarios are based in the main on intuition.   This is why the specialists
prefer other approaches the exactitude of which they can assess better at
each stage.   The method of the trees which we shall bring up gives more
guarantees in this respect.

A powerful tool for analysis: the tree method

The key idea of this technique is that an accident is not due to an

---

*'Sifting' or 'screening' in accepted Anglo-Saxon terminology.

isolated cause but represents the end effect of a series of successive events. The work consists in bringing out the whole of the successive sequences.

The event tree starts from a specific initiating event, such as the rupture of a pipe, in order to explore, downstream, all sequences of events that may follow from it; at each stage forks indicate various possible paths depending e.g. on whether or not the safety system functions normally. In this way, starting from a number of foreseeable incidents, one can bring out multiple sequences which lead to clearly identifiable accidents.

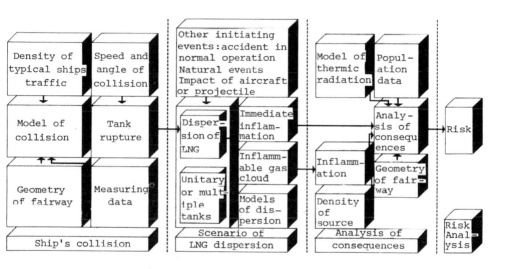

Table 12: How to identify the risks connected with a methane terminal? The so-called tree method was developed for this type of problem. The present case shows the elaboration of an 'event tree', starting from an initiating event, here a ship's collision, in order to study the secondary events that may result from it. If it does not determine the existing risks, this tool permits the assessment of a number of weaknesses in the system. It can be continued with the fault tree which by contrast starts from an accident in order to study the different ways in which this final event could have happened.

(Source: *La Recherche*, Nov. 79, p. 1150).

The trees obtained bring out the points which deserve more complete
exploration: the weaknesses (particularly of safety systems). For this one
also uses the tree technique with a particular tool: the fault tree. One
starts here from the final event — the fault — and this time one carries out
an upstream exploration: what paths can lead to the fault in question?

These descriptions, carried out with the greatest logical effort, make it
possible to show up a very large number of weak points; after the necessary
classifications, some sequences have hardly any significance*, one arrives
at a rather close description of what can happen in a system. The best
known example of the use of these methods is doubtless the Rasmussen report
on the safety of nuclear reactors (16).

For illustration we mention here the work of J. T. Kopecek (17) on natural
gas terminals (cf. table 12).

2. THE ASSESSMENT

This is the quantitative treatment of the qualitative elements that were
brought up in the preceding phase. The pertinent and important questions
have been asked and one can now proceed to figures. This assessment can be
applied to possible final accidents as well as to the sequence of events
that leads to them. One tries to apply quantitative elements to the scale
of the phenomena, in space and in time. The American authorities have for
instance carried out a study on the risks presented by the transport of LNG
in the port of Boston (18).

This study is most often on a par with probabilities affecting each event
in the process. One then obtains paths spelled out in probability and thus
an assessment of the (probable) frequency of faults and final events. An
illustration is given hereafter (Table 13) for the case of an accident upon
landing (19).

3rd. Evaluation

As we have said, the knowledge of the risks (their nature, their size and
their probability) is not sufficient for action and some quarters wanted
to demand of the specialists to go a step further in their work. It is then
no longer just a matter of bringing up simple elements of reflection but
elements of judgement which permit the basing of decisions and their justification if needed. As T. A. Kletz, a specialist on safety issues at
Britain's ICI, has put it very clearly, one needs a method of evaluation that
can be logically defended. The discussion is difficult if I am satisfied
with telling my interlocutor that the risks connected with the transport of
chemical products are no greater than those of them being struck by
lightning; if, however, there is a measuring scale for the risk, a dialogue
becomes possible (20, p. 322).

We want to stress that the methods should not be criticised in themselves;
all scientific investigation brings up interesting elements of assessment
and is a valid help for decision making as the same author remarks:

One cannot do everything at the same time. What should have precedence?
Allocating priorities is always difficult. It is even more complex in the

---

*A tank cannot be empty and full at the same time; neither can the wind blow
simultaneously from the north and from the south etc.

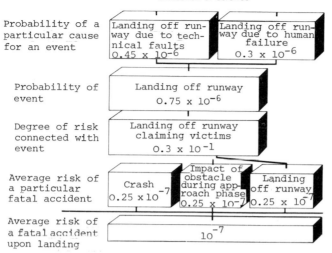

| Probability of a particular cause for an event | Landing off runway due to technical faults 0.45×10^{-6} | Landing off runway due to human failure 0.3×10^{-6} |

| Probability of event | Landing off runway 0.75×10^{-6} |

| Degree of risk connected with event | Landing off runway claiming victims 0.3×10^{-1} |

| Average risk of a particular fatal accident | Crash 0.25×10^{-7} | Impact of obstacle during approach phase 0.25×10^{-7} | Landing off runway 0.25×10^{-7} |

| Average risk of a fatal accident upon landing | 10^{-7} |

Table 13: If the event tree permits the identification of a risk a probability must be associated with each event in order to assess the severity
and the frequency of the identified dangers. This is what is illustrated
in this table in the case of an aircraft landing in conditions of poor
visibility. After identification and assessment of the risk it remains to
be evaluated by putting the above mentioned assessments and the social
practices and preferences in perspective.
(Source: quoted by J. R. Ravetz in *The acceptability of risk*, Council for
Science and Society, Barry Rose 1977).
(Source: La Recherche, November 1979, p. 1152).

field of the uncertain, of safety. In the past one dealt with well known
risks and forgot the rest. This is quite wrong. Whether resources are
substantial or restricted, it is important that they be used in such a way
as to yield maximum benefit (20, p. 322).

In order to spell out this world of risk in which the choices have to be
made, various tools can be used*:

a) Evaluation starting from the cost in human lives. A certain number of
methods aiming at an evaluation of the cost of risk prevention have been
established. Generally speaking, they try to put in money terms the advantages derived from a reduction of risk: this is made partially possible by
the evaluation of the cost in human lives. One can then allocate the
resources in such a way that one maximises the number of lives saved within
the limits of budgetary constraint; it suffices for this purpose to balance
the margin of costs and gains. J. Limerooth has taken it upon himself to
carry out critical reviews of these methods which on the whole result from
the cost advantage approach (21, 22).

―――――――――

*The elements of information which they provide are interesting. But, even
  if this aspect must be dropped later on, it must from now on be stressed
  that the exercise of evaluation often loses much interest: the desire to
  understand all too often surrenders to the obsession with proof, the
  obsession with brandishing in the wake of very questionable examinations
  (or naivety which sometimes contends with the grossest intellectual dishonesty) results which could be classified as 'acceptable'. The scientific
  tool then parts with all its ambiguity and must be examined in the
  political realm.

The approach which considers the individual as social capital (human capital approach): human life is evaluated by actualising the total future net gains of the person; this value reflects the contribution the individual makes to the national product and is therefore compatible with the objective of maximising the national product. This approach could recently be used again for questions of safety in aviation, in health policy, in the nuclear field with evaluations varying between $100,000 and $400,000. Like all evaluations made within the framework of the cost advantage method it has severe limitations by offering always the argument for the absurd: killing the maximum number of old people on the road appears to these calculations perfectly beneficial (22).

- The approach via value insured (insurance approach): the value of the individual is the one at which it shows up in his life insurance. Mishan remarked: Given the motivations that determine the choice of a life insurance a bachelor risks seeing himself ending up with a very low value attributed to his life.

- The calculation of compensation awarded by the courts is a third approach: given the fact that the courts often base their findings on the earnings the victims could have expected, this method comes largely to the same as the first one. The American aviation administration has suggested this method in recent years. The values arrived at, like the ones of the previous method, are in the order of $250,000 to $300,000*.

- The implicit values observed in court decisions are a further approach. It could thus be established that $30,000 had been spent for the prevention of a fatal accident on the roads and between 800,000 and a million dollars in the field of aviation. The values are considered to be the expression of the preference of the community: one still prefers to die in a car. These psychological deductions play a very small part in the social conditions for decision making and the extrapolations which they require in order to be applied to a new problem seem to be particularly perilous.

The willingness to pay is yet another approach to the quantification of the cost of human life; this can be done by using a questionnaire aiming to determine what one is prepared to pay for safety equipment. The results obtained are difficult to interpret because of the obvious influence of variables such as the wealth of the individuals questioned, their mobility of residence, their employment ... In the case of the *Amoco Cadiz* disaster such studies were made in an attempt to evaluate the loss of comfort felt by holiday makers: they were for instance asked how much money they would be willing to spend to have the beaches cleaned up (24). The limitations of this tool are obvious: what if the persons rejects the terms of the question? This was found in enquiries of this type carried out in connection with the planning of London's third airport. Some of the people questioned rejected

---

*This amount is for instance what the insurance companies expect to pay as an indemnity for victims of the DC 10 accident at Chicago (May 25, 1979). However, in the first suit filed (May 29, 1979) an amount of $5.25 million was demanded from each of the three accused (Douglas, the manufacturers; General Electric, the engine manufacturers; American Airlines, the carrier), i.e. a total of $15.75 million for the death of one woman and her two children. For two other victims a claim for more than a billion dollars was made — which is a rather extreme figure (23, pp. 309-310). — E. Mishan, *Cost benefit analysis*, London, Allan & Unwin, 1971, quoted by J. Limerooth (21).

the project out of hand, asked for an infinite amount of money for compensation, the infinite being the only mathematical term capable of expressing their choice. Keener on efficiency than on accuracy or on honesty the analysts corrected these answers and changed them to '£5,000', thus permitting 'acceptable' calculations (25).

One can, therefore, use these tools among others with the greatest caution; it seems at least delicate to use them in simple fashion as tools that lead directly to 'good' solutions.

b) Evaluations starting from a comparison of various already 'accepted' risks. The approach used here consists in putting side by side various risks in order to see whether an examined risk appears acceptable or not. Attention will be centred on levels of risk and no longer on their costs. T. Kletz emphasises the progress made in this approach: the estimates of the costs of a life may be useful in order to decide whether or not a proposition is interesting for the money it costs or whether one should aim at less costly solutions. However they cannot be used as a reason for accepting risk which is often what is required. To start from the network of existing risks appears therefore safer in order to separate the acceptable from the unacceptable (20).

The evaluation of a risk under study is therefore made by comparison with the other known, identified or estimated risks. In proposing this practice Ch. Starr has distinguished a certain number of laws so as to simplify evaluation work further:

- The rate of death due to illness constitutes an upper limit for the acceptability of risks, a bit less than 10^{-2} per annum*which means that out of a population of one million (representative of the whole population) about 10,000 people per year will die from illness.

- At the other extreme natural disasters ('Acts of God') tend to provide a minimum basis; the rate is 10^{-6}**. If the artificial risk is on this level it can be considered as nearly negligible. If it is several degrees lower one can definitely ignore it.

- Social acceptance of risk grows more than proportionally with expected advantages.

- The public seems inclined to accept 'voluntary' risks (mountaineering, tobacco ...) 1,000 times more readily than 'involuntary' risks (industries ...) (26).

These considerations permit the drawing up of a chart in which the *per capita* net benefit is measured on the abscissa (net of costs other than the risk of fatal accident) and the individual risk on the ordinate, which is supposed to be uniformly distributed among the population examined (cf. the following graph as an example).

- This type of graph can be used to compare various projects concerning the same population; one compares the advantages and risks connected with various operations under examination.

---

*This is a very approximate evaluation which must be modified according to age groups (10^{-2} for the forty five-fifty four group, near 10^{-3} for the thirty five-forty four group) (27).

**Rate established for the United States (27).

This method implicitly comes back to figuring out the price of a human life, a perilous exercise which T. Kletz does not want to enter into. This is why the author prefers to start from a much simpler approach which avoids these difficulties: it advocates the examination of risk probabilities of various projects under discussion and starts, with limited resources, by reducing the highest risks through refusal of authorisation for the riskiest projects (in terms of probability).

One starts therefore from risk tables which have become classics, a table supplied by Kletz is shown hereafter.

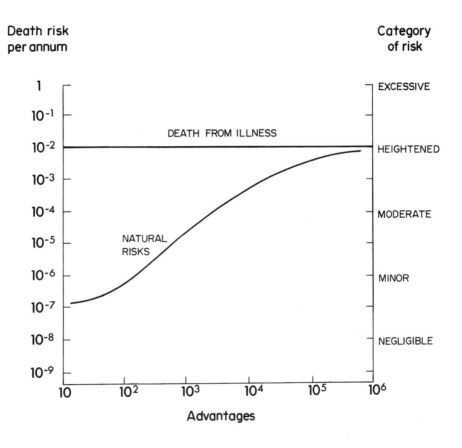

Fig. 27: Scheme for risk vs. advantage evaluation
in the case of involuntary risks.
(Source: acc. to C. Starr)

The conclusions for action are obvious:

We have here a basis for the evaluation of risks of individual origin for the general public. If the risk for those exposed is less than 10^{-7} per person per year the risk should be accepted, at least short term, and no resources need to be allocated to reduce it (28, p. 5).

It follows for instance, if one follows Kletz and his figures, that the risk of a nuclear accident (1 x 10^{-7} per person per year), the risk due to the transport of chemical products (0.2 x 10^{-7}) are very largely acceptable. The same is true for the threshold limit value: they lead to sustaining risks in the order of 10^{-5}, a level comparable to that of other industrial risks*:

The expense for attaining a threshold limit value (TLV) is reasonable; the expense for getting below it is not reasonable at present (28, p. 4).

This practical elucidation permits every decision maker to appreciate quickly whether or not the risk with which he has to deal is acceptable.

T. Kletz gives an example which illustrates the application of the method.

The Dutch factory inspectorate was once called upon to decide whether the storage of toxic chemical products within 2.5 km from a residential area could be authorised. Past experience showed that the frequency of ruptures in a storage tank needed to be watched. Going by the wind force and the distribution of data it was possible to calculate the frequency according to which gas concentrations struck the inhabited area (one in 10,000 years). In certain parts of Holland death from drowning following dike breaks had been a constant risk since time immemorial. The risk of floods could be reduced by the construction of higher dikes but that cost money. The Dutch decided to build dikes at a height that permitted to reach a frequency of one in 10,000 years. Even if the water came over these dikes most people escaped and the risk of drowning was one in 10 million per person per year**.

---

*To shed more light on the practice suggested here: we may recall the explanations proposed elsewhere by H. Otway concerning expressions of $10^{-\alpha}$/person/year:

10^{-3}/p/year: This is an uncommon risk. When a risk approaches this rate of mortality immediate action is taken to reduce it. This level is considered unacceptable all over the world.

10^{-4}/p/year: money (mainly public) is spent to control the causes of the risk: traffic lights, fire police ... Slogans connected with this risk are for instance: "The life you can save is perhaps your own".

10^{-5}/p/year: This level is still taken into consideration by society. Mothers warn their children about these dangers (playing with fire, poisons ...). Slogans belonging to this type are "never swim alone"; "put medicines out of the reach of children". These risks cause certain attitudes and behaviour such as refusal to travel by air.

10^{-6}/p/year: There is little concern about this level of risk. "This can only happen to others".

**It must be understood that this figure of one in 10 million includes the risk of dike break and therefore refers not only to the single risk of drowning in case a dike breaks.

In the same way if a tank holding toxic products were ruptured most people would escape and the risk is of the same order of magnitude.

The Dutch factory inspectorate argued that it was illogical to spend money in order to reduce the risk of poisoning in a part of Holland if the people were not prepared to spend money to reduce the risk of drowning in another part of the country, the two risks being on about the same level; and thus the installation of the storage tank was approved (20, p. 321).

The Rasmussen report has made this type of examination and logic of judgement much more widely known. We show here a table and two graphs suggested in the Executive Summary of this 2,400 page report. The graphs are invariably found in all written work on major risk.

However, we draw attention to the fact that the information they supply on the nuclear industry (the famous 100 nuclear centres considered less dangerous than meteorites) should no longer appear in any writings. The reexamination of the Rasmussen report at the request of the NRC by a group (29) chaired by H. Lewis*has in fact led mainly to the following conclusions in January 1979, conclusions that were adopted officially by the NRC:

As concerns the values given in this report on the overall probability that a specific type of accident occurs, the group declares itself incapable of appreciating whether they suffer from over- or from understatement but it is certain that the margins of error around these values are generally strongly underestimated; too often the lacking data have been figured out with illusory precision or statistical methods have been used in a context in which they could not properly be applied.

The executive summary draws the reader's attention in exaggerated fashion to those results in the report which are the most favourable for nuclear energy; it also gives him too strong a feeling that these results are established with insignificant margins of error. Generally speaking, the executive summary does not give a true impression of the contents of the report (30, p. 40).

Based on these conclusions the NRC was led to: withdraw from the executive summary all approval, explicit or implicit, which previously may have been accorded it; in addition it does not consider reliable the numerical estimation supplied in the report for the overall accident risk of a reactor (30, p. 41).

This scientific strictness, however, does not at all impede the constant use of these graphs. It must, in fact, be understood what propaganda value a syllogism like this may have: a nuclear centre is less dangerous than a meteorite; who finds fault with meteorites? therefore a centre is acceptable. Q.E.D. Even if scientific honesty may be a bit wronged, reasons of state under the banner of science demand that the executive summary of the Rasmussen report continues to be used. How many times will the Lewis report then be forgotten?

It would be regrettable if this 'forgetting' one day justified a complete denunciation of the Rasmussen report which incontestably retains a high technical value. The Lewis committee emphasised this strongly when it recommended

---

*For a detailed examination of the affair cf. Cl. Henry: *'The Rasmussen report A short social history of an important technical document'*. (30).

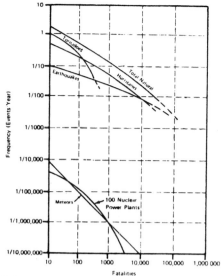

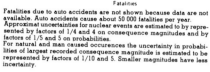

Fatalities due to auto accidents are not shown because data are not available. Auto accidents cause about 50 000 fatalities per year.
Approximat uncertainties for nuclear events are estimated to by represented by factors of 1/4 and 4 on consequence magnitudes and by factors of 1/5 and 5 on probabilities.
For natural and man caused occurences the uncertainty in probabilities of largest recorded consequence magnitude is estimated to be represented by factors of 1/10 and 5. Smaller magnitudes have less incertainty.

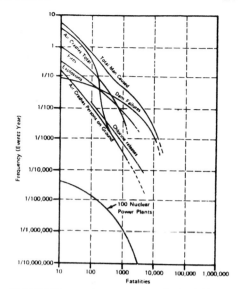

For natural and man caused occurences the uncertainty in probability of largest recorded consequence magnitude is estimated to be represented by factors of 1/20 and 5. Smaller magnitudes have less uncertainty.
Approximat uncertainties for nuclear events are estimated to by represented by factors of 1/4 and 4 on consequences magnitudes and by factors of 1/5 and 5 on probabilities.

Fig. 28: Frequency of deaths due to events caused by human activity. (Source: 16 p. A-2)

Fig. 29: Frequency of deaths due to natural events. (Source: 16, p. A-21)

Accident Type	Total Number for 1969	Approximate Individual Risk Early Fatality Probability/yr [a]
Motor Vehicle	55,791	3×10^{-4}
Falls	17,827	9×10^{-5}
Fires and Hot Substance	7,451	4×10^{-5}
Drowning	6,181	3×10^{-5}
Poison	4,516	2×10^{-5}
Firearms	2,309	1×10^{-5}
Machinery (1968)	2,054	1×10^{-5}
Water Transport	1,743	9×10^{-6}
Air Travel	1,778	9×10^{-6}
Falling Objects	1,271	6×10^{-6}
Electrocution	1,148	6×10^{-6}
Railway	884	4×10^{-6}
Lightning	160	5×10^{-7}
Tornadoes	118 [b]	4×10^{-7}
Hurricanes	90 [c]	4×10^{-7}
All Others	8,695	4×10^{-5}
All Accidents (from Table 6-1)	115,000	6×10^{-4}
Nuclear Accidents (100 reactors)	—	2×10^{-10} [d]

(a) Based on total U.S. population, except as noted.
(b) (1953-1971 avg.)
(c) (1901-1972 avg.)
(d) Based on a population at risk of 15×10^6.

Table 14: Individual risk of death from various causes. (Source: 16, p. 112)

- the use of methods of analysis developed in the report at any time when
such use is logically founded and can support itself on a sufficient basis
of data (30, p. 41).

Generally speaking, the approaches presented above involve a certain number
of implicit hypotheses which must be brought out. They are the following:

One can have great confidence in the established quantitative estimates,
whereas certain authors have pronounced serious doubts about their subject
(31).

- Society has, by trial and error, reached an almost optimal balance between
the risks and the advantages with which it lives (32).

- The acceptability level of risks for an individual result from a rational
approach which consists of a comparison between the new risk and the
habitually known risks.

- Individuals are at all times perfectly informed about the risks they run
(7).

- The approach does not take into account the advantages which individuals
may have from taking a given risk or from submitting to a given risk.

Faced with these approaches the critics set forth that since it is a matter
of evaluation, of taking into account values and social preferences that
attach to a risk phenomenon it seems at least difficult to finesse so
severely on the whole of social sciences: the automatic deduction of a valid
line of action starting from a table of figures cannot suffice. We shall come
back to these critical reflections later on. Here it may suffice to
emphasise that they are serious enough to show that the 'scientific' demonstration of the acceptability of a risk is socially not sufficient: especially
so when liberties are taken with the most elementary scientific exactitude.

2. THE UTILISATION OF SCIENCE IN THE CIVILIAN NUCLEAR FIELD:
   NEW PRINCIPLES IN THE PROCESS OF SYSTEMISATION

Even if it is not the most advanced on the subject, the nuclear field is
and remains a good example of the attraction of science for identifying and
dealing with safety issues. Since 1928 the International Commission for
Protection against Radiation has existed and this in order to respond to a
need which was already evident from a number of accidents involving exposure
to radiation from handling radioactive substances. Towards the fifties there
was concern about the safety of installations while work was done on the
'maximal credible accident'; in 1954 a truly big test was made at Idaho Falls
in the United States: a reactor was deliberately destroyed by fast retraction
of the control rods (33, p. 4). Since then, principles have been defined
and systematised, at least in theory, for all high risk activity.

1st. Knowledge and *a priori* prevention

When risk reaches the scale at which it will be known henceforth it can no
longer be a question of just dealing with the last accident to date. One
also has to pay attention to a very great need which exists from the time of
design: anticipation must take the lead in teaching experience in a big way.

This is what two specialists in the nuclear field have to say. P. Tanguy*:

The methodology of safety analysis (for nuclear installations) constitutes
a case among the methods generally used in industry because it fixes on an
a priori objective whereas safety usually proceeds by successive approxi-
mations, starting from experience acquired i.e. from accidents observed; as
an example one can quote the speed limits for automobiles, the determination
of which, apart from consumption problems, resulted from an analysis of
graphs on the frequency and severity of accidents as related to speed. In
nuclear energy, by contrast, one tries to foresee the different types of
accidents which are likely to occur and to limit their frequency in advance
(34, p. 51).

F. R. Farmer takes up the following proposition made by C. Hinton who was
in charge of the first British installations in the field of atomic energy:

All other technologies have progressed not on the bases of their successes
but on the bases of their failures. The bridges that collapsed ... have
added more to our knowledge of bridge design than the ones which held; the
boilers that exploded more than the ones that had no accidents ... Atomic
energy, however, must forgo this advantage of being able to progress on the
basis of knowledge obtained from failures (35).

We shall later examine the problems which this first requirement poses and
which actually could not be upheld in all its original rigour; the specialists attach henceforth much importance to the lessons from experience,
especially since the shock of the TMI accident. F. R. Farmer himself pointed
out this necessity at the Jülich Symposium in February 1973:

The delays in the completion of large projects ... as well as their
development cost ... considered conjointly with the risks which they impose
on society demand a special effort with a view to increasing the data won
from experience and the means to exploit them whether it comes from the
industry or from elsewhere ... We give too much credit to the 300 reactor
years of safe exploitation which only mean that there has been no accidental
escape of fission products in large quantities, and one forgets conveniently
to see or consider all these small, but sometimes larger, details of equipment and organisation which could very well have led to a disaster (36, p. 9).

Therefore, short of building on ideas, how does the concern for safety
express itself? Various, sometimes opposite, approaches have brought about
the present situation which itself is still in a stage of evolution.

2nd. Determinist approach to safety: the principle of barriers and "defence in depth"

In a nuclear centre safety is a dimension entirely apart from the industrial
project and not a simple rescue arrangement joined on to the installation;
the preoccupation with accident prevention expresses itself from the design
stage in the concern for the proper functioning of the installation and in
its surveillance. This prevention consists in setting up successive independent screens called 'barriers' between the radioactive products and the public
(and the workers). The barriers are three in number as has been spelled out
in the case study on TMI:

---

*P. Tanguy is, as should be remembered, director of the Institute for Nuclear
Protection and Safety (IPSN), the safety and protection organism of the
Atomic Energy Commission.

- The casing for the fuel
- The primary circuit
- The surrounding confinement wall.

The first task for nuclear safety consists in the identification of the various possible situations, in particular accident situations, and the search for their initiating events, internally and externally*. These situations are:

- Category 1: Situation of normal or normal transitory**functioning not requiring an emergency stop of the chain reaction.
- Category 2: Transitory state necessitating a stop of the reactor but without deteriorating effect on any of the three barriers.
- Category 3: Situation that can lead to limited damage to the first barrier.
- Category 4: Situation that can lead to substantial damages to the first barrier but not affecting the airtightness of the third barrier.

At the same time the safety problem has led to demands in matters of equipment reliability (design, control). According to their degree of importance for safety, equipment has been divided into three categories:

Safety Category 1: This is equipment with directly vital functions; its failure is unacceptable for the safety of the reactor and the safety of the population.

Safety Category 2: This is equipment with functions the failure of which causes substantial risk for the population or vital functions in case of failure of equipment in category one or finally functions the failure of which can indirectly cause the loss of category one safety functions.

Safety Category 3: This is equipment with functions the failure of which causes limited risk for the population or can indirectly cause the loss of functions of category two.

In Category 1 we find mainly equipment which guarantees the stopping of the reactor and the cooling of the core: in Category 2 equipment for e.g. the containment of the reactor or equipment the failure of which could fail the reactor's stopping system etc. ...

We can then distinguish three systems categories, defined and dimensioned on the basis of these four situation categories:

a) Systems used in normal functioning. They must be of proper reliability with regard to design, construction, maintenance and utilisation of the apparatus. Not only must a good basic quality of the various elements used be provided (properly laid out to offer the best resistance to demands made upon them) but in addition their quality must allow the reduction of such demands.

The objective is here to ensure the proper functioning of the installation and to minimise the number of incidents that could either fatigue it or break into a process that could, in case of failure of the protections, affect the integrity of one of the first two confinement barriers.

----

*Earthquakes, floods, explosions etc.

**Transitory = change from a state that is considered stable to another state

b) Protective systems for the avoidance of incidents. In case of an event
that involves the normal functioning (electrical fault, incident in the water
supply for the steam generators ...) the systems*called 'protective' intervene
in order to limit the consequences (temperature, primary pressure etc.) and
to maintain the holding-up of the first two barriers. These systems are
designed for very high reliability. The redundancy level of this apparatus
(doubled or tripled) was determined by application of criterion 21 of document
10 CFR 50 of the American regulation called "unique failure criterion" which
can be described as follows:

The protective system must be designed for high functional reliability and
a possibility for checks in operation, proportionate to the safety functions
to be guaranteed. The protective system must be designed to have redundancy
and sufficient independence to guarantee:

that a unique failure does not cause the loss of the protective function,

and

that the breakdown of any component or channel does not entail the loss of
the minimal necessary redundancy unless it can be shown that the functional
reliability of the protective system remains acceptable ...

c) Safeguarding systems to minimise the effects of hypothetical accidents.
These systems are designed to limit the consequences of hypothetical accidents
(category 4) even when the protective systems would have worked. For PWR
reactors these are:

safety injection (which as we have seen at TMI was called upon after the
failure of the discharge valve of the pressure unit);

the spraying of the surrounding wall to reduce pressure and temperature in
in this surrounding confinement wall (safeguard for the third barrier);

Insulation system of the surrounding wall which serves to confine the
dangerous products within the wall.

For fast breeder reactors they are:

the emergency cooling circuits (with sodium in the tank and water on the
outside) which contrary to PWR centres are systems in permanent operation;

the confinement system the function of which is the same as with PWR systems;

the recovery system for melted fuel installed at the bottom of the main tank.

3rd. The probability approach to safety

The safety concept of the American PWR and BWR (pressurised boiling water
reactor) is of strictly determinist origin and has been attacked since 1967
by the systematic probability approach of F. R. Farmer. In 1972 the American
Atomic Energy Commission asked Professor Rasmussen of MIT (Massachusetts
Institute of Technology) to form a team for a complete probability analysis

---

*These systems work by means of the dropping of absorber rods into the core
which stops fission reaction. However, as we have seen in the example of
the TMI accident it remains necessary to continue the extraction of heat
called decay heat due mainly to the activity of fission products.

of the safety of water reactors. This study took three years of work, a total
of seventy man/years of effort and a budget of 4 million dollars. It resulted
in the WASH 1400 report (16).

This imposing work consisted in the study of every important safety system,
the research into their mode of failure, the appreciation of these failures,
the projection of all possible accident sequences (the idea of exhaustive
treatment was not absent) and finally the application of a certain probability
and a certain prediction of the quantity of emitted radioactivity for each
sequence.

According to H. Lewis who later chaired the group appointed by the NRC to
submit an expert opinion on the WASH 1400 report (29):

Only this procedure permits rational devices concerning the measures to be
taken for the improvement of the functioning and the safety of nuclear
reactors (37).

The builders have actually understood the interest which this approach might
sometimes have for them because it could permit optimisation of safety
investments starting from a demonstration of the excessive nature of certain
costly approaches demanded by the safety authorities: the taking into account
of a failure of the emergency stop system, of high intensity earthquakes, of
crashes of large size aircraft etc.

Thus it seems, in France, the designers of fast breeder reactors orientated
themselves towards ignoring the accident in which the rods do not drop for
reactors after the Super Phoenix. For the latter a probability demonstration
given to the safety authorities has led to this accident being "marginalised"
(spelled out in the decree giving the authorisation of construction), an
accident consisting of the stop of the four primary pumps without dropping
of the rods (this accident leads to the release of 800 MJ of energy).

For the safety authorities this approach is of indisputable interest
because it does lay claim to the exhaustive treatment of accident sequences.
In this manner the accident sequence at TMI could in fact have been predicted.
Actually, it was already known that the type of accident that occurred at
TMI was problematic because one of the important conclusions of the WASH 1400
report was from the time of its provisional drafting (1974):

- that the contribution of 'large primary breaches' to the overall risk in
  a PWR centre is negligible;

- that system faults other than those in safeguard systems (small primary
  breaches and human failures), for the management as well as for tests and
  maintenance, contributed substantially to the overall risk.

4th. Connections between the determinist and the probability approach to safety

What has been said shows at which point the two approaches are far from
each other. They are, however, not necessarily opposites and could rather
be of a complementary nature. It remains less so as long as the only
recognised official approach is the first one.

However, a large number among those in charge have been favourable to the
second approach. Thus J. Servant, while he was General Secretary of the
Interministerial Commission for Nuclear Safety, declared in 1976 concerning
the Rasmussen report:

I strongly wish that by absorbing this methodology we could carry out analogous studies on French nuclear centres and perhaps even on other factories (38, p. 391).

P. Tanguy is not opposed to probability methods of analysis for safety:

Personally, I think that (...) in order to try and see clearly the future perspectives of safety with the optics of probability the debate should be calmed down and one should proceed from the simplest to the most complicated. The simplest is the reliability analysis of important systems for safety and the search for the best solutions which when applied at the level of design does not necessarily become more expensive ... At the same time the collection of empirical data must be organised which alone guarantee the results of previous analyses. Obviously, there have to be limits in order not to set up unreasonable organisations which cannot be operated in practice (...) Already worked out, this means the application of probabilities to safety judgements (39, p. 399).

This probability approach has actually influenced the inclusion of the unique failure criterion which as we have seen essentially replaces the determinist approach. In fact the application of this criterion in its strict sense has often proved too restrictive and too penalising. Whence a certain evolution, starting from probability notions, an evolution which actually does not yet deserve this term today*.

Such an evolution would question the officially recognised working methods in the field of safety, to wit the taking into account of the measurability of the pattern of accidents which have been written into the framework of the principle of 'defence in depth'. In fact, in each of the situation categories about which we have spoken the most representative accidents have been mentioned. More specifically, the reference accident (ADR)**has been spelled out in Category 4. For the PWR centres this is the AFPR (accident through loss of primary refrigerant). The probability approach to safety makes this concept obsolete as P. Tanguy remarked in January 1976:

The crucial point that has appeared since 1967 is that such an approach is utterly incompatible with the classical method (...) since one could no longer stop at the analysis of an accident of so called 'maximal credible' measurability but would have to analyse a whole spectrum of accidents that would be more and more improbable. Whence the refusal by certain organisations to accept this approach ... (39, p. 395).

However, this concept of the maximal credible accident remains at such a point that the NRC has not modified its manner of looking at things after the WASH 1400 report as the report by the group for the evaluation of nuclear risk indicates (expert opinion on the Rasmussen report compiled under the responsibility of H. Lewis):

---

*One may ask oneself therefore whether the designers will not be led to review this criterion of unique failure with all its additive prescriptions in a more overall fashion and base themselves more on conscientious probability studies carried out by teams with the best possible knowledge of the functioning of the systems, of the conditions of maintenance and operation of their equipment, of the physical process at work in case of an accident and previous incidents in nuclear centres.

**DBA = Design Basis Accident in American terminology

The contribution made by the WASH 1400 study, as far as the definition of the importance of the various categories of accidents is concerned, is interpreted in an inappropriate fashion in the rules of the NRC. The WASH 1400 report concludes for instance that the transitory phenomena, the small losses of cooling liquid and human errors contribute in an important manner to the overall risk but this analysis does not find adequate expression in the priorities expressed by either the research group or the regulations group (37, p. 89).

and H. Lewis adds:

These three factors (transitories, small losses of cooling liquid and human error) have been the principal causes of the accident at TMI (37, p. 86).

This accident should therefore provoke an important evolution in the approach to safety. An important shock has been felt because of this accident if one follows P. Tanguy:

One cannot deny that Harrisburg calls the approach to the safety of nuclear reactors into question: the sequence in the accident had not been foreseen and the damages to the core have gone beyond what was previously considered acceptable (40, p. 531).

True, it is not a question of turning everything upside down:

the very foundations of safety, the principle of barriers and what the Americans call defence in depth are not called into question (40, p. 531).

Progressive detachment will be possible for designers as concerns the notion of maximal credible accident (which causes an illusion of safety) in order to orientate themselves more towards the study of sequences in accidents. It will still be necessary for these studies to be pushed much further than where they are at present. Coming back to TMI, one may comment that none of the analyses had foreseen the ultimate evolution of the accident in particular the formation of the hydrogen bubble. For the designers the whole field of post-accidental analysis opens up which proves to be fundamental and full of surprises in the study of PWR reactors.

3. ADVANCES AND DELAYS IN THE NON-NUCLEAR FIELDS

The example of the nuclear industry in matters of safety analyses is first rate. It is, however, indicated not to forget what happens in other fields which also call for great vigilance with regard to safety. Two cases attract particular attention: the aeronautics field for its advances in this matter; the classical industry where only recently one notices this type of concern.

1st. The safety of systems in certain activities of the aeronautics and aerospace industries

a) The Concorde case: original safety studies. The probability approach has been developed for the safety of recent civilian aircraft, notably Concorde; this aircraft presented such novelties by comparison with its predecessors that it became necessary to review the habitual approaches in the matter profoundly:

The classical rules worked out for the DC 3 and extrapolated intelligently until they were applicable to the Caravelle, the Boeings 707 and 747 have turned out to be inapplicable, useless and even dangerous for Concorde.

The setting up of certification rules for Concorde has therefore required a long study of the conditions which lead to an accident, a study the philosophy of which is now known in the aeronautics field under the initials ESAU* (41, p. 2).

These works by J. C. Wanner (42) have served as a working basis for a probability regulation which has completed the old regulation under the determinist concept. P. Toulouse reminds us of its principal features (43, p. 402-403):

The classification of breakdowns in four categories: minor, major, critical and disastrous (in this last case the system is destroyed, there are injuries or deaths).

The determination of safety objectives for each category (the probability of any disastrous breakdown must thus not exceed 10^{-9}/hour and in addition the total of all breakdowns of this type must remain below, in frequency, 10^{-7}/hour; this higher demand was adopted because of error margins wanted for such an original aircraft).

The working out of study and security**analysis methods: the aircraft is considered as a system and is divided up into twenty five subsystems which are the object of systematic analysis (qualitative and quantitative); after elementary failures one can spell out, by a chosen regrouping, the list of abnormal functionings of the system.

The definition of utilisation criteria of aircraft starting from specifications of equipment (maximum risk limits per aircraft and for a fleet).

b) The safety of weapons systems. Here, we shall not describe in detail the approaches adopted by the specialists of the "Ballistic and Space Systems" Division of the Aerospatiale company. Rather than add yet another procedural definition (definition of tools, of structuring the safety analysis for the lifetime of the system from its design to its abandonment) we shall shed light on the essential principles which guide analysis here.

The demands adopted by the specialists in this field can be expressed, in the first place, negatively by a series of rejections:

Generally speaking, safety is not just an additional preoccupation: if defined in this way it is rejected by the designers and users of the system because it unsettles the confidence of the teams in the chosen solutions (45, p. 4).

Fundamentally, a dictate of policy by the results of quantitative studies is not accepted: one has to guard against the choice of results obtained through these studies as an acceptability level (45, p. 10).

---

*ESAU — Etude de la securite des aeronefs en utilisation — in English: ISAAC = Investigation on Safety of Aircraft and Crews.

**The term 'safety' is not used in the aeronautics industry which prefers the term security (apparently in order to stress the probability nature of security: nothing is 'safe' but only measurable in terms of security on a scale; this underlines the need for making choices independent of the quantitative results obtained from calculations) (44).

(...) Figuring out the safety of a system precisely was a catch ten years ago, remains a catch today and will be a catch in ten years' time. Not only because of the lack and the imprecision of the probability data but above all because of the intrinsic limitations of human imagination (46, p. 5). The claims of the Rasmussen study in this respect had to be fatal for the method of analysis it set up (indisputably interesting and developed in the Sixties for the American weapons system Minuteman) (46, p. 5).

Analysis must not be carried out (neither explicitly nor implicitly) according to imaginary scenarios; it cannot discard any element *a priori* even if it is judged hardly imaginable. The notion 'beyond measurability' is not accepted (45, p. 9).

The studies must not be guided by a mentality which is apt for basic research and often fond of an exhaustiveness that is useless in industrial management. In a word, it is important not to let oneself be dominated by the complexity of the system under study. The following remark has been suggested in this respect: It is useless to ask oneself in an extreme fashion (10^{-6}?, 10^{-7}?) about the failure frequency of a fuse: this seems to us to make for mathematical amusement if one takes into account the fact that this fuse might, within an operating system, be poorly calibrated or worse still, be replaced by a copper wire that relieves a tired operator of removing it (45, p. 7) — frequency of this type of 'error': 10^{-3}(!).

By contrast the following principles were adopted:

What is required is the setting up of an integral safety programme: its object is the search for the best possible optimisation of a system in terms of performance and cost, the demonstration of the upholding of chosen safety objectives throughout the lifetime of the system (...) The conjunction of these three types of studies in an organised whole (entity) alone deserves the name safety programme (45, p. 6).

A safety programme cannot exist if there has not been a prior choice of policy, defining clearly:

The 'events feared' about the system i.e. the precise definition of the event or events to be avoided during the lifetime of the system.

The accepted probability of occurrence of these feared events: this is the *a priori* choice of the level of acceptability. This decision eliminates the snag which consists in succumbing subsequently to the quantitative results one gets (45, p. 9-10).

The environment to be taken into account for these events. Here again the choice is made *a priori* (45, p. 11).

The safety level sought by putting the programme into operation does not aim at 'absolute safety' — a mythical objective — but to guarantee the absence of risks that are higher than the accepted occurrence level. Quantification all along the line aims only at ensuring a performance level that is higher than or at least equal to the one sought: it does not aim at describing with absolute precision the risk of failure which attaches to one or the other element. The whole analysis is therefore not developed for its own sake but as a function of the policy objectives established beforehand. The authors strongly emphasise this point (45, p. 4).

2nd. <u>Safety studies and classified installations</u>

R. Andurand, safety specialist at the Department for Nuclear Safety (DSN) of the CEA was able to undertake the first in-depth safety studies in non-nuclear industrial installations, at the request of the authorities and with the active cooperation of the operators concerned. He quotes three case studies (47, pp. 117-122).

<u>Grand Quevilly (Rhone Poulenc)</u>. This was a study of the extension of the Rhone Poulenc factory at Grand Quevilly which involved the construction of a synthesis unit for 1,000 t of ammonia per day and the setting up of a cryogenic*storage of 24,000 t of ammonia.

On June 29, 1976 the Higher Council for classified establishments requested a safety study from the operator. We quote a few lines from R. Andurand on the study of the storage:

As concerns the safety analysis for the project of a new ammonia synthesis unit with a capacity of 1,000 t./day the IPSN based its advice on the results of spontaneous releases of ammonia that occurred in large measure at the Mourmelon camp and on the use of the analysis of accidents which had happened (...) In case of accidental rupture of the confinement barriers the liquid is pushed out with brute force. Suddenly carried into atmospheric pressure this fluid is a liquid in a metastable state and overheated by comparison to steam. It boils spontaneously which always starts with a tumultuous emulsion of steam in the form of a flash which produces an aerosol that consists of ammonia droplets at -33°C which in atmospheric conditions can in certain cases travel rather far from the point of emission.

The accepted principle is to estimate the maximum escape that permits the aerosol to be reabsorbed before reaching the confines of the factory. It was in fact expected that the staff inside the factory had been alerted to the chemical risks and equipped to limit the consequences which would not be the case for the public outside the factory.

The application of mathematical simulation programmes for atmospheric dispersion which had been worked out by the nuclear safety department of the IPSN showed a need for splitting up into sections the parts of the installation and of the piping which held pressurised ammonia in order to limit the quantities likely to be set free, taking into account volumes and output.

These limitations of quantities in addition to the sectioning already mentioned implied an effort in the fast and safe transmission of information. Rhone Poulenc increased the number of direct telephone lines and restudied the installation of call posts (call boxes) (...) (47, p. 117).

<u>Pierrelatte (Comurhex)</u>. This was a study of an installation for the manufacture of fluorine based products. At the request of the Industry and Mining service of Rhone-Alpes in 1978 the Comurhex company at Pierrelatte commissioned a safety study to analyse the problems encountered in the reorganisation of a manufacturing unit for fluorine based products.

<u>Mardyck (Stocknord)</u>. In this third case we have cryogenic storage units for

---

*Refrigerated, non-pressurised storage.

ethylene and propylene of the Stocknord company at Mardyck. There again
precise conclusions could be drawn from the study such as the need to be able
to take advantage of the minimum network of the EDF.

The idea which nowadays is valuable to the responsible services is to
generalise, after possibly simplifying them, the methods inherited from the
experiences in the nuclear field*. The industrialists are for instance
required:

- to describe their installations and their projects;
- to record the dangers;
- given whatever they have set up;
- given all possible reactions;
- given the processes in use;
- to record the safety measures adopted;
- to emphasise particularly the most serious menaces connected with the
  execution of the project.

Studies of this kind are developed in general fashion. The chemical
industry in particular has in recent years shown much interest in this type
of analysis, as the recent international conferences at Heidelberg (49),
Mulhouse (50) and Basle (51) show; or studies like the one carried out on a
Scottish port area where very large installations for the storage and the
treatment of hydrocarbons were to be developed (52, 53).

4. A COMPLETELY NEW FIELD FOR APPLICATION OF SAFETY STUDIES:
   LARGE INDUSTRIAL CONCENTRATIONS

The issue of concentrations of dangerous installations in certain areas
('industrial' or not) has begun to attract attention. There again the
nuclear industry was among the first to develop the necessary safety studies.

Electricité de France has thus studied the accidents which can involve its
centres and which have two possible sources (54):

- industrial installations: refineries, chemical and petrochemical complexes;
- communications lines.

However, the non-nuclear sector can also participate in this approach.
There exists one case in the world where a study has been carried out on a
very large scale: it concerns the Canvey Island area in Great Britain (55).

The multiple activities at Canvey Island have been listed previously. May
it suffice here to recall that this is one of the strongest industrial concentrations in the U.K. with, in particular, two of its largest refineries
and a methane terminal with large storage capacity. Since there is the issue
of two additional refineries in this agglomeration local resistance rose and
when the projects were beginning to take shape one of the members of the
enquiry commission recommended strongly what had always been energetically
rejected by the government: the study of the overall risks presented by the
area (2, 3). This perspective was accepted in March 1976: the Health and
Safety Commission was given the task of carrying out a study of the existing

---

*Experience acquired in the aeronautics field could equally be of use in the
industrial field.

risks at the island of Canvey and in the neighbouring area of Thurrock. The technical investigation was entrusted to the Safety and Reliability Directorate, a specialised group of the U.K. Atomic Energy Authority. The mandate given to the group was the following:

In the light of the proposal made by United Refineries Ltd, envisaging the construction of a further refinery on Canvey Island, to study and determine the overall risks to health and safety on account of the fact of any possible major interaction between the existing or projected installations in the area where significant quantities of dangerous substances are manufactured, stored, maintained, treated and transported or used, without forgetting the loading (or the unloading) of such substances on or from ships berthed at the jetties; to estimate the risks and to prepare a report on them for the Commission (55, p. 2).

The H.S.E. emphasises:

To carry out this enquiry (...) it was necessary (...) to estimate the risk represented by each of these installations taken individually before considering any major interactions which might occur, starting from, for instance, conflagrations, explosions or escape of toxic substances into the atmosphere (55, p. 2).

We have specifically requested from the investigating team that it be as realistic as possible in its appreciations and that in case of doubt it must not stray towards optimism (55, p. 2).

The study took two years (1976-78), cost about £400,000 and required the joint efforts of some 30 experts. It permits a safe diagnosis of:

- each existing individual risk in systematic fashion;
- the possible interactions between installations in case of a major accident.

From it results a body of precise recommendations aiming at the limitation of the discovered risks. All this permits a singular reduction of the probability of a serious accident on the site, the diminution of accident consequences if accidents occurred.*

REFERENCES

1. RISK STUDIES IN AN OVERALL APPROACH

(1) Ch. STARR, Social benefit versus technological risk. *Science*, vol. 165, no 3899, 1969, pp. 1232-1238.

(2) H. OTWAY, Societal and scientific causes of the historical development of risk-assessment comments. Battelle Workshop on Society, Technology and Riskassessment, Wölfersheim, June 5-8, 1979 (2 pages).

(3) W. D. ROWE, Application of risk analysis to environmental protection

---

*This 'world premiere', as those in charge stress with good reason, naturally raises many other questions: they will be examined later.

T.N.O. 10th International Conference: Risk Analysis, Industry, Government and Society, Rotterdam 1978, pp. 59-73.

(4) R. W. KATES, Risk-assessment of environmental hazards. Scope-Report no 8, 1979.

(5) R. KATES (editor), *Managing Technological Hazard: Research Needs and Opportunities*. Institute of Behavioral Science, University of Colorado, 1977 (169 pages).

(6) W. D. ROWE, *An Anatomy of Risk*, Wiley, New York, 1977.

(7) H. J. OTWAY, Present Status of Risk-Assessment. T.N.O. 10th International Conference: Risk Analysis, Industry, Government and Society, Rotterdam, 1978, pp. 6-28.

(8) L. P. JENNERGREN, Risk assessment techniques. Technical report series in business administration. Odense University, Denmark, no 6, 1978 (62 pages).

(9) V. C. MARSHALL, Historical and theoretical approaches to the prediction of hazard and risk. *Loss prevention and safety promotion in the process industries*. 3rd International Symposium, Basle, Switzerland, September 14-19, 1980, Vol. 2, pp. 6-395 et 6-408.

(10) Advisory Committee on Major Hazards: First Report Health and Safety Commission. Her Majesty's Stationery Office (H.M.S.O.), London, 1980.

(11) Advisory Committee on Major Hazards: Second Report Health and Safety Commission. Her Majesty's Stationery Office (H.M.S.O.), London, 1980.

(12) Study panel on assessing potential ocean pollutants. Assessing potential ocean pollutants. National Academy of Sciences, Washington, 1975.

(13) Climatic Impact Committee. Environmental Impact of Stratospheric Flight. National Academy of Sciences, Washington, The Academy, 1975. *Science*, vol. 186, 25 October 1974, pp. 335-338.

(14) International Environmental Programs Committee. Implementation of the Global Environmental Monitoring System? National Academy of Science, Washington, 1976.

(15) *Report of the President's Commission on the accident at Three Mile Island*. Pergamon Press, New York, 1979 (201 pages).

(16) Reactor safety study: an assessment on nuclear risks in U.S. Commercial nuclear power plants. U.S. Atomic Energy Commission, report no WASH 1 400, Washington D.C., 1975.

(17) J. T. KOPECEK, Risk assessment of a liquified natural gas terminal T.N.O. *10th International Conference: Risk Analysis, Industry, Government and Society*, Rotterdam, 1978, pp. 86-101.

(18) Final Environmental Impact Statement. For the construction and operation of an L.N.G. import terminal at Everett, Massachusetts (Port of Boston). Federal Power Commission, Bureau of Natural Gas, 3 September 1976.

(19) J.R. RAVETZ, *The use of mathematics in the assessment of risk*. The

acceptability of risks. Council for Science and Society, Barry Rose, London, 1977, pp. 99-104.

(20) T. A. KLETZ, The Risk Equation: What Risk Should We Run? *New Scientist,* 12 May 1977, pp. 320-322.

(21) J. LINNEROOTH, The evaluation of life-saving: a survey I.I.A.S.A., R.P. 75-21, July 1975 (30 pages).

(22) J. LINNEROOTH, Methods for evaluating mortality risk. *Futures,* August 1976, pp. 293-304.

(23) A. MELLY, D. C. 10. *Argus International,* no 13, juillet-août 1979, pp. 304-311.

(24) R. CONGAR, Estimation des pertes de bien-être des touristes en 1978 dans le Finistère. La pollution marine par les hydrocarbures. Colloque international de l'U.V.L.O.E., Brest, 28-30 mars 1979, pp. 181-186. pp. 181-186.

(25) O. GODARD, Environnement et rationalité économique: la prise en compte de l'environnement dans la planification des projets et programmes de développement en pays capitalistes. Thèse pour le Doctorat de Spécialité (3e cycle) en Economie du Développement. Université de Paris I, Paris 1977 (413 pages).

(26) Ch. STARR, General philosophy of risk-benefit analysis. Paper presented at E.P.R.I./Stanford, I.E.S. Seminar, Stanford, California, U.S.A. September 30, 1974, p. 20.

(27) V. PILZ, What is Wrong with Risk Analysis? *Loss prevention and safety promotion in the process industries 3rd International Symposium,* Basle, Switzerland, September 14-19, 1970, pp. 6/448-6/454.

(28) A. KLETZ, The Application of Hazard Analysis to Risks to the Public at Large I.C.I. World Congress of Chemical Engineering Session A.5. Environment and human Activities, Amsterdam, July 1, 1976, p. 1.

(29) Risk assessment Review Group Report to the U.S. Nuclear Regulatory Commission. N.U.R.E.G./CR-0400, National Technical Information Series, Springfield, Virginia, 1978.

(30) Cl. HENRY, Le rapport Rasmussen. Petite histoire sociale d'un important document technique. *Futuribles,* no 28, novembre 1979, pp. 35-41.

(31) H. J. OTWAY et J. J. COHEN, Revealed preferences: comments on the Starr benefit-risk relationship I. I. S. A., R. M. 75-7.

(32) B. FISCHOFF, P. SLOVIC, S. LICHTENSTEIN and R. READ, How safe is safe enough? A psychometric study of attitudes towards technological risks and benefits. Draft, October 1975.

2. THE UTILISATION OF SCIENCE IN THE CIVILIAN NUCLEAR FIELD

(33) Th. MESLIN, Risques sanitaires et écologiques de la production d'énergie électrique. Centre d'Etudes sur l'évaluation de la protection dans le domaine nucléaire (C. E. P. N.), Rapport no 20-2-(2), 1979 (43 pages).

(34) Analyse de sûreté et méthodologie pour les installations nucléaires
Revue Générale Nucléaire, mai 1979, pp. 49-51.

(35) F. R. FARMER, Relationship between risk assessment and reliability
requirements U. K. Atomic Energy Authority, Culchets, Warrington,
England, (Note communiquée par l'auteur, non datée (1976-77) (14 pages).

(36) F. R. FARMER, Development of risk standards. A.I.E.A., Principles and
Standards of Reactor Safety Symposium, Jülich, 5-9 February, 1973.

(37) H. LEWIS, La sûreté des réacteurs nucléaires. *Pour la Science*, mai
1980, pp. 73-89.

(38) J. SERVANT, L'approche probabiliste de la sûreté des réacteurs
nucléaires. *Revue Générale Nucléaire*, no 5, octobre-novembre 1976,
pp. 390-391.

(39) P. TANGUY, La sûreté nucléaire et les méthodes probabilistes. *Revue
Générale Nucléaire*, no 5, octobre-novembre 1976, pp. 392-400.

(40) P. TANGUY, L'accident de Harrisburg: scénario et bilans. *Revue
Générale Nucléaire:* no 5, septembre-octobre 1979, pp. 524-531.

3. ADVANCES AND DELAYS IN THE NON-NUCLEAR FIELDS

(41) J. C. WANNER, Critères généraux de sécurité — Association. Association
Technique Maritime et Aéronautique — 47 rue Montceau, 75008 Paris
Session 1980 (30 pages).

(42) J. C. WANNER, Etude de la sécurité des Aéronefs et Utilisations (E.S.A.U.)
(E.S.A.U.). Octobre 1969.

(43) P. TOULOUSE, Evaluation de sécurité des systèmes d'avions civils récents
Utilisation des méthodes probabilistes dans l'aéronautique et le
domaine nucléaire. *Revue Générale Nucléaire*, no 4; juillet-août 1979,
pp. 399-410.

(44) Terminologie de la sécurité des systèmes. *Société pour l'avancement de
la sécurité des systèmes*. 1ère édition du 1er février 1979.

(45) J. C. DESCHANELS, Le programme de sécurité et ses applications.
Aerospatiale Subdivision Systemes, SYS/P 19258, 28 janvier 1980
(24 pages).

(46) P. LAVEDRINE, Les arbres de défaillance généralisés et leur traitement
informatique Aérospatiale, Division Systèmes balistiques et spatiaux,
Subdivision Systèmes, no SYS/P 19 222, 21 janvier 1980 (36 pages).

(47) R. ANDURAND, Le rapport de sûreté et son application dans l'industrie.
Annales des Mines, no 7-8, juillet-août 1979, pp. 115-138.

(48) Les techniques de la sécurité moderne. Société pour l'avancement de la
sécurité des systèmes en France Société d'etudes sur les "programmes de
sécurité au sein d'un projet" 7 juin 1978, Paris. Actes, mars 1979
(195 pages).

(49) European Federation of Chemical Engineering. 2nd Symposium on loss
prevention. Heidelberg, Germany, 6-9 September, 1977.

(50) Ecole Nationale Supérieure de Chimie et de Mulhouse. Colloque sur la sécurité dans l'industrie chimique Mulhouse, 27-29 septembre 1978.

(51) European Federation of Chemical Engineering. *3rd International Symposium on loss prevention and safety promotion in the process industries.* Basle, Switzerland, September 15-19, 1980.

(52) M. CREMER and M. WARNER, Guidelines for layout and safety zones in petrochemical developments. A report prepared for Highland Regional Council, no C 2056, February 1978 (160 pages).

(53) M. CREMER and M. WARNER, The Hazard and Environmental Impact of the proposed Shell N.G.L. Plant and Esso Ethylene Plant at Mossmorran, and Expert Facilities at Braefoot Bay. A report prepared for Fife Regional Council, Dunfermline District Council and Kirkcaldy District Council, no C 11, May 1977 (2 volumes).

4. A COMPLETELY NEW FIELD FOR APPLICATION OF SAFETY STUDIES: LARGE INDUSTRIAL CONCENTRATIONS

(54) A. LANNOY et T. GOBERT, Evaluation des dangers dus à l'activité industrielle à proximité des centrales nucléaires. Etudes déterministes et probabiliste. E.D.F. Communication présentée au 5e S.M.I.R.T., Berlin, août 1979 (11 pages).

CONCLUSION: AN ARSENAL OF MEANS FOR THE PREVENTION AND THE REPARATION OF ACCIDENTS

The development of sciences and techniques permitted man of the late eighteenth century and of the nineteenth century to achieve greater autonomy from nature. At the same time he experienced accidents, disasters which led him to organise himself, to protect himself. In fact, this evolution took place without the actual intervention of those who were effectively subjected to risk but rather under the impetus of those who held political and economic power for various reasons: economic, social and humanitarian.

At the beginning of the nineteenth century (in France anyway) the risks connected with workshops and factories led those in charge to institute rules and control systems (imperial decree of 1810) and control apparatus (take-over of fire fighting in Paris by the military in 1810).

The development of the chemical industry, of large scale industry, of the petrochemical industry led those in charge at the start of the twentieth century to strengthen the safeguarding and protection systems (law of 1917 in France for instance). By employing the same logic one can also attribute a greater breadth to the means used a century earlier.

The post-war period saw the part-achievement of this edifice: laws took explicitly into account the newly appeared risk of large accidents. In fact, this was done to various degrees. We have seen how in Great Britain, unlike Italy, they managed to give themselves the juridical and legislative means to try and prevent disasters; how France, side by side with a similar effort, has tried to re-adapt its rescue plans (ORSEC, PPI). In other fields, the organisation for the indemnification of damages created national and inter-national pools, compensation funds and conventions of solidarity.

In addition, the progressive change of society towards a high-level tech-nological society has modified the tools and the means of risk management. Science and techniques intervened, essentially on the level of prevention. Thus we have seen the aeronautics and the nuclear industry integrate the concern for safety from the design stage of new projects. The techniques used are various but they all have in common the speciality of taking into account, *a priori*, accidents that are likely to occur. By the same token there has been a change in the nature of risk through the appearance of major technological risk and also there has been a change in working methods. One no longer bases oneself exclusively on experience by drawing lessons from àccidents that have occurred in order to avoid their recurrence; one researches by means of analyses the accidents which one considers both physically possible and relatively probable and protects oneself against them.

However, despite recent improvements in the form of laws and regulations, despite improvements in scientific methods one cannot help raising questions about this inventory of available means and tools. We must therefore critically analyse them and seek out their defficiencies and limitations.

CHAPTER FOUR

Deficiencies and limitations of major risk management

> I Multiple insufficiencies
> II Very serious or even absolute limits

I. MULTIPLE INSUFFICIENCIES

The problems to be controlled have been examined in Part One; the means and tools available have been examined. What appears to make them insufficient? There is a first type of insufficiency which we shall now identify for each of the main chapters we have distinguished: prevention, defence, reparation, the scientific tools belonging to the first term.

The insufficiencies of this first type are related to adaptation. The practice: lack of budgetary means, organisation, laws, knowledge, do not permit defined objectives, no matter what the competence and the determination of those in charge may be like.

We shall try to mark out a certain number of examples of this type.

1. INSUFFICIENCIES IN THE MEANS OF PREVENTION

The difficulties felt on the level of prevention of serious accidents come to the fore in three situations: maritime transport of hydrocarbons and black tides; the general case of classified installations; the more sensitive example of nuclear installations. One notices that the more points of risk of high financial value are reduced in number and surveyed by a tight network of competent people, the more the limitations of management means employed lose their gross and massive nature by comparison with those at the other end of the scale. Restricting thus the field of investigation for these few pages we find on the one hand the automobile risk*and its slaughter, the daily deadlock of prevention; on the other the control of the safety of weapons systems, a difficult but limited and crucial problem which has the benefit of great attention.

---

*Let us remember that there are 250,000 deaths due to road accidents per year worldwide and 10 million injured. In Europe, half or more of the deaths of young men (age fifteen to twenty four years) are due to this cause. The prevention efforts developed in France during the last few years do not seem to yield any longer the very positive results which they did between 1972 (16,000 deaths) and 1977 (13,000 deaths) (1).

We retain here three intermediate cases which we have mentioned.

1st. The prevention of black tides: still too limited means of intervention

The problem of prevention of black tides shall be more broadly examined in the following section as it is related to obstacles. In fact, the situation in this field is so serious that one gets beyond the stage of simply removing the difficulty. We mention, nonetheless, one point upon which action appears easier than upon others: the setting up of adequate means for immediate intervention on the high sea. The setting up of effective means of ultimate rescue would no doubt permit the prevention of some accidents. The enquiry report by the National Assembly on the *Amoco Cadiz* accident was clear in this respect: the situation has been spelled out with regard to international law; it remains to provide proportionate technical means:

It can therefore be affirmed that the governments have from now on greater powers of intervention in maritime traffic, in regulation, control and suppression within their territorial waters and that they can exercise them fully with regard to foreign ships.

Above all, however, sovereignty implies surveillance; and if the exercise of sovereignty is at present limited this is also true and perhaps even more so for the insufficiency of means devoted to ensuring effective surveillance (2, p. 74).

On each occurrence of a disaster steps are taken. Following the wreck of the *Amoco Cadiz* the response was firmer: regulations for maritime traffic, a control centre set up at Ouessant, allocation of a building by the Navy, tugs*based at Brest; clarification of the respective duties of the administrations; structuring of CEDRE**etc. Let us remember here more precisely the recommendations made by GICAMA***on the necessary means for a complete surveillance of the economic 200 miles zone (3, pp. 208-211):

- 8 heavy vessels (1,500 to 1,800 t) equipped with helicopters;
- 8 light vessels (250 to 300 t) for fast intervention
- 1 high sea trawler
- 14 maritime surveillance aircraft
- complete radar control of the channel.

For assistance there must be made available:

- 6 heavy helicopters
- 4 tugs of 20,000 HP for the Channel (moored at Ouessant).

This comes close to 3 billion FF (total expenditure to be foreseen, taking into account battle equipment amounting to 5 billion to be split over several exércises). These funds could best be gathered by means of a 'littoral protection fund' financed by the European community and industry, two bodies for whom the amount needed would not be too significant (the report remarks

---

*This permitted the rescue of the part of the *Tanio* that was still afloat in March 1980.

**Documentation centre for research and experimentation on accidental water pollution.

***Interministerial coordination group of administrations for action on sea.

that the oil companies are not obliged to insure cargoes with a value of more than 100 million FF).

"The French government, more generally: the public authorities, must immediately and effectively draw the proper lessons from the *Amoco Cadiz* disaster", the senators concluded before adding two short remarks: "At once. For tomorrow it would be too late" (3, p. 230).

These funds are not available, and no plan aiming at this objective has been adopted. After the sinking of the *Tanio* on March 7, 1980 some 800 million FF were released to increase the intervention funds of the Navy. But it is only foreseen to make available vessels that had been destined for other purposes. The recommendations by GICAMA that were taken up by the senators in 1978 have therefore still not been fulfilled, either in quantity or in quality.

It is true that there is a financing problem. One cannot see why the recognised principle of the "pollutor=payer" (despite its limits) was not adopted in this field where the economic means of the operators concerned are powerful and they are among the most flourishing in these so-called crisis times.

2nd. The follow-up on classified installations: difficulties

a) Insufficient means. First of all we must remember the enormity of the task: recently, 50,000 installations coming under the authorisation requirement have been counted; 400,000 come under the declaration rule (4, p. 42). To take on this task the necessary means are not available. At the time the law was debated in 1976 a member of parliament emphasised that every inspector was reponsible for more than 1,500 establishments of which 300 are in the first and second category of the 1917 law (5, p. 38). Despite the progress which has since been made the gap between tasks and means remains. The argument regularly raised against this type of criticism is well known: it is not advisable to raise an army of inspectors placed behind every machine twenty four hours a day. The argument is not altogether without foundation but it is hardly a serious one if one looks at the reality: we are far from going to such excesses. Those in charge on the spot agree immediately: they cannot carry out their responsibilities for lack of funds.

This difficulty is not only 'quantitàtive'. The growing complexity of technology, the specialisation which is pushed along ceaselessly in certain fields (chemistry, biology ...), the extreme difficulty of following up on the research carried out in laboratories are an additional challenge for the services in charge of watching over the application of the laws.

b) Are the laws sufficient for the extreme risks? Is the present legislation an adequate tool? The progress brought about by the 1976 law permits us to confirm that the risk of serious accident is now better followed up. But what about the risk of disaster?

We are able to spell out the defficiencies of the available laws. We have pointed out that the serious problems posed by industry should have evoked a more trenchant attitude on the part of the legislature. For instance, Article 3 of the law spells out that an authorisation must only be granted if the dangers can be 'provided for' by the measures specified in the prefectorial order: the word 'overcome' should have been preferred, it has been said (6, p. 33).

It may be retorted that certain decisive means have been put in place, like Article 15:

A decree by the Council of State, adopted on the advice of the Higher Council for classified installations can order the elimination of any installation whether or not mentioned in the nomenclature which presents dangers or inconveniences for the interests mentioned in Article 1 if the measures foreseen in this law cannot eliminate them.

The fundamental problem connected with these debates is dependent on the general policy with regard to the industrial environment. Some call for authoritarian measures to be imposed on the operators; others think that it would not be a good policy to create a relationship of challenge and that on the contrary it is indicated to develop the sensitivity of the operators further, i.e. a greater awareness of risk on their part and more adequate practices. It is a matter of creating a process by which industry itself takes on greater responsibility, spurred on by an intelligent and flexible administration which does not exclude firmness and vigilance.

The debate goes on for ever, one side maintaining that with the question of extreme risk one is bound to adopt much more authoritarian policies, whence their criticism of the law. It is then retorted that the radical and authoritarian approach would not yield results and would not be accepted ..., which gives a reason to the radicals to reject installations that are too dangerous, considering the examples which industry has given of its sense of responsibility. (**Seveso**, **M**orhange, Minemata ...). Thus the debate soon leaves the framework of judicial examination. Later developments will bring some elements of reflection to bear on this debate.

One particular point of agreement, however, comes to light about a deficiency in the law which is nowadays apparent. The law says nothing about the problem of malevolence and terrorism. This question which is very sensitive today deserves the attention of the legislature.

c) Industrial practice. One particular difficulty must be taken into account: the long term maintenance of safety performances. Technological progress makes safety principles obsolete: there is a need for revision, adaptation, change ..., all kinds of things that require more than just the application of regulation standards, which were pertinent. A thorough knowledge of the dynamics at play is needed, judicious arbitration, pauses which permit safety policy to make its voice heard. On the evidence the follow-up on such a complex system calls for substantial capacities.

3rd. The safety of nuclear centres: questions

The prevention of disasters in installations of the latest generation raises completely new questions. A first series of difficulties could be eliminated by using integrated safety programmes from the design stage. The normal functioning is much better controlled than in earlier establishments.

However, the methods used which guarantee better control of risk are not faultless. The scale of events that are likely to occur demand that perpetual improvements are made to existing situations, that discovered deficiencies are treated with the greatest efficiency. What are these deficiencies ? We shall mention a certain number of them here.

In the chapter on the utilisation of science we have seen that the designers base themselves essentially on a determinist approach in order to define the means and the systems to be employed to ensure the safety of nuclear installations.

This deterministic approach is based on a list of accidents to be taken into account from the design stage and on a number of criteria to be applied to each of these accidents. For PWR reactors the French authorities have gone along with the list established by the Americans more than ten years ago and which clearly resulted from a compromise between the demands made by the authorities (the AEC in this instance) and the proposals made by the operators and designers. This approach when applied to the case of fast neutron reactors causes a delicate problem inasmuch as there is no reference to this type of centre, France being in a way the leader in this technology. It is now clear that a good number of those in charge at Electricité de France think that the fast neutron reactor prototypes such as Phénix (in operation) and Super Phénix (under construction) have been designed with too great a luxury of precautions. When one knows that the major problem for the future of this type is an economic problem (the cost of Creys-Malville is estimated to be double that of a PWR reactor of the same size) one can imagine the effort which Electricité de France must make with the authorities to get acceptance for the downward revision of the list of accidents to be taken into account in the lay-out of future fast neutron centres.

One may then ask oneself questions on the independence of judgement of the safety authority. The weight of a favourable policy decision on the development of this type — could it not induce a more flexible attitude of this authority? It would actually be possible for it to justify *a posteriori* such a more flexible attitude by making for instance educated comparisons between the safety arrangements adopted for a PWR reactor and those proposed for the new fast neutron reactors although these two types have very different specific characteristics.

Another weakness of this approach concerns the concept of the maximum credible accident which has proved to be very open to criticism. Thus the reference accident (ADR) of PWR reactors which had been defined as the guillotine rupture of some piping in the primary circuit has been proved not to be the maximum credible accident from the point of view of activation of the safety injection system and that it was necessary to look very closely at the case of small breaches in the primary circuit.

As to the other approach used by the designer, the probability approach, its limitations are well known since they have quickly been pointed out by its detractors who considered this method a dangerous tool because it was demanding. H. Lewis recalls the storm of criticisms raised by the Rasmussen report:

1. The system is much too complex to be quantified.

2. The available data on rates of component failure do not permit such calculations.

3. The statistical methods used are inappropriate and erroneous.

4. The common failures have not been correctly dealt with (this terminology again brings up all breakdowns that affect several independent components and are due to a common cause such as an earthquake).

5. Events of low probability are not intrinsically quantifiable.

6. Human behaviour has not been correctly approached.

7. The importance of quality control has been inadequately evaluated etc. (7, p. 81).

If some of these objections are easily refutable, we shall see later on that some others constitute serious limitations to the use of the method.

Each of these two approaches is actually the subject of substantial improvements. For the first one there is growing concern about the rigour in the application of the concept of defence in depth. This is expressed in the research into successive 'lines of defence' constituted by the whole of the equipment that ensures the surveillance and protection function (first defence line), the safeguarding function (second defence line) or the function that limits the consequences (third defence line).

For the second design method it must be recognised that the Rasmussen report has given it its patent of nobility: the expert report (NUREG/CR-0400) which the NRC requested from H. Lewis also considers that the methodology developed on this occasion should be used for safety analyses and that it constitutes a substantial advance inasmuch as it provides a rational means of risk evaluation. Faced with the criticism quoted before, it must be noted that a rigorous methodology has been introduced which permits the elimination of the inconvenience which the complexity of nuclear installations constitutes. One witnesses here the development of adequate mathematical tools (simulation method, dynamic analysis method etc.) and the introduction of information gathering systems on component reliability which are at present used on nuclear sections in operation.*

This also affirms the willingness of the analysts to deliver the results with a confidence interval that is associated with a given confidence level, which is in fact a *sine qua non* condition for the credibility of these studies.

2. INSUFFICIENCIES OF THE MEANS OF DEFENCE

The *Amoco Cadiz* has clearly shown what the limitations of the means of defence could be; since then these tools have been strengthened. The same type of reflection might be applied to the other arrangements made, particularly where toxic and nuclear accidents are concerned. Certain studies such as the ones recently carried out with competence on the chemical corridor at Lyons show that much — an enormous amount — remains to be done. In the universe of insufficiency which is revealed in the course of reflection now that the risks are with us two complexes may be mentioned for illustration.

How many breathing apparatus (autonomous ones: not gas masks which offer hardly any protection in serious cases) would the authorities have available in case of a large accident? The police who are foremost in warning the people, to advise them to stay at home with doors and windows shut — would they have the necessary means to fulfill their task?

How efficient would the sirens be? The particular sound code for such circumstances**has certainly been published and explained in the local press but the last newspaper article that appeared on the subject dates nearly ten years back (*Le Progres,* Lyon, March 24, 1971, p. 9).

No need to add to the list of surprises that such an examination holds in store. It is sufficient to understand their causes: the awareness of the

---

*SRDF = gathering system for reliability data introduced at Fessenheim and Bugey.

**Three sound signals of twenty seconds, separated by intervals of fifteen seconds which would advise the population of Lyons to listen to radio and television: a few moments later the directives from the Rhone police commissioner would be broadcast.

existence of a large number of extreme risks for which the desirable means are not available is recent. The initiatives taken concerning the particular plans of intervention must urgently be strengthened. However another fundamental factor must also be taken into consideration which also explains to a certain extent the time-lag we have noticed: it is the vague impression which is, however, felt that in case of a really serious large accident there will no longer be a point in fighting it: even less to organise an orderly retreat. This will be examined later but must be mentioned because it doubtless impedes to a certain extent efforts that the introduction of new means could be decided and implemented with the necessary determination.

These fundamental considerations must, however, not hinder a certain adaptation. All those who have examined the presently available means of battle are convinced: the defficiencies are glaring, the delays are heavy. Will it be necessary to have a serious defeat before these means are adapted to the needs of the late twentieth century?

3. INSUFFICIENCIES IN THE MEANS OF REPARATION

1st. Limits of the financial capacities of the operators and their covers

The damages inflicted on equipment are usually well covered. It is different with third party liability insurance which is sometimes still conceived as it was before the massive development of major risk which by definition commits the operator beyond the walls of his installations. In the case of Seveso the strength of the Hoffmann-la-Roche group permitted its president to say "We shall pay for it all; we have sufficient resources for this". But in the hexachlorophene case two of the three companies involved (Setico and Morhange) went bankrupt. In the enterovioform case in Japan the pharmaceutical group involved must have had unusual support from its assessors in order to face the situation: its third party liability cover was notoriously insufficient.

Generally speaking, one may question the ability of industries to face the damage claims which they are likely to occasion nowadays. This is why we see the creation of compensation funds.

2nd. Limits of compensation funds

These are powerful means as we have seen. However, one must have no illusions about their capacities: one must keep in mind the scales attaching to present day risks. We mention here two cases for examination.

a) Funds provided for maritime disasters. In their report on the *Amoco Cadiz* the parliamentarians drew up a list of the deficiencies connected with the 1969 convention (2, pp. 319-323):

Only pollution by hydrocarbons is covered: in case of a chemical disaster or the explosion of a methane carrier reparation would remain derisory;

only pollution by the ship's cargo is covered: no provision is made for hydrocarbons used for the ship's propulsion (case of the *Olympic Bravery*) which nowadays represent large quantities (several thousand tonnes);

the cases of exoneration from responsibility are too generous;

only damages incurred on a state's territory, including territorial waters but not including the economic zone, are indemnifiable;

the limits of liability are inadequate: the amount is about 77 million FF: it was calculated in 1969 for a vessel of 150,000 t which is rather insignificant nowadays.

Even the coming into force of the 1971 convention which took the ceiling to 150 million FF remains very insufficient compared with the amounts that are effectively at stake in case of a substantial accident. The convention of May 25, 1962 concerning the liability of vessels with nuclear propulsion has instituted an integral reparation rule at the expense of the government by whom the operating licence was issued. The parliamentarians suggested that this law be taken as a basis to cover all dangerous transports.

b) Funds provided for nuclear disasters. We quote here the reflections which the TMI incident inspired in A. Melly, a specialist for insurance questions:

The cover. The construction cost of the centre is $2 billion. TMI is insured against third party liability up to $140 million and its equipment for an amount of $300 million ...

The damages. As concerns the damages inflicted on the centre's building they could reach an amount of $60 million i.e. 20 per cent of the risk cover.

The loss of operation amounts to $1.1 million per day, the sum required to supply the region with energy from other sources (*Chicago Tribune*, April 2, 1979). In the field of third party liability things are less clear because it is still difficult to know the number of complaints and claims. Among the numerous claims which followed the TMI accident one may mention:

the creation of a fund to cover the expenses caused by periodical medical examinations for a period of twenty years for all people living within a radius of twenty miles from the centre;

the reimbursement of indirect losses (income) of individuals and losses of operation of companies;

the definite closure of the centre;

the payment of 'punitive damages' by General Public Utilities, the owners of Metropolitan Edison (*Post Magazine and Insurance Monitor*, May 15, 1979).

Six industrial and commercial companies have also brought legal action aiming at compensation for indirect losses suffered in the wake of the accident. For instance: considerable drop in local sales of manufactured products and consumer goods, drop in hotel takings within a radius of 20 miles. Expected losses: more than $560 million, the ceiling amount for liability under the Price Anderson Act. The court will decide whether there was force majeure or not (*The Argus*, May 25, 1979).

Thus, contrary to the statements made by M. Kackson, the spokesman for American Nuclear Insurance (ANI) in the *Journal of Commerce* of April 4, 1979 the centre's cover against third party liability could be insufficient. This is actually the opinion of Robert Lunter, the delegate of the federal insurance administration who affirms in the *Washington Post* of April 18, 1979 that the cover for nuclear risks is insufficient and that in the accident at TMI the insurance will not cover the total of the disaster. Other experts also estimate that the reserves accumulated in the last twenty two years of American underwriting will be exhausted by the TMI accident (*Policy Holder*, June 15, 1979).

There is in fact a problem that can be illuminated with certain figures. The total accidents that occurred between 1957, the start of the civilian nuclear centres in the USA, and March 28, 1979 has cost the insurers $600,000. The most substantial reimbursement amounted to $70,000. The accident at the TMI centre has already cost $1,128,000 in indemnities paid to the people who had been evacuated. In their financial statements for the first quarter of 1979 the New York insurers Crum & Forster have announced net losses before tax which accrued to them on account of the TMI accident and amounted to about $1.5 million. Those of Continental Corporation of New York came to about $3.5 million (*Journal of Commerce*, April 19, 1979).

On the other hand, the authorisation granted to General Public Utilities, the owners of Metropolitan Edison, to increase its electricity tariffs by $49 million has been suspended by the American authorities following the TMI accident. The company has declared that without an increase of at least $33 million it would risk bankruptcy. Taking a rather peculiar point of view, it considers that the cost of the accident should be shared between the investor and the consumer. One would imagine that the latter does not share this opinion (*Chicago Tribune*, April 20, 1979).

The Price Anderson Act. We know that according to the Price Anderson Act the maximum cover for a nuclear centre cannot exceed $560 million. This figure may appear relatively low when one knows that the advertising budget alone of the giant American distribution company Sears Roebuck & Co. (owners of Allstate, the second largest American accident insurance company in the USA) runs to $650 million. One may also mention that the after-tax profits of State Farm for 1978 amounted to $500 million.

By contrast, the ceiling of the Price Anderson Act may seem high compared to the one established by various international conventions for third party liability for nuclear damages, conventions that were signed by most European countries. Thus the Paris convention of July 29, 1969 fixed the maximum liability of the operator at 15 million units of account (equal to about $18 million. The complementary Brussels convention of January 31, 1963, thanks to intervention by governments, carries the amount of reparation to be allocated per nuclear accident to 120 million units of account in total, i.e. about $150 million (*Le Patrimoine*, April 1979).

Out of the $560 million foreseen by the Price Anderson Act $140 million are for the account of the insurer and $335 million to be paid by the pool made up of the 67 American nuclear centres. The remaining $85 million are to be paid out of government funds.

Among the opponents of the Price Anderson Act in the United States we find Ted Weiss, Congressman for New York who submitted a draft law requesting a reform of the limited liability in the case of a nuclear accident. The project HR 789 proposes unlimited liability for the nuclear centres. The part of the liability which falls on the insurers remains the same as under the Price Anderson Act but the new project obliges the centres to share the total cost of damages over and above those reimbursed by the insurers. The government would intervene only with loans to the centres which could not meet their part of the liability. The part of the compensation to be borne by each centre would be determined according to its production capacity, its past safety record and other factors. It is also foreseen in the project that the centre which had the accident would have a liability equal to its total value. This project will be put to the vote in the American Congress before the end of summer 1979 (*Business Insurance*, May 18, 1979).

This is not the first time that the Act has been contested. In 1977 for instance the state of North Carolina had it declared unconstitutional. This ruling was, however, crushed in 1978 by the Supreme Court of the USA. Ralph Nader has also spoken out against the limitations of cover in nuclear insurance as foreseen in the Price Anderson Act. He has in particular reproached the National Association of Insurance Commissioners for not having envisaged the inclusion of a nuclear risk insurance in the multi-risk policies. This question had in fact been raised in 1975 but found no response at the time.

The estimation of the cost of inclusion of such cover in multi-risk contracts remains, however, extremely problematical and depends on technical factors which are still not under control (*Journal of Commerce*, April 20, 1979).

We can hardly say more at this time. The processes set in motion after the TMI accident have not come to an end yet. By the same token, the opposition to the Price Anderson Act will probably cause heated discussions between the authorities, and nobody can tell so far how the matter will finally be dealt with ... (8, pp. 301-302).

II. VERY SERIOUS AND EVEN ABSOLUTE LIMITS

Adaptation problems have been studied. It is a difficulty of a second type, much more serious than the first one, that singularly limits the means and tools of management available for the control of major risk. What is now in question is no longer the need to strengthen the means or better to define the target aimed at but the extreme difficulty, sometimes impossibility, of reducing the gaps that have been noticed even while the size of the existing menaces demand this reduction.

It is now no longer a question of insufficiencies but of obstacles, some of them only temporary, others not, due to history, to technical character-istics that come into play, to theoretical difficulties, to physical problems that must be solved, to the scale of the questions to be dealt with etc.

In the three chapters on defence and on reparation questions of this type appear.

1. The serious limits of prevention

Here we take up once more the three examples chosen in the study of limits to prevention.

1st. Limits to the prevention of black tides

How does one forestall petrol tanker accidents? Progress in many ways may be made but one must understand the seriousness of the situation that has been created in this field. We have seen in the conclusions drawn from the *Amoco Cadiz* case, in the first chapter, which factors block the still timid and laborious attempts at an improvement of the system: the old principles of navigation which are profitably used by those who consider the maritime element only in its narrowly defined capacity as a liquid element that carries tanks; the magnificent success of the North-South dialogue in matters of ship's registrations with the practice of flags of convenience, the increasing traffic density on certain shipping lanes such as the Channel etc.

Here, we shall take up a more technical aspect which it would have been good policy to examine before the disastrous events multiplied: the character-istics of ships. The prevailing situation could not rapidly be changed. This is an example of a severe deadlock in technological development: having been unwilling to integrate the safety issues early enough, this activity has now become extremely dangerous; and contrary to the classical scheme in matters of risk one is here faced with a very large size risk and one with a high probability of occurrence. What, then, are these ships? While being gigantic, they are often fragile, not very reliable, not very manoeuvrable, equipped with inexperienced crews, veritable monsters, they have often been designed' without much concern for safety: some of them are just giant tanks put into motion on the seas of which one hopes that they will safely arrive at their destination intact. May it suffice here to mention a few lines from the parliamentary enquiry reports on the wreck of the *Amoco Cadiz:*

If it is correct that a 200,000 t petrol tanker can be compared to a 35 t truck which has only a 2 CV engine it is no less true that the users are accustomed to this state of affairs and consequently anticipate the manoeuvres (2, p. 32).

In order to avoid or to limit disasters similar to that of the *Amoco Cadiz* it seems necessary that ships carrying hydrocarbons or other dangerous substances should always be able to:

- move, i.e. to have a working system of propulsion;
- steer, i.e. to have a steering system that works (2, pp. 143-144);
- know the immediate dangers of the environment and therefore have valid navigation equipment;
- limit, in the case of damage, the risk of accident or pollution;

It is necessary ... to think without delay of means to be put into operation for the tugging of a giant oil tanker (3, p. 146).

That one should have to announce such recommendations ten years after the launching of the first 200,000 t tankers, eleven years after the *Torrey Canyon* drama and at least fifteen years after the design of such ships seems stupefying*. Inasmuch as a fleet of oil tankers is not changed every few years one has to recognise in this difficulty an extremely serious situation which is likely to be dealt with somehow but not as one would want to be able to deal with it after the repeated black tides, particularly on the Britanny coasts.

The serious faults in options and technological implementations which are so essential have considerable inertia. It remains to be hoped that their impact will be eclipsed before the irreversible damage which they cause and will continue to cause, has too profoundly upset the essential balances of economic and cultural life in the societies concerned.

2nd. Limits to the prevention of large industrial disasters

Here again situations, i.e. tangles of complex problems, have been created which put questions to those in charge that are much more serious than those of the limits examined previously.

There are dangerous factories which are aging, which are located in the vicinity of other factories, at the gates of substantial agglomerations. The control of these heaps of installations is a task with very hazardous results.

One must also consider the great difficulties encountered in the field of product knowledge, of reactions caused and possible after-reactions. Despite laboratory studies, pilot studies, half-size, and years of research (between five and seven years for a petrochemical process) essential points of safety remain obscured. In the case of an explosive toxic product for instance a safety analysis was requested because of extremely serious dangers which the installation apparently causes to the neighbouring agglomeration. One became aware of the difficulties of the work. When one wanted to define the means of prevention one asked for indicators which would raise an alert when a dangerous reaction started; when one wanted to define the means of intervention to be set up one asked about the time one would have to remove and neutralise the product that was not yet in the reaction process ... despite mathematical studies, laboratory studies, document studies one hit on something unknown that was difficult to solve: the kinetics of the reaction in question are not well understood.

---

*Let us imagine a nuclear centre in which the operators are ignorant of the immediate dangers, incapable of dealing with delicate situations, without a chance of acting upon the nuclear reaction, without safety systems ...

In his book *'Of acceptable risk'* W. Lowrance (9) has strongly emphasized this difficulty:

These days we adopt innovations in large numbers and put them to extensive use, faster than we can even hope to know their consequences ... which tragically removes our ability to control the course of events (9, p. 7).

The uncertainties are particularly pronounced in the field of chemical substances as the Food and Drug Administration in the United States has emphasised:

It is particularly difficult to anticipate the long term effect of exposure to a weak dose of chemical products. It can take a number of years before the negative effects become manifest. Diseases such as cancer have an extended latency period. Mutagenic effects will only be apparent in future generations. Finally, it may be impossible to link the pathology observed on man to a particular chemical product because of the incapacity to isolate a control group that is not exposed to the same degree as the rest of the population. (10, p. 70).

There are problems which for the time being do not lend themselves to significant improvements. This was explained for some cases by the group of experts who had to analyse the area of Canvey Island. Recommendations could be made on a number of points of risk; some, however, resisted the attempt.

For example, an explosion in the machine room of a methane tanker moored at the jetty of the terminal could cause an escape of at least part of the content of the vessel which could have serious consequences in the area because of an explosion of the gas cloud that could result from it. No improvement can be suggested by the group. (11, p. 19).

Large gas clouds could develop ... following an explosion on a ship that carries gas to the jetties of Shell, Mobil or Occidental; or if there were a collision between a vessel of this type and another vessel in the vicinity ...

The consequences of these events would be serious and it is thought that little can be done to reduce the probability of an explosion on the basis of presently available data even though it is believed that the strict observation of the speed limit of 8 knots would noticeably reduce the probability of accidents due to ships' collisions (11, p. 22).

The key obstacle amounts to this: in matters of major risk prevention one aims more at a reduction in the number of events than at their elimination. One knows that every effort must undergo a change in size and nature when one wants no longer to reduce such events by 90 per cent but eliminate the rest. This is the essential obstacle to the prevention of large-scale disasters.

To be added is a factor that has been mentioned before: malevolence and sabotage. The obstacle in this respect appears insurmountable. Short of prior reflection on the subject (when essential choices were not given) one must now rely largely on the prudence of men, a quality that is rather absent in those concerned*.

---

*As the latest large-scale criminal attack at Bologna on August 2, 1980 shows: eighty dead and two hundred injured (12).

3rd. Limits in matters of nuclear safety

This field has been the subject of detailed studies with sophisticated means as we have noted. One must recognise, however, that even the best tools remain by nature limited instruments. It is therefore important to understand precisely what at any given time brings about serious limitations or even constitutes sometimes obstacles which one no longer knows how to remove. The next few paragraphs are in their essence applicable not only to the case of the nuclear industry.

a) Key questions for each of the two approaches. Each of the two methods used has its own limitations.

The deterministic approach. Its limitation lies in its essence since it has decided to ignore accidents which do not come within the lay-out of the installation. This must, however, be taken with a grain of salt since we have seen Electricité de France 'parading', under pressure from the safety authorities, accidents outside the layouts (and which are therefore studies). Another limitation comes from the fact that this method does not concern itself with exhaustive treatment and that the surrounding accidents which are taken into account, as we have already seen, do not permit one to be sure about the 'good behaviour' of the installation in accidents the consequences of which have been forgotten or underestimated.

The probability approach. It hits a fundamental difficulty which is the criterion of risk acceptability. For this method to be fully effective it is necessary that a criterion is defined in the sense which Farmer has spelled out. However, to our knowledge no official authority has attempted a clear and precise definition of risk objectives, and this for rather obvious reasons:

- psychological and political difficulties of obtaining a consensus on the level of the criterion of acceptability;

- the need to justify the criterion adopted for the nuclear industry by comparison with the risk levels reached in other industrial fields which would involve that one studies them (one can imagine the surprises waiting for the authorities about the high risks which the population runs on account of certain industrial activities).

b) Three general questions. As concerns safety in general, we shall see three major difficulties which to us seem to constitute absolute limits for several years to come.

Common faults. For the specialists it is a banality to say that common faults constitute one of the greatest difficulties, if not THE greatest, for safety analysis and that they represent the present limits of know-how. In a way it is the main headache of the safety engineer. For the non-specialists it is easy to imagine the difficulties which the designers encounter when identifying these faults which affect, simultaneously or successively, several components or systems. One speaks also of faults of a common cause. It needs much imagination and much effort and rigour in analysis to bring such failures to light without, however, being able to claim exhaustive treatment as these failures can prove to be insidious. They may have multiple causes: external events (an earthquake or a flood for instance), error in design, manufacturing or assembly, error in operation. These faults of a common cause can render all efforts in the design of other systems useless and can be very expensive. The aeronautics industry has had the sad experience of the disaster of the DC-10 at Ermenonville (1974) which saw the defective

La guerre nucléaire pour moins de 100 dollars

La défaillance d'un petit circuit imprimé, de la taille d'une pièce de monnaie et d une valeur inférieure à 100 dollars (environ 420 francs), et le monde avait failli connaître en juin l'apocalypse nucléaire avec la fausse alerte aux missiles déclenchée aux Etats-Unis par des incidents d'ordinateur au quartier général du Strategic Air Command, enfoui sous la montagne dans le Nebraska.

On sait (le Monde des 7 et 10 juin) que, pour la troisième fois en sept mois, des signaux d'alerte erronés, laissant croire à une attaque nucléaire du territoire américain par des missiles soviétiques, ont été transmis, le 6 juin, par les systèmes de surveillance-radar des Etats-Unis au P.C. de la force nucléaire enterré à Omaha. Pour chacun de ces incidents, le 9 novembre 1979, les 3 et 6 juin suivants, les équipages des bombardiers stratégiques B-52 et FB-111 ont mis en marche leurs réacteurs, et les servants des missiles en silos ont été placés en état d'alerte, dite renforcée.

Aucune manœuvre ultérieure n'a, cependant, été décidée : dans les trois circonstances, les anomalies ont été détectées dans un délai de deux ou trois minutes, évitant tout risque nucléaire.

Devant la presse spécialisée, M. Gerald Dinneen, secrétaire adjoint à la défense chargé des communications, du commandement et du contrôle, a révélé que les enquêteurs du commandement de la défense aérienne nord-américaine (NORAD), à Colorado-Springs, avaient identifié la cause des deux dernières alertes : la défaillance d'un petit circuit imprimé, de la taille d'une pièce de monnaie et d'une valeur inférieure à 100 dollars, qui relie, dans un multi-plexeur de télécommunications, un minicalculateur à des lignes de transmissions. Aucun programme d'entretien, si rigoureux soit-il, a précisé M. Dinneen, ne peut éliminer tous les risques de pannes dans un système aussi complexe que celui du NORAD.

Néanmoins, le Pentagone a décidé d'apporter des modifications à la détection et à la correction du système de communications du NORAD, dont les capacités informatiques pourront traiter les données concernant huit mille objets spatiaux différents, permettre d'observer à la fois jusqu'à cinquante satellites et manipuler par jour quarante mille observations de défense aérienne.

Il est vraisemblable que des incidents surviennent, parfois, dans la chaîne du commandement des forces nucléaires, dans les pays qui disposent d'un tel arsenal. Mais, seuls à ce jour, les Etats-Unis n'hésitent pas à en faire état.

En France, il est arrivé, il y a moins d'une douzaine d'années maintenant, qu'un Mirage IV porteur de sa bombe nucléaire décolle de la base d'Orange (Vaucluse), l'ordre d'interrompre sa mission avant son envol ne lui étant pas parvenu à temps lors d'un exercice d'alerte auquel sont régulièrement soumis les équipages. Pour qu'on n'ignore plus cet incident probablement dû à un défaut de transmissions, il a fallu attendre la parution récente d'un livre (1) d'un ancien navigateur de Mirage IV, devenu journaliste depuis, qui relate, sous une forme romancée et de l'intérieur, la vie quotidienne des équipages de bombardement nucléaire.

(1) Les Chevaliers de l'apocalypse, par Germain Chambost, éditions Picollec, env. 39 F.

Fig. 30: Failures in the most perfected systems (Source: Le Monde, September 3, 1980).

functioning of the locking system of the luggage compartment door lead to
the severing of the control circuits which were, however, tripled (redundancy
of the third order). The nuclear industry has also seen a number of common
failures which, however, have never led to such consequences. The best known
accident is the one at the Browns Ferry centre where a candle set electric
cables on fire and disabled several safety circuits. One could thus quote
many examples which show the importance of this category of failure and which
also show at which point it is difficult to know in advance (alas, often one
knows only afterwards). In addition to this difficulty of identification
there is the difficulty of quantification. H. Lewis shows the arbitrary
nature of the procedure adopted in the WASH 1400 report to get a probability
for this type of failure. This procedure was adopted because there was no
better one (7, p. 82).

The failure of a small printed circuit, the size of a coin, worth less than
100 dollars, and the world would have known a nuclear apocalypse last June
from the false missile alert started in the United States by incidents in
the computer at the headquarters of the Strategic Air Command which is hidden
under the mountains of Nebraska.

One knows (*Le Monde* of June 7 and 10) that for the third time in seven
months erroneous alert signals which created the impression of a nuclear
attack on American territory by Soviet missiles have been transmitted on June
6 by the radar surveillance systems of the United States to the subterranean
command post of the nuclear forces at Omaha. For each of these incidents,
on November 9, 1979 and June 3 and 6 of the following year the crews of the
strategic B-52 and FB111 bombers started their jet engines and the missile
crews in the silos were put on alert i.e. reinforced.

No further action, however, was taken: in the three instances the anomalies
were detected within two or three minutes thus avoiding any nuclear risk.

In front of the specialised press Mr Gerald Dinneen, Undersecretary of
Defence in charge of communications, command and control, revealed that the
enquiry officials of the North American Air Defence Command (NORAD) at
Colorado Springs had identified the cause of the last two alerts: the failure
of a small printed circuit, the size of a coin, worth less than $100, which
links, in a telecommunications multiplexer, a minicalculator with transmission
lines. No maintenance programme, however rigorous, Mr Dinneen spelled out,
can eliminate all breakdown risks in a system as complex as the one of NORAD.

Nevertheless, the Pentagon has decided to carry out modifications in the
detection and correction of NORAD's communications system whose information
capacities can deal with the data from 8,000 different space objects, permit
the simultaneous observation of up to fifty satellites and the daily handling
of 40,000 air defence observations.

It is probable that incidents occur sometimes in the command chain of the
nuclear forces in the countries which have such an arsenal. To this day
only the United States does not hesitate to admit it.

In France it has happened less than a dozen years ago that a Mirage IV with
its nuclear bomb started from the base at Orange/Vaucluse because the order
to interrupt its mission before the start did not reach it in time during one
of the alert exercises which the crews regularly undergo. In order that this
incident which was probably due to a transmission fault should not remain
unknown, one had to wait for the recent publication of a book*of a previous

---

The Horsemen of the apocalypse, by Germain Chambost, edition Piccolec,
approx. 39 FF.

Mirage IV navigator, now a journalist, who relates in the form of a novel and with inside knowledge the day-to-day life of the nuclear bomber crews.

Taking the operator into account in the probability analysis *. This also proves to be extremely delicate due to the fact that the behaviour of the operator cannot be compared to that of equipment. Human error, i.e. operator failure, does not have the problematical nature of equipment failure; it depends on the individual, on psychological conditions in which he finds himself etc. One can easily understand that this is difficult to appreciate and quantify.

The existence of events to which a probability can be assigned only with difficulty. It seems to us that the last limit consists in events to which it is difficult to assign a probability. This is then a third extremely serious limit. The probability of the rupture of the pressure vessel in a PWR is difficult to grasp even though experts have thought about this problem. Not to take this event into consideration for the layout of a PWR could only result from a consensus with the safety authority based on specific criteria for manufacture and control. Another problem is presented by earthquakes. For the seismologists earthquakes are not events to which a probability can be assigned because they are not problematical. Since one cannot yet predict earthquakes the designers have accustomed themselves to define a maximal historically probable earthquake (SMHV) to which they assign a probability of 10^{-3}/year and the intensity N of which corresponds to the intensity of the strongest earthquake historically known on the site where the centre is to be set up. As a precaution the designers lay out the installation for an earthquake called 'increased for intensity' (SMS) the intensity of which is increased by 1 relative to the SMHV and to which the probability of 10^{-4}/year is assigned. This approach which is accepted by the safety authority may be contested by some who can always ask for an earthquake intensity probability of $N+2$ (for which the installation is no longer laid out).

2. QUASI-ABSOLUTE OBSTACLES IN THE BATTLE AGAINST DISASTER

Here, justice must be done to the battle plans against disaster. If they are precisely adequate for 'disasters' they can no longer be so, at least not for the time being, in case of disasters of extreme severity.

It had to be admitted already when the *Amoco Cadiz* went aground — which, however, was only a serious warning — as M. Becam put it in parliament:

At present no country in the world has the means to intervene at the same time effectively and satisfactorily on the ecological plane to break up the black tides. We have equipment ... With waves four to six metres high they are clearly inoperative ... The inaccessibility of a wreck and the bad weather at present constitute insurmountable obstacles in the battle against the black tides (13).

This was underlined once more, as we have noted, by A. Achille Fould on the occasion of the consecutive pollution when the *Tanio* sank:

Those who imagine that if all arrangements had been made there would never again be an accident have never seen the sea (14).

---

* These questions will be further developed, pp. 338 ff.

This kind of statement had to be made in the case of the dispersal of a few hundred grams of dioxin into the sky at Seveso. It would be the same for a number of menaces mentioned before. We know already that nobody can prevent a large layer of gas from catching fire, that nobody knows how to extinguish a large conflagration of liquified gas or of liquid sodium, that nobody could fight a fire storm, that in case of a chemical or nuclear pollution of the Channel or the Mediterranean, for instance, one would be caught short.

One knows how to deal with large accidents. It is no longer a question of expecting much from the rescuers in the case of a disaster caused by one of the great menaces just mentioned. Yesterday one healed, tomorrow one will lighten the burden a bit by actually making the 'prevention of the fire spreading' the essential part; in certain cases one will have to be satisfied with the diagnosis that one can establish with the means available. Certain areas will have to be condemned, like the 45 hectares condemned for an indefinite time at Seveso, while waiting to discover how, within a few generations, the accident wll enforce its consequences.

3. ABSOLUTE OBSTACLES TO REPARATION

The largest reinsurance groups in the world know their strength. They know that they were able to face up to the Ford accident in 1977 ($200 million) or to the damages caused by hurricane Capella ($500 million) that struck northern Europe in 1976. They are in a position to insure installations of still much higher cost: the largest oil platform in the world is in the North Sea: it represents a value of 2.1 million DM. Another one, bigger still, is under construction: its value will be double or nearly double (14, p. 43).

However, these insurance mechanisms are not without limits. We have already mentioned some. There are other more serious ones which one can hardly refrain from calling absolute.

The Muenchener Rueck, the foremost reinsurance group in the world, however, offers glimpses on some doubts about the market's capacity to follow suit on similar escalations:

... never before in the history of mankind have so many individuals lived so close to each other; never before have objects of such high value been found concentrated in so many limited spaces and never before have so many people had accidents simultaneously and of such severity and such frequency. For to the natural causes of accidents and disasters have been added those which man himself created, causes which must be attributed to techniques and to civilisation (15, p. 35).

Its opinion? It goes beyond the technique of insurance — largely. For from the thresholds which are today about to be cleared the question rebounds to a new plane:

The existence of insurance companies will not change much in this respect. The foresight, the preventive measures against damages are only too often caught up with or surpassed by the appearance of still greater perils, by still greater concentrations of values. We should not forget, however, that the worst dangers threatening human existence come from man himself.

The institution of insurance results from human reason. In large measure it permits the material reparation of the consequences of human failure.

But it will also logically find its limits at the time when mankind no longer has the capacity to settle the problems of its existence in a reasonable manner (15, p. 36).

It is the same situation for private or even public compensation funds. True, it would not cost the oil market or the nuclear profession much to set up funds much larger than the ones presently available. However, one must see the size of the menaces of our time. Evacuation of 2 million people for ten years from an urban area: what would it cost? Does the question even make sense? By the same token, what sense does 'compensation' make in case of substantial pollution following a severe accident in the Channel or in the Mediterranean?

REFERENCES

(1) *Le Monde*, 9 août 1980.

(2) *Rapport de la Commission d'Enquête de l'Assemblée Nationale*, présenté par H. Baudoin. Première session ordinaire 1978-1979, novembre 1978, no 665, Tome 1 (333 pages); tome 2 (94 pages).

(3) *Rapport de la Commission d'Enquête du Sénat*, présenté par A. Colin. Seconde session ordinaire 1977-1978, juin 1978, no 486 (289 pages).

(4) J. COLLIOT et B. FONT-REAUXL, La prise en charge de l'inspection des installations classées par les services de l'Industrie et des Mines. *Annales des Mines*, Juillet-août 1979, pp. 41-46.

(5) *Journal Officiel*. Débat Assemblée Nationale, 15 avril 1976.

(6) C. LEPAGE-JESSA et Ch. HUGLO, La législation sur les nuisances indus-trielles. *Annales des Mines*, Juillet-août 1979, pp. 29-40.

(7) H. LEWIS, La sûreté des réacteurs nucléaires. *Pour la Science*, no 31, Mai 1980, pp. 73-89.

(8) A. MELLY, Three Mile Island. *L'Argus International*, no 13, Juillet-août 1979, pp. 297-303.

(9) W. LOWRANCE, *Of acceptable risk*. William Kaufmann Inc. Los Altos, California, 1976 (180 pages).

(10) E.P.A. Consolidated DDT hearings, 37 Federal Register, 13 369-13 375, footnote 19, July 7, 1972. (Cité par W. LOWRANCE, cf. (9), p. 70).

(11) Health and Safety Executive, Canvey, an investigation of potential hazards from operations in the Canvey Island — Thurrock area, Health and Safety Commission, London H.M. Stationery Office, 1978.

(12) *Le Monde*, 3-4 Août 1980.

(13) Les conséquences du naufrage de l'*Amoco Cadiz* ont été longuement évoquées au Sénat. *Le Mois de l'Environnement*, no 21, avril 1978.

(14) *Le Quotidien de Paris*, 14 mars 1980.

(15) Centenaire de la Münich Rück: 1880-1980 Münchener Rückversicherungs Gesellschaft. Münich, 1980 (102 pages).

CONCLUSION: FROM TECHNIQUE TO POLICY

The two developments in this fourth chapter have permitted us to get a better measure of the reality of the forces available in the face of major technological risk. Without this knowledge of the limits to the means and tools available it would be difficult to avoid disenchantment and deadlock which is always savage because it is a matter of major risk.

These two successive examinations lead the specialists to turn to the political field. First and foremost in order that the means be adjusted to the objectives, a basic condition for all planning. Secondly and above all in order that the totality of questions raised by major risk be reexamined in terms of options and not just in terms of means for managing them.

The essential point is this: the disruption which has appeared in the field of potential menaces due to technological and industrial development no longer permits us to look at the tools and means of management as one did in the past.

Given what is at stake, given the impossibility of working at zero probability for accidents, policy must provide options to be followed up and for the approaches retained, the requirements to be respected. It is no longer good enough that the technician does his best to reduce the failure probability of systems which have not been the object of any explicit device on the political plane.

Accidents of extreme severity no longer permit the same results from the battle against them; one could circumscribe, isolate, evacuate, heal, repair; one will now sometimes have to be content with diagnosis (if that is possible) and with lightening the burden (within the limits of what is possible). One cannot decontaminate an area affected by a dioxin cloud as one would extinguish a fire; one cannot give a beautiful rose colour back to rocks that are blackened to the core; one cannot extinguish a large fire of liquified gas; one cannot avoid that a layer of gas derivatives catches fire or causes a conflagration; one cannot extinguish a large sodium fire etc. In a word that is well known in the field of natural risk: "one cannot stop an avalanche". This new problem for policy must be recognised: one can no longer count on means for battle that are adequate to the accidents which nowadays threaten if prevention is at a dead end. The break is clear. One can use 100 times more rescuers than yesterday but to what good if it is only putting the lives of 100 times more of these brave rescuers in danger — and this for a very meagre result. A hundred more decorations do not give one single satisfactory answer.

By the same token, as the world's foremost reinsurance company emphasises, the compensation of victims reaches its limits and this brings us back to the need for looking at economic development 'reasonably'.

These thoughts therefore bring us to policy. It is a matter of providing general options in full knowledge of the issue. Who will provide the options when, as was discovered in the military order of things during the first world war, it is no longer just the professionals who are in the firing line but the whole of society?

We shall take these difficult questions up in the fourth and the last part. However, before that we must undertake a further exploration. Before acting in the face of the problem of major risk, in full knowledge of the limitations that affect the available means of management, it is important to understand how these questions are dealt with by the body politic. The human and the social factors singularly complicate the problems which we are examining.

It is necessary to study them in order to get the proper measure of the challenge presented by major technological risk.

PART THREE

The social regulation of major risk

CHAPTER FIVE

Operators, public authorities, citizens facing major risk

I The operator
II The public authorities
III The citizen

The official report on the TMI accident includes 19 pages headed "Recommendations by the Commission". Only two of these pages show technical appreciation. The seventeen others deal mainly with procedures concerning organisation and training.

The official report on the accident of the Bravo platform (at Ekofisk) includes three and a half pages of conclusions. They identify seven main reasons for the accident. Only one of these seven reasons raises a purely technical question. The others refer to organisation and training problems.

E. Bjordal
"Is risk analysis obsolete?"

Loss Prevention and Safety Promotion in the Process Industries.
3rd International Symposium, Basle, September 15-19, 1980, vol. 2, p. 643.

The vessel Baltic Star, registered in Panama, ran aground at full speed on the shore of an island in the Stockholm waters on account of thick fog. One of the boilers had broken down, the steering system reacted only slowly, the compass was maladjusted, the captain had gone down into the ship to telephone, the outlook man on the prow took a coffee break and the pilot had given an erroneous order in English to the sailor who was tending the rudder. The latter was hard of hearing and understood only Greek.

Extract from the protocol of a Swedish maritime tribunal.
Le Monde, November 15, 1979.

The administrative court of Rennes, applying Article 64 of the law on the sea ports has sentenced Mr Papadopoulos, captain of the Greek freighter Irinikos, to pay a fine of 36 FF for pollution by releasing hydrocarbons in the port of Lorient on July 27, 1977.

Until the finance law of December 29, 1956 came into effect, which raised the rate of fines by 50 per cent the polluters were fined only 24 FF.

Le Monde, May 19, 1978.

I. THE OPERATOR: PRIME RESPONSIBILITY
FOR THE CONTROL OF MAJOR RISK

As creator of risk, the operator is the first actor on which one could count
to control danger. The law, the policy of companies (which provide that that
company will always and everywhere behave as a 'good citizen'), rationality
itself or quite simply commercial interest (what sacrifices would not be made
to ensure a good brand image?) should lead to good risk control by the company.
And therefore: to a good policy of prevention and in case of an accident to
immediate protection for the staff, to an efficient battle and an equally
immediate information of the public authorities.

If we restrict ourselves for the moment to the case of France we notice
that in fact safety is an important concern in the nuclear field. Electricité
de France and the Atomic Energy Commission have integrated this type of concern into their way of operation from the beginning. In the private sector —
the classified installations in general and the chemical industry more
precisely — we notice certain important developments during the last few
years. Safety is in the process of winning a better status. There was and
there often remains a serious backlog in this respect. It became necessary
to adapt the organisation charts, to put into the newly created jobs highly
qualified engineers (which was hardly the custom as all specialists agree).
Sometimes energetically pursued for two or three years, this movement towards
control of major risk goes further in some cases: companies join their efforts
together (as in Lyons and Rouen) in order to get to know the risks of a
common area or a common field of activity better.

One must pay attention to these new experiences or these new arrangements:
they provide a valuable basis for action to be taken with the actors concerned.
Much does, in fact, remain to be done in order to set up better apparatus,
adequate to the control of extreme risk. Many factors hinder the development
of this apparatus. The 'imperatives of production', the dogma of the industrial secret, strengthened by that of the freedom of enterprise, the keen
competition for a share of the market or for regional or worldwide supremacy,
the lack of internal information, the lack of knowledge, irrationality, error
most certainly, are among the dark features in the picture. These are the
general conditions which will later be examined because they reveal more than
the general policy in industrial development.

Despite these fundamental difficulties much can be done by the industrialist
who cares about safety. In order better to put in evidence the degrees of
freedom (which most often remain to be won) we shall successively examine a
certain number of weaknesses which show the gap between the desirable and
the actual. For each of these discovered insufficiencies on the level of the
operator we shall submit some illustrations, frequently taken from cases
previously examined. These cases used as examples come from the whole of
industrial activity and not just from one specific branch: the messages they
carry are in fact of the order of logic: it is of little importance if a
regularity shows up in one field or another (railways, shipping, chemical
industry ...); the essential point is to bring out structural information.

Here then is a series of reflections that suggest themselves through the
observation of various accidents.

1. AN ASSEMBLY OF CLASSICAL FAILURES

1st. The general attitude of the mind: excluding the extreme risk

In order to develop a prevention strategy effectively, the very first

difficulty to overcome is the one of the general mental attitude of the
actors involved. The mental images can in fact constitute absolutely airtight barriers which reduce all efforts in matters of prevention to nothing,
even if they are massive. The man in the street often uses the expression
"this happens only to others" to describe this shield which wards off all
attention to the problems. It is the same with organisations as we may recall briefly with three examples which clearly show the type of weaknesses
that one encounters in this matter.

The Titanic. How could one have even the slightest apprehension? Why go for
the usual rescue exercises? One knew perfectly well that the *Titanic*
was unsinkable. "Not even God could make this ship sink". The monster with
its 55,000 HP, its three propellers, would make child's play of the waves and
the wind. Warning messages? Ice all around? Full speed ahead! The
reputation and the power of the company demand punctuality. Titan does not
bend to the laws of the world. Seen from a distance — in time, after the
drama, or in space, from the ship that saw the *Titanic* pass by at full speed,
the blindness and deafness seem pure folly. For those who had the honour of
running the *Titanic* nothing was abnormal. The prism through which they read
all information prevented them from looking reality in the face. Icebergs?
In the mental image of the actors in the drama at worst the *Titanic* would
knock the wall of ice over like a heap of straw. This, of course, was not
explicit but it ruled behaviour. The curtain was not rent until the impressive mass of the iceberg rose out of the night.

Flixborough. The report by the court of enquiry spells out:

> The enquiry has clearly shown that nobody among those in charge of the
> design or the construction of the factory foresaw the possibility of a
> major accident that could happen suddenly (1, para. 218).

Three Mile Island.   The Kemeny report dwells a lot on this question of
general mental attitude: 'mindset'. The term recurs incessantly in the statements of the agents interrogated during the enquiry. The commission writes:

> After years of operation of nuclear centres without any accident for the
> general public the feeling that the nuclear centres are sufficiently safe
> has become a conviction. One must recognise this in order to understand
> why a large number of key actions which could have prevented the accident
> at TMI were not taken. The commission is convinced that this attitude must
> be changed: one must say that nuclear energy is by its very nature
> potentially dangerous; consequently one must constantly question the safety
> systems that are installed and ask oneself if they are sufficient in order
> to prevent major accidents.

Canvey Island. The study group installed by the British administration has
clearly brought out the fact that major risk is sometimes left out of the
concerns of industry.

During preliminary visits to the installations selected for detailed study
the investigation team was relieved to find that wherever dangerous products
were treated, handled or stocked the companies generally paid much attention
to operational safety questions. Where appropriate codes of conduct existed
they had been taken into account in the design and construction of factories
and equipment.

However, these visits also established that none of the companies concerned
had made systematic attempts at examining and documenting those potentially

serious events which could cause accidents among the population in neighbouring communities (3, p. 8).

The case of Canvey Island is no exception. Whence the anxiety of certain
industrialists to fill in the extremely worrying gaps that have recently been
discovered in this field.

2nd. An approach to the safety of insufficiently integrated systems

Safety is not just an additional task, an element added to the life of the
organisation. It is a dimension entirely apart from the workings of the
system and must be treated as such. As the saying goes: a chain is as strong
as its weakest link.

In other words: it is not very judicious to plan good investments without
adequate means of maintenance; inversely, it will not do to provide such
means if the basic design of the installation evidently needs to be reviewed.
However, one must go beyond this merely technological aspect of safety: it is
a socio-technical system that must be managed. The British advisory committee
on major risks has taken note of this when in 1976 it recommended in its
first report a detailed and multiform evaluation of the safety of the most
dangerous installations. This evaluation, it wrote, must be applied to
design, construction, operation and maintenance; but in addition managements
would have to satisfy the administration that they have:

- an appropriate management system and approach in general;
- competent staff;
- effective methods for the identification and evaluation of risks;
- an installation that is designed and functions in accordance with the
  appropriate rules and regulations;
- adequate rescue procedures;
- an independent control system where it seems to be required.

This overall approach is essential. It may be that at a given time the
most critical question has nothing at all to do with equipment but with
people, with social relations inside or even outside the organisation. One
must be able to conceive the conduct of safety in new terms. One must also
know that a system is the least subject to disaster when limited incidents
cannot find a 'sounding box' or that they cannot escalate into large-scale
accidents or disasters. In order that a localised defect does not develop
into a general collapse one must make sure of the quality of the whole
(equipment, people, communications, social relations etc.). Deadlock in
prevention must not become the rout of the safety function as a whole. This
presupposes good options from the start, a good quality system in general, a
well established safety function.

Often the operators are not at all aware of these points.

Flixborough. Without much risk of error one can say that the system installed
by Nypro could not have escaped serious accident. Everything led up to it:
from the design of the factory (the most populated buildings — canteen,
administration, control room ... being in the most dangerous area, the operation (a patent that was more dangerous than others), the maintenance
("escapes that would absorb themselves"), the staff (insufficiently competent),
the organisation chart (an engineer who considered himself to be in charge of
safety) to the economic difficulties of the company (which hardly encouraged
circumspection, neither in normal times nor after an incident had occurred as
we have seen early in 1974).

<u>Three Mile Island</u>. The case gave P. Tanguy cause for these interesting reflections:

It is not good for safety to pursue the operation of an installation that is not in an excellent state. At TMI, before the accident, there were various defects; on the secondary side: with pluggings of transfer lines for concentrates; on the primary side: with the escape from the discharge valve and a problem of heterogeneity of concentration of boron on the treatment side of effluents; with a non-airtight tank the repair of which had been put off (4, p. 531).

In the analysis of an accident one must always guard against the search for one cause — the cause — which on its own explains the accident. In the case of TMI operator errors were found. The haste with which these scapegoats were pointed out quickly appeared suspect. In fact, the whole of the system had, here again, to be considered with all its faults as P. Tanguy stresses. The Kemeny report remarked actually, in order to terminate a faulty debate, that the operators in the accident installation were better than the average operators in the United States (2, p. 49).

Without waiting to find a single cause (we shall deal with the question of the human operator later on) the commission appointed by President Carter brought out a whole lot of deficiencies: (2, pp. 10-11).

- The training of the operators was insufficient.
- The prescribed procedures were very confused.
- The preceding incidents in similar centres had not been brought to the knowledge of the operators.
- The general running of the factory was hardly satisfactory: there were permanently some fifty alarm systems in operation in the control room; maintenance left things to be desired.
- The control room did not permit to face an accident: hundreds of alarm signals were released simultaneously, key indicators in places where one could not see them, ambiguous signals, maintenance labels all over the place hiding sometimes important luminous signs, instruments foreseen only for normal operation, an inadequate computer etc., providing in the end little chance of controlling a significant failure.

Finally, the commissioners write:

... whether or not operator error 'explains' this particular case, given all these faults, we are convinced that an accident of the type that occurred at TMI was inevitable in the long run (2, p. 11).

One could multiply these examples. The two cases quoted are no aberration. Even if the one at Flixborough looks like a caricature, one must nevertheless remember the opinion of the court of enquiry:

The normal practice at other chemical factories was not sidestepped at Nypro (1, para. 204).

This opinion is doubtless too severe if read out of context. In fact, one reads elsewhere that the court seems to have had some difficulty choosing between its judgements. We shall reexamine this point later on.

At no time during the enquiry has there been any proof that the chemical industry or Nypro in particular was not aware of its responsibilities in matters of safety. On the contrary, there are indications which show that

conscientious and positive measures have continuously been taken for this
purpose (1, para. 201).

3rd. Preceding events — the return of experience

It is always difficult to identify the vulnerable points of a system
properly. It is therefore all the more important not to neglect the means
which can help with identification. Among these means is the close examination of previous events. The cases examined are eloquent on the insufficiencies in this respect. The typical example in this matter is the case
of the landslide at Aberfan which was mentioned previously. We shall bring
it out here before showing with other examples that this failure to take
notice of the lessons of experience is very frequent.

Aberfan. We have mentioned already that at 09.15 h on October 21, 1966
140,000 t of material from a slag-heap at Aberfan covered a school and
eighteen houses, claiming one hundred and forty four victims of which one
hundred and sixteen children. A few summary explanations can be given before
we examine the facts, the basic lesson which this disaster teaches all
industrialists.

a) The problem of the slag-heaps. Various slag-heaps had successively been
piled up on the Aberfan site, above the town, along the slopes of Merthyr
Mountain (270 metres). The acute question was the one about rainfall: the
area has a lot of rain which sinks in and resurges in springs and rivulets on
the flanks of the hill where the waste from the mine is accumulated. If
these resurgences had been paid attention to one would, in principle, not
have located the slag-heaps just above Aberfan as was the case. There could
be a landslide on account of this and other reasons: the rain drags the clay
in the slag-heap downwards where it forms an impenetrable layer which
activates the trickles; water falls on the slag-heap and not only on the hill,
accumulates in the soil and soaks it. The type of material in the slag-heap
may aggravate the situation.

The site has been used for a century; a first slag-heap was piled up.
Between 1914 and 1918 it had reached a height of 30 metres, and a second one
was started but this time on the flank of the hill. It was piled up to 30
metres before piling a third one to 40 metres. A fourth one reached 45 metres
within eleven years; but in November 1944 the fifth one was started and piled
up to 50 metres. Slag-heap Number 6 was started in 1956 but abandoned two
years later because it infringed on agricultural land. Thus in 1958 the fatal
Number 7 was started. It was 35 metres high at the time of the drama eight
years later.

Slag-heaps Numbers 1, 3 and 6 were between the trickle lines. They caused
no worry. This was not so with the others: the local authorities were
constantly worried about them. The coal mining officials were affected only
after the deaths of one hundred and forty four people

The slag-heaps came under the management of the Merthyr Vale Colliery of
group 4, zone 4 of the South Western Division of the National Coal Board
(NCB — an establishment created in 1947 during the nationalisation of coal
mines).

b) Total blindness and deafness.

1927: Warning about landslides from slag-heaps. Professor Knox holds a meeting in Cardiff at the Institute of Engineers of South Wales. He issues a
warning about the threat presented by the infiltration of rainwater into the

slag-heaps. He adds that if the companies do not pay for the installation of
the necessary drainage systems they will have to pay for the effects of the
landslides. A participant in the discussion warned that the cost would drive
the companies into bankruptcy and that research into the location of slag-
heaps is required.

It so happens that the person who was to become production director of the
South Western Division of the NCB had studied this pronouncement by Professor
Knox when he was a student.

1933: Start of slag-heap Number 4 — first slide. Slag-heap Number 4 is
started. Very quickly a first slide occurs. This causes a layer of material
which was not very different from that of another accident (1944).

December 5, 1939: the accident at Cilfynyd (180,000 t slide down). The
company which owns the slag-heaps of Aberfan (nationalisation commenced only
in 1947) had a serious accident eight kilometres away at Cilfynyd, 180,000 t
of material slid from a slag-heap and blocked a road, a canal and a river.
There were neither deaths nor injuries but the accident caused a strong feeling in the world of coal mining companies. A report on previous incidents
was requested by the company. A study was prepared: "The slide of slag-heaps".

One finds that the chief engineer of one of the divisions of this company
sent a copy of this report to his son who had taken part in the excavation
work. One also finds that this son, twenty seven years later, was Divisional
Mechanical Engineer at the South Western Division of the NCB. Further on one
finds that the future production director of this division had also been a
witness to the 1939 accident; he had called it the worst he had ever seen
(he had seen two others). A further engineer (the Mechanical Engineer) of
group 4 of the coal mines on which the whole of the Merthyr Vale depended was
also familiar with the accident.

All this was forgotten, and the report of 1939 remained in a drawer.

February 1944: local worries. The company gives assurances. The local
authorities, being worried, are reassured by the company: there is no
immediate danger, drainage will be done.

Late 1944: Major slide from slag-heap Number 4. For its defence in the wake
of this slide the company blames the rain; it even claims that the drainage
system has permitted a limitation of the damages. Alarm is again averted.
It would have been necessary to bring up the problem of rain and to question
the piling of new layers on old ones. Twenty years later slag-heap Number 7
had come to cover the material of slag-heap Number 4. Number 4 had become
unusable and one started Number 5. The latter which was started by covering
the trenches dug for drainage did not cause any reaction from those in charge
who had recommended the trenches.

1950: Local worry. The NCB gives assurances. The local authorities are again
worried. The person in charge of the installations at the NCB brushes these
fears aside and stresses that the slag-heaps will be abandoned when the switch
is made to underground storage: this would happen within a year. He adds that
the slag-heaps are under constant supervision.

In fact, there is no control at all. The slag-heap concerned is kept in
use for another five years. There has never been underground storage. On
the contrary, there are even studies for the extension of this slag-heap;
but this turns out to be more expensive than making further slag-heaps.

Slag-heap Number 5 is started in 1955 but abandoned later because it affects agricultural land.

1958: Site choice for slag-heap Number 7 without either study or visit to the site or consultation with those locally in charge of the coal mines. The site for slag-heap Number 7 was chosen without study, without boring, without any limitations other than those of the property of the NCB. The Colliery Manager and the Group Mechanical Engineer who took the decision did not consult anybody else in charge. As a result, the scheme, based on a map from 1919, was introduced without a visit to the site. It had to be acknowledged after the disaster that the scheme was a sheer absurdity. It is not known, however, whether this scheme could have misled those in charge because the latter knew on what inexact data it was based. It must also be said that the scheme was the best available which indicates the competence level of the existing management.

1959-1960: Local worry: the NCB gives reassurances. The council and its *ad hoc* sub-committee for slag-heaps complain repeatedly to the NCB about floodings caused by the slag-heaps but also about the risk of slides in case of rain. The local section of the Labour Party supports the requests. The only response from the NCB consists in arranging stakes: they were to be covered over within two months.

1962: Concern about the residues from (coal) washing: the NCB ignores it haughtily. There appears a new worry about the "residues from coal washing". These are materials that have been chemically treated to permit a better extraction of coal. This type of waste is hard when it is dry and becomes mud when it is humid. There are great fears that it may contribute to a slide. At division level it is decided to interpose these materials into disused pits in order to separate it from other waste. This decision is not passed on to group 4; the director general of production at the NCB is not informed either of the dangers connected with this material.

From 1962 to 1965 these residues therefore continue to be poured on slag-heap Number 7. Independent experts said after the disaster that this material played no part in the drama. One may think that the complaints which had been raised should have started serious studies. These studies could have brought other risks into view.

The area engineer says in reply to complaints that he will consider the use of another slag-heap for the residues. Nothing will happen.

November 1963: Slag-heap Number 7 slides: worries; the NCB does not move. Slag-heap Number 7 slides. Local reaction is strong. The local man in charge of coal mines decides to stop the pouring on of the residues.

However, the pouring continues, true: in reduced quantities. What is more serious: this slide does not stop the use of slag-heap Number 7, contrary to what happened in 1944 with slag-heap Number 4. This incident does not appear in the archives of the NCB. The people in charge at the NCB say they have no knowledge of it.

January 1964: "If the slag-heap moved it could put the school in danger". At the town's planning council a councillor declared as the *Merthyr Vale Express* reported on January 11: If the slag-heap moved it could put the school in danger.

March 29, 1965: accident at Tynawr; the report has no effect — like the one of 1939. On that day the division had the accident at Tynawr which cost it

£20,000; headquarters, however, was not alerted. The accident occurred precisely on account of residues from coal washing. In order to interpose them a sort of lagoon was created after a dam had been built from waste. The semi-liquid material was spread out inside. On that day the dam gave way and the mud flowed over a road and a railway track.

The Division Chief Engineer made a report which takes up the one of 1939 and adds thoughts about the residues from coal washing. This report of 1965 too could have avoided the Aberfan tragedy. It explained clearly why there could be slides from the slag-heap and it detailed the precautions to be taken.

The report had no effect. The Area Mechanical Engineer in charge of the slag-heaps of Merthyr Vale (Aberfan) who had very bad personal relations with the Division Chief Engineer replied that the pouring of "residues from coal washing" had stopped and that, actually, the slag-heaps had been quite stable.

1965: Visits from people high up in the NCB; nothing changes. During the year preceding the disaster top people from the NCB visited the sites because an oil company was interested in treating coal from Aberfan. The problems caused by the slag-heaps, the local complaints had nothing to do with these visits.

The last six months: an advance of the slag-heap and runnels are noticed. During the last six months of its existence the slag-heap advanced by 7-10 metres. There were runnels.

October 21, 1966 08.00 h: Runnels are reported. The surveillance team reports runnels. It is stated that the slag-heaps will be abandoned. So: was it known that the slag-heaps presented dangers? This could only be a high-level decision. A year earlier there had been talk of a new site, between slag-heaps Number 4 and Number 7 but the decision was postponed on the grounds that the material delivered was not right. This site was hardly safer than the existing ones.

October 21, 1966, 09.15 h: the "incident" this time causes 144 deaths. Preceded by small movements inside the slag-heap 140,000 t of material started off runnels. Material of muddy consistency hit the base like a wave. Other debris cleared the lanes opened up and reached the school and the houses. The slides broke two large pipes which caused a complementary flooding.

August 1967: The enquiry report — Ineptitude. The tribunal establishes that the disaster could and should have been avoided. It speaks of ignorance, ineptitude, faults in communications. On the evidence, as far as the school is concerned, one could speak of fate:

- the phenomenon had been known scientifically for half a century;
- the measures to be taken were known;
- multiple runnels had been experienced previously;
- the one on October 21, 1966 was not the biggest one.

Such failures have also been noticed in other cases. In addition to the one at Seveso there were the precedents of BASF (1953), Dow Chemical (1961), Grenoble (1968) etc. which were ignored. One remembers those of Hixon (a railway disaster in Britain) and of TMI which are equally exemplary.

Hixon. On January 6, 1968 an express train running at 120 km/h hit an exceptional road convoy of 162 t at the level crossing at Hixon, Great Britain.

We shall also study this exemplary case from another point of view later.
Here, let us just remember that the automatic barrier system installed in
Great Britain at the time did not take the safety of exceptional convoys
properly into account. A telephone was put at the disposal of the drivers
but most of them ignored the usefulness of this telephone and above all the
need to have recourse to it: if they obstructed the track for too long nothing
warned the train of the fact that the level crossing was obstructed (there
was no return signal from road to rail apart from the telephone which permitted to join the signal service).

On January, 1968 this fault became evident when a heavy convoy crossed the
Hixon level crossing at 3 km/h. A train arrived. The collision occurred:
the trailer had not cleared the track between the time of the alarm bell and
the arrival of the express train.

What is important for our purpose is that a similar incident had occurred
14 months earlier. On the Leominster crossing a vehicle carrying a crane
came to a halt on the rail track following an incident. The driver was
surprised to see a train pass by when he had only just managed to remove his
vehicle. A serious accident had only just been avoided. Intrigued by the
affair, the road-to-rail safety mechanism seemed strange to him, the driver
reported the incident to his employer who wrote to British Railways. In their
answer they said:

- that Leominster conformed to the standards approved by the ministry;
- that vehicles must not stop on level crossings*; it did not mention the
  existence and the function of the telephone.

Chance had it that it was the same transport company that was involved in
the Hixon accident. Neither British Railways nor the road carrier knew how
to draw lessons from the previous event (5, 6).

Three Mile Island. The Kemeny report makes interesting observations on the
subject of previous events:

In September 1977 an incident occurred at the David Besse centre, also
equipped with a B & W reactor. During this incident a discharge valve
remained open and the level in the pressurizer rose while the pressure sank.
While there were no serious consequences to this incident the operators
acted improperly with the high pressure injection system by apparently
basing themselves on the rise of the level in the pressure unit. The
David Besse centre was running only at 9 per cent of its capacity and the
insulation valve was closed about twenty minutes after the blockage of the
discharge valve in an open position. This incident was the object of a
simultaneous enquiry by B & W and the NRC but no information to draw
attention to the expected correct actions by the operators was sent to the
installations before the TMI accident. A B & W engineer had stated in an
internal B & W memo, more than a year before the TMI accident, that if the
David Besse event had occurred in a reactor running at full capacity "it
is quite possible, perhaps probable, that there would have been inadequate
core cooling and even damage to the fuel". (2, p. 29).

In January 1979 an official from the NRC stressed the probable nature of
erroneous operator action in an accident of the TMI type. The NRC did not
alert the centres of this before the accident. (2, p. 29).

---

*Of course, a level crossing is not a parking lot; a year later it was
to become a cemetery.

An engineer from the Tennessee Valley Authority analysed the problem of
the rise of level in the pressurizer and the drop of pressure more than a
year before the accident. His analysis was sent to B & W, to the NRC and
to the Advisory Committee on Reactor Safeguards. Once again no notifi-
cation on it was sent to the centres before the accident (2, p. 29).

4th. The problem of modifications in technological processes or the life of systems

We know that technological systems present risks that accrue in the phases
of transition: starting and stopping. This problem is well recognised. It
calls for a broader examination of other questions of similar nature. Among
the changes that occur during the lifetime of a manufacturing system one
must pay attention to:

- changes in the size of the installation;
- changes of process;
- important internal innovations of a general nature;
- changes in the technical environment (installation of dangerous factories);
- changes in the most general aspects of the life of the system
  (particularly economic and social crises etc.).

The cases examined provide some illustrations of failures of this origin.

Flixborough. The process was changed and the production capacity of the 1972
installation tripled. But safety questions were not rethought.

Seveso. The process was changed to manufacture trichlorophenol but the
safety measures which could have been taken were not reviewed. However,
there again, it was a switch to the use of a more dangerous patent.

Roman Point. This case has not yet been mentioned. It is the case of a
partial collapse of a highrise building in the suburbs of London in 1968.
The principle of the accident is the following: safety had not kept pace
with the innovations adopted in the field of construction. The standards were
no longer valid because a highrise building is not just a somewhat larger
than usual individual house. In short, industrial techniques bearing strong
imprints of the 'systems' mentality had been employed to facilitate the
construction of the tower. Safety did not keep pace with the changes of
logic which this evolution implied. What was crucial was henceforth, less
the safety of each cell of habitation than the safety of the connecting
elements (supports, bases, independences and dependences etc.). This was not
understood. The first limited incident (explosion in an apartment) caused a
card house effect (5, 6).

Summerland. The leisure centre installed on the Isle of Man under an
enormous plastic bell was not a usual building. The permits were, however,
issued on the basis of standards applicable to theatres. The administrative logic required the discovery of the category in which Summerland
belonged; the theatre category was adopted even though on the evidence it
was not suitable. Administrative logic was satisfied. However safety was
in no way ensured: a new problem had been treated in an old framework (5, 6).

Nowadays the problem of rapid transition from Phenix to Super Phenix (the
prototype having been too successful as some specialists think) is quoted as
an example to give cause for concern. Or again, in general terms, the end
of the industrial era in which there were foremen with long-standing knowledge of a process, who knew how to anticipate a serious accident because
they had seen several of them: with today's mobility and above all with the

fast change in processes and technologies one can be much less sure. By
the same token one must pay attention to the changes which the economic
crisis might bring about.

5th. When safety gives way to the requirements of production or to safeguarding investment.

The profit motive is banal and far from sufficient to explain on its own
the disasters recorded but it is nevertheless present and sometimes
decisively so. The typical example is the one of the *Amoco Cadiz*:
"Negotiations which I do not hesitate to call sordid", according to the
pronouncement of the French Interior Minister previously quoted. When the
TMI accident occurred the French officials also stressed in broadcasts that
a nationalised organisation like Electricité de France could not be compared
with the American companies and their greed for profit. This kind of
reflection certainly needs to be developed for all high risk enterprises and
this before disasters happen.

The power motive must also attract attention when it becomes too exclusive.
What pride to see one's name on the biggest ships in operation! To have
the most impressive installations! To make the biggest investments etc.
What is the price to be paid for this in terms of safety?

Much more difficult are the cases of judgements to be made in a situation
of foreseeable or immediate crisis: a possible breakdown or the restarting
of units that have certain faults. The respective discussion of this matter
developed around the fissures found in the casings of the tube plates of
steam generators and the tubes of nuclear water boilers.

In all of these cases the responsibility of the safety authorities is of
critical importance. Without their fast intervention and their competence
of judgement, conditions which cannot be met if there is no real independence
between these services and those in charge of controlling the normal workings
of productive activity, accidents are aggravated.

A few illustrations taken from the maritime field — which has no monopoly
on these questions — deserve to be recalled.

Torrey Canyon. Captain Rugiati, the skipper of the *Torrey Canyon* is under
an obsession when he enters the Channel in the morning of March 18, 1967: to
make up for a delay which threatens that he might miss the tide at Milford
Haven which would mean a forced stop of more than a day and a half. He
gives the order to steer an extremely dangerous course: in addition he omits
a reduction of speed and even leaves the rudder on the automatic pilot. The
red alarm signals launched by bewildered observers who see the ship enter
such waters do not alarm anybody on the tanker's bridge. The ship hits a
reef at 15.75 knots, at full speed; the hull is ripped open over a length
of 100 metres. A last minute manoeuvre, hampered by the automatic steering,
does not help. The game is already over (7).

Amoco Cadiz. Eleven years later Captain Bardari is also gripped by an
obsession: to have the endorsement of his ship's owners for the tugging
contract which he must accept if he wants to get the *Amoco Cadiz* out of its
bad position. During the eleven and a half hours which his ship drifted he
has had hardly any time to think of the threat he causes to the coasts. He
does not send a distress signal to the authorities: all is well on board.
He receives the offers from the tug *Pacific* with the greatest reticence.
Even in a storm and without steering the 'commander' of the oil tanker does
not dare to decide on his own authority: everything must go through the ship's

owners in Chicago; the referring reinforces the difficulties in communications
which are due to a fatal time difference. The ship's officer on watch standing in for the skipper foregoes ship's owners' privilege at 23.18 h and sends
an SOS call: the ship is aground and has been polluting the area for two hours.
The bravery of the air-sea rescue pilots is called upon and, in a subsidiary
fashion, the resignation of the coastal inhabitants. In this affair the main
preoccupation has been with the cost of tugging; the rescue of the ship and
its cargo (actually: not insured) took second place and the lives of the crew
members came last. There are no referees.

Three Mile Island. We have also said that at TMI the operator had thought
more about limiting the destruction of his equipment than of limiting the —
admittedly: minor — radiological consequences for the people in the neighbourhood.

The important point is to provide structures which constrain the operator
to take the exceptional into account and to renounce certain possibilities,
no matter how immediately advantageous they may be, in the case of imminent
or potential danger. This is not always understood. The following illustration, supplied by the French association of ocean pilots is very interesting in this respect. One may retain its spirit even for non-maritime
situations. It is here the question of the safeguarding part the pilot can
play in the relationship between the ship's owner and the skipper:

In the case of an incident we are not too much at odds about command. The
skipper remains the skipper; the pilot is only a councillor. But I have
for instance found myself on a mineral carrier where after having dragged
the anchor which had not caught properly under strong wind we found ourselves grounded north of Dick. I had gone to sleep when we were at anchor.
The lieutenant woke me up: Pilot, pilot, something is wrong; it was so
"wrong" that we had, at 500 metres, shallows of 4 metres depth while we
had 14 metres draught. No VHF. Passing ferries could have alerted Gris-
Nez but the weather was too bad. A decision had to be taken. We had to
get under way as fast as possible. But the engine took three hours to
become operational. When we were about to swing around the windlass jammed:
at sea incidents often happen in series. What could we do with a 220,000 t
mineral carrier of 14 metres draught? There is no means of raising an
anchor without the windlass. One had to cut the chain to get away. The
Brasilian skipper did not want to sacrifice the chain: "You put in the log
book that this is done on the advice of the pilot". We got under way and
moored a bit further away on the other anchor. The following day, back at
Dunkirk, I had the pleasure of receiving congratulations from the insurers.
The skipper then admitted that alone he would never have dared to sacrifice
his anchor: he would have risked "getting the sack" from the ship's
owners on arrival (8, p. 69).

6th. Infringements.

Seveso, Flixborough, TMI, Amoco Cadiz ... the infringements were manifest
and sometimes decisive. One may retort that no installation can exist without constantly infringing. But the argument is not valid: when it comes to
it, it is not a question of exceeding the product quantity used by a litre
or of forgetting to fill in a monthly report form. At Flixborough, let us
remember, they had installed a storage capacity which exceeded the authorised
capacity forty three times; at Seveso Icmesa was infringing with regard to
the control services. It is here not a question of "admissible largesse with
regard to a meddlesome regulation". When one comes, as in the case of Seveso,
to requesting a permit for the construction of a waste treatment unit to
eliminate the residues of a dangerous production which had been neither
declared nor authorised then there exists again an intolerable delinquency.

We shall guard against hasty conclusions. These infringements do not necessarily reveal conscious strategies. In the case of Seveso, the Hoffman-la-Roche group remarked that Icmesa had not continued with the processes tested in Basle and that headquarters had not been kept in the picture. The errors, the lack of information between parent company, subsidiary and subsidiary may be critical and explain, more than a deliberate choice of contravention, the attitudes observed.

Whatever may be the case, one must know that an important gap may exist between what the law prescribes and what the operator effectively does.

7th. Faced with danger or disaster: the behaviour of the operator

It is hardly easy to face an exceptional situation which in an extreme case is liable to wipe out the organisation of which one is in charge or from which one derives benefit. Quite a variety of behaviour can therefore be expected from an operator in such circumstances. A few examples may illustrate these various attitudes:

Seveso. We have already acknowledged the courage and the lucidity of G. Reggiani, the director of the Medical Research Board at Hoffmann-la-Roche. We may add the determination of the president of the group who did not hesitate to step into the front line and take charge of the affair: he who had opposed the acquisition of Icmesa but whose hand had been forced; he who had been absent at the time of the drama and who was informed in an erroneous manner. This man in charge was to announce: "We shall pay for it all".

On the other hand one must also remember the start of the drama, between July 10 and 12, 1976. While the technical director of Givaudan immediately thought of the terrible hypothesis of the dioxin, from July 11 (9, p. 107), the company was satisfied with the admission of the escape of a "cloud of herbicide". It took 10 days, of going to and from Geneva since the company had no laboratory on the spot, for Hoffmann-la-Roche, after having obtained the proofs requested, to admit the presence of the poison July 20). Roche insists on always having measured up to its duties: one had to know precisely the extent of contamination before speaking out; because the band of uncertainty was extreme, either a few hundred houses or the whole of Milan might be involved, for some time or for years. Headquarters at Basle was, actually, only belatedly alerted by its sub-subsidiary: On July 14, it is said by the management of the group which repeats unflaggingly that everything was done to deal as best as possible and with the greatest responsibility with what had happened; and this despite a large number of difficulties in communications with its sub-subsidiary and the Italian authorities*which must not be alienated.

Nevertheless, much is open to interpretation and judgement as the Italian side has not failed to demonstrate. The lawyer for the plaintiffs speaks of criminal hypocrisy (10, p. 106). For others there is conjecture. Roche would not have admitted the presence of the poison unless the Milan experts had uncovered the situation.

As we can see in such an affair there is nothing automatic. A large number of factors intervene which, depending on the precise contexts and circumstances, can turn out to be minor or decisive.

---

*One example among others: how many high-level experts who spoke Italian fluently could one have at Basle at the height of the summer season? This is the type of question one is confronted with in such circumstances.

<u>Manfredonia</u>. In September 1976 between 10 and 30 tonnes of arsenic salt escaped from the ANIC factory (100 mg are sufficient to kill a man); a report from the Italian Health Ministry indicates that the first reaction from the company's management was to deny: the cloud contained only steam made up of water and anhydride of arsenic, absolutely inoffensive; there was arsenic only in 'minimal quantity'. Actually, if everything had been done to protect people one could understand this caution.

The mayor was alerted by telegram that in case of a stop to the activities at Manfredonia 3,000 workers would be made jobless in other factories; therefore nine hundred people continued to work on the site and to take their lunch at the company's canteen for several days until an official ban was sent to the company (9, pp. 462-463).

<u>Three Mile Island</u>. In this case the company did announce that there had been an accident. It started a general alert. But very quickly the NRC officials were to declare that the company had not informed them properly: the operator tried to minimise the importance of the accident on the first day despite evidence which showed its seriousness (9, p. 18). The company also lost all credibility with the press by declaring on March 30 in the course of a conference: I do not see why we should have to ... tell you in detail everything we do (2, p. 120).

<u>Ixtoc 1</u>. In October 1979 in his annual speech to the Mexican people President Lopez Portillo under the applause of Congress declared that "Mexico will not pay" (11).

<u>Givaudan and hexachlorophene</u>. In its ruling of February 11, 1980 the court of appeal at Pontoise stressed "the insufficiency of the information released by Givaudan-France". It noted (12, pp. 65-71):

Givaudan-France had (nevertheless) a real knowledge of the toxic nature of hexachlorophene and this from the large number of studies which the company had carried out since 1939, as well as from the studies published in the specialised press following the accidents and intoxications stated (p. 65).

- the "absence of warnings in technical bulletins" (p. 65).
- the incomplete nature of the indications included in the technical bulletins (p. 66).
- the maintenance of recommendations that were manifestly outdated and dangerous (p. 66).
- the fact that at no time Givaudan-France had spontaneously addressed circular and individual letters to its customers or spelled out the data published previously in its technical bulletins or even to inform of the results obtained in current work (p. 68).
- the fact that by contrast the company undertook to brush aside all concerns which its customers had reported:

    ... Givaudan-France had been set upon by a certain number of its customers who had been alerted to the potential dangers of hexachlorophene be it by foreign scientific reviews, be it by the general press (*Le Figaro*, January 2, 1972, *Le Monde*, February 16, 1972, *Le Nouvel Observateur*, February 21, 1972, *Chimie Actualites*, April 20, 1972, *Le Parisien Libere*, April 14, 1972), be it even by consumer journals (*Que Choisir*, February 1972, *50 millions de Consommateurs*, April 10 and 16, 1972).

Givaudan-France replied to all its customers in very general terms, limiting itself to indicating the recommendations established by the FDA for the use of hexachlorophene.

To some customers, nevertheless, Givaudan-France replied more explicitly, criticising mainly the articles which to them seemed unfavourable to the use of hexachlorophene.

Thus, in a letter dated April 15, 1971 and sent to customers, it presented these articles as "pseudo-scientific, in search for headlines".

In a letter of April 29, 1971 Givaudan-France indicated to Laboratoires Midy that "the sensational articles which have appeared here and there in the press should not be taken into consideration. They exploit, the company declared, fragmentary data of problems which, taken out of their context, no longer have any scientific value.

On September 6, 1971 Givaudan maintained "that hexachlorophene is a product without any danger". It added, "those who contest the inoffensive nature of the product are ill-informed and incompetent journalists".

On December 30, 1971 Givaudan-France wrote to M. Niama, the director of the Derno Vitta company in equivalent terms: "Hexachlorophene G 11 has for several months been the object of more or less violent attacks in the press following research work carried out in the USA. However, the extrapolation of these results to man in normal conditions of use has no scientific value". (pp. 68-69).

In a similar vein Givaudan tried to play down the importance of the FDA recommendations*.

On April 18, 1971, in its bulletin No. 820 to its customers over the signature of Vicklund, in charge of press relations, it was spelled out:

"that hexachlorophene is without danger and that it is effective". This bulletin, taking up for its own account the affirmations of the Wall Street Journal, adds "taking into account the absence of replacement products with comparable effectiveness and safety, the manoeuvres of the FDA are regrettable".

On September 19, 1972, when since August Givaudan had had knowledge of the accidents imputable to talc Morhange polluted by hexachlorophene, Mr Vater-laus, the head of scientific services of the Hoffmann-la-Roche group did not hesitate at the scientific congress in Hamburg to pass these accidents over in silence and to declare "that the results of studies have shown that the fears of the FDA were without foundation" and that "the good tolerance of hexachlorophene has recently been demonstrated to be clear to a child". He added that in the food studies no change in the central nervous system that could be compared to the cerebral oedema in rats and in newborn Rhesus monkeys had been found in observations which, as he added, had caused the FDA to formulate the proposals published in 1972 and put into their proper context (p. 69).

The court, before remarking further that Givaudan has not alerted SETICO to the dangers which the handling of the product presents puts on record that the company has been more preoccupied with its commercial problems than with the concern for information:

---

*Food and Drug Administration, American control organism that had warned against the product in December 1971.

Thus, Givaudan at the time published articles favourable to hexachlorophene, criticised the FDA, claiming particularly and contrary to all evidence "that there are no documents with proofs to support them which would permit to declare a possible damage by hexachlorophene".

Sterling Winthrop for its part, forgetting its communication to the FDA of September 18, 1971, published, on November 14, 1972, a declaration qualifying the decision by the FDA as premature and illogical and threatening even to attack this organism in the courts if it persisted in its intention to classify Physohex on the Rx list. Givaudan seemed more preoccupied with the commercial problems which the decision by the FDA raised for it than with the concern for complete and regular information of its customers.

This curtailed information has had a repercussion on the attitude of the Givaudan customers (pp. 69-70).

We end our illustrations here. It is obvious that a large number of other attitudes are possible: from the preventive introduction of trucks equipped with the necessary apparatus for eventual decontamination of territory (a commendable effort by ICI of Great Britain) to the most determined denial of the very fact of an accident and through half-truths (such as when one calls a 20 t wagon the very toxic contents of which had erroneously been poured into a river* — 'recipient'), the hanging of the scapegoat and the threat of general lock-out as at Manfredonia.

One understands the interest taken by French legislation which obliges, as we have noted, the industrialists to declare any accident or incident that has occurred immediately.

8th. The economic problem

An essential point is to know whether the cost problem which a more adequate prevention of major risk and the introduction of means for the battle against accidents would raise proves to be decisive for the industrialist. For lack of a general study in this field it is hardly possible to answer with assurance.

In the opinion of certain specialists, however, it would seem that in the majority of cases the cost of safety does not represent a decisive obstacle. This would be the case for the safety of storage; the question is doubtless more debatable in the pharmaceutical and nuclear industry. One may still bring up some questions for discussion in precise and detailed studies.

- Does one not in a situation of normal operation, in the absence of accidents, notice a tendency to sacrifice safety?

- In case of economic difficulty, is not safety one of the items most affected?

- In investment decisions, are there not various factors (insurance, safeguarding equipment, brand image) that push in the direction of taking the safety problem into account?

- But, from a certain cost level on, does one not find a less determined

---

*The case of leakage of acrolein on July 10, 1976. Two days later the management of the company declared an incident with a 'recipient' (13).

attitude? Are there still mechanisms which permit taking risk into
account? Or does one feel that one is entering the field of 'fate' and
that it is then the whole of society which is involved, no longer just
the operator who has done what is 'reasonable' on the economic plane?

9th. The attitude of industrialists faced with major risk

Can one draw a synthetic view of the attitude of industrialists who are
confronted with the question of major risk from all the observations made
above? This seems very difficult. An understanding of the problem is no
doubt in the process of evolution, varied according to public and private
sectors, the lines of activity, the companies, the services, the teams and
the men.

In the field of energy the problems of safety have for a long time been
the object of definite concern. But does one not see a certain weariness
dawn? Weariness to see oneself imposing demands which have nothing in
common with those which are imposed on other high risk sectors, a real
problem but one which must not lead us to take this as a pretext for lowering the level of completely justified demands. Weariness to find oneself
constantly beset by the safety authority who, one thinks, demands too much
and on points which are not valid, which becomes counter-productive for
everybody. Here we have, actually, a difficulty of size for the follow-up
on major risk: high risk leads to an inflation of studies on points which
are not necessarily the most critical ones; mechanisms get inflated while
high risk slips through the meshes.

In the chemical field the topping up requested by the authorities in
matters of safety seems to some people rather demanding. Witness for instance these few lines from the European council of chemical industry
federations:

We have recently seen a proliferation of legislation on safety in all industrial countries which affects quite particularly the chemical industry
which many consider to be unclean, cumbersome, polluting and dangerous.
It is usually forgotten that the development we have seen depends for a
large part on progress contributed by the chemical industry. It is true
that the rapid development has not happened without some inconveniences;
but the intention to impose on this industry in the name of safety of
products measures that would mean intolerable burdens the usefulness of
which is not evident would paralyse it (32, p. 235).

Confronted with citizens' groups who are worried, the chemical industry
painfully feels the grievances of which it is the object. From it follows
in certain cases a fall-back on very unconvincing positions which are closer
to a "Maginot Line" strategy than to a dynamic and modern policy as
evidenced by the language of the chairman and chief executive of PCUK in an
interview with the magazine *Decideurs:*

Decideurs: You approach here one of PCUK's problems in the region of
Rhone-Alpes, the one of its relationship with the environment.

A. du Fretay: Yes, and this is to put on record that the chemical industry
which has contributed massively to the economic development of the region
does not develop in a favourable environment, less so in the Lyons agglomeration in a broad sense. The example of acrolein is quite telling in this
respect. This is a product which is no more dangerous than many other
chemical products, its manufacturing conditions are very safe despite some
incidents which have been systematically exaggerated. However, its

manufacturing technique has been called into question while, curiously, the manufacturing technique for acrylonitrile at the factory at Yvours which is chemically and technologically very close to it was considered very safe by the same people even though acrylonitrile itself is certainly no less danger- ous than acrolein.

We find ourselves faced with people, media included, who reproach us vehemently for having stopped acrylonitrile which no longer had a future and for pursuing the manufacture of acrolein at Pierre-Benite even though the product has an important development potential. There is a lot of incoher- ence in this attitude.

I want to add that not only did we take all practical precautions for this manufacture of acrolein but in addition we have undertaken an exhaustive study of the risks it may present by submitting ourselves to the judgement of an expert who was approved by the administration.

Finally, I want to recall that the reliability of our technique has been confirmed, if that was needed, by the sale of a licence for the process we use at Pierre-Benite to China.

Decideurs: You seem to feel the consecutive reactions to the incident which occurred on October 12, 1978 at the acrolein unit of Pierre-Benite as some- thing serious?

A. du Fretay: In fact, this is very serious in the present international context where our best established commercial positions are called into question.

Because, after acrolein, what manufacture will the anti-chemical-industry lobby attack next in equally arbitrary fashion? But it is also and above all serious for the Lyons area because elsewhere we find more sense of measure, more responsibility and less demagogy in the defence of the environment.

Have I not been told in some schools in the area that the students had been given courses on the mortal dangers which the activities of our factory present?

Along this line one risks a dramatic decline of the Lyons chemical industry over the next ten years.

I hope, however, that those in charge in all fields will reassess their positions in time to reverse this tendency, and I shall be happy if our discussion could, no matter how little, contribute to their necessary aware- ness of the problem (33, pp. 9-10).

In other directions we have seen groups launching themselves into very profound safety studies. However the reticence towards the development of analyses in such depths as those carried out in the nuclear industry is still strong.

The idea prevails that the private sector cannot afford the luxury of the means employed by the state. It seems that one is equally far from having understood that major disaster is a distinct part of existing possibilities and not only a sign of misplaced pessimism and treachery to the cause of progress.

The reaction to disasters seems to have developed slowly since the Seveso affair. The discussion has become more cautious even if it remains, to the public, a bit woolly and almost completely self-assured.

To witness, these statements made on Europe 1 by the president of the Union of Chemical Industries at the time of the Love Canal affair:

Europe 1: But then, at the present time, M. Achille, what do we do with our chemical waste in France?

UIC President: Since 1976 waste which is considered to present some possible danger is destroyed. There are some fifteen installations, I believe exactly fourteen in France, which permit the carrying out of the destruction of, say, all waste which one thinks could present some danger. In fact, at present they exist perforce ... before legislation was introduced there was waste from products which had been made here and there but all waste is controlled, supervised, recorded and, we may say, regularly examined.

Europe 1: There are, then, nevertheless residues which you deposit in known and recorded places ...

UIC President: ... and supervised, you understand ... by the industrialists who in turn are supervised by the public authorities.

Europe 1: But then, in France ... You cannot tell us tonight for certain that we don't have a comparable situation ... to wit: a dumping of chemical products and later on top of them eventually some buildings.

UIC President: A. Listen ... A-hm ... I believe. What I can tell you is that ...

Europe 1: You get me a bit worried M. Achille ...

UIC President: You have nothing to worry about ...

Europe 1: One must still know where the waste from chemical products is hidden ...

UIC President: It is known where the waste from chemical products is hidden but one can never be absolutely sure that in the course of time, say, tens of years, something will not have been done in some place, in particular during a time of war or other. This is the only qualification I make on this subject

Europe 1: So you have no particular worry?

UIC President: Ah. Listen ... no. In the end, really, one always has worries in life but this is not one of the subjects one might worry about in our country (34).

Thus it seems that a first move could have been made. Doubt in certain cases could win the day, timidly, of course. Without this and its translation into actions and determined commitments the industry would doubtless head for serious rebuffs.

2. THE PROBLEM OF THE HUMAN OPERATOR

1st. 'Human error': simplistic dodging of the real problem

The expression "human error" is used without much discernment. In the accident at TMI it was used without much descretion. True, J. C. Wanner remarks, 'human error' is not 'human fault', the latter expressing voluntary transgression of the rules while error is only an involuntary transgression which therefore does not involve culpability. (14, p. 2).

But behind the word 'error' the idea of fault could have been communicated, a fault against the technique. Since in work relations man complicates all existing systems just as in matters of safety the operator could have been presented, implicitly, as a saboteur*who made the beautiful edifice of the technicians collapse.

Pierre Vianson Ponté in the front-page article which *Le Monde* devoted to the event was aware of this sudden attraction of the 'human error' — which was actually the title of the article in which one could read:

On the evidence, and it has immediately been confirmed, what happened at Harrisburg, what was decisive, was not a technical failure, not a scientific error, not a faulty construction and operation: for that we can have confidence in the technicians. They know their job, their calculations are correct, their apparatus in proper shape, at least one wants to believe so. What was decisive was human error, actually: in the event repeated two, three or four times.

"Everything is provided for", the famous experts sound haughtily and with unfailing reassurance. "Everything, even the quasi unforeseeable incident, even faults in handling, even error. Whatever the cause of the breakdown or accident, one, two three safety apparatus come automatically and instantly into play. One after the other, all possible dangers are eliminated, the most severe measures are taken. Be quiet, sleep tight, nothing can happen, never: we guarantee it."

They had provided for everything: that is true. Except for human error (15).

After this example of recourse to the argument of human error, abusive and suspect as it is, we must do justice to the operators concerned: working permanently with fifty alarm signals in operation (there were continuous breakdowns), facing at the time of the accident the setting off of more than one hundred alarm signals (2, p. 11), managing with an indicator that showed the giving of an order (closure of the discharge valve) and not its execution (the valve was blocked in open position, information on its "state" was therefore needed) etc. could only lead to rebuffs and errors ... If one adds still the absence of post-accidental procedures or the arrangement of procedures the results of which one might fear (transformation of an accident of a small breach into an accident of a large breach?), very few operators would have emerged well from such a situation.

---

*Or explicitly as quasi-saboteurs — cf. the recent article by R. Latarjet: One cannot say that there has been sabotage at TMI but one can say that saboteurs could hardly have done better! (30, p. 502). The latter had foreseen everything but not this 'unpardonable' slip-up. The idea of the "bad student" was underlying the comments in April 1979. Condemnation and culpability were not far away.

<u>Three Mile Island</u>. In our first meeting, four weeks after the accident every-
body said that this was a simple case of operator error. The operators had
failed to recognise that a certain valve had remained blocked in open position,
had failed to recognise other symptoms which led them to stop the safety
injection system. In fact, these omissions have without any doubt transformed what should have been a minor incident of which we should never have
heard into a truly major accident. However I remember very well, during one
of our first hearings, the deposition under oath by these operators who
insisted on the fact that they had never been trained for anything comparable
with what they had to face. I did not believe them at the time but before
the end of our enquiry I learned that they were telling the truth. In fact,
they had not been prepared.

This has convinced us that we must examine the training programme for
operators very closely. A part of the training, as it seems, is done on the
job and under contract with the manufacturer. A witness, the person responsible for training at the manufacturer's was very proud of the last five
years of his company's programme. What was the first important progress
achieved, we asked him. "When I arrived, he said, many courses had been
given by engineers. But the engineers don't know how to talk in a way which
people can understand. Consequently, the first rule I introduced was that
no engineer was authorised to participate in the training of operators." We
were not completely convinced by this answer, particularly after we analysed
things more thoroughly and found that all theory had been taken out of the
operator training programme.

They were trained to be button pushers, quite adequate for normal conditions
of operation but they had really never been prepared for a serious situation.
This was quite legal with regard to the NRC requirements and had become
standard practice: The operators were only trained for an accident in the
course of which only ONE thing went wrong.

They were never trained for a situation in which TWO independent things
could go wrong. And in this particular accident THREE independent things
went wrong. J. Kemeny and human error (Source: 29, pp. 65-66).

In short, 'human error' can only constitute an excuse which permits to
close the case for good. This preoccupation was too strong in April 1979 to
let it be passed on immediately for more serious analysis by specialists.
This having been spelled out, it is indicated to approach the real problem
raised by this man: machine interaction.

The experts know, in fact, very well that the operator must be understood
as a factor within the whole set-up; this set-up has been designed, rules
and limits of operation have been fixed, the piloting has been entrusted to
a group of people. The error, if it exists, must therefore be studied in
a very broad context and the idea of fault put aside, as the title of an
article put it perfectly: "Track down the causes, not the people" (16).

This perspective has for instance been developed by the INRS*which suggests
to analyse work accidents by means of a fault tree which was named "tree of
causes". The effort of scientific study in this field is particularly
important to be undertaken as the fault is habitual, the immediate and ultimate reference providing the explanation. The diagram following which is
suggested by the INRS (31) for illustration shows clearly how this method by

---

*National Research and Safety Institute.

developing an implacable logic permits to reveal the important failures in
the systems involved. One can actually conclude from it a certain number of
particular and general lessons as the table below also shows.

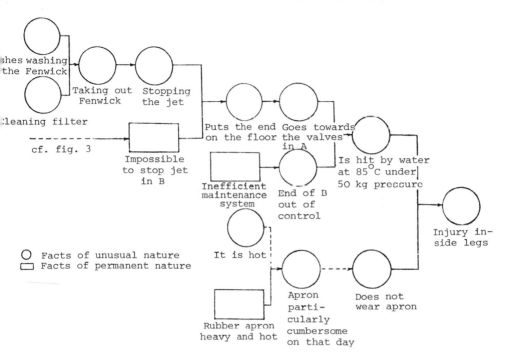

Fig. 31: Hot cleaning under pressure: tree of accident causes.
(Source: 31, p. 28)

Before doing so we might remember that the Rasmussen study had already used
this type of analysis. The question of the human operator can thus be
suitably approached. We shall also stress the following recommendation by
P. Tanguy in the wake of the Harrisburg accident:

We must, and this is doubtless the most important point, study the safety
of installations by putting ourselves in the place of the operators;

- check that the instructions they have are compatible with the accident
  studies and the behaviour of the systems;

- analyse all cases in which the operator can inhibit a safety action and
  make sure that he makes the necessary checks beforehand;

- supply the operator with clear and unambiguous information which permits
  him to know at all times the real state of his installation;

- make sure that the training programme for operators is adequate to the
  actions which they will have to perform in case of an unforeseen event.

In the long term one must no doubt undertake a more fundamental research
into the problem of interfacing between man and machine, the design of control
rooms, the diagnostic apparatus and decision making aids etc. (4, p. 531).

Station	Description of accident factors	Particular immediate curative measures	Potential accident factors
Equipment washing	Washing machine available to various factory departments	•Assigning and training of one person in charge of washing • Entrusting him with current main tenance of the washing machine • Giving him a washing machine in good condition	Anarchic utilisation of a piece of equip- ment by several departments, teams or individuals
	Elimination of spray gun. New mode of operation	Installing a spray gun of which the lock also stops the water supply of the machine	Equipment modifica- tion eliminating a safety feature
	The absence of the spray gun was not announced	Encouraging staff to report anomalies in matters of safety — 1. by inviting them to do so 2. by following up pertinent sugges- tions	Poor circulation of up-line information
	Heavy and hot apron	• Making more comforta- ble aprons available to operators • Thinking of this comfort when buying aprons	Individual protection uncomfortable
	Mural coil system unreliable	Strengthening the sys- tem by putting in a gu- ide pointing downwards and placing it near the floor (cf. Fig. 1)	Safety apparatus in- sufficient in design

Table 15:   Study of the tree of causes.
(Source: 31, p. 29)

Along the same lines the Central Safety Service for Nuclear Installations has requested Electricité de France to take a certain number of actions after the Pennsylvania accident, among them:

- Reexamination in depth of information given to operators with regard to the state of the installations and the general rules of operation.

- Reflection on training and qualification of staff.

- Studies concerning control command instrumentation and display of infor- mation in the control room which is likely to improve the means at the disposal of operators (4, p. 529).

The aim having thus been corrected or at least the positive part of it re- membered, we can now explore the problem of the human operator more precisely. We shall do this in three successive points which will show the scope of the questions to be explored and the importance of the questions asked on account of this about the security of the systems.

2nd. Man in accident situations: the limits of adaptation

The most advanced studies take human limits in high risk situations into account after the level of abnormality is passed which can be defined in terms of work load. (14, p. 13). R. Andurand spells it out:

... The studies of human reliability have shown that men have a heightened performance level for a long time and, put into a frightening environment, 'crack' suddenly. The times for recuperation are long and could be figured out in certain cases; one must take this into account, especially where measures destined for the limitation of incident or accident consequences are concerned and in order to get the installation back into a safe state.

This is certainly a borderline aspect but one must know that panic is only the visible manifestation of the total disorganisation of a man suddenly placed in a different situation. In the first minute which follows a very serious accident man, no matter what his specific training, either panics or takes useless decisions or decisions contrary to safety. Five minutes after such an event 90 per cent of people are still in this state; half an hour later the figure is about 10 per cent and 1 per cent after several hours (18, p. 131).*

One must take these data into account. In particular for post-accidental situations one must provide 'instructions' of extreme simplicity. An instruction must never be presented as a procedure. Being concise is of decisive importance. Thus one will spell out e.g.:

"In case of a gas escape leave as fast as possible and in passing hit the red button next to the lock-chamber door with your fist. Do nothing else." For this one must, of course, have set up 'positive' security measures i.e. systems which for lack of fluid (water, compressed air, electricity) or for lack of the human order automatically put the circuits or the systems in a favourable position; and one must also have good barriers, good fall-back positions.

A few more lessons are contributed by the studies of human reliability:

Human failure. The average rate of human error committed by the staff of a nuclear reactor in a very frightening situation like for instance a LOCA is estimated at 0.2 to 0.3.

Generally speaking, one considers that the rate of error for a given task is described by a curve which is the function of the rate of fear felt (19, p. 5).

Incredulity. Following a LOCA, human reliability would be low, not only because of the stress which has been caused but also because of a probable reaction of incredulity.

In the minds of the staff the probability of a large incident, a LOCA, occurring is considered to be so low that for a few moments the probable potential reaction would be not to believe what the panels show. In such conditions it is estimated that no action can be taken at least for a minute

---

*In the French nuclear centres the protection and safeguarding systems are designed in such a way that in case of an accident the installation can be kept in a safe state without human intervention for a period of 10 minutes.

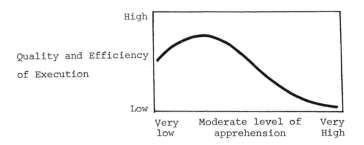

Fig. 32: Quality and efficiency of execution

and that if nevertheless some action is taken it will probably be inadequate
(19, pp. 6-7).

Reliability and return to normality. If one refers to the curve which shows
the quality and efficiency of execution as a function of stress ... the rate
of overall error is valued at 0.9 (9.10^{-1}) five minutes after a major incident
of LOCA, at 0.1 (10^{-1}) after thirty minutes and at 0.01 (10^{-2}) after several
hours. It is certain that seven days after an important LOCA there would be
a complete return to a normal stable state and that the normal rates of error
could be applied to individual behaviour (19, p. 7).

Drastic reduction of reliability in case of error. A theory of human behaviour under the constraint of time maintains that the rate of normal error
for each corrective action which follows another doubles when an error has
been committed in the preceding corrective attempt or when the preceding
attempt has not had the foreseen corrective effect. So it is that if one
starts at an error rate of 0.2 it is theoretically sufficient to have just
three supplementary attempts at a corrective action in order to reach the
borderline case of error rate of 1.0. This limitation corresponds to the
fact that the individual becomes completely 'disorganised' (19, p. 7).

Learning and the exceptional. The operators questioned could explain in
general terms what they would have to do if they were put in hypothetical
emergency conditions but they did not always seem certain about the location
of buttons, the texts on the posters concerning actions of manual operation
which would be necessary to take if the automatic security system broke down
(19, p. 7).

The danger of habit which may veil the alerts. ... in one of the cases
analysed while using the operating modes for the calibration of an apparatus
correctly a technician approximately anticipated the reading that was going
to appear on the instrument for each operating stage. He had done this
calibration for a long time and so often that he knew what he expected. This
knowledge coupled with a very low frequency of discovering a reading out of
tolerance created a state of fact in which he intensely expected that every
reading would be within tolerance range. In these conditions a certain probability exists (estimated at 10^{-2}) that the technician 'sees' a reading outside tolerance as being inside the tolerance range (19, pp. 17-18).

The importance of the 'insignificant'. The quality of written instructions
has been evaluated in an estimation of error rates. Interest was centred on
factors like the ease with which an operator could find the written operating

modes for cases of emergency, the importance of presentation and the ease of comprehension of non-routine procedures. The style of written instructions plays a 'material' part in the estimated error rate. The written instructions should not conform to the well-established principles of good literature; they should rather resemble the operating mode type of maintenance which has been practised in the army for about twenty years. Other insufficiencies which have contributed to relatively high estimates of error rates are the bad quality of printing or again the fact that there is no distinctive binding and storage space reserved for the instructions for cases of emergency. This is also the case with the lack of thumb-indexes or with inappropriate indexing which has made it difficult to find the specific operating modes and also a poor presentation (19, p. 18).

The coupling of human errors. One must know that when too closely coupled valves must be handled an error is likely to be made with the second one if there has been an error with the first one. It is important to track down this type of 'common modes' which intervene through human operation. This question recalls the case of Feyzin: the safety valve is not treated as such but together with the cleaning valve as one and the same. The operator, on account of this, destroyed the existence of a safety device. Since then it has been decided to separate the two valves with their distinctly different functions very clearly. (19, pp. 19-22).

Redundancy of staff. It is certain that if one person only is given a critical task one must expect an increased error rate; a simple inspection per round will not radically attenuate this rate (the error is 0.5 for this type of control). In the same way, the signing of a register by the agent indicating that he has carried out the operation must not inspire too much confidence: the action can quickly be considered by him as "carried out as a formality" without any more significance than the ticking of an item on a check list (19, pp. 23-24).

3rd. The piloting of systems

We have just examined the problem of the operator in a borderline situation. This is a large field for research to explore. However as a complement another field needs to be studied: the one which stretches from the normal situation for the operator to these borderline points where he might breakdown suddenly. Reducing the probability of too large a gap to the normal position is a very essential management task. J. C. Wanner was able to clarify this problem from his work in the aeronautics industry. Here are some of the essential results he has presented (14).

Generally speaking, J. C. Wanner defines the accident in these terms:

A system is usually a set of sub-systems and elements of which the function of each can be characterised by one or several parametres ... The field of variation of each of these parameters of operation is equally limited. It is thus possible to represent the operation of a system by a "point of operation" in a space of n dimensions, each dimension being devoted to one of the parametres of operation. We shall say that there is an accident when following a succession of incidents the point of operation of a system has gone beyond one of the limits of the authorised field. Usually the going beyond a limit is due to a succession of incidents of which none plays a preponderant part (14, p. 3).

The different incidents the succession of which leads to the accident divide up into three categories of which only the third brings the operator into play:

- incidents of manoeuvrability: the system, especially after a breakdown, no longer responds correctly;

- incidents of sensitivity to disturbances: an external event (wind for example) or an internal one (breakdown) projects the point of operation outside the authorised field;

- incidents of pilotability: the operator has the necessary controls to maintain the point of operation within the authorised field but the work proves too difficult to carry out.

The prevention of the incident of pilotability leads therefore to questioning oneself about the work one demands of the operator and above all about the way in which the human operator behaves. Summed up, the characteristic points are the following:

- The operator does his work in sequence: the brain treats difficulties not simultaneously but successively (be it as a voluntary act or as a reflex); on account of this the work load increases with the number of operations to be carried out.

- The operator cannot stand the absence of information; if he does not receive it he creates an internal image and this means the loss of vigilance.

- The operator compensates, up to a limit, the variation of difficulties by an increase in work load without variation of performances; when the limit is reached, as we have seen, an abrupt degradation of performance appears (14).

These statements permit the definition of a certain number of management rules:

Three Mile Island. A second field was the information available to the operators in the control room. I shall never forget our visit to the sister unit (TMI-1) which operated relatively without problems ... All was extremely impressive (in quality) until we came to the control room, where I had rather unusual experiences.

When you enter you see a large wall entirely covered with small panels which serve as alarms. I think there were nearly a thousand if my memory is correct, and as a complete tour of the control room did not interest me much I just looked at what was going on. After about five minutes I heard a bell ring. An operator did something just as the guided tour arrived at the spot. I questioned the guide and he asked if it would inconvenience the operator if he told us what was happening. He replied that an alarm had rung but that an alarm does not necessarily indicate something to cause alarm. It means only that the operator must look at something and take corrective action. Behind every piece of plastic there is a light, he said, a blink-light which tells the operator the problem. The operator pushes the right button. The bell stops ringing and they take care of the problem. Thus, ten minutes later when another alarm started, I had become an expert, I fixed the blink-light. The operator came, looked at it and did the things necessary and all was in order again.

Fifteen minutes later I again heard an alarm, I looked everywhere but there was no blink-light. I really saw the operator gesticulating and calling one of his assistants who started removing those small pieces of plastic one by one. I raised my hand once again and said: Sir, would you mind asking the operator what is going on now? They had a discussion in low voices and said:

Oh, this is nothing important. It is just that when you hear an alarm ringing and there is no blink-light this indicates that the light bulb concerned
has burned out. Then I looked on for ten minutes while he assiduously removed the small pieces of plastic in order to find in the end the burned out
light bulb. It was just an ordinary flash light bulb. They replaced it and
when it began to blink they knew what was not working.

Now I fear that I have put myself into serious difficulty after this because I remarked in the presence of journalists that I did not think that
this particular control room was a masterpiece of modern technology. In
fact, I said that it was at least twenty years behind the times. I was
heavily criticised for this. And with good reason, because my statement proved
to be wrong: we discovered later, in the documents of the NRC, a report
written ten years earlier in which an expert had said that the control rooms
were, then, twenty years behind the times. J. Kemeny and the control rooms
(Source: 29, pp. 66-67).

Providing an information system that is reliable and credible for the
operator which implies:

inhibition of the alarms (e.g. the alarm of a disconnector of a pump
designed for maintenance should be inhibited) without weakening their
reliability;

categorisation of the alarms in order of seriousness (which is already done
in the nuclear centres);

equipment of excellent reliability;

making truly useful information available.

Providing an information system that is of help to the operator:

providing information that can be quickly interpreted;

providing information to avoid the loss of vigilance*;

providing information which permits precautionary measures.

The recommendations by J. C. Wanner must be reported in detail.

In all this design work prior to the task of the operator it is important
not to forget the mental image. If an operator, following an ambiguous information, has interpreted a complex situation in a certain way it will be
difficult to undo his mental image subsequently. J. C. Wanner takes up the
case of the operators at the Harrisburg centre in this respect: the operator
had nineteen other parameters that permitted him to conclude that the discharge valve was not closed; has on the one hand interpreted across some of
these parameters in order to justify his hypothesis of a shut valve and on
the other hand he has not consulted the others (this consultation did not even
enter his mind: why verify a hypothesis that is already largely confirmed!)
(14, p. 17).

---

*Solutions have been found e.g. in the aeronautics industry for automatic
approaches: a cathodic screen representing the runway which is 'enlarged'
gives the information while all the rest is obliterated in the case of an
incident the image is dislocated and the pilot takes over manual control with
the advantage of already having the analysis of the situation (14, p. 19).

In a study similar to the one by J. C. Wanner, H. Otway (20) clearly
stresses the paradox encountered in the control rooms of centres: except for
about ten times a year the operator does not have much work to do*: one is
below the necessary vigilance level. Abnormal situations do not arise more
than once a year. However if there is a problem one expects the operator to
understand immediately what is happening, without a diagnostic error, and to
act perfectly from the time the automatic mechanisms, which are provided for
the first ten minutes or the first half hour (depending on the country) cease
to respond to the situation on their own.

Simulators permit the training of staff. However the procedure suffers
from certain insufficiencies (unreal nature of the situation which one can,
however, correct with various gadgets) and above all from serious limitations:
one can simulate only what one has designed .. while serious accidents can
result precisely from design problems; in addition, the simulator by training
the operator in the best responses may reduce his aptitude for improvisation.
Some people therefore ask themselves about the ultimate yield of this tool
(21). One may at least hold on to the observation by J. C. Wanner: the
simulator does not permit the training of an operator but it is useful to the
already well experienced operator (14, p. 23).

We remain with the question H. Otway asked:

... it is when a totally unexpected situation arises that the operator is
most necessary, i.e. when he must quickly diagnose and respond to a situation which was not covered by his training (20, pp. 28-29).

We add to this only the words of confidence from the specialists. It is
precisely in the face of the unforeseen that the human being must draw on his
capital of creativity and ingenuity. H. Lewis who does not ignore this trump
(22, p. 83) concludes without optimism or pessimism: In this situation human
action can be beneficial, ineffective or aggravating (22, p. 84).

4th. The operator placed in networks of a social nature

An important step is taken when one recognises the complexity of the situation created by the interaction between man and machine: there is no longer
the 'good machine' on the one hand and the 'bad technician' on the other.
Other considerations of the same type are needed. It is a matter of bringing the whole of the social components of the task to light and the rules
which structure it. We shall distinguish here two levels of reflection: the
first will be mentioned to be remembered (and will actually show the path
chosen by the large units of advanced industries); the second, more subtle,
will be more interesting for the comprehension of present and future
situations.

*a) About the broad notion of 'apathy' in the analysis of the operator's
social situation.* The report by the committee chaired by Lord Robens in
Great Britain (1972) thought it could attribute work accidents to apathy.
From this it draws the conclusion that by reducing the number of rules and
regulations accidents could better be prevented because the individual would
be more responsible. This point of view was criticised by various studies
which the unions carried out.

Two university scholars, T. Nichols and P. Armstrong undertook for instance

---

*A case also for the conduct of large ships but not for aircraft, J. C. Wanner
stresses (14, pp. 29-30).

to follow accident cases in respectable companies which had safety officers, satisfactory equipment, staff representatives for safety etc. They drew several conclusions (23):

- Superiors do not at all prevent subordinates from taking non-scheduled measures to restart production (in case of an incident).

- This is particularly apparent with workers: production must be restarted; all their work makes sense only when production works, this is instinctive. One has already jumped over a dangerous barrier to carry out a repair before one could think of danger.

- The worker knows that all delays must be made up.

- In all cases there is pressure from the organisation to strengthen production (questions, attitudes taken by superiors, bonuses etc.)

- This pressure from management and from foremen is quickly interiorised. The worker anticipates from it the demands made by his superiors.

- Why do foremen, no matter how trained in matters of safety and used, like the workers, to hear management speak of the requirements of safety, exercise such pressure? They know where management's fundamental preoccupation lies. They see the man in charge of production burst out of his office like a cannon ball when the conveyor stops.

- Naturally, from time to time management and the foremen have preached about safety; there are even sanctions in case of an accident. But in day-to-day operations one sees clearly what is important. (Nichols, Armstrong, pp. 16-20).

Thus the two researchers write:

Every one of the accidents we have described has occurred in the context of an interruption of production and while the man involved tried to maintain or to restart this production. In every case the dangerous situation was created on account of the objective to gain time and to ease operations. In every case the company's safety rules were violated. The interruptions of the process were no isolated events. Nor were the dangerous means used to deal with them. The men acted as they did in order to face up to the pressure exercised by the foremen and the management who aimed at maintaining production. This pressure was continuous; the interruptions of the process were rather frequent and equally hasty methods used to deal with them were employed repeatedly (23, p. 20).

How, asked the authors, do these accidents square with the scheme of the Robens report? How can the notion of apathy account for these events?

On the evidence, they do not fit the scheme at all. In every case the dangerous situation had been created ... by the production problems and not by a problem of apathy. It was precisely those who had made efforts to continue work who were injured. And the pressure to act in this way came from the foreman and from management. Risks were taken not because the workers did not care to know whether or not they were taking risks but for a very obvious reason: Maintaining production. 'Apathetic' men would not be so easily disturbed (23, p. 21).

The impressive study by the National Institute for Industrial Psychology

(NIIP) which reported on 2,000 industrial accidents seems to point in this direction. As P. Kinnersly*emphasises, this NIIP study calls a number of all too convenient ancient myths into question when it spells out in particular:

Even though the companies studied were probably among the better ones in terms of attitudes and results where safety is concerned ...

The risks were such an integral part of the working system that the more work was done, the more accidents there were ...

One expected employees to repeat complex actions without stopping with perfect dexterity. And one expected from them that they avoided danger with the greatest regularity by instant reactions to alarm signals. When they could not deal with one or the other difficulty they exposed themselves to danger. And yet these are two activities for which the individuals are slow and not very sure (24, 25, pp. 200-201).

As soon as one leaves that woolliness of 'apathy' one encounters difficulties, one gets hurt by the problem of ill-conceived systems as the Kinnersley report does (25, pp. 203-205);

- work places designed for people of 1.30 m height with arms of 2.40 m length;
- brake handles which are inaccessible in case of emergency;
- non-homogeneous**control systems.

Kinnersly finally stresses that this easy recourse to 'apathy' to 'human error' for explanation of accidents has been abandoned at Volvo for example. He quotes the opinion of an engineer from this company, an opinion that is quite alien to the philosophy of the Robens report:

Our philosophy is that every operator or maintenance mistake that can be made will be made sooner or later. Therefore we have taken all safety measures against human error (25, p. 206).

This approach is difficult to develop. It is much more convenient to stick to the idea of negligence. Kinnersly vigorously attacks the action by associations working on work safety, above all the Royal Society for the Prevention of Accidents (RSPA) which remains faithful to the old approach as the following lines show which have been approved by this institution:

Accidents do not happen, they are caused ... by lack of precaution, reflection, concentration, regard for oneself and for others, familiarity, drink, fatigue, hatred, working conditions, frustration, irritability and boredom (25, p. 197).

They could have added, Kinnersly comments, infrequent attendance at church services. The economic factors and the employers' attitude are not mentioned (25, p. 197).

---

*In his book: *The Hazards of Work: how to fight them*, a kind of Robens report presents the point of view of the wage earners who were recognised as 'excellent' by the minister, Michael Foot, in parliament.

**As the economic study of fifty seven cranes and operators of such has shown (Glamorgan Polytechnic's Department of Civil Engineering and Building) there were fifty four different control systems on the fifty seven cranes, 72 per cent of the operators forgot at least once that the controls were different from the ones on the crane they had last used; 95 per cent of the motors could be put into motion accidentally.

Other authors (Grayson and Goddard) also bring up another point concerning the responsibility of the individual at work. Often a relationship is established between those who have accidents and certain characteristics which are said to be typical of the 'unlucky' individual. Some people would have accidents no matter what precautions the employer took; these were always the same people who had accidents ... From there it is only one step to involving 'individual destiny'. Our authors declare war on these ideas (26, p. 16):

In its study the National Institute of Industrial Psychology sets aside all theories which tend to link certain personality features to accidents. The NIIP studied age, sex, size, weight, marital status, the people in charge, transport habits, medical history, absenteeism, late reporting for work, eating habits, models of rest and concluded:

"Personal characteristics have very few significant effects" (24, p. 22).

The relevant factors have proved to be: the time a man has worked on a job, the number of different jobs he has done, the effect of surprise, the dangers connected with the work, the processes associated with the work. In fact, the, foreseeable, conclusion is as follows:

"The work situation is the source of accidents, not the individual".

b) The operator, the rule and the institution — questions about the safety of large advanced systems. The accident due to an unfortunate intervention by the operator is taken into account by organisations which issue a certain number of rules and instructions. In the first analysis the situation may appear simple and clear. P. Mayer, in a recent study, has shown, however, the real complexity of the problem. Here again, we have nothing mechanical. Man can definitely not be isolated from the social context in which he lives and works. Let us take up some observations from this examination.

About the evidence in an interrogation. The evidence can be read in the classical discussion of the institutions; P. Mayer sheds light on this habitual discussion:

The regulation safety is founded in the law and in fact on the concern for maintaining safety; in order to avoid incidents the regulations must be respected, and there are always ways of making good regulations which are not contradictory; if the regulations are bad it is because they are badly made and one must make new ones (27, p. 46).

If people respected regulations better there would be less incidents.

... it is a matter of human error, of individual error, human nature is made in such a way that people don't want to follow regulations ... there are two types of agents: those who follow the rules twice over and those who don't want to know about them ... (27, p. 73).

But observation denies this apparent simplicity. If the framework remains quite clear for all repetitive situations, it is not the same when one leaves the field of quasi-automatisms. In the first case it is easy to follow stable instructions, issued *ex ante* on the basis of satisfactory theoretical knowledge. For work which offers no relief from routine (checking for instance, a part of improvisation is sometimes necessary. At the other extreme, experimentation in particular, one no longer finds precise general instructions; theoretical knowledge is absent or reduced to external experts; the exact carrying out of the work is accessible only to the specialists directly involved. From then on a large number of factors enter the behaviour as parameters for these non-repetitive tasks. One must look at their diversity.

The 'social' situation of behaviour in a non-routine situation. P. Mayer
stresses in his study which was done on a large R & D* centre that it happens
that a production or experimental process requires, given its technology,
going beyond the standards set for a much larger group of processes. In
other words, unless one changes the process or the equipment one cannot respect the standards of the centre or the installation. Certain departures
from the rules are therefore foreseen. They are negotiated between the
safety engineer of the installation, the safety engineer of the centre ...
(27, pp. 16-17).

The rule such as it is finally fixed, looks then like a compromise between
the safety requirements and production and research requirements (27, p. 17).

This is not to say, the author stresses, that safety has necessarily given
way to the requirements of production if the design of the installation which
departs from the standards has been specially studied from the point of view
of safety (27, p. 17).

It is clear that one has left the classical framework that can be taken up
without hesitation.

How are safety rules modified? P. Mayer identifies several components of
the decision to be worked out:

- arbitration and compromise between taking safety requirements into account,
  the convenience of work and the requirements connected with the production
  and research activity.

- arbitration and compromise between the economic and financial viewpoint
  and the foregoing considerations.

- the power of the safety authorities over the day-to-day organisation of
  work (27, p. 20).

In a more general way a certain number of explicit and implicit considerations will weigh on the devices such as those linked with:

- the life of the institution and its rule books;
- the investment choices for safety;
- the division of decision making power in this matter and more generally
  the organisation of work;
- social relations at work in a still more general way (27, p. 14).

Once again, from here on it becomes difficult, at least for these non-
repetitive and therefore not strictly regulated situations, to lean on a
single rule for judging an operator's behaviour. The rule, fruit of nego-
tiation and compromise, no longer offers, either to the hierarchy or to the
operator, a single recourse that permits him to discharge his responsibility.
Having been negotiated, it has involved the various parties in a complex
manner; worked out by compromise between various requirements it presents itself as an attempt at optimisation under constraint and not as the universal
'truth' and can easily be opposed as 'faulty'.

A characteristic example of this type of evolution has been related in

---

*Research and Development.

Le Monde of November 17, 1978, P. Mayer recalls. It was an incident at the
retreatment plant at La Hague:

For management, poor individual control has led to a spread of contamination
in a lock-chamber and in neighbouring corridors through disrespect for instructions.

By contrast, for the CFDT the accident or incident only put the inadequacy
of certain work zones in evidence, the difficulties of operation and the
decay of the installation and, the CFDT adds, one may say that the obligation
put on the operator to transfer heavily contaminated pieces from one zone to
another had inevitably to lead to this type of accident even if it is easy
to conclude, as COGEMA does, that it was human error (27, p. 30).

It follows that in these installations where the quasi-automatism are not
the rule one is faced with regulations whose operational value is very uneven. The author thus distinguishes:

- operating regulations, safety regulations which are at the time operating
  procedures; they do not contradict production and productivity
  considerations;

- prescriptive regulations, regulations which correspond to compromises
  made by the organisation between two points of view, regulations which are
  accompanied by their means of application. These are recognised and institutionalised in the organisation;

- symbolic are covering regulations, regulations which testify to a loosen-
  ing of the impact of forces which go for the establishment of internal
  regulations and the impact of those which effectively go for safety in
  decisions about and execution of production. These regulations are therefore not applied;

- opportunist regulations: regulations corresponding to the known rules of
  the art but which appear only when the activity which they regulate can
  be avoided and carried out differently;

- phantom regulations corresponding to known rules of the art or to manufacturers' rules which are not institutionalised as internal regulations
  of the company or the production unit but which could be or are treated
  thus in other companies or other production units; rules of the art the
  organisation of which tends sometimes to force out even knowledge and
  which are generally not respected.

The last three types of regulations are regulations of which it is usually
said that they are 'made in order to be violated'. One finds the first two
of them where safety has been particularly institutionalised. The third one
is seldom if at all institutionalised (28, pp. 10-11).

REFERENCES

(1) The Flixborough Disaster. Department of Employment. Report of the
    Court of Inquiry, Her Majesty's Stationery Office (H.M.S.O.), London, 1975
    (56 pages).

(2) *Report of the President's Commission on the accident at Three Mile Island.*
    Pergamon Press, New York, October 1979 (201 pages).

(3) Health and Safety Executive, Canvey, an investigation of potential hazards from operations in the Canvey Thurrock area. Health and Safety Commission London, H. M. S. O., 1978.

(4) P. TANGUY, L'accident d'Harrisburg. Scénario et bilan. *Revue Générale Nucléaire*, no 5, septembre-octobre 1979, pp. 524-531.

(5) V. BIGNEIL, Ch. PYM and G. PETERS, *Catastrophic Failures*. The Open University Press, Faculty of Technology, 1977.

(6) P. LAGADEC, Développement, environnement et politique vis-à-vis du risque. *Le cas britannique*, tome 3. Laboratoire d'Econométrie de l'Ecole Polytechnique, avril 1979 (115 pages).

(7) *Le Monde*, 5 mai 1967.

(8) La pollution marine par les hydrocarbures. Union des Villes du Littoral Ouest-Européen (U.V.L.O.E.) Colloque de l'U.V.L.O.E., Brest, 28-29-30 mars 1979 (227 pages).

(9) Camera dei Deputati VII Legislatura. Commissione Parlamentare di inchiesta sulla fuga di sostanze tossi che avvenuta il 10 luglio 1976 rello stabilimento Icmesa e sui rischi potenziali per la salute et per l'ambiante derivanti da attivitá industriali (Legge 16 giugno 1977, n. 357). Juillet 1978 (470 pages) (Notre traduction).

(10) C PECORELLA, Qui va payer? *Survivre à Seveso*. Presses Universitaires de Grenoble, Maspero, pp. 105-117.

(11) S. CROSSMAN, Ixtoc 1, la marée noire du siècle. II. Bataille politico-juridique *Le Monde*, 3 avril 1980.

(12) Le talc Morhange, *Tribunal de Grande Instance de Pontoise*. Jugement du 11 février 1980 (102 pages).

(13) Dernière Heure Lyonnaise, 13-14 juillet 1976.

(14) J. C. WANNER, Critères généraux de sécurité. Association Technique Maritime et Aéronautique, Session 1980 (30 pages).

(15) P. VIANSON PONTE, L'erreur humaine. *Le Monde*, 4 avril 1979.

(16) *La Technique*, no 25, février 1978.

(17) E. QUINOT, Méthodologie d'étude des accidents du travail. *Annales des Mines*, janvier-février 1977.

(18) A. ANDURAND, Le rapport de sûreté et son application dans l'industrie. *Annales des Mines*, juillet-août 1979, pp. 115-138.

(19) R. ANDURAND et F. NIEZBORALA, Analyse de fiabilité humaine (traduction de *Nuclear Safety*, vol 17, no 3, mai-juin 1976). Commissariat à l'Energie Atomique, Département de Sûreté Nucluéaire, décembre 1976 (29 pages).

(20) H. OTWAY and R. MISENTA, The determinants of operator preparedness for emergency situations in nuclear power plants. Technology Assessment Sector, Systems Analysis Division, Joint Research Centre, Commission of

the European Communities, Ispra, Italy. Invited for presentation at the Workshop on Procedural and Organizational Measures for Accident Management: Nuclear Reactors, International Institute for Applied Systems Analysis, Laxenburg, Austria, January 28-31, 1980 (34 pages).

(21) J. R. HINRICHS, Personnel Training. *Handbook of Industrial and Organizational Psychology*. Edited by M. D. Dunette, Chicago, Rand Mc Nally, 1976, pp. 829-860.

(22) H. LEWIS, La sûreté des réacteurs nucléaires. *Pour la Science*, mai 1980, pp. 73-89.

(23) T. NICHOLS and P. ARMSTRONG, *Safety or Profit. Industrial Accidents and the Conventional Wisdom*. Falling Wall Press, 1973 (13 pages).

(24) National Institute of Industrial Psychology. 2000 accidents: a shop floor study of their causes, 1971.

(25) P. KINNERSLY, *The Hazards of work. How to fight them*. Pluto Press, London, 1973.

(26) J. GRAYSON and Ch. GODDARD, Industrial Safety and the Trade Union Movement. *Studies for Trade Unionists*, vol 1, no 4, December 1975 (29 pages).

(27) P. MAYER, A propos des règlementations de sécurité. Rapport de recherche Centre de Recherche en Gestion, Ecole Polytechnique, avril 1979 (63 pages).

(28) P. MAYER, Techniques de gestion et analyse des institutions: l'exemple des règlements de sécurité. Centre de Recherche en Gestion, Ecole Polytechnique, février 1980 (30 pages).

(29) J. KEMENY, Saving American democracy: the lessons of Three Mile Island. *Technology Review*, June-July, 1980, pp. 65-75.

(30) R. LATARJET, *Malpasset, Amoco Cadiz, Three Mile Island ... Le poids des risques dans le choix des sources d'énergie*. R. G. E. Tome 89, no 7/8, juillet-août 1980, pp. 497-504.

(31) E. QUINOT, Méthodologie d'étude des accidents du travail. *Annales des Mines*, janvier-février 1977.

(32) L. VILLEMEY, Exposé au Colloque sur la sécurité dans l'industrie chimique, Mulhouse, 27-29 septembre 1978, pp. 235-241. Conseil européen des Fédérations de l'Industrie Chimique, Bruxelles.

(33) *Décideurs*, no 15, 16 février 1979, pp. 7-10.

(34) *Europe 1*, 22 mai 1980 (magnétothèque personnelle).

II. THE PUBLIC AUTHORITIES

The public authorities are, as we have seen responsible for making sure
that the standards set in matters of industrial safety are respected. The
difficulties due to the insufficiencies of available means, to the real
novelty of the problem of major risk, to the insufficiencies of certain laws
have been noted previously. To be added are certain difficulties which have
just been examined in the case of the operator and which certainly also
affect the public services: the general disposition of the mind which can
lead to misinterpretations of a situation, a too partial approach to
phenomena, negligence towards preceding events, insufficient follow-up on
systems changes that have been made etc. The problem linked to the human
factor cannot, of course, leave out the administrative safety services either.

These difficulties lead to deadlocks similar to those noticed at Flixborough
(apparently total absence of inspection services), at Seveso (insignificance
of measures taken) or in the case of the *Amoco Cadiz*.

We shall not resume the examination from this angle. We shall much rather
ask about two more fundamental difficulties which can limit the effectiveness
of public action in matters of safety. They have to do with the way in which
two fundamental functions of the state find expression in the field of major
risk control: the stimulation of economic development on the one hand, the
maintenance of social cohesion and public peace on the other. These two
requirements deserve to be reexamined in the light of the new stakes of major
risk.

1. PRODUCTIVE ACTIVITY AND SAFETY CONTROL: AN ORDER OF RANK
SELDOM DENIED

The fate of the most useful establishments, I shall go further: the very
existence of several arts, has depended so far on simple police regulations,
and some which were pushed away from supplies, labour or consumption by
prejudices, ignorance or jealousy continued to battle in disadvantaged conditions against countless obstacles put in the way of their development
(1, 1474).

These few lines from the report dated 26 Frimaire year XII were recalled in
the general debate of the law submitted to the senate on July 11, 1975 which
was to replace the law of 1917 on dangerous, unhealthy or inconvenient
establishments (1)

It is true, as we have indicated before, that this situation of the 1800s
is no longer with us as we search for a certain harmonisation between
development and environment. Nevertheless, as regards major risk, which is
often connected with key operations, one cannot be sure that this first concern for productive activity is not still effective. This is not necessarily
to be criticised as such but deserves careful examination as soon as risk
becomes extreme and can involve, even in the name of a blockage to economic
development, serious setbacks for a large community, or even for the country
as a whole.

Three points attract attention.

1st. A series of cases

Flixborough. The official enquiry commission did excellent work as we have

seen. However when it became a question of passing from the simple statement
of discovered ineptitudes to calling the industry concerned into question the
commission showed a strange timidity. One really must read this mixture of
precision in the examination of facts and the more than cautious reserve in
assigning responsibilities. One has some difficulty finding in these lines
the coherence which distinguishes the rest of the analysis:

We completely absolve everybody of any grievance according to which their
desire to restart production could have led them, in full knowledge of the
facts, to get a dangerous process going without paying attention to the
safety of those who did the work. We have, however, no doubt that it was
indeed this desire which led them to neglect the fact that it was potentially
dangerous to restart production without having examined the remaining reactors and having determined the cause of the fissure in the fifth reactor.
In the same way, we have no doubt that the error in appreciation as concerns
the problem of connecting reactor No. 4 to reactor No. 5 was largely due to
the same desire (2, para. 57).

At no time in the enquiry has there been any proof that the chemical
industry, and Nypro in particular, was not conscious of its responsibility in
matters of safety. On the contrary, there are indications which show that
conscientious and positive arrangements were continuously made in this
respect (2, para. 201).

Nypro was conscientious with regard to safety issues (2, para. 202).

We repeat that there is no proof whatsoever of any kind which would permit
(us) to say that Nypro had put the production target above that of safety.
(2, para. 206).

This attitude is taken up by several observers who object to the attitude
of the commission. Two main reproaches are formulated: no all-out effort was
made to shed light on the case, quite on the contrary, efforts were made to
limit the commission's field of investigation.

Brigadier R. L. Allen thus notes that the court could have appealed to the
witness by the Explosive Storage and Transport Committee (ESTC) which already
had detailed knowledge of the explosive and inflammable substances, of what
they involve in terms of immediate protection, of loca(lisa)tion etc.
(R. L. Allen, 3, p. 76). This was not requested for months. Still more
serious, Brigadier Allen writes (3, p. 78) that a committee official let him
know that instruction had been given to the members of the ESTC to abstain
from all comment on the Flixborough disaster. The ESTC had voiced its
interest and had offered its help to the court. These offers were not taken
up.

Two specialists who are close to the unions develop the second reproach:

The Government rejected a large-scale enquiry into the petrochemical industry after Flixborough because its whole economic strategy is based on the
development of North Sea Oil and on the inevitable supremacy of British
industry thanks to the petrochemical sector. Any criticism aimed at petrochemical factories which are dangerous in their existing form both for the
workers and for the neighbouring communities would call this strategy into
question. Immediately after the explosion at Nypro Sir Derek Ezra*declared:

---

*Chairman of the National Coal Board.

"after having expressed his deep regrets, that the disaster would seriously affect the balance of payments situation" (4, p. 24).

Canvey Island. For ten years the public authorities have refused any overall study of the risks of the area. Two additional refinery projects were discussed. The conservative MP for the constituency, Sir Bernard Braine (6, 7) tried to have the idea adopted that such a study was indispensable: until 1976 he found himself opposed with the idea of 'national interest' by successive governments. The 'national interest' seemed to forbid this type of analysis which could have led to changes, to a ban on activities, alternative locations. The MP finished by putting the dot on the "i":

... If by 'national interest' it is understood that Great Britain must have additional refinery capacity nobody in South Essex pleads against this proposition. It is manifestly in the interest of the oil companies to build their refineries and have them working close to the market on which they hope to sell their product. There is no objection to this from a commercial point of view. It is also true that the most lucrative market and the most expanding one for these products is the already overcrowded area of South Essex.

Should one authorise them to locate their refineries where it seems right to them without taking into account other factors or should one have a policy for the whole that would ensure that the companies are asked to go where they cause minimum damage to the community?

To argue that it is in the national interest to build these new oil refineries in the Thames estuary where one lives already in a dangerous environment is not only stupid and irresponsible: it makes complete nonsense of what successive governments have said for tens of years about the need for regional planning and the reversal of the insane attraction to South Essex (5, pp. 1467-1468, 6, p. 25).

In the same sense, by the way, if one follows one of the ministers whom the conservative MP had the honour of fighting in Parliament, he writes in his memoirs: We cannot permit ourselves to ruffle a foreign oil company (8).

Three Mile Island as seen from France. In a very clear fashion the French Prime Minister drew the line of conduct for the public authorities during his visit to the 'prestigious' installations at Tricastin on April 9, 1979:

"Our will for economic development determines the scale and the rhythm of our electronuclear programme; our safety imperative determines its modalities of execution" (9).

This expresses perfectly the policy of the great industrial countries in the face of major risk: the imperative of production comes first; sacrifices are then consented to in order to respond to the requirements of safety according to what is at stake but also according to budgetary possibilities are of the kind "as far as is reasonably possible" (10)*. We have already noted the vote in the Italian parliament imposing regulations to be respected with regard to safety in the framework of the 'requirements of production' (11)

---

*While previously laws were, in their working, much more strict, particularly for mines. It was, however, a matter of limited risk compared to today's.

Ermenonville and the American authorities 1974. This is an example of the
opposite kind. It would, of course, be necessary to look at all the background of the decision to ban all flights of all those big three-jet Douglas
aircraft but one can nevertheless note this case in which the authority chose
not to respect the, undeniable, requirements of economic activity* (12, pp.
306-307). The American authorities proved therefore quite a different
determination from the one which had prevailed in the affair of the luggage
doors of the DC-10 which had led, for lack of decisive measures, to the
disaster of Ermenonville.

The Senate Trade Commission has concluded on Thursday, June 27, that the
federal civil aviation administration (FAA) and the manufacturer McDonnel
Douglas Aircraft did not in time take the appropriate measures to avoid the
disaster of the DC-10 of Turkish Airlines which crashed last March 3 near
Ermenonville, causing three hundred and forty six deaths.

According to the commission the accident has apparently had the same
cause as an accident which occurred with another DC-10 near Detroit two years
ago: the malfunctioning of a luggage door. "We believe, the authors of the
report declared, carelessly or not, the manufacturer of the aircraft and the
American administration while seeking a solution of the problem wanted to
hide the bad news from the public in order not to draw attention to the reputation of the DC-10.

According to witnesses' statements presented to us, the members of the
commission added, we believe that McDonnel Douglas has not acted as the
circumstances required. After the Detroit accident the FAA has not taken
"the measures that were needed. Rather than presenting legal depositions
concerning the locks of the luggage compartment, the report continues, the
FAA and McDonnel Douglas agreed on the publication of a 'service bulletin'
which explained the problems and the measures taken to solve it".

"The Paris accident could have been avoided, the Senate commission concludes, if the service bulletin had insisted on the urgency of the problem
and if the aircraft had been completely modified" (AFP) (25).

We do not discuss the well-founded technique of these decisions here.
However, it is important to note that the public powers are sometimes able
not to follow the habitual lines of compromise. They are actually so
habitual that this type of decision caused a very big surprise: as if an
unwritten rule of benevolence had to lead the safety authorities not to go
beyond the stage of firm recommendation. As the example comes from the
economically strongest country, the one most attached to the rules of
liberalism, the surprise turns into stupefication, into secret worry to see
rules which were implicitly held to be sacred, undermined.

2nd. Key arrangements

To permit the development of economic activity the public authorities have
had to provide various types of arrangements without which this development
would have been blocked.

In matters of protection, i.e. minor contamination, the governments have
set, either in a joint fashion (in the case of ionising radiation) or
separately, levels below which the effects are held to be tolerable.

---

*In the same way the NRC stopped five nuclear centres in 1978 because of their
excessive vulnerability to earthquakes.

In matters of safety one has limited the requirements by defining the design criteria.

In matters of indemnification one has also limited the possibilities by providing ceilings.

Without this triple action major risk could not have been taken on which would have led to a blockage of technological development in its existing form. What should we think, for instance, could the arrangements of the French Civil Code have led to:

Article 1382: Any human act which causes damage to someone else obliges the one by whose fault it occurred to repair it.

Article 1383: Everybody is responsible for the damage he has caused not only by his action but also by his negligence or by his lack of caution.

Article 1384: One is responsible not only for the damage caused by one's own action but also for that which is caused by the action of persons for which one is responsible or of things which one is entrusted to guard.

Nuclear legislation has clarified the application of these arrangements (for instance by determining clearly who is the guardian of the 'nuclear object', a question which has already given cause for endless casuistry). The legislation also aimed at permitting the development of industry despite the rigours of these three articles. The experts from the Atomic Energy Commissariat note it without ambiguity and on several occasions in their recently published work of reflection and synthesis:

On the other hand one had to avoid shackling the development of nuclear industry with excessive burdens (13, p. 326).

For lack of appropriate legislation the liability incurred by the fact of a nuclear accident risked to be excessive to the point of shackling the desired development of atomic energy (13, 331).

For lack of appropriate legislation the government risked to hurt itself on popular opposition to the development in the use of nuclear energy, for legal and psychological reasons simultaneously (13, p. 332).

The first two of these conventions (Paris 1960 and Vienna 1963) ... contain the same exorbitant principles of liability as the common law ... in particular rising limited liability for as long as it remains bearable and insurable (13, p. 333).

The forfeiture period is fixed in the convention at ten years from the date of the nuclear accident taking into account that the insurers have made it known that it would not be possible for them to accept charges for indemnifiable consequences which came to light only after this period had passed* (13, p. 341).

---

*The Paris convention leaves, however, some latitude for national legislation to fix a period longer than ten years on condition that appropriate measures are provided to cover the search for liabilities during this supplementary period. Thus France has extended the period to fifteen years and the U.K. and Germany have held on to the thirty years of common law. (13, p. 341).

In the British case one also had to depart from 'Common Law'; the ruling issued by the House of Lords in the Ryland vs. Fletcher case (1865) said:

We think that the rule of the law is the following: the person who for his own ends brings to his land or there gathers and keeps no matter what that is likely to cause damage by escaping keeps it at his own risk and danger ... and is prima facie responsible (i.e. without need for other proof) for all damages which are the natural consequence of the fact that it has escaped (14, p. 160).

In the Aberfan case one clearly sees the limits of such principles; the Lord Justice spelled out:

Apart from the requirements of mankind the law does not recognise any general obligation to protect others against misfortune (15, 16, p. 433).

3rd. The problem of the independence of safety authorities

The insufficient independence of the authorities responsible for the safety of installations is often lamented. This is a difficult subject inasmuch as the balance between competence and independence, effectiveness and independence is hard to strike. It is, however, evident in certain cases that choices have been made which clearly express the primacy accorded to production over safety.

Generally, we may mention a certain number of examples.

a) The creation of the Health and Safety Executive in Great Britain. In Great Britain the Robens committee had advocated the setting up of an independent administration for risk management. When the HSE was created in 1974 it was, however, put under the Ministry of Employment. It was argued that this would facilitate its relations with parliament and the executive. In the Canvey Island affair one could see its limits: the HSE had apparently some difficulties making its views felt even if it finally obtained good results from British Gas (a nationalised establishment which in this affair appeared more influential than the organisation in charge of safety).

b) The creation of the NRC for the nuclear safety of the United States. Until 1974 nuclear safety was the responsibility of the AEC (Atomic Energy Commission), an organisation which gathered together the fundamental research functions, research and development and safety.

In October 1974 the Energy Reorganisation Act eliminated the AEC and replaced it by three orgnisations which divided the competences between them. The Nuclear Regulatory Commission (NRC) became responsible for regulation and the issue of authorisations. Control and promotion of nuclear energy therefore came under two different authorities.

Independence on the level of organisations is, however, only one necessary condition. President Carter's commission on the Harrisburg accident echoes this: "It has become clear to us that part of the old philosophy connected with production still influences the practices of regulation at the NRC. While certain compromises between safety requirements and the requirements of an industry are inevitable the facts suggest that the NRC has sometimes sided with the interests of industry rather than pursued its primary mission which is to ensure safety ... The old attitude of the AEC is also evident in the reluctance to apply the new safety standards in centres previously authorised (2, pp. 19-20.

c) Two French cases. In 1971 the service of the industrial environment being in charge of control of classified installations left the Ministry of Industry and went to Environment. The two functions of promoting productive activity and of controlling safety where thus separated on the level of organisation charts.

The second case, concerning the nuclear industry, is different. The organisation chart following, published in *Revue Generale Nucleaire*, would not in itself be sufficient. One does not put the Council of State in direct dependence of the Interior Minister.

As a complement we mention these few lines from the CFDT trade union of the atomic energy industry which adds hardly anything new to what has already been said and said again in the United States on the same type of question:

In these circumstances, can one consider the Minister of Industry independent in his judgements concerning the safety of nuclear installations when he himself, through the Directorate General for energy and raw materials, sets up the French electronuclear programme in agreement with the EDF which also depends on him?

Is it normal that EDF requests authorisations when the permanent group in charge of reactors, the control organism for the safety report, includes four experts among its fourteen members which are appointed on the recommendation of the EDF?

An even more flawed situation exists in the control of the safety of fast breeder reactors set up by the CEA and EDF and judged practically by people from the CEA and EDF about whose advice one will not be surprised.

All these entanglements within a hermetically closed circle risk generating a large number of compromises and it seems to us extremely damaging to the quality of work on the whole of nuclear safety problems which are not slight as we have seen ... (18, pp. 401-402).

J. L. Masson*, RPR member of parliament was worried about this type of question in May 1979. Concerned about the capacities and the independence allocated to the safety authorities he had tabled a 'Resolution Proposal' in the National Assembly:

... tending to create a control commission in charge of studying the conditions under which the administration services in charge of nuclear control and safety are organised and function.

In his paper on motives he writes:

The control of nuclear safety is at present entrusted to the Central Service for the safety of nuclear installations. This service does not have the means in terms of equipment and staff to assume its functions in a completely reliable manner. In fact, the agents in charge of control, in this case the inspectors of nuclear installations are few in number and seldom available; they have rarely the adequate training to carry out their duties and in addition the independence of the controllers from the controlled is not complete.

---

*Member of the Mining Corps and previously inspector for basic nuclear installations.

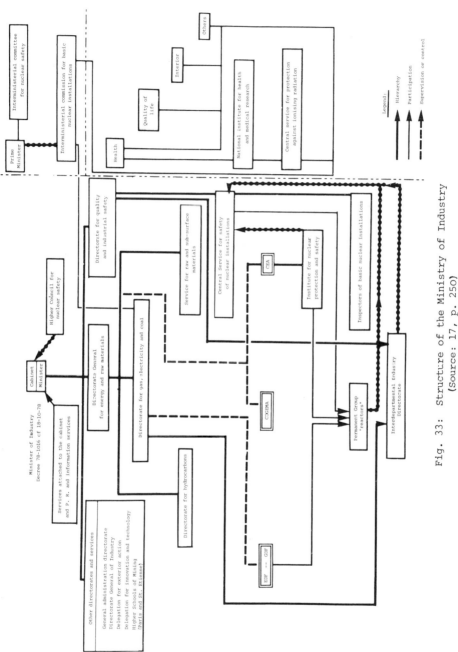

Fig. 33: Structure of the Ministry of Industry
(Source: 17, p. 250)

The training of inspectors of nuclear installations:

The inspectors are usually chosen from among the large technical bodies of the state (mining corps and bridge and roads corps mainly). However, and without calling the scientific knowledge of these officials into question, it seems on the one hand that in nearly 90 per cent of the cases they have at the time of their appointment no technical knowledge of what the problems of nuclear safety are and on the other hand the only experience of the matter they have are the memories of a visit to a specific nuclear centre during their student times. One can certainly proclaim the advantages of on-the-job training but in a field of such technical level it is legitimate to ask one-self whether this solution is realistic.

The number and availability of inspectors. The number of full-time inspec-tors is not large; the essential rest of them consists of part-time inspec-tors. The latter maintain their previous activity (without any connection with nuclear safety: road construction, mine control ...) The inspection of nuclear centres is for them supplementary work and recent statistics show that the part-time inspectors carry out between three and eight inspections per year. They devote therefore on average less than 10 per cent of their time to nuclear safety. Nobody doubts that at this rhythm it would hardly be realistic to speak of competence obtained from on-the-job training.

The independence of the inspectors. It follows from what has been said that the most competent agents of the Central Service for nuclear safety are pre-cisely those made available full-time by EDF and the CEA. As for the other inspectors, in order to mitigate their lack of technical knowledge they usually carry out their inspection with an EDF or a CEA engineer 'doubling' to guide their work, which is not entirely satisfactory either.

This situation is indisputably regrettable, and it is important that parliament assumes its responsibilities in this matter by informing itself about measures foreseen to remedy certain deficiencies (19, pp. 2-3).

The objective of the Proposal was judged 'absolutely pertinent' in that it responds to "considerations which are at present very preoccupying" (20, p. 2) The argument was put forward that one must not "excessively complicate the work of the National Assembly" by creating thus multiple commissions on particular subjects; one had to use as much as possible existing organisms. The law commission rejected the Proposal (20, p. 5). One is left with the procedures rejected in Great Britain and in the USA.

2. MAINTAINING PUBLIC PEACE

The public authorities feel the critical nature of the crises caused by large scale disasters. Faced with the drama while questions are asked about the technological options, the organisation of production (and its control) one sees a great upsurge of disquiet among the people in charge: fundamentals are called into question. The decisive fear seems to be related to the effects on production which public feelings may cause. Public feeling is not understood as a symptom of real failure but only as a symptom of passing psychological difficulty. When the feeling is exhausted by a month of communiques in the media economic life can resume its normal course; adjust-ments are made in calmness and serenity. This scheme is fine, up to the time when adjustment is no longer sufficient or the 'psychological' difficulty felt by the citizen (and also by quite a number of the experts) spells out something other than incompetence and irrational fear.

For the time being the mechanism described imposes itself with great
regularity. It is an essential element in the regulation of major risk. In
these circumstances every 'serious warning' is presented and finally treated
as a simple incident of the road which has the great advantage of improving,
in the end, the arrangements made in matters of safety. Is that enough?

A few examples may be recalled.

Seseso. This case, the most serious that ever occurred, shows clearly the
bouquet of mechanisms put to work in such circumstances. Let us remember:

The silence at the start of the affair

The still 'official' denials brought out in record time of information revealed by newspapers and experts ... It is a matter of reassuring others
and oneself towards and against everybody that dioxin is not dangerous.

The optimism in the various communiques, particularly those from the police
commissioner's office, when in fact the situation had completely passed
beyond control. One does not find the energy to act but enough to declare
on TV: "everything is under control", according to the classical expression
which does not fool anybody.

One reassures oneself about the toxicity of dioxin, about its hypothetically
teratogenic nature, about its geographical stability, about its natural decay.
All that connotes dangers but it is not proven, it is held to be doubtful,
even non-existant.

The refusal to follow up seriously on the effects of the poison, the
reluctance to listen to certain scientists who deviated further towards
reality than they observed, the need for reassurance was held to have
priority.

The fetishism of measuring and the recourse to extreme sophistication:
long, costly, hardly instructive, when simple observations permitted more
effective measures but also constrained reality more.

The escape reveals itself with surprising transparency, one is close to
the deadlock of the mechanism which must remain invisible to be effective,
with the affair of the statements by Roche's research director who launched
the alarm call. The regional health assessor threatened then "to report him
to his superiors" and asked himself not whether the ideas of the researcher
might be well-founded but whether the latter was speaking "in an official
capacity": this gentlemen will have to answer for his statements. One understands that the sovereign and absurd denials of the ideas of such a highly
placed person from Roche gave no hope for a less mistaken attitude when it
came to simple citizens, Italian researchers or parliamentarians who ventured
on to the road of lucidity.

Priolo. There was also much emotion in July/August 1976 about the strong
pollution caused by the local industrial complex in this suburb of Syracuse
(Sicily). Would one have to evacuate part of the suburbs of Syracuse? No.
The police commissioner soon closed the affair, affirming that the pollution
was within reasonable limits. Reports from the ministries of health and work
(attached to the Seveso report), however, make one think again: the values
measured are within the tolerance limits ... But the measuring apparatus,
the reports say, are not in a sound state. The reports suggested that one
should try and get better instruments (21, p. 426, 22, pp. 244-246).

The Amoco Cadiz. The seriousness of the situation can lead to the asking of certain essential questions: setting up a genuine navigation system*(not just improved apparatus), financing of this system etc. Here again the people in charge try mainly to reassure the public:

all measures have been taken**... All available means have been put to work ... We still have barriers in reserve***... The dispersants permit the dispersion of the oil, to make it disappear****... The coast roads are closed to traffic during the Easter weekend but these measures must not be interpreted as aiming at a cover-up of the situation*****...

Three Mile Island seen from France. The Prime Minister had the difficult task of bringing it home to the French people that they must not let themselves be overwhelmed by emotions; this was at the Press Club a few days after the accident:

Francois-Henri de Virieu: Prime Minister, for five days now an event has dominated the news by its concrete consequences, by the emotions it raises, by the fantasies it awakens in men all over the world. I want to speak, as you will have understood, about the accident at the nuclear centre of Harrisburg in the United States. We have now been talking for an hour and five minutes and I am actually astonished that we have not talked about it until now. Well, what lessons does the French government draw from this accident? When will it draw them? And does it foresee a change in its policy of building nuclear centres?

M. Barre: You are right in saying that this is a considerable event; it is more considerable for its psychological fall-out than for the technical reality which we can observe. About this technical reality I still cannot tell you anything. You know that studies are still in progress in the United States; we have sent missions to study the problem on the spot and we are awaiting the results of these missions to draw conclusions.

I can tell you at once that even if the centre in question at TMI is of the same type as the reactors which are built in France this centre has very different features******, and the scenario which unfolded in the United States could not happen in the same way in France. We have in fact safety systems which take the possibility of such accidents into account, and this shelters us from the consequences which could be considerable.

F. H. de Virieu: But, Prime Minister, the problem is not just technical, it is also psychological.

---

*According to P. Kerdiles, Chief Commissioner of the Navy in the colloquium of the UVLOE at Brest in 1979 (23, p. 76).

**The Prime Minister at Landeda, Saturday, March 18, 1978 Europe 1, personal tape recording).

***The Interior Minister on Europe 1.

****The Minister of the Environment on Europe 1.

*****Communique by the police commissioner for Finistere, Radio France.

******The steam generators in France have indeed different technical features which is what the authorities are emphasising; but this difference is not comforting as one so hotly felt one could affirm; the generators did not play the central part which was immediately attributed to them.

M. Barre: I was coming to that. You have asked me about the technical problem and I gave you a technical answer.

But I believe that the problem is psychological and there must be broad information for the people about the technical conditions under which the centres are built, about the safety measures taken, in such a way that the mysterious halo, which actually goes back to the Hiroshima bomb, cannot bear down on behaviour and mentality.

I can tell you that in the light of the conclusions which we shall have, all arrangements will be made so that in the centres which will be built the safety apparatus will be arranged accordingly.

You have asked me a third question: has the government decided to call its electronuclear programme into question? My answer is: No. Because even if France has to multiply the safety measures it cannot renounce electronuclear energy. But we shall carry the programme out with explanation, information and the development of safety measures.

F. H. de Virieu: You will continue building centres near heavily populated areas? Because it appears from the example in the United States that the big problem is not the seriousness of the accident, it is rather that people evacuate themselves very quickly and consequently cause panic problems.

M. Barre: M. de Virieu, I have read the dispatches which say that people went off; today I received dispatches saying that they are coming back. If you will allow me: we are drawing no premature conclusions. Let us see how this thing develops because its development has not come to an end. Don't think that we take this lightly because it is a matter of the safety of the French people.

Josette Alia: Exactly. I wanted to tell you that I think you are throwing out the consequences a bit quickly, which are not only psychological; the fall-out can be much more technical and much more serious than the merely psychological fall-out.

Well, since you say that the government is taking care of this, when will the ORSEC-RAD be published for example? Will the French people not only be informed but consulted which would in any case be useful before new centres are built? One could imagine a suspension, not for very long, of the nuclear programme in order to organise a consultation on a national scale which would mean taking into account the serious psychological fall-out of which you have spoken just now and those worries which I believe to be quite real and justified.

There is another question which is not negligible. If one ends up with the conclusion that one must in any case suspend or slow down the construction of nuclear centres, if the American example proves to be so dangerous that one comes to this conclusion, must not France from now on and in any case develop the new industries about which it talks a lot but to which it gives only very few means and very few credits? I have become aware of this during this week because I have made an enquiry into what French woodland would present as an energy potential. There is a deficit which approaches the oil deficit; it comes to 7 billion FF per year. The forest industry is not negligible. I believe there are forty seven reports in this respect at present at the Ministry of Industry. None has been followed up so far, nobody talks about them. When one enquires about them the public authorities begin to smile and then one is told: "Certainly, we are looking into this;

new energy sources are very interesting". Subsequently they tell you: "Now let's talk about serious matters".

M. Barre: There is much in what you are telling me.

Josette Alia: This is a very important problem.

M. Barre: In the first place, I have not evaded the technical problem. What I have said is that at the present time we do not know what the technical aspects are of what has happened in the United States and we have people on the spot to study them. Therefore I cannot tell you what technical conclusions we are going to draw from them.

Secondly, as concerns the psychological fall-out, I want to tell you that we have made a very big information effort in France. EDF has made a very great effort towards the population and the mayors of the local communities concerned, we have created the High Committee for Information on Nuclear Energy which is chaired by Mme Veil and which allows us to pursue carefully the studies which give us the possibility to know the safety problems better and at the same time to inform the French people better.

Josette Alia: But who does not publish these studies?

M. Barre: Reports will be published ...

Josette Alia: We have not seen any for a year.

M. Barre: ... when the High Committee decides to do so. It has been a year since it was created.

Josette Alia: There is no publication on this.

M. Barre: Perhaps a certain amount of time is needed in order to prepare a good publication. The High Committee for Information on Nuclear Energy does not write articles for weeklies.

Thirdly, you have talked to me about the forests and the forest industry. Madam, I know the problem well since in 1977 I launched a certain number of studies and work on the issue. I can tell you that your expectation will not be disappointed and that decisions will be taken in the matter. But we are far from nuclear energy when we talk about the forest industry. As for solar energy, you know that we are making efforts but that this is not for the near future.

Flora Lewis: Still concerning the nuclear question, there have been, yesterday and today, demonstrations in Germany which have gathered hundreds of thousands of people. With regard to the accident in the United States the problem in Germany is fundamentally more serious and more difficult to solve than the technique of centres. It is the problem of waste. This has never been talked about in France. What are you going to do with your waste? Where are you going to put it? In fact, the more atomic energy is produced, the more waste there will be.

M. Barre: Well, madam, until now we have solved the problem of waste without it causing dramas and we shall continue to do so.

Flora Lewis: Where do you put it?

M. Barre: It is put in various places. (24, pp. 52-61).

The Tanio. In case of deadlock in the discussion there remains the strong arm method. The Brittany members of parliament from all sides who had come to express their despondency in Paris in the wake of this latest black tide knew something about it and are not prepared to forget it. The public peace risks again being endangered on account of this brawny reception, and those in charge must reconsider their attitude. Repair has not completely wiped out bitterness.

REFERENCES

(1) *Sénat*, séance du 11 juin 1975 (p. 1474).

(2) Department of Employment: The Flixborough disaster. Report of the Court of Inquiry. London, HMSO, 1975.

(3) Brigadier R. ALLEN, Some Comments on the report of the Court of Inquiry into the Flixborough disaster. *The Acceptability of Risks*. Barry Rose, in association with the Council for Science and Society, 1977.

(4) J. GRAYSON and Ch. GODDARD, Industrial safety and the trade union movement. *Studies for Trade unionists*, vol. 4, September 1975, p. 24.

(5) Sir B. BRAINE, House of Commons, Official Report, Parliamentary Debates. Wednesday 24 July, 1974, Vol. 877, no 82, HMSO.

(6) P. LAGADEC, Le problème de la sûreté d'un grand complexe industriel: le cas de Canvey Island. Textes de base. Service de l'Environnement Industriel, Ministère de l'Environnement, Laboratoire d'Econométrie de l'Ecole Polytechnique, septembre 1980.

(7) A. MARSDEN, Planning, Health and Safety: a case study of the Canvey Island Petrochemical installations. Leicester Polytechnic, School of social and community studies, April 1979.

(8) R. H. S. GROSSMAN, *The Diaries of a Cabinet Minister*, vol. 1, 1964, 1966. Longmans, 1975 p. 366. Cité par A. MARSDEN, cf. (7).

(9) *Le Monde*, 11 avril 1979.

(10) Health and Safety at Work. London, H.M.S.O., 1974.

(11) Senato della Republica. VII Legislatura, Disegno di Legge, (Articles 20 et 24), approvato dalla Camera dei deputati nella seduta del 22 giugno 1978. (V Stampati nn. 1252, 971, 1105, 1145 e 1271) risultante dall'unificazione.

(12) B. MELLY, D-C 10. *L'Argus International*, no 13, juillet-août 1979 pp.

(13) M. PASCAL, *Droit nucléaire*. Commissariat à l'énergie Atomique, Eyrolles, octobre 1979 (462 pages).

(14) P. M. DUPUY, *La responsabilité internationale des Etats pour les dommages d'origine technologique et industrielle*. Pédone, Paris, 1976 (309 pages).

(15) Report of the Tribunal appointed to inquire into the disaster at Aberfan on 21 October 1966. H.M.S.O., 1967.

(16) Department of Trade and Industry, Safety and Health at work. Selected
     Evidence (Report of the Committee), vol. 2, London, H.M.S.O., 1972.

(17) M. BURTHERET et Mme De CORMIS, Le régime administratif des centrales
     nucléaires. *Revue Générale Nucléaire*, no 3, mai-juin 1980, pp. 249-258.

(18) Syndicat CFDT de l'Energie Atomique, Le dossier électronucléaire. *Seuil*,
     Paris, 1980.

(19) J. C. MASSON, Député, Proposition de résolution. Assemblée Nationale,
     no 1031, Enregistré à la Présidence de l'Assemblée Nationale le 2 mai
     1979, Annexe au procès verbal de la séance du 2 mai 1979 (4 pages).

(20) A. RICHARD, Rapport fait au nom de la Commission des Lois Constitution-
     nelles, de la Législation et de l'Administration Générale de la
     République sur la proposition de résolution (no 1031). Assemblée
     Nationale, no 1213, Annexe au procès verbal de la séance du 28 juin 1979
     (5 pages).

(21) Camera dei Deputati VII Legislatura. Commissione Parlamentare di
     inchiesa sulla fuga di sostanze tossi che avvenuta il 10 luglio 1976
     rello stabilimento I.C.M.E.S.A. e sui rischi potenziali per la salute
     et per l'ambiente derivanti da attivitá industriali (Legge 16 giugno
     1977, n. 357). Juillet 1978 (470 p.) (Notre traduction).

(22) P. LAGADEC, Développement, environnement et politique vis à vis du risque.
     Le cas de l'Italie. Laboratoire d'Econométrie de l'Ecole Polytechnique,
     Avril 1979 (280 pages).

(23) La pollution marine par les hydrocarbures. Union des Villes du Littoral
     Ouest-Européen (U.V.L.O.E.) Colloque de l'U.V.L.O.E., Brest, 28-29-30
     mars 1979 (227 pages).

(24) Le Club de la Presse. Europe 1 Dimanche, 1er avril 1979 (Texte diffusé
     par Europe 1).

(25) *Le Monde*, 29 juin 1974.

III. THE CITIZEN

Nowadays faced with major industrial risk the citizen as a wage-earner in an installation, living in the neighbourhood or simply as a member of the community of the nation, organised in a trade union or an association, member of a political party or independent of these orgnisations, finds himself involved. How does he see the situation? Most often as one who is left out.

1. IGNORANCE

This is the situation in which most citizens find themselves who are absorbed by too many tasks to be able to devote the efforts necessary to understand, to inform themselves, to train themselves, even if everything is done to encourage such a quest (which is extremely rare).

2. ACCEPTANCE AND IMPOTENCE

Passivity, like apathy, are only symptoms. It is not a question of stopping somewhere and being content with rapid condemnation. We shall therefore examine, in three points, what may contribute to create this balance between opposing forces which are called passivity or apathy.

1st. Passivity in general

Several phenomena come into play here: the idea that one cannot in any case change the situation in any way; the importance of mental dispositions or determining values such as work (case of Seveso where dioxin was accepted so that the workshops should not be closed); fear felt while those in charge show optimism on every occasion and lucidity would be too heavy a burden to carry for the citizen who would find himself alone in his quest for truth.

This leads the citizen to accept the situation or the event, not to be a driving force, or even to feel contempt for the problem.

This passivity is not without effect on the general regulation of major risk. Pressed on one side to give or refuse the necessary authorisations or to withdraw authorisations already given those in charge within government are hardly pressed by the other side to act with determination. The defence of jobs impedes in many cases the application of the necessary rigour and the stimulation of the imaginative effort of the operators (one can often do without a product, a storage, a transport as many examples have shown). Without strong pressure, concerning critical questions, of course, things stay as they are and the administration is lead to tolerate more (and more) in its function as arbiter. The extreme case was doubtless the one at Livorno in the affair of the red mud: the workers occupied the court building to put pressure on the judges who risked impeding the operation of the company (1, p. 56). In the case of Seveso a member of the work's council, if one follows C. Rise, put the population on the alert against the "campaign aiming at the closure of the factory" (2, p. 67).

Of course, many actions go in the opposite direction. But, generally speaking, one cannot say that the citizen (doubtless for lack of information, training, habit of reflection on collective choices and their inflection) is an actor who is present in these debates.

2nd. Passivity: alienation beforehand

Laura Conti emphasises this aspect of the problem in the case of Seveso when examining the behaviour of the future victims before the drama.

Seveso. Despite certain denials and without neglecting the combative attitude
of the unions in the wake of the escape of dioxin it is useless to hide from
oneself the important fact: the workers and the population knew before the
drama that Icmesa was a dangerous factory; one benefited from it:

I know the lies of the Right and I know the lies of the Left. They knew
very well the people who lived near the factory had rabbits and chickens.
When an animal died because of air pollution they took it to the factory; the
animal was deposited and one received 10,000 Lire. Who gave the 10,000 Lire?
Mr Waldvogel? No! It was a worker. And this worker knew therefore very well
that the atmosphere was poisoned. In addition, a magistrate came to see me:
"Madam, I am bound by professional secrecy, but I can tell you that there has
been contamination of the Seveso (the small river) which came from Icmesa
and which killed sheep that drank from it. There were complaints and I asked
the carabinieri to come with me and take water samples. We, myself and the
carabinieri, went to the river bank. We noticed the odour. One of the carabinieri took the tube and bent down, choked and collapsed. I bent down and I
too felt the danger. There were people from the village who saw the episode.
Then came the ambulance the people from the village had called in to help
the carabinieri". Well! They knew very well that it was polluted. Forty
sheep died. Everybody knew it.

Naturally the work's council said "No" because the image of the working
class had to be kept irreproachable.

It does not see the truth in all its obstinacy. What does that mean:
alienation? It means being one's master's accomplice. That is alienation.
If the working class had been at all times as sound as the work's council
said it was, it would not be alienated ... Marx said: the working class
fights by "embracing" its enemy. This is what alienation means. It is not
(just) being the victim of La Roche but being La Roche's victim while, in
fact, "embracing" La Roche.

The work's council is not marxist. It sees itself as the archangel with
the flaming sword. They got furious with me. I always told them. The truth
is sufficient ... It needs no imagination, no factastic dream ... And here
is the truth: La Roche is exploiting you. They exploit you to the point of
buying your complicity. Naturally, there are workers who are not accomplices;
they are ashamed of those who are ... which is a moralist way of' putting the
problem; one must understand the mechanism which determines this complicity
(3).

In France one finds cases in which the citizen did not know how to play a
particularly active part. We mention two examples.

Public enquiry into a chemical factory, Spring 1980. This was the case of a
public enquiry as prescribed by the authorisation procedure for classified
installations. The installation was destined for the manufacture of a dangerous chemical product. Within a radius of five kilometres about 10,000 people
in nineteen communities were involved. The enquiry went on for thirty days
during a favourable period. There was hardly any discussion in depth on the
project but instead on the level of the municipal councils:

- 8 favourable opinions requesting only that precautions be taken;
- 7 debates without giving an opinion, however, demanding all guarantees;
- 3 cases of no debates;
- 1 unfavourable opinion, taking into account the already existing pollution.

Public enquiry into an acrolein installation, June/July 1980.   In the habitual
terminology, this 'went very well'.   One recorded only:

- reservations from the CFDT and the CGT trade unions at the factory;
- rejection by a consumer association;
- a letter with remarks from one of the mayors involved.

For a large community, this is not much.   It is true that the publicity
(organised by the police commissioner's office) did not mention acrolein but
only an 'incineration plant'.

3rd. Passivity (as a psychological and political mechanism) "after the event"

In the case of a disaster this first attitude may even be strengthened as
we have seen in the example of Seveso: "You see it, this poison, do you?"
"You believe in dioxin?"   L. Conti develops this point in her analysis: if
the psychological need is intense the mechanism of denial can reveal itself
with great force, particularly if the political and social context lends itself to it; particularly also if the measures adopted by the public authori
ties are poorly thought out and managed.

A large part of the population did not criticise those who denied the
danger of dioxin but on the contrary had unlimited faith in them while the
same population criticised severely and sometimes even subtly those who insisted on the toxicity of dioxin but had not known how to take rational
measures.   This caused a very peculiar experience such as taking part in
popular meetings in which the directive to shut the air inlets of automobiles
while driving on the polluted road was discussed.   It was remarked quite
sensibly that to shut off the air intake ducts was a precaution if one did
not clean them when leaving the polluted area because in that case the reopening of the car windows would have carried the polluted dust at once inside the cars.   Well, instead of concluding that whoever had advocated the
shutting off was not a far-sighted or realistic person one concluded with a
real logical 'jump' that dioxin was inoffensive.

I have even seen people who had photos in their wallets that were taken in
the period from the end of July to early August of officials in charge of a
strange service.   They sprayed automobiles which came from the polluted
section of the main road with solvent.   The normal conclusion from this
practice would have to have been that if there was dioxin on the wheels of
cars the solvent would not only make it fall to the ground but by dissolving
it would make it penetrate to deeper layers.   Strangely, instead of concluding from it that these arrangements were faulty, they concluded that whoever
had made these arrangements "knew" that dioxin did not do any harm (49,
p. 249).

Laura Conti, however, goes beyond a simple statement:

This way of thinking (on the victims' part), is it really strange?   Doctors
for instance know that it isn't.   There are seriously ill people who seem to
infringe deliberately the prescriptions and consciously behave in a dangerous
manner (forgetting to take the medicines, neglecting diet rules etc.), simply
in order to measure, according to the energy and vigour with which the doctor
calls them to order, the real seriousness of the danger that is threatening
them.   The fact of a violation of the rules is a way of questioning with
actions rather than with words in order to get an answer to which one is
anxiously listening (4, p. 50).

After the doctor it is the politician in charge who gives his opinion: the

incongruous arrangements and the reasonable ones which, however, were not
adhered to caused confused messages and the population concluded from them
that the statements about the danger of dioxin were only made to "discharge
all responsibility" but one did not really believe in them. This is the behaviour of a nation that is badly ruled by tradition. It is used to see laws
promulgated which nobody can observe, to see terms and periods fixed of which
one knows in advance that they will not be respected, to see tax rates fixed
which would starve any citizen who declared his real income etc. The regional
government has shown that it was as far from popular reality as the central
government had been since time immemorial. It has shown that it knows neither
how to get nor how to transmit information (4, p. 50).

Regional deputy (and previously surgeon) Laura Conti emphasises also how
the authorities by their behaviour and their actions induced passivity among
the majority:

In order to reduce dangers, a series of extremely rigorous attitudes where
needed, especially in order to teach people that one had to think of dioxin
as a menace without ever reducing vigilance. One would have needed to behave
as one does in the operating room of a hospital: the surgeons know very well
that if one tolerated a doctor entering without sterilised coat and mask one
could expect a few months later that someone would approach the operating
table while nibbling a sandwich. The simple infraction is not dangerous in
itself if it does not lead to successive infractions.

People laughed when they saw that the soldiers who supervised the sealed-
off areas were not fully protected. From that moment on all precautionary
rules seemed bizarre and like the numerous bureaucratic dictates which are
formulated exclusively to "avoid (escape) responsibility".

That traffic was not banned on the main road which crosses the polluted
area removed all credibility from the arrangements which for months followed
each other. It is not the fact that some motorist could have breathed in
polluted air that was preoccupying but rather the spreading scepticism which
led many people to live for months without taking any precautions. Writing
circulars and making arrangements which recommend the strictest body hygiene,
taking a bath or a shower every day, does not serve much purpose if one does
not check that every house and flat really offers the possibility to have
(at least) a shower. (4, p. 48).

The same author quotes this opinion of Marisa Fumagalli:

Everybody is put on alert but one does not exactly know why; nobody has
been informed, nobody has received explicit information (5, p. 25).

3. DISCUSSION AND REJECTION

Despite frequent absences of the citizen (mostly interpreted as a sign of
social health by the previously mentioned agents) he still contributes his
part in certain cases. He sometimes does it violently (when it is a matter
of setting up nuclear installations in particular); the examples are well
known, the death of a demonstrator at Creys-Malville is a resounding memory
in this field. He sometimes does it with reports, books, films, petitions
etc. Above all he does it with his poor means: in terms of time, money,
expertise. This therefore often leads him into errors, into mistaking the
target. Everything is then set for the neutralisation of his intervention:
an 'expert' will quickly demonstrate to some citizens 'of good will' that a

good heart cannot mitigate a flagrant lack of technical competence and that therefore abstention would be the best proof of civil conscientiousness.

However, this image of the 'public'*is more and more in need of revision: certain associations (such as the GSIEN**have enough competence and access to information to make the authorities jealous. Discussion may then start even if equality of access to information and to expertise, which are a fundamental requirement for real involvement of the citizen, remain an unsatisfied demand.

We shall see later how this situation can develop into conflicts: the effective neutralisation of the citizen or the radicalisation of his actions through increased impotence. We shall also examine the developments noticed, mainly abroad, which tend to strengthen the citizen's position in the debates which the problem of risk raises. For the time being we stick to one example. It is the case of Canvey Island. It shows how citizens who were able to gather competence and to choose pertinent approaches to express their opposition managed to bear down on a situation and to create the conviction (together with others) that studies must be carried out on the subject that preoccupied them. From this resulted in the end the risk analysis about which we have already spoken and of which the authorities are proud with some good reason: 'a world first' which no doubt could never have seen the light of day if it had depended exclusively on the good will of the authorities or the industry (the latter being still more reticent if not opposed in certain cases).

The ferocious determination of the citizens in this case, no doubt the price of many difficulties, succeeded in dragging along the interest of the authorities and the acquiescence of industry (which sometimes is still hostile). This is a way of citizens' participation in the social regulation of risk which seems more attractive than plain passivity or impotent violence.

Canvey Island. The remarkable study by the Health and Safety Executive of the existing risks in the Canvey area did not emerge spontaneously from the offices of the experts. It was the fruit of a process in which the citizen played a preponderant part (6, 7). From among their actions (there were demonstrations, petitions, resolutions by local communities, compilation of technical files with the help of independent experts, particularly J. Fay, MIT Professor, D. J. Rasbash from the Department of Fire Safety Engineering at Edinburgh University) we choose as an example the intervention by their MP, Sir Bernard Braine, a Conservative. His passion in the House of Commons for a settlement of the problem of major risk which affected his constituency was constant. He started his battle in 1964: it was then a matter of opposing the arrival of an additional refinery in the area and to demand an overall study concerning the existing risk. The risk of having this refinery and another one installed was a burden for many years during which Sir Bernard Braine put as much relentlessness into the defence of his idea of a general risk analysis as the government put into its rejection. In 1976 at last Sir Bernard Braine won his case. He was helped in this by various factors: the

---

*This term is always being employed. As various specialists who took part in a meeting organised in Berlin by the EEC commission emphasised (April 1-3, 1979): the word 'public' must be banished; it does not have any precise meaning. Most often it is used to describe a group of people held to be 'incompetent'. The one who uses the term usually considers himself to belong to the opposite group: the group of 'experts' who 'know' and therefore can judge, evaluate, decide, justify, ...

**Scientific Group for Information on Nuclear Energy.

refineries asked themselves how well-founded, economically and politically, their choice was; certain experts, such as a well known university scholar who blocked an enquiry procedure in order to enforce a more far-reaching risk study, supported his idea; the population backed him up, confounding all political affiliations (and Sir Bernard kept his pressure up throughout the affair, no matter which political party was in power: political colour did not change the line pursued by London in this matter at all*). In 1976, then, the idea of an overall risk study of the area was adopted and the necessary means made available. According to the MP it was still necessary to keep up pressure so that the 'complexity' of the subject should not get lost in the report. The latter was at last published in June 1978. Sir Bernard continued his battle, opposing the political conclusions of the report (one did not know how to justify the closure of specific installations for reasons of risk, the HSE concluded).

Let us mention here some of the interventions of the MP in the House of Commons.

December 21, 1964: a fait accompli. Sir Bernard Braine: last June 28, my constituency and I for the first time heard about the project by AGIP Ltd., the subsidiary of ENI ... to build a £15 million refinery at Canvey Island. According to the press the project had been approved by the Board of Trade and an 'industrial development certificate' (IDC)**had been issued to the company. The refinery was expected to be completed by 1967, and it seemed to me and my fellow citizens that we were faced with a *'fait accompli'*** (11, p. 1011; 7, p. 1).

April 2, 1974: Overall analysis still rejected. The minister for local government: we do not foresee at this time an overall study such as the honourable member suggests but we are examining the situation (12, pp. 110-111); 7, p. 12).

April 3, 1974: Rejecting the argument of 'national interest'. Now we are getting the truth: national interest refers to the economic demands made by oil companies to locate their refineries in the proximity of the market (13, p. 1188; 7, p. 12). I warn the Minister: this is a situation which I shall never accept. It is a situation which my fellow citizens will not accept. I warn the Labour government as I have warned the Conservative government which I supported: I shall not tolerate the situation. It cannot count on my support for the proliferation of industrial installations of this type in an area which already has too many of them (13, pp. 1189-1190; 7, p. 12).

April 8, 1974: Rejecting the argument 'one more refinery is not so serious'. Sir Bernard Braine: I don't know how many times I have pleaded in the House for this simple proposal which the government should adopt and which has a bearing on the environment and on safety. All I ask for is an overall view ... We have had this ridiculous affirmation last week from the promoter of the Burmah-Total Refinery Trust Bill (2): because there are already other refineries and installations near Cliffe (in the Thames estuary), adding

---

*In each case there was actually benevolent understanding, even commitment before elections and firm rejection after electoral success (8, 9, 10).

**First administrative action leading to the process of authorisation.

***French in the text.

others does not make much difference. The whole problem is that it does make much difference (14, pp. 105-106; 7, p. 13).

April 18, 1974: Rejecting an ostrich policy. Sir Bernard Braine recalls what he said in the House in 1970 already: I have no desire to be an alarmist. I choose my words carefully. But the Aberfan disaster has fallen upon us ... precisely because nobody in charge ever thought that he had the responsibility of calculating the risks which were taken. I beg (the Minister) to make sure that his colleagues know well that before taking any firm decision which would increase the fire risk at Canvey Island they must remember that the safety of more than 25,000 living beings is at stake (15, p. 378). And Sir Bernard adds: All that has happened since is that the population has increased by about 5,000 and that we have seen two refineries establishing themselves (16, pp. 252-253; 7, p. 18).

June 26, 1974: The possibility of reexamining the course chosen. Sir Bernard Braine: Too often in major planning decisions does a bureaucratic law establish itself once the decisions are taken. But this must not cause any doubt: when one has events like the Flixborough disaster on one's mind one must have the possibility of thinking things over again. Planning permission has been granted: does this mean that inexorably we shall make the error more and more serious? This defies common sense; society must have the opportunity to think things over once more (17, p. 1642; 7, p. 21).

July 23-24, 1974: The discussion is started by Sir Bernard who wants to create a shock. Sir Bernard Braine: This is a sad story. It concerns the ministers of successive governments whose inability to understand the situation is only matched by their arrogance. It also concerns the Prime Minister who thought it right to exploit the anxiety of my fellow citizens by making a promise to them during the general election (campaign), the promise to reexamine the whole question, the promise which unfortunately he omitted to carry out. This is a story with a message for every community that considers its environment to be in danger (18, p. 1437).

... I am proud to have represented the people of South Essex for more than a quarter of a century. I feel I would be failing in one of my prime duties to these people if I did not pursue my proposal by using all means at my disposal. When a problem of this kind appears and a community has exhausted all other reasonable and constitutional means of protest it turns with good reason to its MP to pursue the battle in the House. I shall not permit this privilege to be denied (18, p. 1438; 7, p. 24).

My aim, starting from the beginning, is to show how, little by little, ignoring the small nuisances and the small risks has now created a situation of extreme seriousness (18, p. 1446; 7, p. 24).

A planning enquiry commission had been demanded for the area by the local authorities: the demand was not upheld (18, p. 1464). When in mid-June we were received by the Secretary of State we asked him to undertake an independent enquiry in order to examine the implications of the planned projects: We did not get a satisfactory answer. The feeling of bitterness, despair, and frustration is becoming all the stronger (18, p. 1466; 7, p. 25).

... We have had to wage a continuous war against these ministries. It has been difficult to obtain the information requested on which to base a judgement. This is the more serious since it is a matter concerning the safety of a large community (18, p. 1472, 7, p. 25).

Sir Bernard Braine is not listened to. In the end, the camp of those who
called the member for Canvey to order and common sense in the debate won the
day; one must not go into fundamentals in the third reading of a law (p. 1473).
In the examination of a law such as the one under debate which refers to railway lines (true: mainly on Canvey) one should not talk about refineries and
shipping conditions on the Thames (p. 1479). As for risks, why not concentrate them all in one place? (p. 1467).

May 25, 1975: In the wake of the enquiry report on Flixborough: a new charge
by Sir Bernard. Sir Bernard: Truly, the situation is worse than it was at
Flixborough. At Flixborough, there was one factory in a sparsely populated
area. At Canvey, there is a concentration of risks in a densely populated
area (19, p. 1832; 7, p. 27).

... Listen carefully to what the report says about risks of this kind (explosive gas clouds):

"While explosions of non-confined gas have occurred in other parts of the
world, there is a notorious lack of information on the conditions under
which a cloud of non-confined gas can explode or on the type of mechanism
that leads to such an explosion".

What an extraordinary admission! We have here a group of distinguished experts who are unable to give information on a risk to which thousands of
people in this country may be exposed at any time (19, p. 1836; 7, p. 27).

... There is a rescue plan for Canvey but I dare say it is totally unrealistic. An orderly evacuation is foreseen which even under ideal conditions
would take more than three hours in a flow of uninterrupted traffic, using the
same approach road as the one which the rescue services would have to use in
trying to get to the island (19, p. 1836; 7, p. 28).

April 3, 1978: Debate of the HSE report on Canvey. Sir Bernard congratulates
himself on the technical results of the report which show at what point one
was in the dark before the experts had done their work. But the MP keeps up
his pressure (20, pp. 938-958):

Even if the report had been limited to these statements and suggestions it
would already have been useful. Improbable as it may seem, it goes further
and concludes that the proposed development of the oil refineries, on condition of certain improvements in safety would not significantly increase the
risk level and can therefore be envisaged. This is a stupid conclusion. To
say at the same time to a community which has been subject to such terrifying
risks that its life expectancy may now be increased by fifty or even seventy
five per cent and then add, however, that two new risks may be introduced does
not only comfort nobody but defies all logic and in large measure renders the
investigation laughable.

Sir Bernard considers the analysis given on the risk presented by liquified
natural gas (LNG) especially inadequate:

It is extremely odd that a study which cost the taxpayer £400,000 should
end up with a document that lacks a vital annex. Yet this is the case. One
will search in vain for annex No. 10. I was obliged to table two questions
in parliament in order to establish that the missing annex was a text prepared by the British Gas Corporation on "the possibility and the consequences
of an explosion of non-confined LNG"; a subject which no doubt is not without
importance or pertinence in the context of the study. But the minister's
answer was that it's inclusion would have been "inopportune".

I quite understand the position taken up because the annex, which I have now seen, shows that the British Gas Corporation greatly underestimates the dangers of LNG if one refers in this context to the attitude taken by the United States for example. The experts from the British Gas Corporation think that LNG cannot explode in an unconfined state and that this risk is hardly worth thinking about. It would certainly have been embarassing to expose the nonchalant attitude of the Corporation to the public and I quite understand why the annex was omitted. All I can say to (my fellow) MPs is that the criteria of risks for life, as established by the Federal Power Commission in the United States in a refusal of a site for LNG are such that if they were applied in this country the methane terminal at Canvey would be closed down tomorrow morning. I therefore find the attitude of the Health and Safety Executive towards the activities of the British Gas Corporation at Canvey extremely disappointing.

Sir Bernard continues by bringing up a number of requirements for the effective application of the report before concluding:

Nothing less than seeing at last the approval of what I have brought up will satisfy me or reassure the inhabitants of my constituency. I demand common sense actions now. I have no desire to be the man who was proved right by a disaster to support his views. I demand from parliament to remember Aberfan where one hundred and sixteen children and eight adults perished in 1966. These deaths would not have occurred if those who knew that the slagheaps at the colliery of Merthyr Vale were dangerous had acted. The report on this tragic event shows clearly that this accident need not have taken place.

The tragedy of Aberfan and that of Flixborough were the result of a single neglected danger. At Canvey, the sources of danger are innumerable, each one aggravating the others. Therefore, what I still have to tell the minister is this: act now, act before it is too late, act firmly and with determination; don't wait any longer (20, pp. 938-958, 7, p. 42).

French information in the electronuclear field: intervention by the CFDT. In our country one may mention, among other examples, the determination of the CFDT trade union for the atomic energy industry to take charge of the problem of citizens' information; successive publications, considered to be of high quality, have been made available to the public. The justification of this work, of this non-renunciation, is given in the introduction to the work which was published in 1980:

One may ask oneself the question whether this type of information agrees with the role and the responsibility of a trade union organisation. Normally that is: in a democracy, the organism in charge of research, development, production and operation of energy resources should also be of service to the community by informing it in order to allow it to make its choices; these choices are a political responsibility and do not only concern technicians. In France, the people in charge of the responsible organisms such as the atomic energy commissariat and EDF have chosen to sell nuclear energy, and the information they offer is often nothing but disguised advertising. As for the government, the documents distributed by the Ministry of Industry and Research or by the delegation for information show rather the concern for confirming their choices than a desire to provide genuine information.

All those who work in the field of nuclear energy and do not consider it right to keep silent in the name of 'reasons of state' have therefore a responsibility towards society: it is this responsibility which we have decided to assume collectively in the framework of our trade union and within the limits of our means (21, p. 14).

REFERENCES

(1) Ch. HUGLO et R. CENNI, *Une société de pollution.* J. C. Simoen, Paris, 1977.

(2) Cl. RISE, V. BETTINI et C. CERDERNA, Derrière l'I.C.M.E.S.A. *Survivre à Seveso?* Presses Universitaires de Grenoble/Maspero, 1976, pp. 61-68.

(3) Entretien avec Laura Conti, Secrétaire de la Commission Santé-Ecologie du Conseil Régional de Lombardie, Député de la Région de Lombardie.

(4) L. CONTI, Trop d'écheances manquées. *Survivre à Seveso?* Presses Universitaires de Grenoble/Maspéro, 1976, pp. 45-58.

(5) L. CONTI, *Visto da Seveso.* Feltrinelli, Milano, 1977

(6) P. LAGADEC, *Le problème de la sûreté d'un grand complexe industriel: le cas de Canvey Island (G.B.).* Service de l'Environnement industriel du Ministère de l'Environnement, Laboratoire d'Econométrie de l'Ecole Polytechnique, 1978 (98 pages).

(7) P. LAGADEC, *Le problème de la sûreté d'un grand complexe industriel: le cas de Canvey Island (G.B.),* tome 2: textes de base. Service de l'Environnement industriel du Ministère de l'Environnement, Laboratoire d'Econométrie de l'Ecole Polytechnique, 1978 (98 pages).

(8) Lettre du Premier Ministre travailliste, James Callaghan à l'Association locale de défense, Citée par Sir Bernard BRAINE. House of Commons, Official Report, vol. 871 no 20, Tuesday 2nd April 1974, col. 1164-1196. Burmah-Total Refineries Trust Bill.

(9) Lettre du Cabinet de Mrs Thatcher, le leader de l'opposition à un habitant de Canvey, 19 avril 1979.

(10) Lettre de Mrs Thatcher, Premier Ministre, au secrétaire de l'Association locale de défense, 12 juillet 1979.

(11) House of Commons, Parliamentary Debates, Official Report (Hansard). London, Her Majesty's Stationery Office (H.M.S.O.), 21 December 1964, Col. 1011-1022.

(12) House of Commons, Parliamentary Debates, Official Report (Hansard). H.M.S.O., vol. 870, 20 March 1974.

(13) House of Commons, Parliamentary Debates, Official Report (Hansard). London, H.M.S.O., vol. 871 no 20, April 1974, col. 1165-1196.

(14) House of Commons, Parliamentary Debates, Official Report (Hansard). H.M.S.O., vol. 872, no 24, 8 April 1974, col. 95-114.

(15) House of Commons, Parliamentary Debates, Official Report (Hansard). London, H.M.S.O., vol. 807, 24 November 1970.

(16) House of Commons, Parliamentary Debates, Official Report (Hansard). H.M.S.O., vol. 875, no 56, col. 241-348.

(17) House of Commons, Parliamentary Debates, Official Report (Hansard). London, H.M.S.O., vol. 877, 24 July 1974, no 82.

(18) House of Commons, Parliamentary Debates, Official Report (Hansard).
     London, H.M.S.O., vol. 877, no 82, 24 July 1974.

(19) House of Commons, Parliamentary Debates, Official Report (Hansard).
     London, H.M.S.O., vol. 892, no 132, 23 May 1975, col. 1831-1846.

(20) House of Commons, Parliamentary Debates, Official Report (Hansard).
     London, H.M.S.O., vol. 955, no 168, 3 August 1978, col. 938-958.

(21) Syndicat CFDT de l'énergie atomique, Le dossier électronucléaire.
     Le Seuil, Paris, 1980 (531 pages).

CONCLUSION: MAJOR RISK OUTSIDE THE ACTORS' FRAME
OF REFERENCE

Despite the limits to means and tools available for managing major risk or
more precisely: for mitigating insufficiencies one might see the social
capacity for the regulation of this problem strengthened. It is no aberration
to think that operators, government and citizens contribute actively to give
full effectiveness to the cadres appointed by legislation and to the possibilities offered by the tools.

We are still wide off the mark. For each of these great agents in the
situation there exist other demands which are hardly conducive to taking up
the problem of major risk resolutely.

Industry will install three valves instead of two in a system but it is much
more rare that it will examine what can so conveniently be left outside the
frame of reference; in case of a serious fault one can always invoke fate
and the impossibility of finding alternatives. This was the case in Seveso:
Industry is by its nature a risk enterprise. The parliamentary enquiry group
replied:

> The Commission does not ignore the fact that it is practically impossible
> in the chemical industry to work at zero risk but it is (also) very
> conscious of the fact that this does not diminish the responsibility of
> industry but on the contrary increases it.*

One may get to the final argument that industry can only be held responsible
for 'normal' accidents. Would all others be force majeure? The Givaudan
company seemed willing to adopt this theory for its defence in the hexachlorophene (talc Morhange) court case. The court at Pontoise did not follow
it as its ruling indicates:

> The defence wants to establish in this respect that 'the dangerousness of
> a product can only be appreciated within the framework of normal use'.

With this suggestion it concludes "that in the case of abnormal conditions
of use it is not the product which is at fault but the user". It concludes
(further) that at the rate of 0.50 per cent recommended by the manufacturer
hexachlorophene could not be considered dangerous.

This is a case of petitio principii (begging the question).

In fact, it is always the product that is at fault even if the conditions
of use, the importance of which cannot be denied, are of a nature as to
displace, attenuate or make the responsibilities incurred disappear.

A product is dangerous because it contains in itself the very properties
which bring about damaging consequences and on account of this requires precautions in its use.**

The public authorities have always had their mission: to protect productive
activity, to guarantee public peace. The citizen must work to live. The
scheme has no drawbacks ... as long as major risk does not appear. But when

---

*Parliamentary enquiry commission, cf. p.   (21, p. 62).

**Ruling by the court of Pontoise, cf. p.   (2, p. 63).

it appears one finds, as in Seveso for a while, an industry in a critical position, public authorities incapable of guaranteeing the economic and social order, citizens deprived of their work.

Because, surely, it is not a question of choosing between prosperity with risk and the candle and cave (backwardness) without risk. With the problem of major risk it is now rather a question of avoiding the scenario that combines all disadvantage.

We have seen how every one of the main actors in economic and social life can take his part in what an active regulation of major risk requires. We must carry this examination still somewhat further because, naturally, none acts independently of the others. It is not actors in isolation who intervene but networks with complex dynamics. How do they affect the general regulation of major risk?

CHAPTER SIX

Social situations to be understood and mastered

I   Very complex networks of agents
II Complex dynamics

I. VERY COMPLEX NETWORKS OF AGENTS

The preceding chapter might give the impression that the deficiencies in the social regulation of major risk are caused mainly by negligence, even ill will or at the limit by deliberate decisions taken by agents who are only after their own interests. This would be quite insufficient.

In order to understand the difficulty of the question asked one must get the measure of the extreme complexity of the systems involved. Systems which, as J. Forrester has emphasised, can react brutally to actions of apparently mild criticism while at the time displaying exasperating inertia when faced with desirable corrective measures (1).

In short, the search for the 'one' responsible party becomes largely insufficient or must at least proceed from an approach that is more subtle and in depth than yesteryear's tribunal. For the guilty party one must substitute participants who are more or less involved in situations which are brought about by negligence and deadlock in matters of risk control.

1. THE WORK TOOL INTRODUCED INTO NETWORKS

The outstanding example is maritime transport. In the case of the *Torrey Canyon* the vessel flew the Liberian flag. The shipowners, Barracuda Tanker Corporation, are domiciled in Bermuda, a subsidiary of a big American company. It had been chartered by a Californian company, Union Oil & Co. It carried oil for the account of BP. Its crew was largely Italian. The shipwreck occurred outside British territorial waters (2). This is the classical puzzle repeated eleven years later with the *Amoco Cadiz* under Liberian flag, with an American shipowner, a British charterer, a crew of various nationalities, Italian skipper etc. However these are still simple puzzles. The case of the *Tanio* leads into quite a different world.

The Tanio. Under the headline "But to whom does the tanker belong" Francois Grosrichard made a first attempt at untangling the story: The *Tanio* had been built at Schiedam in the Netherlands in 1958 for a group of companies called Worms, the French oil transport company.

The *Tanio* which in its youth was called *Lorraine* was 66 per cent owned by
the French Maritime Transport Company (another company of the Worms group)
and 33 per cent owned by Pechelbronn at the time of its assignment, five and
a half years ago, to the Malagash company Petromad which was created in 1966.
As Petromad and the Malagash banks had no substantial financial resources
(and there was probably fear of nationalisation by the Malagash authorities)
it was the Panamanian finance company Cruz del Sol which was created for the
occasion and in which Worms had shares which bought the *Tanio* in July 1974.

In Petromad, 60 per cent of the capital was Malagash and 40 per cent belonged to Worms and to Petromer, a company domiciled at Bordeaux*. Cruz del
Sol had leased the *Tanio* to the latter. The Malagash national flag had been
obtained thanks to the shell-only method of chartering, i.e. on the spot.

From 1975/76 onwards the increasing difficulties of the Madagascar nationalised oil industry which had lost the Reunion market progressively led its
owners to withdraw the *Tanio* from its usual coastal service in the Indian
Ocean and to assign it to the Mediterranean for the account of Elf.

Thus the vessel underwent a double recasting, financially and technically.
The Panamanian company Cruz del Sol mentioned above sold the shell of the
Tanio to Locafrance International Leasing (a company under Swiss law domiciled at Lausanne) on July 23, 1979. Locafrance chartered the *Tanio* for
three years to another Panamanian company called Gardelia. The latter sub-
chartered the ship's shell only to Petromad, the Malagash company which in
turn, after having equipped the ship with sailors chartered it again to
Gardelia. (3)

2. INTEGRATED PRODUCTION IN DEPENDENT ENTITIES

This specialty in modern business quite singularly reduces the margins of
freedom. One can see this very clearly in the affair of the red mud of
Montedison. In Italy, one could not easily decide the closure of the
Tuscany mines the survival of which depended on the titanium dioxide factory.
In France, one could not easily accept the problem because of the existence
of similar factories on national territory; so, the Italians were implored:
Just do it in your territorial waters!**

In the case of the DC-10 and its grounding after the Chicago accident one
clearly sees the difficulty of taking and upholding brutal decisions. Most
of the big airlines (UTA, Swissair, Alitalia, Lufthansa ...) found themselves
in a very difficult situation. On account of this tough action is only
possible in the presence of extreme, evident and imminent danger***.

One may also still recall a case which shows how a production intrudes
into the economic, cultural and local or national political life. What, for
instance, does one think of the acrolein factory at Pierre-Benite in the
Lyons suburbs? It is part of the Lyons chemical industry (for which the
Feyzin refinery was installed); it provided jobs with regard to which the

---

*Little by little the French shareholders had reduced their holdings and
  Petromer's had disappeared by 1978.

**According to a pronouncement by the French environment minister on the pro-
  gramme "Files of the screen (TV files) of February 26, 1974 (4, pp. 60-61).

***Which is not always the case with pre-accidental situations.

municipality and the trade unions are particularly vigilant. It supported
a whole downstream sector: animal feeds (since the main aim of acrolein
production is the manufacture of méthionine which serves as a component of
animal feeds). The downstream sector developed on the basis of this supply:
less need for cereals, purchase of American soya, very intensive type of
stock breeding. Still downstream, a particular type of nutrition developed
in the country because of these possibilities in matters of stock breeding.

All this would be involved in an action from the top of the edifice. One
can understand that in many cases determination appears necessary only after
the accident, especially if as is often the case in matters of major risk the
elements of decision are not neatly divided into black and white.

3. A MULTIPLICITY OF AGENTS AROUND THE SAME PROBLEM

A great many examples could illustrate this point. We mention one which is
absolutely clear. It is the railway disaster at Hixon in Great Britain which
was already mentioned. It shows, in fact, very well how a whole lot of
agents may remain impotent. The safety of level crossings is not one of the
most complex problems but still ...

Hixon. Let us recall that the automatic barriers at Hixon which were at the
time just newly introduced in Great Britain had been well designed from the
safety point of view: half-barriers to permit a possible escape in case of
a very late crossing (actually: an infringement), short-interval alarms to
discourage motorists from crossing at all cost, signal panels etc. The
problem of exceptional convoys had been foreseen: a telephone had been put by
the barrier indicating that one must ring in case of an accident or cattle
crossing or heavy vehicles. This telephone was the only means of stopping
the train: there was no return signal road-to-rail that would indicate to
the train driver that the passage was not free. This was the price to be
paid for the new automatic system which saved the motorist a lot of time and
thus dealt much better with traffic.

The telephone was not used when the road convoy with a heavy transformer
arrived at the Hixon level crossing. One could close the case with this
statement and put the whole responsibility on the road haulier. However, one
must go further. The culpable omission by the road haulier, and the police
patrol which accompanied the exceptional transport, is only the beginning of
a process which in many points shows the signs of insufficient attention to
the eventuality which manifested itself tragically on January 6, 1968 at
Hixon. Let us examine the system which produced the disaster in some more
detail (5, 6).

a) The Ministry of Transport

The authorisation given to the haulier. In usual fashion an authorisation
was requested from the service of bridges for the exceptional transport of
the transformer. The complete routing was spelled out: the difficulties
were pointed out. The Hixon level crossing was not mentioned; the carrier
was not warned that he would have to use the telephone.

The publicity. Second responsibility of the Ministry of Transport: the
haulier would have thought of the telephone himself if it had figured large
in publicity. The haulier was no amateur: it was a big company with one
hundred and seventy vehicles and specialised in transport of exceptional
loads; the question of the telephone had not become a routine issue for it.

The highway code. Of course, the highway code mentions level crossings.
However it did so mainly to emphasise the dangers of the zig-zag between the
barriers and the duty to respect the warning lights; it did not mention the
telephone.

In addition, the use of the term 'automatic' risked to induce inappropriate
attitudes; it could create the impression that the train was 'automatically'
warned of danger; perhaps it led the truck driver to feel no longer responsible since the word 'automatic' means, for instance in the expression
'automatic gear change', that the driver can take his attention away from a
particular mechanism. Here, by contrast, the word 'automatic' transferred
all attention on to the truck driver.

The official documents of the Ministry of Transport. A document from the
Ministry (July 1966), distributed by the Royal Society for the prevention of
accidents to the police in particular, went into details on the crossing of
railway tracks. About the telephone it said only that it "could be necessary"
without going any further which would have permitted bringing home the reason
for this necessity. In addition, an "explantory note" (July 1966) emphasised
that the document did not have the nature of a regulation but only that of
an advice; the note did not mention the absence of a road-to-rail signal nor
the telephone.

The inspections of the Hixon level crossing. An inspection took place in
January 1966. The road was narrow. One might not expect exceptional convoys.
There was no telephone installed.

In June 1967 there was a further inspection; the road had been widened and
a telephone installed. However at the time of the control the eventuality
of the passage of an exceptional convoy was not discussed; nor was the
purpose of the telephone or of the panel describing its use in certain
circumstances.

b) British Railways

Publicity. With the telephone system British Railways had certainly installed
a return loop from road to rail. Not everything had been done to ensure that
this system would be effectively used. For the Hixon level crossing 1,000
brochures and fifty posters were distributed to the local authority. These
documents, like the highway code, were not sufficiently explicit. Here again
one could misunderstand the term 'automatic'.

What would have been needed was a message of this type:

ONLY a telephone call can stop a train. When the lights start flashing you
have only eight seconds to pass under the barrier and sixteen seconds to clear
the level crossing. If you cannot cross within twenty four seconds, don't
try! TELEPHONE FIRST.

What is more serious, the brochures distributed after 1966 don't mention
the question of a telephone. The road transport companies were not directly
informed of the new responsibility incumbent upon them.

Telephone indicator panel. The panels showed the following text:

    - IN CASE OF EMERGENCY (BOX)
    - or before passing with heavy or exceptional vehicles
    - or with cattle
    - TELEPHONE THE SIGNAL STATION

This panel drew attention to the case of an emergency while there could be danger in other situations such as we have seen at Hixon. British Railways made no studies to find out whether the panels effectively attracted attention to the telephone.

An alert ignored: the incident at Leominster (November 8, 1966). The incident and the response it raised from British Railways have previously been noted as an example of the inability to register precursory events. Fate had it that the same transport company was involved in the Hixon accident.

c) The transport company involved. The company made the point that it had no knowledge of the telephone system. However, it was the same company as was involved in the Leominster incident, and the details of its letter to British Railways show that it knew the impossibility of stopping a train automatically. The company did not bring this to the general attention of its truck drivers.

d) The owner of the transformer. The company had received no special advice concerning the Hixon level crossing in the various authorisations received from the Ministry of Transport. This time there could have been a chance of localising the danger because the carrier was new to it; the carrier, in fact, asked the owner for specific instructions to be followed. However the exact risk of Hixon was not mentioned to them.

e) The truck drivers. They took no notice of the panel. The driver was concerned that the permitted height was only 16 feet 6 inches while the transformer stood 16 feet 9 inches high, (the permitted maximum heights are often reduced for safety reasons). The driver should have handled the system so as to lower the height by six inches. He took the risk of getting stuck on the level crossing. Like all other members of the convoy, even the one in charge of safety who let the driver act as he liked, the driver being above him in rank, he relied in the end on the policemen for questions of safety.

f) The police. The patrol which escorted the convoy, like those who had drawn up the instruction, paid attention mainly to the clearance available underneath the electric wires. The policemen paid no attention to the telephone. They were surprised that nothing stopped the train.

Generally speaking, the police authorities had instructions which did not alert them sufficiently to the problem caused by exceptional convoys on automatically controlled level crossings.

II. COMPLEX DYNAMICS

1. CONFLICTING INTERESTS

It is a banal observation which only serves to emphasise that ordinary conflicts between interest groups all play a part in safety issues. For simplicity's sake we have previously distinguished three sub-systems of actors: operators, public authorities and citizens. Each one of these subsystems has internal conflicts and the relations between components of subsystems are often of a conflicting nature. Each cell is guided by the satisfaction of objectives and interests of its own. It is in the thick of these conflicts that safety issues are raised, discussed, solved or ignored.

In the case of Canvey Island for instance the citizens demanded another siting for the planned oil refineries, the public authorities wanted them set up in the areas foreseen for this purpose in the planning policy, the companies had their commercial interests and succeeded in bending the administration to their view. One also finds internal conflicts in this affair: conflict between the safety authority and the Department of Energy; conflicts between the ministers and their inspectors who did not follow the recommendations; conflicts between the companies involved etc.

In each case the play of forces can be shown. Let us take up just two points: The conflicts may be settled or not. If not, they can degenerate into outright opposition.

No group is ever a compact block; in the Seveso affair for instance, if one considers the Hoffmann-La Roche group, one must distinguish between the parent company (Roche), the subsidiary (Givaudan) and the sub-subsidiary (Icmesa) each of which has its own and to some extent diverging interests. Even inside the parent company one must distinguish between top management, the research departments, the remainder of the company etc.

2. CONTRADICTORY REFERENCES FOR ACTION

The conflicts are relatively simple when they concern the same objective and refer only to modalities to be adopted: division of profits, power, stability of the organisation or the company. They are usually more complex because they occur in non-homogeneous groupings of preferences. Two illustrations will make the point clear.

In the first case a certain scientific idea about the freedom of research is opposed to the wishes of the community. In the second one it is a matter of diverging attitudes to the utilisation of the ecosystem.

Harvard and genetic engineering in 1976. In 1976, Harvard University (USA) wanted to pursue its researches in genetic engineering: it based itself on the freedom of scientific research. In June 1976 the *Washington Star* asked on its front page: Is Harvard a good location for Frankenstein operators? In mid July 1976 the city council of Cambridge imposed a three month moratorium on these researches; it based itself on the safety of its citizens. The scientific commentators decried the measure: the incompetent man-in-the-street must not be allowed to intervene in science policy (7).

Amoco Cadiz. In 1978 after the wreck of the *Amoco Cadiz* two objectives were confronting each other: reestablishing the cleanliness of the beaches, the hotel owners' point of reference, and safeguarding the fundamental qualities of the environment by using only very small amounts of dispersant: this was the point of reference of the sea fisherman and the biologists.

The period of time chosen for reference in an action is thus very often a bone of contention. For some agents the most important thing is the next commercial season (such as the tourist season in the case mentioned above); for others it may be a period of several years while still other agents may think of a period of several decades. In each case the motivations and actions are different.

In the same way, depending on whether one refers to the imperatives of siting or the demands of fairness one sees further opposition appear. One sees national development and regional autonomy hurt, national development and the equality of citizens in the face of public burdens etc.

At Canvey Island where so many risks are concentrated the local MP asks: Why have we been given such treatment? It is the same question that the riparians of the Loire river raised when they were upset by the siting of nuclear centres (13). For liquified gas terminals the challenge seems to be still more acute: there are only between three and five sites envisaged in France.

Added to these questions are social and ethical ones when one approaches the development of biology.

The difficulty is that there is hardly any simple solution possible when the choices of reference are so heterogeneous. From here on one understands why magical entreaties calling upon 'the common interest' and 'the national interest' are so ineffective.

3. MULTISHAPED CONDITIONS

The social universe of a technological project or an industrial establishment is comparable to a kaleidoscope: very many factors enter into it, constructions that are surprising in their outcome, likely to be overturned under the influence of rather weak impulses. Two examples may again illustrate this additional difficulty.

Seveso. The Italo-Swiss group opposed to the Italians. The Democracia Christiana had to stand up against the communist party. The central government controlled the region. The regional executive tried to limit the prerogatives of the regional council. The church tried to control the decisions etc. But one must not forget: the historic compromise between the PCI and the DC in Rome; the desire of the Lombardy communists not to foil an attempt by the DC (whose liberal and open-minded figure was V. Rivolta, the regional health minister) and avoid another attempt which the PCI did not like; the impotence of the Italian authorities which permitted Roche to appear as a safety factor in the disaster (the group undertook decontamination, promised integral indemnification, showed its determination in action).

The aborted project of the oil complex anchored off Brest, 1967-1974. This was a question of setting up a refinery in the sea area of Brest, an installation to receive large ships and to have equivalent storages.*

---

*Six oil companies had been 'invited' to realise the project.

The risk was on the one hand the possibility of black tides inside the area or the slow pollution of this environment which was extremely rich in and used for breeding shellfish.

On the side of the protagonists one found the municipality of Brest supported vigorously by the Departmental Directorate for Equipment and in particular by the local Director (the latter had actually chosen a different political party from that of his minister, Olivier Guichard, the mayor of La Baule, who on account of this latter function was explicitly suspected by the Brest parliamentarians to prefer the development of his region to the detriment of Finistere). Equally protagonist of the project was the local UDR MP, vice president of the national defence commission in parliament (the Navy was actually very reserved about the project). Finally, there was the chamber of commerce and industry supporting the project. The communist party and the CGT came out in support at least for a while.

On the opposite side the network was more complex: the cooperatives of the sea fishermen and the oyster breeders and the CDFT. Allied with them were, not clearly recognisable but nevertheless very useful, five multinational oil companies (the sixth one favoured the project). The local farmers found themselves in the fray as well without close links to the seafarers which was not surprising. There was also a long standing adversary of both: Edouard Leclerc whose ardour was disquieting. He had nothing to do with Gabriel de Poulpiquet, the UDR MP for Finistere who opposed the project more and more and by the same token with the Brest deputy of the same party. The regional party, the Union Democratique Bretonne, represented a further dimension in the opposition to the project. The local press soon made no secret of its hostility.

At the police commissioner's office the representative of the government tried to follow Paris' political line which came to light in a delicate exercise: DATAR, the ministry of the environment and equipment hesitated between reserve and opposition; the ministry of industry did not hesitate in its determination to see the project through. The Prime Minister had little room for manoeuvre because of a promise by the President of the Republic, General de Gaulle*(8, 9).

The decision to site a refinery unit (with a capacity of 3-4 million tonnes) and a landing site for oil tankers (with a capacity of 200,000 tonnes) in the Brest area was taken on October 9, 1968 within the framework of a series of measures in favour of Britanny. The Government confirmed its commitment on February 2, 1969 through the President of the Republic. The studies started then at the end of December 1971 when the main technical features could be spelled out. To carry the operation through, the complex was planned to go into operation at the end of 1974, a mixed syndicate combining the county, the Brest chamber of commerce and the municipality was constituted on June 21, 1971. It was then that the project in its broad outlines was presented to the public. Financing would largely be taken care of by the oil companies. Until the beginning of 1972 everybody tried to find out more about it, to get information particularly about the protection of the environment. Meanwhile, the project had been confirmed by G. Pompidou who had taken up his predecessor's commitment and who had invited the oil companies, who were very

---

*A short summary of the affair:
  Promised by the Government, studied, prepared, the project of siting an oil complex at Brest was finally abandoned; the problems of the environment had played an essential part in the agony and death of this operation.

reserved about the economic viability of the project, to overcome their
reluctance. They did so, but only apparently, while braking the operation
which they considered uneconomical, with all their might.

Local reluctance soon turned into opposition (1972/73). It hardened as
the months passed by, while one side affirmed the necessary and irreversible
nature of the project and the other demanded the reopening of the case
because of the danger which the project caused for the off-shore area of
Brest. The following table shows a simplified map of the conflict which did
not develop along the classical lines of division.

Faced with the seriousness of the situation that had been created, the
government finally urged all interested parties to reexamine the usefulness
of the operation on October 12, 1973. After the decision, the studies, a
start of preparations, there was thus basic agreement. From being explosive
in summer 1973 the situation then became only very confused.

In the end, the 'energy crisis' permitted an honourable 'shelving' of a
project which had become dangerous on all levels (Source: 14, p. 52).

Groups or actors favouring the project	Groups with non-clarified positions	Groups or actors opposed to the project
• Prime responsible: Municipality of Brest (afterwards the urban community) Chamber of Commerce and Industry (CCI) Equipment services of Brest • The UDR MP for Brest, general councillor • The young economic chamber of Brest and its region	• General council • CGT, communist party • The Government	• Sea fisherman and oyster breeders of the Brest off-shore area • Farmers • E. Leclerc • G. de Poulpiquet, UDR MP for Finistere, General Councillor • Persons and organisations from the sciences • Socialist party • Local press • Oil companies

Table 16:   Simplified map of the conflict
            (Source: 14, p. 51)

4. THE OPERATION OF NETWORKS IN A DISASTER SITUATION

A disaster is usually the result of a series of failures involving a large
number of actors. The situation to be controlled appears in the form of
complex interrelated networks reacting upon each other even if they ignore
each other. Also usually, however, one looks for the 'one' in charge, a
chief who can give orders. To have a coordinator is certainly necessary,
but to what end? What should be his task? Applying the decisions better or
animating the networks, or a still more complex task?

D. Fischer has asked himself about the problem and took two observations to

start (10). The one of the Royal enquiry commission which had to analyse
the accident on the Bravo platform at Ekofisk: The underlying cause of the
accident was that the organisational and administrative systems proved to be
inadequate to ensure safe operation; even if there were certain technical
weaknesses they were of secondary importance in the course of events (11,
p. 4).

The one of the presidential commission set up to enquire into the Harrisburg
accident which concludes: changes will be necessary in the orgnisation, the
procedures and above all on the level of attitudes at the National Regulatory
Commission and inasmuch as the institutions we examined are representative,
concerning the nuclear industry in general (12, p. 7).

The statements invite a close examination of the networks concerned. With
D. Fischer we shall examine the cases of Ekofisk and TMI after some more
general observations.

1st. General model of the network of the intervening parties

In a simplified manner one can enumerate the various intervening parties
as follows:

- the operator
- the safety authority
- the authority in charge of rescue operations
- the official experts
- the victims or groups involved
- the outside groups such as political personalities, independent experts,
  press etc.

A first 'social' map of the network of actors involved can thus be drawn
up. In order to understand the dynamics of the network one must still spell
out for each intervening party:

- the various actors involved (there may be several authorities for example)
- their essential objectives
- their secondary objectives
- the actors associated with them
- their criteria for decisions
- their uncertainties
- their unresolved internal conflicts
- the lessons they draw from the event.

2nd. The network involved in the Ekofisk accident

a) The intervening parties

The operator. The company involved (Philips) did not at the time have a
rescue plan in case of an accident; it had neither the organisation nor the
necessary resources to manage the accident and its consequences. But it knew
(how to): close the wells; evacuate the staff, start the battle against the
fire; alert the authorities an hour after the start of the accident; obtain
the resources of outside experts.

The safety authority. There were two organisms involved: the directorate for
hydrocarbons (in charge of technical control and safety) and the pollution
control authority (in charge of following up on oil treatment operations).

The authority in charge of rescue operations. An *ad hoc* organisation was
set up: the Action Command. It was supported by the cooperation of the safety

, organisms, various experts, foreign countries; it was open to representatives
from the oil company.

The official experts. These were various military experts or people from
research institutions.

The victims and groups involved. In a very general way these were the central
government, coastal communities, fishermen, the public in general (but in the
end the oil did not reach coasts).

The outside groups. Environmental experts got in touch with the Action
Command. Foreign governments offered assistance. There were also the media,
the national and international press; the Norwegian journalists had easy
access to aeroplanes and provided news quickly, sometimes before the Action
Command had been informed; this kind of problem was finally resolved by the
setting up of an experienced Action Command press team on the shore.

The following table shows a scheme of the organisations involved.

b) The dynamics of the network. The second table is an application of the
scheme of analysis to the Ekofisk case. We recall the following questioning
by D. Fischer. The Action Command (AC) managed to put its foot down: the
administrations involved made themselves directly available — without going
through their usual channels; the political world abstained from criticism
of the Action Command on the spot during the operations. For how long could
such a situation prevail? What would have happened in the case of a long
drawn out disaster?

In fact, depending on duration, the setting up and the running of the networks must follow different rules. It would certainly be necessary to spell
them out in advance.

3rd. The network involved in the TMI accident.

a) The intervening parties. We have already presented them in the analysis
of the case. D. Fischer suggests the following presentation.

b) The dynamics of the network. We add only a few complements or reminders:

- Absence of a precise rescue plan at the operator's as well as at the
  authority's.
- Absence of a plan for the information of the public.
- The obtaining of reliable information soon proves to be difficult for
  everybody. Much confusion results from this.
- The network is very complex inasmuch as each intervening party is itself
  composed of various organisations and pursues sometimes multiple aims.
  This is for instance the case with the NRC which with its federal, regional
  and local offices must at the same time take on the technical operations
  and retain its role as a control organisation (which involved actions
  aimed at improving its image with the public or avoiding seeing it
  degraded).

One notices here that the dynamics of the network would necessarily have
changed if the accident had proved to be more serious. The organisation in
charge of evacuations (PEMA) in particular would have taken the foreground
of the stage.

A more serious situation would also have put into clearer focus the lack
of integration of the various components. H. Denton, when he was appointed

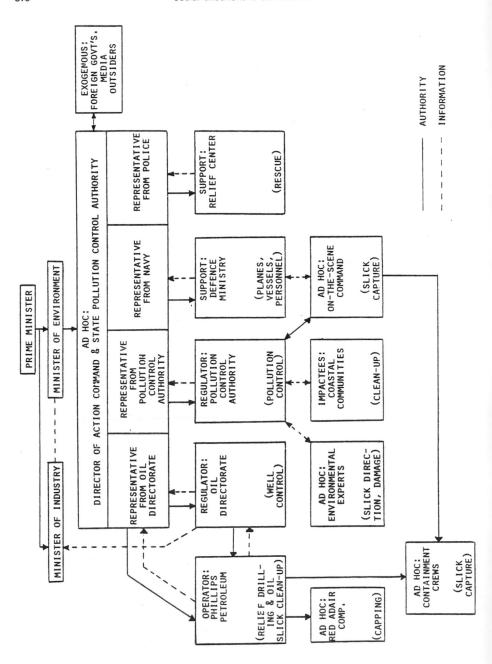

Fig. 34:   Ekofisk: Networks of actors involved in the accident
(Source: 10, p. 8)

SUMMARY OF ACTOR RESPONSES TO BRAVO ACCIDENT. EKOFISK

	Accident Response Group(s)	Normal Regulator(s)	Operator	Experts	Impactees	Exogenous
Key Actors	Action Command (ad hoc)	Oil Directorate, Poln Control Authority	Phillips Petroleum	Military, Inst. for Cont. Shelf Research, Marine Research Inst., Saga Petroleum	central government, coastal communities, fishermen	Nordic countries, media, other oil companies
Key Areas of Choice	supervise Phillips, support Phillips, contain conseqs., public information	support Action Command, assess Phillips plans, obtain polln oil eqipm., alert communities	prevent blowout, stop blowout, control pollution, protect life and property	provide transport, personnel, forecast oil slick, assess damages	display good response, protect shore, protect livelihood	aid to Norway, responses if prolonged acc., understanding of and access to info., cooperation with Phillips
Associated Choices	internal organiz., assess alternatives, take over operation	responses under prolonged accident	prod. on other wells, find and use equipment, determine procedure	responses to prolonged accident, assess alternatives, research	future oil prod., remain in power, preparation for slick, remove vessels	gain new information, public right to info.
Associated Actors	Ministries, Politicians	Min. of Industry, Min. of Environment	Red Adair Company, Poln Control Equip. Manuf.	other research Inst., foreign colleagues	general public	other North Sea govts., NATO, outside experts
Choice Criteria	retain estab. principles, ensure public prot., show govt. in command	ensure competent mgt., ensure readiness to meet pollution	guarantee good relations, minimize liability, minimize losses, costs	proven research results, try new research model	economic base, future of party, resource alloc., future of catch	contain polln to North Sea, test own polln control equip., headlines, heighten public awareness
Uncertainties	impact of slick (polln, public), disposition of slick, management of slick, timing of well control, use of research	timing of well control, performance of equip., coastal protection	performance of equip., government demands, future govt. responses	performance of equip., crews, damages from slick, predictions of drift, acceptance of results	harm to party, harm to oil program, harm to fish, shore	harm to North Sea, harm to oil program, meshing of equipment
Unresolved Conflicts	alternatives to polln control, actions for prolonged accident, degree of advance preparation	role in accident, role in inspection and enforcement	responses to govt. investigations, cost-effectiveness of preparations, liability limits	coord. performance of seperate units, researched advice vs. best guess	ability to prevent acc., oil vs. fish, degree of future damages, northern drilling prog.	separate info gathering, international vs. national, ltd vs. unltd liability, use of chemicals
Lessons Learned	Action Cmd. worked, advance prep. necessary, enforce prep. and coop.	need for apriori roles, Information split, human and oil safety, new appr. to polln ctl., more information on operators	strong govt. response, greater attn. to acc.pot., more emphasis on crew supervision and training	integrate research results a priori, greater training, coord., more applied research, more safety research	damages nil due to timing and location of acc., accidents are part of dev., more resources for acc. prev. and mgt.	nations interested and responding, media relations important, central media info releases, coord. necessary apriori, enforced company coop.

Table 17: Ekofisk: Dynamics of the network
(Source: 10, p. 12)

coordinator, managed to establish more coherence. But could he still have
controlled the system if the accident had been more serious and more prolonged?
On the evidence one could not expect the situation in itself to have led to
the desirable integration of the multiple intervening organisations. It would
have to have been prepared.

The 'defence in depth' would also have to be applied in terms of organisation and social capacity. Once more one discovers with major risk data which
have long been known in the military field: an event of vast impact is no
business for specialists used to do the same manoeuvres, communicating in the
same terms and sharing the same objective. From here on a 'battle' is no
longer won simply by the capacity of staging a fine isolated operation thanks
to some act of heroism but by the aptitude to manage very complex networks in
which information, specific policies and conflicts are the main factors.

The following table gives a certain number of factors of the dynamics
observed in connection with the TMI accident.

4th. The need for anticipation in setting up networks

With D. Fischer we can make the following observations:

Given the fact that an accident is always possible (no matter how optimistic
the people in charge may be), it is necessary to have plans of intervention
which foresee the institutional arrangements to be made.

These post-accidental conditions must be precisely analysed in order to
integrate all key actors in case of an event (and not only the official
experts) into the framework that has been worked out.

The reference values and mental inclinations of each of the intervening
parties must be clarified in advance. If this is not done communications will
prove difficult in a crisis situation.

The clarifications permit working on alternative response scenarios —
which are no longer just uses of organisations each of which has and pursues
its own objectives and policies despite all efforts by the coordinator.

These perspectives open an approach to a large number of difficult tasks.
These must be undertaken because, as D. Fischer finally emphasises, one
cannot say that after TMI (in particular) one 'knows' the problem and that
the administration will henceforth be better prepared. On the evidence,
intense reflection combined with effective practices are necessary. The TMI
accident is thus not in itself a lesson but a serious warning.

REFERENCES

(1) J. FORRESTER, La planification sous le règne des influences dynamiques
     des systèmes sociaux complexes. In *Prospective et Politique,* OCDE,
     Paris, 1970.

(2) *Le Monde,* 13 novembre 1969.

(3) *Le Monde,*

(4) Ch. HUGLO et R. CENNI, *Une société de pollution.* J. C. Simoen, Paris,
     1977.

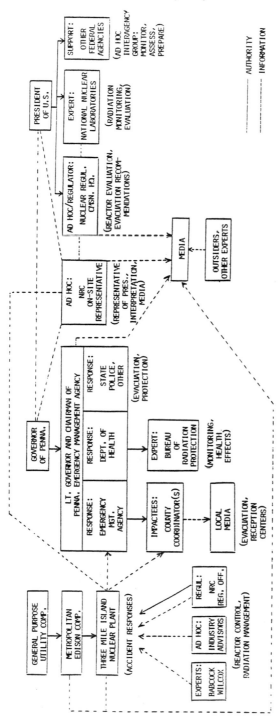

Fig. 35: Three Mile Island: Network of actors involved
in the accident.
(Source: 10, p. 26)

	Accident Response Group(s)	Normal Regulator	Operator	Experts	Impactees	Exogenous
Key Actors	Penn. Governor's Office Penn. Emerg. Mgt. Agency Nuclear Regul. Cmsn.	Nuclear Regul. Cmsn.	Metropolitan-Edison	Dept. of Energy National Laboratories Penn. Bureau of Rad.Prot.	residents counties, public facilities	U.S. President media
Key Choice Areas	evacuation preparation and execution direct involvement in accident management on and off-site	supervision of operator public image gathering and dissem. info. coord. with acc. mgt. actors	notify others control accident control radiation releases	monitor radiation protect public health stop explosion	evacuate/stay	political concern national responses understand and inform public
Associated Choices	planning for protection, reception, communi-cations obtaining good, timely information	understanding nature of accident degree of centralization coord. of research	protect workers reach cold shutdown gain timely advice public image	forecast drift of rad. handling of hydrogen bubble	Information sources protective measures personal/public evac. property/family	demonstrate action centralize info sources
Associated Actors	counties other Penna. State Depts. federal support agencies	National Laboratories Environ. Prot. Agency Dept. of Energy	Babcock and Wilcox Ad Hoc Industry Group	federal agencies	non-affected counties relatives general public	outside experts foreign observers
Choice Criteria	radiation doseages public health and safety panic	protect public health & safety protect nuclear reactor & workers	understand accident technical feas. of options minimize plant damage	necessity of rad. releases weather monitoring physics of reactor, boiler systems	personal health & safety costs of evac./stay anxieties	political responsibility alloc. of fed. resources info mgt.
Uncertainties	understanding nature of accident understanding radiation releases distance of evacuation, if needed	seriousness of accident emph. on worst case lack of previous exper.	seriousness of accident reactions of NRC future of plant commun. to media	criteria for where to monitor weather forecasts future rad. releases nature of hydrogen ques.	nature of radiation need for evacuation conflicting info.	severity of accident conflicting info. technical understanding
Unresolved Conflicts	criteria for evacuation good, timely information sources of information liability for advice planning vs. ad hocracy	acc. preparation + role public info releases reliance on regional offices radiation release mgt.	radiation release mgt. relation to media future of plant future of company costs and reimb.	criteria for evacuation worst case vs. expected research vs. best guess	liability of officials criteria for evac. trust in experts, media, govts.	national vs. international political roles media roles sources of info.
Lessons Learned	need for NRC acc. role need for commun. and preset information roles new accident management required	raise priority of acc. mgt. merge acc. prep. and mgt. with licensing enhance acc. reporting and inspection	enhance oper. training design plant for acc. mgt. use of checklists of exper. from elsewhere	need for new criteria for evac. need for pre-analyses of expected accs.	nuclear accs. a reality unreliability of media, oper., experts	info mgt. apriori integrated responses and info. needed

Table 18: TMI: Dynamics of the network
(Source: 10, p. 28)

(5) V. BIGNELL, Ch. PYM and G. PETERS, *Catastrophic Failures*. The Open
University Press, Faculty of Technology, 1977.

(6) P. LAGADEC, *Développement, environnement et politique vis à vis du
risque, le cas britannique: cinq catastrophes*. (tome 3) Laboratoire
d'Econométrie de l'Ecole Polytechnique, avril 1979. (115 pages).

(7) E. MENDELSOHN, Frankenstein at Harvard: the public politics of recombinant DNA research. The social assessment of science. Wissenschafts-
forschung, 13 Report, Universität Bielefeld, 1978, pp. 57-78.

(8) P. LAGADEC, L'impact des grands projets de développement sur l'environnement. Le cas de la raffinerie de Brest. Contribution à la théorie
de la planification. *Ecole des Hautes Etudes en Sciences Sociales*,
Paris, 1976.

(9) P. LAGADEC, L'étude d'impact: instrument d'évaluation des décisions
lourdes. *Futuribles*, no 9, hiver 1977.

(10) D. W. FISCHER, Organizing for Large-Scale Accidents: Experiences from
the Bravo and Three Mile Island Accidents. Institute for Industrial
Economics, Bergen, Norway. For the workshop on Procedural and Organizational Measures for Accident Management: Nuclear Reactors International
Institute for Applied Systems Analysis. Laxenburg, Austria, January
28-31, 1980.

(11) Royal Commission of Inquiry, *Uncontrolled Blowout on Bravo*, Preliminary
Edition, Oslo, 10 October 1977.

(12) *Report of the President's Commission on the accident at Three Mile
Island*. Pergamon Press, New York, October 1979 (201 pages).

(13) *Le Monde*, 27-28 janvier 1980.

(14) P. LAGADEC, L'étude d'impact: instrument d'évaluation dés decisions
lourdes. *Futuribles*, no 9, Hiver 1977, pp. 23-52.

CONCLUSION: SITUATIONS WHICH FURTHER COMPLICATE THE TASK BUT GIVE NO REASON TO DESPAIR OF THE USEFULNESS OF ANALYSIS AND ACTION

The inertia which all of the structures examined in this chapter add to the social insufficiencies studied before gives a measure of the difficulty of action. The actors, like the structures, are there so that this 'dynamic conservatism' functions of which D. A. Shon*speaks:

The mobilisation of the forces involved in the maintenance of existing situations. Social regulation hardly compensates for the insufficiencies and limitations of technical management: much more often it reinforces these difficulties and obstacles.

However, even if the analysis of social systems appears still more complex than that of techniques we have noticed that much remains to be explored, to be presented in models; that certain possibilities could emerge from analysis which would permit to counteract habits that can no longer prevail when major risk becomes acute.

If possibilities exist it is indicated that one should know in what directions action can be taken. Here we enter, more directly still, into the field where choices must be made: the political field.

---

*D. A. Shon, *Beyond the stable state: public and private learning in a changing society*, Penguin Books 1973.

PART FOUR

Politics

CHAPTER SEVEN

When politics discard the question of major risk

I   Defending progress, getting projects accepted
II Possible results of this first scenario

*Sire, these are the motives which have guided me in the
preparation of the work which I have the honour of presenting to your majesty.*

*At first I thought it was indicated to order the display of
posters every time that a request is made for the establishment of a factory which exudes unhealthy or inconvenient
odours; but further reflection has made me change my mind.
Such an arrangement would have raised much opposition, often
illfounded, and subsequently impeded the founding of
chemical factories which merit all the protection and benevolence of your majesty since they supply us with products
for which we depended before on foreign countries.*

*It seemed preferable to proceed with information on what is
'convenient' and what is 'inconvenient' which will provide
all the guarantees one can wish for.*

> Report of October 9, 1810 which led to the imperial
> decree of October 15, 1810, quoted by G. Gousset in
> *"The law on classified establishments"*, 1968, p. 9,

*What already characterises the administration in France is
the violent hatred with which, quite indiscriminately, all
those, noblemen and commoners, who want to take part in
public affairs outside the administration inspire it.   The
most insignificant independent body which seems to have a
desire to establish itself without its concurrence causes
fear; the smallest free association, no matter what its
purpose, annoys it; it permits only those which it has arbitrarily set up or over which it presides ...*

> Alexis Tocqueville
> *The Ancient Regime and the Revolution**

*Which amendment (of the Constitution) guarantees the freedom
of the press? I am against it.* — The President of the
National Regulatory Commission**

---

*Quoted in the Delmont report: "The participation of the
French people in the improvement of their conditions of
life", January 1976, unpublished report.

**Recording — at the time of the TMI accident, *Le Monde*,
April 14, 1979.

I. DEFENDING PROGRESS, GETTING PROJECTS ACCEPTED

Can one permit oneself to be troubled by the risks one runs by the limitations of means which exist to cope with the dangers and menaces or even by
the difficulties of an organisational and social nature which one encounters
in the management of risk?

A first response is very rapidly supplied to end these questions: on the
one hand, the insufficiencies stated are largely contingent, arrangements
will very largely reduce these inconveniences; on the other hand and above
all, one must not be misled by partisans.

There, in fact, goes our collective future. Even before and above urgency
and necessity the technical people in charge have one absolute and sovereign
duty: to guarantee progress and to authorise the industrial projects
necessary for this progress. In this constant effort one must not permit
oneself to be discouraged by opposition or give in to the obsolete and
dangerous illusions of certain groups.

This is the first line of defence which tends to guide reflection and
action as soon as the question of major risk surfaces.

1. THE ARGUMENT FOR NECESSITY AND COMPETENCE

1st. The fantastic leap forward achieved over three centuries thanks to industry

Could one have any doubt even for a moment when faced with such a technological option? The evidence demands respect for the achievements of
industrial society in terms of living standard and type of life to realise
to what extent the renunciation of the scientific and technical adventure
would be pregnant with consequences. As a plea the following lines by
J. Fourastie are sometimes quoted: They are, in fact, very pertinent:

At the end of the seventeenth century the life of the head of a medium
sized family, middle class, first married at the age of twenty seven could
be described like this: born into a family with five children he saw only
half of them reach the age of fifteen; he himself has had five children like
his father of whom only two or three were alive when he died.

This man, if he had lived to be fifty, which was rather rare, about one in
five, would have seen the deaths in his immediate family (leaving out uncles,
nephews and second cousins) of an average of nine people of which one of his
grandparents (the other three would have died before his birth), both his
parents and three of his children. He would have lived through two or three
famines and in addition three or four periods during which cereals were
expensive because of poor harvests which occurred on average every ten years;
in addition to the deaths he would have survived the illnesses of his brothers
and/or sisters, his children, his wives, his parents as well as his own, he
would have experienced two or three epidemics of infectious diseases without
even mentioning the quasi-permanent epidemics of whooping-cough, scarlet
fever, diptheria ... which claimed victims every year; he would have suffered
physical ills such as bad teeth, injuries that took a long time to cure; the
sight of misery, malformation and suffering would have been constantly before
his eyes (1, pp. 234-235).

One can sum up the influence of technical progress on the material life of
man by saying that technical progress frees men from slave labour; it increases

the length of their lives; it increases their autonomy with regard to physio-
logical needs and with regard to their environment; it initiates the passage
from a vegetative state of life to a speculative state; it allows the average
man to have access to higher education and opens up for him the road to
intellectual civilisation (1, p. 242).

2nd. The central position of the scientist and the engineer in these achievements

Would one, even for a moment, be taken in by the mode doubting the seriousness of the experts and trusting the incompetents? Here again, common sense
must take over. Technological progress being a necessity, who can best solve
the question of modes of action? The most competent specialists, on the
evidence. It is to them that the fantastic leap forward mentioned above is
due; it is they on whom depends the prosperity of tomorrow or at least the
maintenance of our well-being and of our place in the world.

In this respect it becomes more and more intolerable to see the expert
constantly questioned about his choices, to see the incompetent preferred to
the specialist, almost as a matter of principle. Thus to see, if one pursues
the principle, the options taken incessantly debated, called into question in
public is distressing where effectiveness of action is concerned.

The following text expresses with perfect clarity this feeling of uneasi-
ness — even revolt — in the face of a development aiming at participation of
the public in technological decisions:

Decision making in technological matters in this country has reached a
point where the guiding principle seems to be: never mind the decision,
whether good or bad, provided there is maximum "participation of the public".

What I object to is the imposition of such laws (on safety, environment),
of tools by which uninformed people can indefinitely prolong the process
of decision making. The tool I am referring to is called adversary hearing.

... I think that the present situation is in large measure the result of
the action, or rather: the inaction, of the technical community itself.
The technical community, without the least organised protest, has permitted
Congress, the states, the administrations to pass amendment after amendment,
just to keep the lawyers employed, statutes, rules, procedures causing more
and more public controversy.

I will tell you frankly that the technical decisions to be taken in the
fields of environment and safety are too important to be left to the lawyers.
Yet the technical community makes no effort to keep control of these issues
I had engineers telling me that this was the fault of the media who are
incapable of understanding technical issues and who on account of this have
adopted the practice of printing or distributing exclusively simplistic
opinions in excess and for sensational effects. Lamenting about the press
will not solve the problem. The right of the media to be stupid is
guaranteed by the First Amendment to the Constitution and, alas, thrives
on it.

The problem is that you must educate the media. You are the only ones who
can do it ... (you are) the creators and the guardians of technology. The
engineers and the masters of applied science have done more to relieve the
sufferings and burdens of man than anyone else, whether they be esoteric
thinkers, lawyers, politicians or demagogues ... I respectfully suggest
that you go back into the public arena and take up your legitimate position
of leadership (2, pp. 25-26).

2. THE ESTABLISHMENT OF RATIONAL OPTIONS DETERMINED BY EXPERTS:
   THE PROBLEM OF ACCEPTABILITY

The decision on the adoption of a project has been made; the general outlines of operation have been established. It remains to get these options
off the ground. The main difficulty is likely to come from certain opponents.
Necessity and urgency lead the promoters to be firm and often also defensive,
not knowing how to convince the opposition. They prepare, any way, to be
under attack: every project is subject to criticisms. These criticisms are
the harder the more they voice the ecologist point of view. It is therefore
a matter of 'passing' despite the barriers erected against the efforts of the
experts. For this, skill and firmness are needed; if these conditions are
not met there remains forcing the issue.

Whatever the unpleasant nature of the modalities of the scenario, they are
held to be necessary because of technical, economic and social imperatives.

The problem of acceptability of the risk for the public, according to
hallowed terminology, is the most delicate point in the whole edifice. Will
the population accept what one wants to do? And if there was an accident,
would they not profit from the passing crisis caused by such a disaster to
raise unreasonable questions? Here again one must 'pass mark' by avoiding
the calling into question which would lead to lots of disagreements, no doubt
even to 'economic decline and the end of social progress'.

1st. Getting projects accepted

a) *Skill in dodging, firmness in discussion*. A bouquet of means may be used.
Here is a sample:

Denial (Refusal). Refusing to examine the issue of risk is the most obvious
attitude. The decider will have a tendency in such a case to affirm that
the absence of danger has been 'established at scientific level'. The theory
may be supported by the argument according to which 'exceptional safety
measures' will be taken; and to close the debate he will add, in any case,
that 'the decision is irreversible'.

Displacing (shifting) the question. If the questioning persists a new way
will be to explain that there are other risks and to lean on the most impressive excuse of common sense: "my risk is acceptable because there are other
risks; I too have the right to take risks". And once more it will be possible
to get around the risk.

Refusal of studies, refusal of their publication. This may also be, for a
while, a dodging solution; it is first of all a matter of not starting a
debate. However this position may quickly become untenable. One will have
to find new lines of defence.

Publication of a large number of approximate data, even false ones. With a
flood of more or less exact data one will be able to try and discourage those
who attempt to shed light on the real risks connected with the operation;
they will first have to unravel the true from the almost true. If there is
still doubt which might come down on the side of the supporters of the project
one will clarify the debate with a simple, clear and well presented document
that will be widely distributed. One will take up those factors of the in-
extricable case which are the most 'reasonable'. One will fight those who
'pollute the minds' and will bring straying opinion back to 'reason'. The
decider will also find new means of strengthening his position in the conduct
of those who are most aggresive and determined.

<u>The solemn promise, the affirmation beyond doubt</u>. Conviction and determination may be used in the attempt to eliminate doubts, questioning about the risks involved leads to denials or 'states of mind' that are unfavourable to continued action. The sovereign assurance that there is no danger is only an intermediate means, and if further clarification is demanded one will find other lines of defence.

<u>Calling scientists and university scholars into question</u>. This is an approach that can be used but is difficult to maintain. It will be possible to quote multiple errors that were committed in this field*.

<u>Exhortation</u>. If one puts oneself 'above polemics', if one really wants to be 'serious' and 'honest', one must consider the actions undertaken to be 'irreversible' if not: 'that amounts to renouncing all hope for growth', to adopting an antinational attitude, to proving oneself 'irresponsible' etc.

<u>Assurances</u>. One gives assurances that everything has been studied by eminent scientists and technicians for years and that everything will be done to eliminate danger, if need be thanks to "world firsts". H. Green thus takes up the chain of arguments that may be used to reassure and manage the risk (of seeing the opposition close ranks):

The dangers are not as serious as they seem to be; at least it has not been shown that they are really serious.

Should there really be some danger or something disagreeable for the population this may be tolerable or acceptable given the enormous advantages it will draw from the use of technology.

We are undertaking research in order to understand the potential risks better and to find the adequate type of R & D to come to a technological solution which eliminates or diminishes the danger.

In any case, there is no reason to worry about unfavourable consequences of the technology as long as its feasibility has not been established.

We shall only permit the use of the technology if the appropriate measures have been taken to avoid any accident; if these measures do not ensure sufficient protection the government will not permit the use of the technology (4, p. 55; 5, p. 380).

<u>The appeal to man's abilities, to history, to the citizen</u>. Man will always manage to face difficulties and will always know how to adapt**.

One can also call risk into question in an everyday universe which inspires no fear; the fairy tales proposed as arguments are well known: there is the

---

*In the case of the debate on France's nuclear option one finds two beautiful examples of argument: the opinion of the scholars, you know ... let me tell you that the Academy of Sciences declared in 1829 that the railway was a dangerous monster, a criminal folly, and it suggested that it should immediately be banned in France. Well? ... I can tell you that Louis Lumiere himself declared that the sound film would never speak. Well? (3, p. 78).

**R. Dubos emphasises that man's possibilities for adaptation are immense but that he surely cannot adapt to everything and anything (6, p. 77). This opinion, however, may be contested.

car accident which must not lead to the renunciation of this precious means
of transport; there is the aircraft accident which cannot condemn flying etc.
This type of mental slip which dispenses with all examination on the level of
questions asked can also be taken up in case of a deadlock: it will be argued
that the whole thing "will have cost the taxpayer no more than x packets of
cigarettes".

Finally, one can still praise risk as a courageous attitude, the supreme
human attitude: the grandeur of man and nations consists precisely in launching forward into uncertain adventures. One will tell the citizen: the
challenge will be taken up by a society, a region which has always shown its
care for the safety of its inhabitants and for the preservation of its
environment; the difficulties will be overcome and one will learn lessons
from them which will definitely permit to fend off menaces of the same kind
better in the future. The operation, despite its risks, will therefore
prove to be beneficial: running risks will thus, finally and paradoxically,
be the best way of not running them*.

The use of force on the spot

In the debate: necessity will become (the) law. Given that the project
responds to 'higher' interests no prevarications will be permitted. Whoever
calls the choice into question marks himself out as a dangerous deviant.
Words become incisive, bare of even the least hesitation. The necessary consultative procedures for issuing the expected authorisation to undertake or
pursue the work must take their course without a hitch. Any delay would be
serious for the national economy.

On site: force if needs be. If there is still opposition despite the official
information, despite assurances given, despite approval by the parliamentarians, there remains only the strong arm method: quasi-military occupation
of the area. The most extreme example is the one of the international airport of Tokyo-Narita where the confrontations were violent and drawn out.
In France, Creys-Malville and Plogoff are the main references. Whatever
regrets there are: the idea takes hold so that in such a scenario there is
no other choice.

2nd. In a disaster situation: standing firm

While in the defence of progress the realisation of the project has been
ensured as perfectly safe and invulnerable, the disaster comes as a brutal
shock to the credibility of the experts and the people in charge. It is then
a question of 'standing firm'; for a week or two, the time when the news
chases up the spirits of disappointment and aroused feelings are dominant in
the groundswell.

Two lines of approach may then be taken, depending on whether in the minds
of the people in charge there has only been hot alert or very serious
discomfiture.

a) In the case of hot alert: talking big and strong. As we have said before,
one will then denounce uncontrolled feelings among the population, alarmism
of the 'so-called experts', the disaster mood of the media, who are always
keen on the sensational, the pollution of minds etc. One will give assurances
that the problem is not really a technical one but much rather psychological.

---

*The final argument used by the protagonists of the oil complex project at
  Brest (a case to be taken up later).

It is all a matter of information. Efforts already undertaken in this field
will be strengthened. One must not give way to panic. There is no objective
reason that could cause stress. Therefore, one must not be afraid. The
experts have the situation in hand. Commonsense can still be used to smooth
out the debate and lead to reason:

> The nuclear industry has caused less deaths than any other industry; it is
> less dangerous than the DC-10s or, for that matter, Central Africa — and I
> speak in my capacity as Minister for Cooperation (7).

b) In case of serious discomfiture: appealing to fate. When one can no longer
deny the evidence, when the latter is really dramatic, one changes tune
radically: the scholars are no gods. Thus the help of God was invoked at TMI
in an official poster. At Seveso "one let nature run its course". The
scholars are powerless against fate: and one invokes the natural disasters
which have always happened. Finally and at any rate, there is no choice and
there was no choice. One must lick one's wounds and rebuild with courage,
solidarity and dignity. A last piece of advice to the victims: if they do
not want, on top of it all, to be put in quarantine they must be discreet
about their fate, ask only for reasonable compensation for immediate damages,
avoid all political polemics; actually: decency demands silence.

As an illustration, the official reactions to the black tides are
interesting.

Torrey Canyon.

The Interior Minister (in the National Assembly): So, the oil reaches the
beach. M. Max Lejeune and other speakers reproach us now for having delayed
the ordering of barriers; but this disaster was without precedent.

M. Pierre Cot, in his admirable eloquence, I regret that he uses it
politically but this is a magnificent voice in parliament, has almost convinced the House that white is black and black is white, that it is the very
beginning of the art of government to foresee such events and that one is no
minister, nor governing, if one does not know that oil tankers will spill
their oil on the Brittany coasts.

M. Pierre Cot, Japan for example does not ignore the existence of typhoons.
Or will you tell me that means for the defence of the Japanese coasts have
been foreseen by that country? By the same token, Italy knows that volcanic
eruptions may occur but can you tell me what measures to fight them have been
foreseen?

Quite true, this is a disaster (Shouts from the benches of the communist
group and the federation of the democratic Left and the socialists — applause
from the benches of the Democratic Union for the Fifth Republic).

M. Pierre Cot: ... I just want to emphasise that you are making a mistake.
In fact, I did not say that you could foresee the details; I have pointed out,
and this is quite a different thing, that you have not studied them.

It is quite obvious that from the time when oil tankers appeared on the
scene, I am not saying that they are uncontrollable but difficult to control
and these are the same words that Commander Ropars used as everyone knows,
and when their numbers increased certain shipping lanes in the world became
particularly tight, such as the Channel and its approaches. It was therefore
necessary to examine the measures to be taken more carefully and to draw up

a plan of defence. It does not seem, at least you have not said so, that an arrangement in this respect has been studied.

The British, you tell me, are to be blamed for the same negligence. However one man's fault does not excuse the other's.

Right or wrong — if it is wrong it is much too serious, and I reproach your general policy ...

The Interior Minister: Don't talk of the atom bomb!

M. Pierre Cot: ... you started by underestimating the danger, voluntarily or unconsciously, it is up to you to say which, and above all you have given the impression that no means were provided. A newspaper has even claimed that you have then somehow taken up the famous formula: "there is not a gaiter button missing". But when one had to use the gaiters the buttons were on one side and the gaiters on the other. It is the impression of deficiency, of insufficiency in study and preparations which, with all due respect, I permit myself to reproach you for ...

The Interior Minister: M. Pierre Cot, I acknowledge your courtesy, your dialectic talent, but nothing else. (Applause from the benches of the Democratic Union for the Fifth Republic).

M. Lejeune has reproached us for having ordered the barriers too late. Well, no company had any.

<u>Torrey Canyon</u>. *The Interior Minister* at Lannion at a press conference:

In no country in the world has anybody ever imagined that an oil tanker could sink near the coast. In the first place, there must be an end put to that veritable piracy which is a challenge to maritime law: the concession of flags of convenience. This is a challenge to the rules of the civilised world. Obviously, nobody could have thought that 100,000 t of oil could be spilled one day in the Channel. Now the oil is there. We must face it. But there cannot be a political issue of it. Political oil does not exist*(9).

<u>Amoco Cadiz</u>. *The Interior Minister* (at Brest at a press conference):

I have been impressed by the outsize amount of media coverage of the *Amoco Cadiz* affair. Nothing was worse for Brittany than the "Save Brittany" campaign because only a small part of the region was affected. On the northern coasts there are no more traces except for one single beach. There has been a failure in terms of information which must be attributed to various people in charge and it has been prejudicial for Brittany.

In a family, when a child is ill, one keeps it to oneself: one does not shout about it in the streets (11).

<u>3rd. To comfort everybody's confidence: the cement of common sense</u>

---

*This comes close to the statement made by the hospital at Desio in the Seveso case: "We do not carry out political abortions here", or to the leaflets distributed when the *Amoco Cadiz* ran aground in which the population was asked not to "politicise" the black tide.

In an unruly fashion, the following thoughts serve as a justification of
risks taken:

- All human actions involve risk ...
- Life means risk ...
- All men are mortal, this is the only certainty ...
- One rejects a risk that is estimated at 10^{-6} per year; however, this means
  the risk that I could be dead in five minutes time. So, ...?
- One crosses the road ...
- One uses one's car ...
- At Flixborough, no more people died than did on the roads the same day ...
- There are aircraft accidents; it is no reason to reject flying ...

One accepts natural risks: the flooding of London (£3 billion) without
doing anything about it (probability: 10^{-2} per year); there are definite
earthquake risks in Tokyo ($250 billion) or San Francisco ($50 billion) without it stirring emotions. So, ...?

One accepts a risk of 10^{-2} per year in construction (earthquake) but one
rejects the 10^{-6} per year risk of the nuclear industry!

- The people are badly informed ...
- The public does not understand ...
- It is a matter of national interest ...
- The people always reject industrial developments ...
- Man has always accepted the price of progress: one cannot see why this
  should change ...
- True, there has been an accident, but this cannot happen here ...
- We shall take all necessary measures ...
- Radiation? One goes for winter sports. One lives in towers. One wears
  watches ...
- Mutagenic effects? You are wearing underpants made from synthetics ...
- Opposition groups lead systematic campaigns which are most often
  orchestrated. They are manipulated and of an obvious political colour.
  Some are of a truly subversive kind. It is important to control them;
  this is actually being done, and it works* ...
- The media are often irresponsible ...
- This is all politicised ...
- It is a very serious matter to cast doubt on the sciences and techniques ...
- A risk? If you were an expert you would know that there is no danger and
  you would not ask the question ...**

Since the start of industrialisation, the balance sheet has been in credit.
So, why quibble about any specific point? It is a matter of take it or
leave it. If one does not want to go back to the age of the candle one must
take it all, the bomb and medical surgery ...

- Without risk the economy would die ...
- It is still better to take a big civilian risk than to have an atomic war
  because of economic crises ...
- Actually, we are in the middle of an economic war ...

---

*Opinions held at a big international seminar by a specialist on communi-
cation problems. The Anglo-Saxon experts were surprised by them.

**Answer given to the author during a talk at a large R & D organisation.

The argument of war opens up large perspectives in matters of social acceptability. Every year the nations celebrate the memory of difficult times when they had to give the blood of their sons to avoid losing them. Should we be less brave today? Less responsible in the face of history?

REFERENCES

(1) J. FOURASTIE, *Machinisme et bien-être. Niveau de vie et genre de vie en France de 1700 à nos jour.* Ed. de Minuit, Paris, 1962.

(2) Th. G. DIGMAN (Attorney), The technical community must speak out. The Adversary proceeding: is it compatible with technology? A panel discussion. Chemical Engineering Progress (C.E.P.) March 1977.

(3) P. DELOUVRIER, Une semaine nucléaire: le grand débat. *L'Express,* 17-23 mars 1975.

(4) H. P. GREEN, The adversary process in technology assessment. *Technology Assessment.* R. G. Kasper ed, Praeger, New York, 1972.

(5) F. HETMAN, La société et la maîtrise de la technologie OCDE, Paris, 1973.

(6) R. DUBOS, Environnement: les dangers de l'adaptation. In *Dialogue,* vol. 3, no 1, 1972.

(7) *Le Monde,* 12 juin 1979.

(8) Assemblée Nationale, Séance du 27 avril 1967.

(9) *Le Monde,* 14 avril 1967.

(10) *Le Monde,* 11 juillet 1978.

III. THE POSSIBLE RESULTS OF THIS FIRST SCENARIO

1. SUCCESS

1st. When the decisions 'hold', when the projects 'go through' in spite
of everything

This is the case with the French electronuclear programme, the most
generous and the most flourishing in Europe. True, there has been in all or
nearly all cases vigorous opposition. It was hoped in each case that
lassitude and resignation, for lack of comprehension, would finally leave
the field open to the technicians. Some tariff arrangements might help in
this respect. If one avoids indecision and disregards the state of mind, if
one strengthens the programme each time opposition hots up one may hope to
achieve the objectives. The opposition will give up.

Surely, there will be hostile articles such as the following one which
shows the discontent of the local inhabitants concerned.

Golfech or "what are public enquiries good for"? The question was raised in
Le Monde by M. Marc Ambroise-Rendu when the Mayors of Golfech and Auvillar
protested against the way in which the construction of the local nuclear
centre was decided:

For fifteen years EDF have had plans to install electricity generating
reactors at Golfech on the banks of the Garonne river, 20 km from Agen. At
the start it was a question of two reactors of 800 MW each. But as the time
went by the project was considerably enlarged: with its four 1,300 MW
generators and their gigantic cooling towers the centre has tripled in
capacity.

Anxious to get the opinion of the electorate, the parliamentarians of the
counties of Valence-d'Agen and Auvillar organised a referendum in June 1975.
The results were very unambiguous: 82.3 per cent of the voters said NO to
the project. In July 1978 the regional council for Midi-Pyrenees unanimously,
with two abstentions, also rejected it. This line was also followed the year
after by the general council for Tarn et Garonne with twenty one votes
against, two votes in favour and three abstentions.

Faced with such opposition one might think that the technicians in the
south-west would look for less 'difficult' sites. Not at all. EDF began to
buy land, then the necessary gravel pits for the extraction of the millions
of cubic meters of building material it would need. The parliamentarians
from the area revealed that teams had already been assigned to work on the
building site.

Realising that the decision had already been taken 'elsewhere', the
parliamentarians did not welcome the public enquiry which was opened in the
town halls of sixteen communities, from October 22 until December first.

However, this was still an improved procedure compared to yesteryear's
habits. It now stretches over two months (instead of three weeks). A study
of the impact on the environment is added to the package. The enquiry
commissioners are chosen more carefully: for Golfech one thought it right to
assign a socialist mayor, an ex-colonel and an ex-police subcommissioner.
The commissioners went to every community to register the objections directly.

These precautions are no longer of any use once distrust has set in. Nine
mayors refused to give shelter to the enquiry registers in their municipal

offices. "We don't want to be accomplices to a parody of consultation".
they explained. To save legality, the authority responded by sending nine
vans which for the purpose were baptised 'town hall annexes'. They were
protected by four squadrons of mobile constabulary who arrived in the villages
with their barred-window cars, their car radios and their rifles. Anybody
who wanted to enter an observation in the register had to do so under the
eyes of several uniformed men. One can imagine the atmosphere.

It got worse still when on the first day three registers were stolen and
burned in a public place. It was again aggravated when at Valence-d'Agen the
opponents organised a sit-in around the vehicles. The constables answered
with tear gas and truncheons. A number of onlookers who did not know what
to think of it all were first flabbergasted and then scandalised. Stupefied
they were also when during other demonstrations the police commissioner spoke
of "specialists in violent agitation" when everybody had perfectly recognised
friends and neighbours in the crowd.

The blunders accumulated throughout the enquiry, the population from the
neighbourhood of Golfech turned from a resigned "No" to the most determined
opposition. During the last few weeks no less than four hundred and fifty
people symbolically tore up the enquiry registers. Everybody knew that they
would immediately be seized and later on charged. Yet some of these people —
mothers of families, farmers, doctors, local craftsmen — had never taken
part in any demonstrations until then.

One event occurred like an exclamation mark on the scene. On November 17,
at Mazamet the President of the Republic on a visit to the Grand Sud Ouest,
declared that the centre at Golfech would be built and promised two more.
At the time the enquiry was not complete and obviously one was still far
from the declaration of public usefulness.

This was, however, the same Valery Giscard d'Estaing who told us on
January 26, 1978: "The apprehension must be answered in depth and not with a
propaganda campaign. It cannot be a question of imposing a nuclear programme
on the French people to which they would be profoundly opposed after having
been completely informed." The public enquiries should be the occasions to
give the citizens this complete information and this open debate to which
our neighbours in Germany, the Netherlands, in Denmark and Sweden are
entitled. The French people are offered soothing proposals which one contra-
dicts shamelessly in the subsequent debate and, on the ground, it is a
comedy of consultation. Does not the nuclear industry risk the demise of
democracy with these detestable practices?

This is only mishap. The essential thing which will be remembered in
history is the overall unfolding of the programme, and there is hardly any
worry. The precedent of Malville may actually be taken as encouraging. A
resolute policy, powerful enforcement — and victory is achieved.

Creys-Malville or the time of disillusionment. We shall come back every
summer; Super Phoenix must go, the 60,000 people gathered from all over
Europe affirmed on July 30/31, 1977, proclaiming their opposition to the
construction of the fast breeder reactor at Creys-Malville (Isere). This
was at the time of confrontation between a fraction of demonstrators and the
forces of order which had ended with the death of an ecologist, M. Vital
Michalon, and which had claimed a large number of injured people. This year,
on July 31, some people will simply place a sheaf of flowers to the meadow
where M. Michalon fell while a mass will be read at Mepieu ...

Behind the iron grills and the barbed wire which last year only protected huts and a concrete slab some ten giant cranes continue the creation of the 1,200 MW centre. Sixty three thousand tonnes of concrete have already been poured. Malville in 1977: has it killed off the anti-nuclear challenge?

July 1978. The farmers are in the fields, the people from Lyons and Grenoble in their second homes. "The summer will be calm, the anti-nuclear protesters have gone elsewhere to raise their voices", an inhabitant of the village of Bouvesse (2) remarked.

2nd. When simple alerts or even disasters do not cause very strong crises

Various examples may reassure the protagonists of this scenario of the closure of the issue and of firmness.

Incidents of chemical origin in the Lyons area.
The multiple incidents that occurred in the Lyons area on account of the chemical industry have not caused large protest movements. It is certainly necessary to take all necessary precautions but it is not imperative to review such and such choice fundamentally. The question of employment is actually an essential point here: if there is a risk of job losses, social acceptability is largely achieved.

The 'fissures' which affect certain elements of nuclear boilers (generators).
Surely, a lot of 'persuasion' was necessary in facing the trade unions but a bit of reassuring information was enough to silence the public. The fact that EDF could be held on tenterhooks for a year by the private construction company, that the safety authorities could be kept ignorant even longer, did not cause questions that were too difficult to control. In this case — where after all many specialists declared they were at their wit's end in following the phenomenon*and even more serious limitations were encountered during a shut down after the reactors were started up, there was no major alert: some solemn undertakings were sufficient, and the reactors were started up.

Torrey Canyon, Amoco Cadiz.
In the same way, can it not be said that the accidents of the *Torrey Canyon* and the *Amoco Cadiz* were successfully overcome? True, the battle was long and expensive but it was possible for the government to limit the indemnifications. True also that new regulatory measures were taken but none of them called any powerful interest group into question. Certainly, there was local grumbling but the spirit of solidarity was magnificent and on the electoral plane the events had no consequences. Of course, new means of surveillance had to be installed but the recommendations of parliament (a structured plan for the protection of the coasts) did not have to be enacted which latter would have been an effective burden on public finances and would have required the application of the principle "polluter = payer" which is adopted everywhere else.

Seveso.
The majority of the population did not seriously call the economic and political choices made by public and private figures into question. The company involved was not rejected locally: its financial and technical aid was sought. Various measures were put up for study, of course, in various commissions but no major change surfaced. The most acute difficulties were encountered when the incredulous evacuated inhabitants forcibly retook possession of their land. But that was rather a symptom of extreme success: the victims themselves pushed the risk aside. Repression actually helped

---

*Particularly for the 'knee' parts where the casing is of lesser quality.

success along; the moral question "Has one the right to abort or not? took
the front of the stage and quickly masked other problems which were more
critical for those in charge. The final question, for the second time
round, also ensured the *status quo*. Indemnification is a minor evil.
Finally, last pay-off to success: oblivion was demanded. This present could
be granted to the population: it is the most beautiful demand the worried
people in charge could receive in order to overcome a crisis of this size
at minimum cost.

Thus, achieving this 'social acceptability' does not seem to be beyond
reach. In a disaster situation the public has enough to do to survive and
live again to overcome its fears: it no longer has the energy for political
battle which one would actually have quickly denounced as unseemly. Faced
with simple menaces, the problem of jobs and judicious information, so much
easier when one has tight control of the media, are precious trumps for those
in charge.

2. DIFFICULTIES

Getting things passed and tenacity are not always achievements which one
can win without fear. The opponents, the parliamentarians, the local
figures can considerably impede the unfolding of operations. There, despite
determination and even force, history hesitates. It is no longer possible
to convince with some simple formula, no longer possible to silence by
moderate recourse to legal violence. A good illustration of difficulties
thus encountered can be given with the case of the aborted oil complex at
Brest. All the previously mentioned means were used; they were not
sufficient to win the battle.

BREST OIL COMPLEX

PEOPLE IN CHARGE   OPPONENTS

The presentation of risk

The installations of the refinery
integrate themselves into the land-
scape without being camouflaged or
disguised. The colours will mix
with those of the sky and the earth.
Trees will make the oil monster invisible from the neighbourhood.
Even next to the giant towers with
their piping systems the swallows
of the field will spread their
melodies and display their acro-
batics ... Departmental equipment
directorate (3,p.10).

The decision having been taken, the
seafishermen have tried to get information from the files on the preparation of the project in order to
examine the conditions in which the
battle against pollution was foreseen.
They found very quickly that these
documents did not exist according to
the statements by the promoters of
the project themselves.

The only document to which one could
refer is a brochure by the departmental
directorate for equipment in which the
passage on the protection of the environment is certainly poetic (seafishermen and oyster breeders) (4,p.4).

Refusal

The police commissioner considers that the whole of the scientific studies carried out to this day tend to show this compatibility which will be guaranteed by exceptional technical protection measures. At this stage it is therefore indicated to continue the research and mutual information efforts within a working group open to the parties concerned. *La Prefecture* (5).

M. Maurin, director of ISTPM:
My experience has shown me that there is incompatibility between the presence of oil industries and oyster breeding (6).

Professor Lucas, Director of the Marine Zoology Laboratory of the Faculty of Science at Brest:
As a biologist I know that all the promises of guarantees against pollution are a decoy.

M. Le Faucheux (CNEXCO):
This offshore area is sound but it is sensitive. At the landing stage there exist 'definite dangers' of pollution which could exceed an annual spillage of twelve tonnes of hydrocarbon (8).

Dr M. Aubort from CERBOM:
At the present stage of the technique it seems impossible to us that this pollution will not occur in the wake of the setting up of these installations (4).

Begging the question

The offshore area of Brest becomes polluted, which is the normal lot of a receptacle basin for several thousand hectares on which a population of more than 300,000 people live and work.

In particular, hydrocarbons from combustion wastes, losses through escape, (oil changes) and from the activities of the Navy, the commercial port, the fishing fleet, the pleasure boats, clandestine discharge, service stations ... represent a volume which is difficult to measure but certainly substantial.

These same people pollute the offshore area unconsciously. One needs only walk along the waterside to see them throw left-overs of fish into the water, to look at some of their premises around which the rats swarm ... In short, everybody (to start with: those who speak of the defence of the environment); pollutes, and one cannot see

When we ourselves realised the risks, we ran into a total refusal of the examination of the question ... and for obvious reasons since the risk is so enormous to guard against that one asks oneself how it can be so. The business turnover in the offshore area is 50 million FF and the stock is valued at 100 million FF. Moreover: how could one exactly determine the causes in case of slow loss? *Seafishermen* (10).

... we would have wished that such a problem, on account of its seriousness and its importance for the future, were approached in a serious manner. *Seafishermen* (12).

why the setting up of the refinery
is refused; if one refuses it one is
against industrialisation, against
progress. *Chamber of Commerce* (11).

Refusal of studies and their publication

The city of Brest let it be known
that it cannot distribute the report requested from CNEXCO on
account of reproduction difficulties.

The studies can of course be communicated to the interested parties,
this, however, on condition that
they are completed. Information has
a tendency to change into a pollutions of minds, and the "university scholars" are unfortunately no
strangers to this phenomenon. This
pollution develops along with press
campaigns. M. de Bennetot, UDR MP
(14).

To say that there has been no information is untrue. Information has
been given, and it was important,
but it was refused. Could it have
been more extensive? Yes, certainly,
if it had led quickly to an arrangement which we were refused at the
time when it would have been most
useful. *Mixed Syndicate* (9,p.21).

The senator and mayor have said in
public that the study would, of
course, be widely distributed. Well,
when it was sent to the mayor's
office (first half of April 1974) he
claimed difficulties with reproduction
to avoid any publication. *Seafishermen* (10).

Avoiding all publicity about the refinery, letting things run their
course as if nothing was happening,
this seems to be the present line of
conduct of the protagonists of the
oil complex. However, we have had
this report ... *Le Télégramme* (local
daily) (13).

I have been obliged to say that a
barrier has been erected against information meetings in order to delay
them and that in the course of these
meetings the scientists who developed
unfavourable theses could hardly
express themselves; their report was
distributed only parsimoniously. One
went as far as to say that there was
no reproduction facility. G. de
Poulpiquet, UDR MP (15).

Seeing how offhandedly all this was
dealt with, I concluded that the
operation would be carried out with
a minimum of risk. The errors committed, I was still inclined to let
things go and I even defended the oil
mini-complex in an election campaign.
"The choice has been made", I said,
"I believe it is a mistake but we
have reached the point of no return.
It remains only to get assurances
and guarantees".

I insisted that studies be carried
out and above all communicated to all
elected representatives of the political and economic scene of the province of Brest. This was done only
very partially and reluctantly. G.
de Poulpiquet, UDR MP (15).

Publication of a large number of approximate and even false data

The bay of Hiroshima

The refinery of Marifu in the bay of Hiroshima deserves some attention. It is one of the largest refineries in Japan ...

Fishing, a very important activity in Japan continues in the bay without damage. *Mixed Syndicate* (9,pp. 16-17).

It seems that in the bay of Hiroshima oyster breeding is condemned to disappear progressively for the benefit of industrialisation. ISTPM*(5).

The pond of Thau (Frontignan refinery)

At Frontignan near Sete there has been an oil refinery for fifty three years. The basin of Thau is surrounded by an oil and chemical complex which includes in addition to the refinery some ten companies which; employ several hundred people. The breeding of oysters and mussels in the basin has never been impeded by the industrial installations. A world renowned vineyard produces Muscat wine at Frontignan. The vines grow on the immediate borders of the oil refinery, some less than two hundred metres from its enclosures. M. de Bennetot, UDR MP, (16).

Through the ISTPM we have seen the Frontignan refinery. The director thought we were parliamentarians. When we raised the problem of setting up a refinery near an oyster breeding centre his reaction was: "And they let you do that!" His stupefaction, however, soon gave way to confusion when we told him that he was mistaken, that we were not the promoters of the project but the representatives of the seafishermen. *Seafishermen* (10).

We shall never let the offshore be polluted. *Chamber of Commerce* (18).

It (the Chamber of Commerce) feels sure that the oil complex can develop without affecting other activities and it undertakes to this effect to do everything to ensure the protection of the environment and to facilitate development. *Chamber of Commerce* (18).

On the level of the landing-stage there will certainly be escapes of hydrocarbons which will cause a slow death of the livestock in the offshore area; there might even be an accident which would have a more radical effect. Nobody can guarantee that there won't be pollution; we cannot run this risk and play the promoters' game. *Seafishermen* (10).

---

*Scientific and technical institute for maritime fishing.

Calling scientists and university scholars into question

A Professor at the College de France had shown 'scientifically' that trains travelling at more that 25 km/h in tunnels would be fatal for the travellers ... Scientists can be mistaken and above all they are not competent to talk about techniques; one may be a specialist in sea urchins and know nothing about anti-pollution barriers. *Mixed Syndicate* (19).

The section objects to the malevolent affirmations and to certain local dignitaries who are casting doubt on the competence of marine biologists some of whom are internationally renowned and to seeing scientific research and the profession of researcher discredited in the eyes of the public. It is, on the contrary, through the development of basic and applied research ... that one will be able to control certain problems caused by industrial development. *Union of Scientific Researchers* (20).

Assurances

The men who have given us their counsel are all scientists, highly qualified engineers; their high level of culture and their knowledge guarantee that their moral and intellectual qualities cannot be doubted by anybody. *Chamber of Commerce* (19).

Exhortation, appeal to the citizen and to history

In order that our youngsters shall not reproach us tomorrow for the underdevelopment of the Brest region ... *(Chamber of Commerce* (18))

In 1920, the Engineer of Bridges and Roads, Coyne, had tried unsuccessfully to get the authorities in Brest to permit the installation of a refinery and an oil port. The requesting companies thus fell back on Donges.

We reject betting, lotteries, insufficiently coherent projects which as they grow collide with existing activities without sufficient compensation. We are for precise and consistent projects. We are for a clear language which is the best way of respecting people. Let it be said clearly what is at stake and what is the seriousness of the choice to be made. *Seafishermen* (22).

It is sufficient to compare the economic upswing at Nantes-St. Nazaire with the Brest area to see how loaded with consequences the decision was. Mixed syndicate (9, p.11).

If one supposes that the existence of a risk which is inherent in all human activity must lead to the rejection of a project, no action in any field is possible any longer. *Mixed Syndicate* (9,p.39).

If the existence of these risks causes the rejection of the project there is no further action possible. It amounts to transforming Brittany into a national park. *Chamber of Commerce* (18).

The Brest offshore area constitutes for the industrialised countries an extremely rare example of preservation of the ecology of such an area; it seems strange that one is getting ready to upset this ecology while other countries go to enormous expenses to recover natural spaces of a similar quality. *Seafishermen* (10).

Debate without appeal

The land for the storage area of Caro has been bought completely on amicable terms one year ago. The land for the refinery at Lanvian has been bought partly on amicable terms, partly expropriated ... The detailed studies for the landing stage as well as for the refinery are in progress and the deadline for the work has been set. The definite projects for Lanvian shall be completed by May 1, 1973, the date by which consultations with the companies will be started and who will then have four months in which to reply.

By now we have gone largely beyond the stage of study, several hundred million FF have already been invested. The financing of the whole operation is assured. *Mixed Syndicate* (9,pp.7-8).

The present situation is irreversible: the government's decision is final and has materialised in the commitment of funds; the invitation for offers for the first (stages of) work has been launched. Henceforth the application of these decisions must be provided for on the land and of their consequences on the level of safety. It is up to the professionals of the offshore

area to decide on the technicians
to be associated with them who will
take charge and not to get themselves involved in these questions.
Mixed Syndicate (5).

The refinery will be built in the
place chosen or it will not be
built; the decisions taken in this
respect after long studies are
irreversible. *Mixed Syndicate* (23).

We ask you solemnly to review your
positions and to appoint a different
site before the irreparable has
happened. *Seafishermen* (24).

On the verge of confrontation

Well, is it serious and honest to
call everything into question as
one has heard? The refinery of
Brest has become a reality, and
the process started must be considered irreversible if one is not
to give up all hope for growth.
Mixed Syndicate (9,pp.6-7).

In the name of my fellow fishermen I
turn to all of you, parliamentarians
and administrators of the county:
above polemics, you must know that our
fight has no other cause but the defence of jobs. Be well aware of the
risks we shall run because you will
share in the responsibility.

Hitherto we have shown restraint.
Should we be proved wrong? It is true
that we are not in the habit of storm-
ing police commissioners' offices ...

Let nobody be mistaken: our tranquil-
ity and sense of responsibility is
only matched by our determination to
defend what is our most elementary
right: our right to work. *Seafishermen*
(24).

For how long have we been told it is
too late? *Farmers* (25).

Wanting to annihilate the offshore
area of Brest and its natural riches
for a non-existing future is aberration and hysteria. It means pro-
voking our revolt, and we are asking
Paris coldly: Do you want Brittany to
become a detonator? *E. Leclerc* (26).

3. DEADLOCK

In a large number of situations, forcing things through may 'work'; but
deadlock lies in wait for those in charge. At Brest, as we have seen, the
local refusal substantially strengthened the position of industry who was
not keen to obey government orders and the project was abandoned. In the
case of Canvey Island the language used for ten years had to be abandoned;
risk studies had to be launched; one had to debate with and finally constrain
the main industry of the island (British Gas) to face the abandonment of its
LNG installation.

In the same vein, the Kemeny report in the wake of the TMI accident anticipates: if those in charge do not substantially change their practices they

will destroy the public's confidence and will be responsible for the elimination of nuclear industry as an energy source (27, p. 15).

For the time being nothing obliges us to take the risk of deadlock into
consideration. One can get by and by contrast redouble determination and
firmness. One can then count on the resignation of the majority and the
outmanoeuvering of some which is not very serious. One also risks refusal:
be it by some who will show a determination equal to that of those in charge,
be it by the majority.

As a beginning of the first scenario we have just quoted the warning
addressed to the public authorities in the affair of the oil complex at
Brest.

Examples of the second scenario were found in the case of the *Tanio:* the
passive resistance of all littoral communities (no matter what their
political colour) in the wake of the reception the Republican Safety
Companies offered them in Paris; or again at Plogoff where the forces of
order were opposed (who tried to ensure in usual fashion the execution of
the public enquiry), not as small or specialised interest groups but, much
more worrying for a political authority — as a culture with all that involves in terms of contradictions but also with underlying powerful forces.

REFERENCES

(1) M. AMBROISE-RENDU, A quoi servent les enquêtes publiques? *Le Monde,*
    10 janvier 1980.

(2) Cl. FRANCILLON, Un an après, Creys-Malville ou le temps des désillusions.
    Le Monde, 30-31 juillet 1978.

(3) Raffinerie de Brest. Département du Finistère, Direction Départementale
    de l'Equipement, 24 décembre 1971 (14 pages).

(4) L'avenir de la rade de Brest: Pêche, Ostréiculture, Aquaculture ou
    Raffinerie. Union des Coopératives Ostréicoles de l'Ouest-Bretagne;
    Comité Local des Pêches Maritimes de Brest; Syndicat CFDT des Marins
    Pêcheurs de la rade de Brest; décembre 1972 (57 pages).

(5) Préfecture du Finistère; Complexe pétrolier de Brest. Procès-verbal de
    la réunion du 28 avril 1973 à la Préfecture.

(6) C. MAURIN, Directeur de l'ISTPM, Lettre au Président du Comité Local
    des Pêches, 4 avril 1973.

(7) A. LUCAS, Déclaration in Le Télégramme, 9 janvier 1973.

(8) Implantation d'un terminal pétrolier en rade de Brest. Examen pré-
    liminaire des problèmes éventuels posés par la pollution de la rade.
    Centre National pour l'Exploitation des Océans (CNEXO), avril 1973.

(9) Complexe pétrolier. Syndicat Mixte pour la Création et l'Aménagement
    de Zones Industrielles et Maritimes dans la région de Brest, avril 1973
    (43 pages).

(10) Entretiens avec des représentants des marins-pêcheurs. Octobre, novembre, décembre 1973; janvier, avril, mai 1974.

(11) La rentrée à l'Ecole Supérieure de Commerce: la séance solennelle a été marquée par une lecon inattendue sur la pollution. *Le Télégramme*, mardi 3 octobre 1972.

(12) Les pollueurs ne sont pas ceux que l'on croit. Réponse des Marins-Pêcheurs au Président de la C.C.I. *Le Télégramme*, 5 octobre 1972.

(13) Le rapport du CNEXO sur la pollution de la Rade: "Il faut prévoir plus de 3 accidents par an au niveau de l'appontement pétrolier". *Le Télégramme*, 29 juin 1973.

(14) Extrait du procès-verbal des délibérations. Département du Finistère, Conseil Général. *Séance* du 12 janvier 1973.

(15) G. de POULPIQUET, Lettre à Monsieur le Président de la Chambre de Commerce. 27 novembre 1973.

(16) M. de BENNETOT, Lettre à Monsieur le Rédacteur en Chef du Télégramme, "L'exemple de la raffinerie de Frontignan". Publiée le 26 février 1969.

(17) Le *Télégramme*, 19 décembre 1973.

(18) Pour que nos jeunes ne nous reprochent pas demain le sous-développement de la région brestoise. Chambre de Commerce et d'Industrie de Brest, tract accompagné d'une lettre du Président de la C.C.I., 15 novembre 1973.

(19) Débats communautaires. Comptes rendus. *Le Télégramme*, 18-19 novembre 1973.

(20) Déclaration du syndicat F.E.N. des chercheurs scientifiques, Section du Finistère, 27 novembre 1973.

(21) M. de BENNETOT, Lettre à Monsieur le Rédacteur en Chef du Télégramme, en réponse à la déclaration des 120 scientifiques. Publiée le 16 novembre 1973.

(22) "Raffinerie et Cultures Marines: "Ne déformez pas nos propos s.v.p. ..." demande le Comité Local des Pêches à la Chambre de Commerce". *Le Télégramme*, 14 novembre 1973.

(23) Débat au Conseil Général sur la raffinerie de Brest; déclarations du Préfet et de Me Lombard. *Le Télégramme*, 15 mai 1973.

(24) M. KERVELLA, Président du Comité Local des Pêches Maritimes de Brest et Président de la Société Coopérative Ostréicole du Tinduff (S.C.O.T.) Intervention lors de l'inauguration des installations portuaires du Port du Tinduff, vendredi 13 juillet 1973.

(25) Note des Agriculteurs sur l'enquête d'utilité publique, juillet 1973.

(26) E. LECLERC, *Le Soleil de l'Ouest*, no 7, octobre 1973.

(27) *Report of the President's commission on the accident at Three Mile Island*. Pergamon Press, New York, October 1979 (201 pages).

CONCLUSION: A FIRM AND AUTHORITARIAN EXECUTIVE
FOR THE DEFENCE OF 'PROGRESS'

Urgency and necessity leave no room for doubt, respite and demagogy.
Commonsense demands to go ahead without ever giving in to hesitation. A
simple glance back will, if necessary, reassure the well-foundedness of the
determination that one is showing.

Whatever the problems or the oppositions, one must defend the march of
technological progress. The solutions when they have not yet been achieved
will in any case be achieved in the future: it has always been like this.
The challenges, even if they are lively, will only be passing if those in
charge do not allow themselves be impressed: it has always been the same.
If by chance a disaster exacerbates the difficulties one will hold on to the
steering wheel and round the Cape.

Such is the first type of response in defiance of major risk. We have
asked ourselves if it had a chance of success and in the first analysis we
saw that it did. A ferocious will permits to sweep off the opposition. We
have, however, raised questions: there may be difficulties, even deadlock.

Opposition may harden. What is more serious: the show of strength may
awaken forces that have so far been dormant. We mentioned it in the cases
of the *Tanio* and at Plogoff. When one brings up a culture against oneself
one can no longer be sure of success. One does not disturb the peace of the
sea with impunity.

CHAPTER EIGHT

When the body politic opens up to the problems raised by major risk

I   New directions for the socio-technical
control of major risk
II Innovations in the relationships between
the citizen and the decisions involving
major risk

I. NEW DIRECTIONS FOR THE SOCIO-TECHNICAL CONTROL OF MAJOR RISK

The perspective provides a contrast to the preceding one. It is no longer
a question of making sure, first of all, that usual activities go on in
order to think subsequently about possible adaptations in the field of safety
(while always trying to make these options sustainable for the citizen).
Here, one no longer overcomes the obstacle by acceleration. One tries to
define policies which respond to the new problem that arises, knowing that
the social body as a whole is at stake.

Safety, from being a secondary concern, becomes the object of a central
policy. The management of major risk then calls more for the redefinition
of fundamental principles, in matters of prevention, of battle and of reparation, than for the simple setting up of complementary means. It is no
longer the exclusive business of safety specialists (who are sponsored by
those who hold political power and are in charge of keeping the institutional
debate going). It becomes truly a political question.

The problem no longer exists: what attitude to adopt so that risk does not
impede development policy. It becomes: what policy to work out in order to
deal with the problem of major risk.

1. A DIFFERENT STATUS FOR THE SAFETY FUNCTION

1st. Safety, a preoccupation for general management

A first question arises for major risk. How is the 'safety' function to
be placed in relation to the task of piloting the organisation? The answer
to this preliminary question depends later on the very possibility of setting
up or not setting up an adequate management system for major risk.

If one considers what is at stake in this field one immediately understands
the need for a substantial change in the status usually accorded to the
'safety' function. When on top of the physical integrity of the installation,
the lives of a very large number of wage-earners, the safety of a town or an

area are at stake; when on account of this the very existence of the industrial group (perhaps of the multinational kind) may be affected*; or when
for the same reasons the capacity (or even the fall) of a government could
be up for discussion ... then one can no longer consider safety a question
of secondary importance.

On account of this safety becomes a general management function. It does
not just involve the setting up of means for fire fighting for instance: it
must be accepted as an important criterion among the big options of organisations: choice of investments, markets, products, technologies, objectives
in the field of social relations, brand image, more generally: the social
image of the organisation. Even the very purpose of the enterprise might be
affected by taking the question of major technological risk into consideration.

More generally, the whole life of the organisation and its relations with
the environment may be affected by this taking charge of the problem of risk.
Safety is no longer a particular function when it comes to these major risks
but a dimension completely apart from the life of the system, and it comes
to the cell that pilots the organisation to give the necessary great
directions.

We may explain in a few points what this approach covers.

a) A commitment by general management. Given what is at stake, management
at the highest level cannot relegate the conduct of risk management to a
lower echelon.

b) Choices of a political, not a technical nature. A supporting technical
service has little room for manoeuvre. Given the choices that have been
taken and the constraints imposed it has to set up the best (possible) construction. In matters of major risk the difficulty that must sometimes be
met in satisfactory manner, the task of management and the menaces which
result from it force a reversal: given the possibilities that exist in
reality for the control of risk, it must be decided in the very first place
whether the intended operation shall go ahead, be modified or abandoned.
This is the fundamental choice which it is up to general management to make.

This point may cause surprise inasmuch as the habit is different. In the
name of economic imperatives, imperatives that are considered 'natural'
('going ahead') one usually practises another approach: first the choices
and the irreversible commitment to them, then their management of which the
panoply of means for safety form part.

John Dunster, Deputy Director-General of the British Health and Safety
Executive, has expressed this need to make a choice in matters of major risk
very well:

There are three categories of risks: 'unjustifiable' ones when the risk
is too high to be accepted no matter what the benefits; 'justifiable but
non-justified' ones when the advantages are too small to compensate for
the problem of risk level caused by them; and 'justified' ones when the
risk is worth being accepted because of the benefits associated with it
(1, p. 13).

---

*No doubt, production would not be stopped for good (it would start again
 later, eventually, the company being changed into a national enterprise for
 instance); but the financial interests would undergo substantial changes.

It belongs specifically to general management (after a process of decision making open to agents other than itself, as we shall see) to set the desired objectives in matters of accident probability. This has been done in the field of weapons systems. We have seen the difficulties encountered in this respect in the use of probability methods by the safety services of the civilian nuclear industry. Without precise and explicit definition of the accepted risk levels, the quantification is difficult to carry out*and serves no longer to ensure that the objective is attained but to dictate the choices with a risk of fixing objectives as a function of the results obtained.

2nd. The interpretation of policies for prevention, battle and reparation

a) The perspective. Up till now we have followed the logic imposed by current practice in matters of risk management: we have distinguished the three phases of prevention, battle and reparation. The sequence seems even natural: one must do everything to avoid the accident. If it occurs, a known telephone number must be called, i.e. No. 18; when the fire brigade have finished their work the insurance pays. This, for me, must be reviewed for major risk.

Prevention cannot fix its rules and objectives without looking at the capacities and limits of the means of battle and reparation. The policy of battle cannot, for its part, just base itself on the courage of the fire fighters whose motto (in Paris) shows the unchallengable abnegation:

Save or perish.

With major risk safety requires even more than heroes; it demands, because in the situation it will often be too late, determination beforehand.

Whence perhaps the new keynote of policy, if one is needed for the political organisms in charge of the safety of the population:

Prevent before you forbid.

This is actually understood without any ambiguity by many of those in charge who care for the lives of the men who are committed to safety and to the importance of the mission with which they are entrusted.** As an example we may quote these lines which were written less than two years ago by someone in the hierarchy of the rescue services of a large European agglomeration (he was addressing a local parliamentarian):

The risks caused by today's chemical industry are important in many fields and, given the long-standing existence of the factories, it is very often difficult to reconcile the needs of their development with the safety that one legitimately desires, for the staff as well as for the neighbourhood ...

These difficulties are one more argument in favour of still more careful studies, for better knowledge and greater respect for the limits which must

---

*Specialists for weapons systems spell out that it is quasi impossible to supply exact results in terms of probability; but it is possible to certify that the system has a degree of safety at least equal to that established by the political authority.

**Respect for this mission demands warnings when needed. A disaster avoided is always better than a posthumous medal.

not be exceeded, given the real possibilities for prevention, protection and intervention.

Of course it cannot be a question of 'debating' when one is faced with disaster: the rescue services will always act, no matter how serious the situation. This courage and this heroism at the time of drama does not prevent, on the contrary: it authorises, those in charge to take a decisive part in the working out of decisions which may one day involve their administration, their staff, their mission. Even if the fire brigade must always answer calls, those who are politically in charge of the safety services must still have a more important part to play than they are at present given in the definition of industrial and technological development strategies.

The same is true of the financial reparation function. It has certainly been protected by the ceilings which have often been set for its commitments. Even while maintaining this shield it could be more directly associated with the choices so that no illusion remains as to the capacities of this ultimate rescue in case of disaster.

b) Lost battles; battles not to be lost. The typical example of what not to do is the one of the oil tanker fleet. Once this fleet was constituted, the degree of freedom of choice being close to zero, conventions for the indemnification of victims of black tides were signed*. Once the obvious limits of the funds available had been duly established the question of the means for prevention and battle was raised ... to realise some arrangements could certainly be put in place but one could no longer set up a genuine and necessary prevention system.

As for the automobile, one has long settled for a highway code which as J. C. Wanner emphasises, contrary to the rules for aircraft certification, aims at defining the 'guilty party' but does not try to prevent accidents**. (2, p. 7).

Half a century after this means of transport was developed which is so essential for the economy, for social life, for the structuring of space etc., one began introducing some safety considerations. However this is a further battle lost. One cannot counterbalance huge political failures with technical patch-ups (like the safety belt): even the victims jib or oppose them (as the in-town rules show which have become a major affair before one could really judge the usefulness of the measure).

A third deadlock of this kind (but outside the industrial framework) are the fires in the mediterranean countryside. In the absence of a country planning policy, a large expanse ready to catch fire develops; the fighting forces are largely impotent, even with the most sophisticated means. The course of events is accelerating when malevolence is added to 'natural' causes.

One might mention other battles that are going on at present which are not

---

*The signing must not be confused with application.

**Whence the unwarranted rules such as right of way (a source of serious errors: who has not sometimes confused right and left) or absurd practices which oblige construction companies to decorate their road work sites with posters which are useless for safety (and inauspicious because one no longer pays attention to useful posters) (2, p. 7).

yet lost. However the lessons of deadlocks that have already occurred should
suffice. Without integration of the various components of the safety function, without a new definition of the place of this function within the
policy of development, which is very upstream in the study of projects, one
must fear new deadlocks of unsuspected seriousness.

c) A ministry for safety. The question of the status to be accorded to the
administration of industrial safety has at last been asked. Civil safety is
too often thought of from the outside, as a sort of 'after-sales-service' on
which one can always count when an operation runs into trouble; on a service
to whose advice one does not listen beforehand and the demands of which one
listens to even less; a service, at last, which is useful as a scapegoat that
is immediately available in case of serious trouble. Major risk leads us to
rethink this kind of approach completely.

The previously mentioned developments give cause to rethink the idea which
some put forward: one must give the industrial safety organisms a different
status in order to confirm their political purpose clearly.

The creation of a Ministry for Safety, for instance, would give the administration the power to intervene more effectively on the level of development planning and to exercise to the limit a right of veto very much upstream
in the choice of projects and technological options.

This right of vetoing deserves some explanation. It cannot be a question
of going from one extreme (as spelled out very well on the front page of a
daily newspaper: "If all goes well, nothing will go wrong") to the other
which would be a fall-back on the 'caution' positions which on the strength
of being amplified would themselves become risk factors. The concern for
safety must not become an obsession. A responsible service should not go for
extremes. It would have the possibility in really serious cases not to
permit the creation or the continuation of situations which forbode too
great an upheaval. Of course, such an innovation would not go through without difficulties. But the stakes of major risk may lead to a strengthening
of the demand for further change rather than choosing the _status quo_.

Inasmuch as this integration of the various components of safety could be
realised (civil safety in particular would thus gain the political weight
which it has lacked) one could think of setting up systems and means
adequate to the needs. Here again the first effort to be made is of a
structural nature. Even if strengthening in quantitative terms is necessary,
it is important above all to define new logics of action.

2. A DIFFERENT STRATEGY FOR SAFETY

1st. Prevention adapted to the problem of major risk

The essential point in matters of prevention is the recognition of the
phenomenon of major risk. From there derives the need to mark out the existing dangers and to define clearly: the political objectives (in terms of
probability of occurrence), the agents concerned, the means made available
to ensure prevention. In every large field completely clear thresholds must
be fixed. One does not take on just any danger even if the calculated
frequency of deadlock is very low; one does not take on just any risk based
exclusively on a consensus within the regulatory authority; one does not
accept those confused situations where the centres of responsibility clash

in an ungraspable fog*; one does not tolerate flagrant insufficiency of the means of prevention.

This kind of minimal precaution having been ascertained, one can then examine the quality of the control one has over the system. The analysis to be carried out is social as well as economic and technical. If on this level serious insufficiencies appear they must lead to an immediate return to the first stage because the minimal conditions are unsatisfactory. It is then important to remedy them, which can involve substantial cost, or, if this is not possible, to reexamine the choices made. This reexamination may call much into question; one will have to judge in the light of risks involved. If a city of a million inhabitants is endangered by a factory fundamental changes in the installation should not be excluded. As this is already the case for toxic products the ban may be upheld.

This type of extreme measure, when it is well-founded, is not necessarily against existing interests at an economic level, neither is it at medium term because an accident could have much more serious consequences, nor at long term because the disaster could affect more than one installation: the whole of an economic activity, as the commission appointed by President Carter on the accident of Harrisburg has pointed out (3, p. 25).

The analysis of the systems involved in major risk is as difficult as the complexity of those very systems. We know that accidents usually result from a number of 'minor' faults (and not from the sudden fall of a meteorite, to mention the most explicit example). From here on a different alarm system (and no doubt a different stopping system) must be foreseen: it concerns uncertainty. If from a certain moment onwards the analysts see that their results enter too much into the field of uncertainty they must go back to the basic options. Many have asked themselves whether the uncertainty which exists concerning the development of fissures affecting certain elements of nuclear centres should not start this kind of process.

2nd. Faced with the disaster

a) The exercise of authority. The protection of the population in times of crisis was organised for the first time in the middle of a world war: it was strengthened during the 'cold war' mainly because of the fact that the Korean war caused fears; first the ORSEC-RAD was worked out to respond to the risk of the transport of fissurable material (creation of the Force de Frappe and the civilian nuclear industry). It is therefore not surprising that military logic has determined the adopted organisation of the rescue operation. The problem of secrecy is often stressed. There is another one which is no less important: the social systems, in peace times, are much more complex to manage than a well organised army. The issuing of orders is replaced by a communication between sub-systems each of which has its own aims, objectives, interests. The networks, as we have seen, are therefore very complex.

In the second place, a military model is often built on a scenario that is alien to a disaster situation. The civilian disaster is not an operation limited in time and space, a prepared engagement. It rather resembles a 'phoney war', a collapse. If one were to look for lessons to learn from

---

*As in a cloud of fog: the more light one puts on it the more the diffraction of it blinds. If at the same time the risk increases there comes a time when a halt must be called to restrain circumspection.

military sources the most useful ones would be found in the management of
defeats, and retreats. What does one do when communications no longer work,
when the command posts have been destroyed, when the units have been broken?
When information turns out to be fragmentary and false? When the available
plans are useless, even dangerous to follow? Exceptional personalities may
then emerge who take command out of the hands of those by whom it is normally
held. Depending on how the crisis develops and on the personalities concerned the disaster may be used to confront the existing political society
and challenge it for power.

The nature of disaster is to be extraordinary: in overthrowing plans,
calling into question both the means available and the organisation provided.
If one claims, as is always done, that "everything is provided for", that
the "ORSEC plan has been put into operation" etc. one is heading for bitter
disappointment. Between principles and an integrated action in a concrete
situation there is a gap which undue confidence and certain political
imperatives often try to overlook imprudently. The battle won on paper is
lost on the ground.

Thus again, to provide a command authority is right. Wanting to clarify
information is also right. However going to extremes and translating
principle into authority without participation, into a general restriction
of information, is inadequate. This point is crucial. There is even the
idea, especially since the TMI accident, that in case of a serious event the
hierarchical powers must be tightened, based on a quasi-military model and
again on an elite unit. Messages must be simple and reassuring: all is well,
there is no problem; everything is under control; the authority is looking
after you ... Authority will be concentrated in the hands of four or five
people, a single one is given power to supply information. This scheme may
be adequate for a parachute unit which is getting ready for a raid; but is it
adequate for a large scale disaster? Without rejecting it altogether, an
exceptional situation calls for clarity and firmness, one may nevertheless
raise some questions.

Restricting information? At TMI the news went on the air before the local
authorities had been alerted: a journalist had been connected, by mistake,
directly with the control room and learned the news (3, p. 103). At Seveso,
the independent doctors were also soon aware of the presence of a formidable
poison; and they were faster than the official doctors in searching the
libraries. On the *Titanic*, a dumbfounded steward starting shouting: "We
have hit an iceberg" at the very time when the crew tried to get everybody
back to their cabins: "Nothing has happened, all is well".

One cannot determine beforehand that information will be strictly channelled. Trying to stick to it, one risks running into trouble and losing one's
credibility quickly.   -

Highly concentrated authority? If the principle is pushed to the extreme
this also leads to disaster. In a situation of widespread uncertainty a
sovereign attitude can cause incredulity and at the first error the chief is
discredited. In a confused situation, however, it is difficult not to set a
foot wrong. When a highly placed personality, such as Dr Reggiani, the
Director of Medical Research Laboratories of Hoffmann-la-Roche, in the Seveso
case, goes to say that in fact the situation is extremely serious and the
official error is interpreted as a deliberate attempt to cover up the facts,
the authorities lose all credibility.

The principles of preparation for battle and of a unified command are good

and right. Their descent into false reassurance and into intransigent
authoritarianism can only have fatal consequences. A disaster, let us stress
it again, resembles much more a collapse (potential or real) than a commando
raid. Therefore different rules of action are required.

b) The management of information. The normal idea in this field is that the
authority must be well informed, 'on the hour', about the situation; thanks
to experts and competent services the information available will be the best
and the safest. On this basis it is possible to act with determination,
without hesitation, always on the understanding that not everything will be.
revealed to the public so as to avoid confusion.

One must approach the problem differently and begin to ask the question:
as far as information is concerned, what is a disaster? With Y. Stourdze we
shall say that a disaster is characterised by two factors:

- the generation of information at great speed;
- the accumulation of information which one can no longer deal with and of
  which one does not know the limits of veracity.

The following lines from the same author explain these two realities: The
contemporary question of disaster asks in turn the question of the make-up
of the network of information. The vulnerability of the classical systems
comes more and more to light. They prove to be not only inoperative in
exceptional situations but they also prove to be dangerous amplifiers. Why?

The disaster crisis expresses itself in a fantastic increase in the mass
of information produced; it gears down the speed of its emission and dissemination. Such information then constitutes a multiform pile ...

The classical systems of interpretation which hold up, provided they have
not been swept away by the shock wave of information, try to agglomerate the
information in order to justify the well-foundedness of their previous
strategies. In short: the information camp shows quite an unusual configur-
ation: an enormous mass, confused, fractured ... and the classical treatment
instruments continue logically (mechanically) to treat information according
to the lines of interpretation which they consider most favourable for their
own defence.

Summing up, the situation is catastrophic because the mass of information,
and the emergency, imply means of rapid gathering and interpretation of
information when even the networks are smashed and even the poles of
analysis are naturally inclined towards projecting in their analysis their
own retroactively protective coordinates.

Must one not think after the fashion of military strategy that it would be
possible to distinguish from here on two aspects for analysis:

a) the first features of traditional strategies for minor disaster. The
   classical official response to such affairs.
b) in the second one by contrast the major disaster immediately encounters
   the problem of complexity. The traditional hierarchical procedures
   applied to its solution lead inevitably to discomfiture ...

Here, facing a crisis situation means first of all catching the multiform
signals which emerge from the critical zone. One cannot stress strongly
enough the difficulty of controlling a multitude of signals which have been
generated at great speed. One must get hold of them, retain them, then treat

them in layers, progressively make sense of the strata of information by
sorting out, distilling and isolating up to the proposal of an outline of
the 'core of usefulness'. How does one achieve such a process? By setting
in motion large series of joint gathering and analysing operations, themselves paralleled by a set of simulations which permit the application of a
(reliability) coefficient to each hypothesis of meaning, then by coupling
and aggregation from these 'particles of sense'. This is a difficult job of
understanding the complexity which is more exasperating when one must act
fast.

Whence the need to have structures that will bend to a discipline of this
kind ... Either communication configurations of the more inert structures
solidify little by little or they propose flexibility and potential inter-
connections ... In the first case, any catastrophic avalanche will cause
irreparable collapses; in the second one the established networks will intervene in order to produce transitory and intermediary states of equilibrium which permit the neutralisation of the 'shock wave' (4, pp. 128-130).

These observations constitute an interesting warning. The problem is not
whether or not to let out the good information one has at the command post.
It is a matter of managing the information networks in a much more complex
fashion, to absorb a great mass of data, to know how to read them, to know
the credibility limits of each message and again to coordinate a public
reading of the information, knowing that in any case everybody has information, 'true or false', sometimes even before the authorities have it; the
non-official networks function more subtly and faster than the classical
hierarchical lines of communication.

These remarks lead to the linking of battle and prevention because the
networks to be set up for gathering, handling, evaluating and giving information must be there before the event. In a crisis situation it is too late.

3rd. Disaster management, management of a rout

The previous points may surprise but is not the classical representation
the right one on the evidence? The police commissioner has a plan. He gets
the news of the disaster. In line with the government he orders the ORSEC
plan to be put into operation, ensures respect for the authorities, forms a
general staff for giving information and supplies at the right time the sort
of information that is considered useful and judicious in order to avoid
panic. The implicit representation is finally the following: technicians
will soon have the situation in hand; it is above all a matter of dealing
with the psychological problem created by the event. A good plan will help
the technicians, well proportioned information will permit taking care of
the citizen in suitable fashion.

These mental representations (as expressed, actually, in certain written
work are of great importance as we have seen before. We shall therefore
insist on this other image which is much closer to what must be envisaged if
one wants to be better prepared for really serious disasters and understand
the disorganisation which may follow a modern disaster. Sending 200 ambulances to the end of a runway at Charles de Gaulle airport is a matter of
organisation: a little authority can see the operation through. Finding an
answer to a major escape of toxic gas in an agglomeration is another problem.
What order would the police commissioner want to give? In the first place
one would not easily reach one's correspondents: a number of intermediaries
would already have fled, having heard that there was serious danger. One
would not have a sufficient number of gas masks and the rumour would soon
spread that the available masks offer hardly any protection against the

substance spread about, that respiratory apparatus is indispensible*. The
authorities would make an appeal over the radio and give orders. Being ill-
informed on the latest events, or on the latest rumours, they would leave a
lack of certitude in their message or sound inadequate in their communication.
Disbelief would be on the march.

The situation is in fact too complex to permit the authority to take charge
of everything, to rule everything by demanding understanding, calm and
passivity of the citizen.

The problem must be framed in quite different terms as it must be for the
purpose of information. The first task for the authorities is to draw the
outlines of a response and to go along with them, coordinate afterwards a
general effort to put the adopted policy into action. There is hardly any
other policy in a situation of great danger (of an inestimable menace such
as a toxic cloud unless one is willing to go to extremes, to resort to
armoured vehicles, a rather unglorious perspective for societies which claim
to be democracies).

It is therefore a question of exercising one's authority by showing one's
ability to coordinate and direct a collective movement. This too will have
to be prepared in advance. One remembers here D. Fischer's reflection on
networks.

As an example and to take up once more the essential point of the information we shall give below, the case of the use of 'citizens band radio' in
the USA as an organisational support in a disaster situation. The resources
of the volunteer organisations are then used (instead of being neglected)
and attached to the official services (instead of being rejected). To accept
this scheme takes no doubt much intelligence and strength on the part of the
authorities: they accept to lose their power of facading in order to gain in
efficiency and to share the responsibility for the battle with others, the
pride not to be useless for the community. In a word, the authorities take
on what is the most delicate: strategic leadership of the action and delegation of some of the ground work; which again is not comfortable: one can
no longer lose oneself in mere activism to cover up strategic incapacity.

Citizens Band: the case of its use in the 1978 spring floods in North
Dakota and Minnesota. Citizens Band means a linking of citizens, largely by
means of radios which are mainly installed in cars. At present there exist
more than 35 million sets (transmitter-receivers) of this type in the USA.
This communication system permits, particularly in rural areas, to make up
for the lack of telephones, to coordinate activity during peak periods (such
as harvest times). It is a valuable system in case of road accidents: the
first car that stops can alert the rescue services (this has for instance
permitted the reduction of intervention time for rescue from fourteen to
eight minutes in the state of Missouri). One also notices the cooperation of
citizens band operators with the police services**.

What could be the use of citizens band in a disaster situation? It could
be multiple:

----

*This kind of problem was found in the case of the disaster at Texas City
  in 1947.

**Contrary to what one might have feared the citizens band people do not just
  concentrate on impeding the police.

- giving information to the authorities;
- help with calls for assistance;
- supplementing or making up for the inadequate telephone network which may be out of operation or overloaded;
- offering to the authorities groups that are set up, organised, equipped with apparatus that is most valuable in a disaster situation, i.e. transmission apparatus, and ready to be used as a support that is immediately operational (for non-specialised tasks) in conjunction with the existing official services which thus can concentrate in their operational command posts on the direction of the battle.*

The usefulness of CB is above all important in case of a drawn-out crisis. In the case of the 1978 spring floods in North Dakota and Minnesota the following phenomenon was noticed:

the police attached volunteer groups to itself which were registered as "official emergency workers", given ID cards, orange jackets, numbers etc. They were used as dam inspection patrols and as traffic police;

the Red Cross used the support of another CB organisation. The latter took on a food distribution and sandbag supply function and general assistance to the command post. At the very beginning CB supplied the Red Cross with the radio facilities it was lacking;

in addition the Salvation Army who had radio facilities supplied assistance with food distribution, search for shelter and making a labour force available.

Thus conveniently attached to an administrative structure CB proved to be a very valuable communications channel: mobile units, support missions which permitted the achievement of certain objectives without having to dismantle the command posts of the police or the Red Cross. CB provided equipment which would neither have been given to nor bought by them and expertise acquired over a long period in the use of this equipment.

Some difficulties came to light. In this case there was insufficient preparation: relations with the official services had not been sufficiently studied; the problem of children returning from school at the end of the afternoon had to be sorted out as they disturbed radio communications for a while; the problem of irresponsible use of the channels was, however, resolved.

These few difficulties did not discourage the officials at all who found the experience extremely positive. There was very little irresponsibility; CB was a gold mine. On condition that the relations to be developed between organisations are well studied in advance and rules of utilisation for each case are provided (according to the local sociology and not along the lines of bureaucratic federal rules) the system proves itself useful and keeps the officials free to take care of the most difficult tasks.

REFERENCES

(1) J. DUNSTER, Virtue in compromise. *New Scientist*, 26 May 1977.

---

*The confidence given to these non-official organisations may even be an absolute necessity: there has been a case where the aerial of the police station was put out of action by the event. The CB network was most welcome.

(2) J. C. WANNER, Critères généraux de sécurité. Association Technique Maritime et Aéronautique Session 1980 (30 pages).

(3) *Report of the President's Commission on the Accident at Three Mile Island.* Pergamon Press, New York, October 1979 (201 pages).

(4) Y. STOURDZE, Hypothèses sur la relation catastrophe/réseau. *Futuribles,* no 28, novembre 1979, pp. 126-130.

(5) Th. RABEK, D. BRODIE, J. EDGERTON and P. MUNSON, *The flood breakers C. B. use during the 1978 flood in the Grand Forks region.* University of Colorado, 1979 (123 pages).

II. INNOVATIONS IN THE RELATIONS BETWEEN THE CITIZEN
AND THE DECISIONS CONCERNING MAJOR RISK

Major risk, as we have just seen, challenges technological and industrial development policies. In the second scenario that we are studying here the response is not limited to a strengthening of the means of persuading and coercing someone in charge into the defensive: the policy, set up to guide the event, adapts itself and takes charge of the problem of major risk. It is the same where relations between the authority and the citizen are concerned. In the approach set out in the preceding chapter a social 'technology' drawing on the whole arsenal of manipulation was the weapon of the decision maker. There exists another perspective: we shall examine it hereafter. It carries the imprint of quite a different spirit: no longer to get projects and risks accepted which are considered non-negotiable but to work out together with the social body the approaches to and the means of its way of life of which technology is one of the main parametres. The manager of necessity hands over to the politician. His main task is to construct possibilities for choice between variants, to put the exercise of this social choice into operation and finally to guide its realisation.

This is a situation which strongly contrasts with the preceding one; we shall study one aspect of it. Various ways of bringing it about will be presented subsequently. Of course, some of them are more timorous than others; some of those mentioned in the following pages also still conform, in their spirit and still more in their effective application, to the perspective set out in the preceding chapter. They present, however, a first step, even if sometimes hardly perceptible, towards an attempt at opening up the procedures of decision making.

1. 'RISK ASSESSMENT' OR THE POLITICAL EVALUATION OF MAJOR RISK

1st. The typical aspect of 'assessment'

The term 'assessment', in the sense in which we use it here, appeared for the first time in the expression 'technology assessment' (evaluation of broad technological options) as used by the senator Daddario, President of the subcommittee for Scientific Research and Development of the American Congress in October 1966. The use of this neologism answered a twofold need: to express the urgency of much broader studies of technological choices and their immediate and deferred effects on society and environment and also to stress the urgency to reconcile the social body as a whole with the tool set up to ensure its well-being: technology (1-7).

The translation of the term 'assessment' (into French) is delicate but one often uses the formula 'evaluation of the broad technological options'. Some additional remarks will permit us to understand its scope.

What is fundamentally at stake, and what the Americans had perceived since 1966, are the following questions. One can no longer escape the questions which arise more and more acutely and incessantly within our industrialised societies: who may legitimately take a specific decision and involve a specific risk? What, within a choice to be made, constitutes a 'given' constraint? Who shall be the arbiter in terms of style of development, pulling back a specific constraint, necessarily at a cost, in the broad sense of the term?

Until now it was up to the decision maker to define at the same time the margins of freedom and to design the optimal actions to be undertaken. Now

one sees more and more the appearance of two social demands. There is in
the first place the examination of everything that was traditionally presented as 'necessary' which explains itself by the fact that the customary
ways of development are called into question. Secondly one wants to have
the possibility of altering the choices that have been studied which requires greater participation in the capacity for information, expertise,
decision making and control.

This, then is the spirit of technology assessment. Let us recall the two
complementary definitions which were formulated by F. Hetman and J. C.
Derian-Staropoli respectively: It is a question of seizing hold of technology, society and the natural environment conjointly (7, p. 82). What
defines technological assessment in the last resort is the association of
the interested social groups during the phase of enquiry ... (3, p. 65).

A. Staropoli adds further that it is a matter of retrieving the idea of
the 'assizes' (3, p. 27), where the decision must be taken, if there is no
simple solution which a code of law prescribes beyond any doubt, on the
conscience of the jurors. One can easily see the perspective suggested by
the notion of assessment: a society must choose 'on its conscience' what
shall determine its development; it is not at all a question for it to bend
simply to the orders given by the experts. In matters of technological
development freedom and creativity must be given precedence over fate. The
experts have only the task of preparing the choices: it is not for them to
decide what is best for a community.

The same demands must be made in what is called 'risk assessment'. The
fundamental question is the following: given the potential dangers attached
to a specific style of development, to an available or envisaged technological option, what policy, involving such a future and such risks, shall be
adopted? This is the question.

The word 'policy' covers two kinds of problems. On the one hand a problem
of arbitration to be exercised between various approaches open to society:
this presupposes a good knowledge of the constraints and the margins of
freedom that exist. The question is one of working out projects which are
technologically, economically and socially coherent, that are clearly
explained where the 'risk' dimension is concerned which they necessarily
involve. On the other hand, there is a problem of legitimacy: it is a matter
of using proper institutional means to ensure that the arbitration (from the
initial study to the decision, the application and control) rests on a
clearly political choice (in the sense of general interest). In other words,
it is recognised, all along the arbitration process, that this is an exercise
by which the whole citizenry with all its components works out the means and
approaches to its development, that it is not a matter of optimisation under
arbitrarily appointed constraints to be realised by experts who find legit-
imisation in science or in their position of power within society.

We have recalled the two key demands of the assessment approach, planning
of the whole, search for legitimacy, in order to avoid all misunderstanding:
risk assessment is fed by all the achievements of scientific study, management and general risk control. A social consensus on an option devoid of
interest would be nonsense. The developments of the preceding point in this
chapter, new directions for the socio-technical control of major risk, must
therefore be taken into account in the search for legitimacy, a point which
we shall now examine more closely.

2nd. The irreducible need for social choice

   The tendency which consists in making tools to help in the decision on
tools while avoiding the crucial problem of choice and imposing just one
solution, 'the one', dictated by rationality, is energetically rejected by
those who clearly recognise the highly political nature of decisions which
affect major risk. If one can speak of counter-attack one might say that
the latter takes place along three lines of argument. The first one intends
to ascertain for policy the control of a field which the technician would
like to appropriate for himself. The second aim is to show that ambiguity
bears down heavily on scientific results, and that this is the more so when
it is a matter of little known major risks. The third one consists in a
political analysis of scientific practice: it still broaches the assurances
of objectivity and rationality with which scientific analysis most often
wants to bedeck itself. This kind of questioning is certainly not new but
it acquires an unexpected strength in the field of major risk.

   Before examining these pleadings one by one let us stress once more that
the aim is not at all to reject the tools which help decision making as such,
they are recognised as absolutely necessary as we have already said, but
only to make better use of them.

a) The scientific tool must not obscure the political choice. The main rejection concerns the notion of acceptable risk. *'Acceptability vs. Democracy'*
(9) is the way L. Mc Ginty and G. Atherley refer to it. They reexamine the
propositions made by experts like T. Kletz and develop their criticism. Can
one talk of an 'acceptable' risk? Can one establish comparisons between
different risks to define a unique measuring scale for the acceptability of
risk? What is in the end the function of this notion? We mention the
following reflections:

It is wrong to think that there is only one acceptable risk level. One must
ask these questions: acceptable for whom? Who is going to judge? The
promoters, the potential victims, the experts, the representative organisations, the community in general, the government ...? In fact, every point
of view defines its own level of acceptability and each one differs largely
from the others.

The idea that one can establish correspondences and comparisons between risks
of different nature, measured on a single scale and deduce from them a safe
evaluation must be rejected. To sum up one could say that the parametres
which lead to estimation*cannot be a basis for evaluation. Mc Ginty and
Atherley develop this and stress particularly the seriousness of a risk
expressed in a frequency figure is only a dimension of this: it is not
because the risk connected with the chemical industry is lower than that of
being struck by lightning that the chemical risk is 'acceptable'; the risk
of being struck by lightning is accepted because the remedy against it would
be too expensive, the authors say. In additions one should ask oneself what
this idea means: "accepting to be struck by lightning". What means have
people been given to express their preference in this respect? Has it been
explicitly discussed?

   These two questions show already the banality of these shortcuts**which an

_____

  *The notions of 'estimation' and 'evaluation' have been set out in chapter
   three.
 **Of the following type also: 'people accept the automobile'. When has the

approach concerned too much with appearing 'objective' permits itself.

One finds a similar banality in the argument: "accepted risk in a specific industry": is there really acceptance? or simply a state of fact against which the victims are largely impotent? The argument resembles the one which consists in measuring disturbances from noise (airports are the classical example) on the reduction of rents. Without reentering into the whole discussion, let us say that it is difficult to pretend that one is dealing with a social question, the acceptability of risk, by skimming over the socio-logical realities (powers, interests, economic capacity of the groups involved, social representation, values etc.). As Mc Ginty and Atherley say in picturesque style:

comparing specific risks in different fields amounts to comparing apples and oranges. One may be able to demonstrate statistically that the apple is heavier than the orange but the statistics are of little use in determining which fruit the people prefer (9, p. 324).

In fact, as they emphasise with good reason, there is no quantitative scientific methodology which could be substituted for political judgement. This is where we find the essential criticism in the encounter with the so-called method of acceptable risk.

The notion of 'acceptable risk' is unacceptable largely because of the role it has been given. The idea that one can in quasi mechanical fashion pass from the estimation of a risk to its evaluation is in fact unacceptable. As always in such cases there is great temptation to substitute for the political process, for the assumption of responsibility in decision making, a logic that has a scientific appearance, an appearance of solidity. When it is actually a case of a mirage — Mc Ginty and Atherley remark — the image of precision and veracity in the measurement of risk is sought for and serves to bring about conviction. Here again we come across the criticisms made with regard to the cost-benefit analysis:

The cost-benefit analysis has the tendency of converting the political, social and moral choices into a pseudo-technical choice. Whence the attraction it has for administrators, whence also its obvious logical fault for those who are used to the analysis of choice (10, p. 53).

Our two authors make an identical analysis concerning risk:

Because the (British) governments have no risk policy they turn more and more to strategies drawn from the philosophy of the "acceptable risk". This approach is profoundly undemocratic and must be abandoned (9, p. 324).

This demand was already formulated by K. W. Kapp for instance in the debate on the political economy of the environment:

We shall have the task of introducing (earlier) politically formulated standards into the socio-economic process. In short, we shall have to

---

citizen been asked, while being offered a variety of transport systems, what he would like to see developed? The fact that there has not been a general uprising against the automobile does not mean that one 'accepts' the 13,000 deaths per year on the French roads. But, of course, the argument is too nice, even if it is not well founded: if one 'accepts' 13,000 deaths on the roads, that leaves a good margin in other sectors.

operate with positive and socially acceptable criteria which have received
political sanction (11, p. 124).

This is precisely what the method of automatic estimation of acceptable
risk leaves out of consideration which has become so dominant in thoughts
on safety that it is no longer questioned as Mc Ginty and Atherley also
remark and continue:

There is no overall policy for risk management. One ministry has its
standards, another ministry has different ones. The regulation of risk is
rarely, if ever, discussed in the council of ministers or in parliament.
The political parties show little interest for it. As a result the safety
rules and general risk policy are not subjected to any of the traditional
democratic controls or constraints. In the absence of any coherent policy
for the regulation of risk, the ministries turn more and more to the
philosophy of the acceptable risk which is favoured and promulgated by
officials and experts who take the place of more open and democratic debate
... These experts have no special qualification to declare which risk
level is acceptable (9, p. 324).

Thus the hypotheses necessary for research into acceptable risk are called
into question; we quote here H. Otway and then (again) Mc Ginty-Atherley:

The implicit hypotheses of these approaches to risk evaluation are that
the preferences which society has shown in its acceptance of risks in the
past can be drawn from statistical data, extrapolated for the future and
compared with other types of risk (12, p. 8).

We think that in practice ... it is impossible to compare risks of a
different nature, risks run for different reasons and in different social
circumstances (9, p. 323).

Thus the process of decision making based only on quantitative analysis is
condemned: a calculation, an opinion poll, a psychological analysis etc. are
not acceptable as the only operators of social choice.

b) Science in the grip of ambiguity. The idea according to which the
scientists give the 'facts', the indispensible rational basis for the social
debate which latter is 'passionate', 'subjective', 'irrational', 'politicised'
— has been questioned for a long time. Rather than 'facts' the scientist
presents 'documents'. This questioning of the status of science is
strengthened in a new way when one approaches the problem of major risk:
science proves then even more impotent to say what is 'true'. One must in
fact consider, as has been emphasised in the study on limits of scientific
tools, a certain number of factors which gain particular importance when one
moves away from normality:

Differences in measurements and methods of measuring can lead to the
generation of bodies of different data which rapidly produce very diverging
results when one is concerned with phenomena as sensitive as major risk.

Practices such as extrapolation can be called into question as soon as one
concerns oneself with risks of disasters.

The practice of verification is often impossible. The frequent absence of
experience makes the production of data difficult. The high degree of uncertainty requires working with such large reliability intervals that the
data produced lose their significance.

Many risks can remain undetected for a long time, cause and effect relationships occur in unusually complex contexts which can condemn even science "to
get there always too late".

H. Novotny makes quite a number of these observations: from the study of
the case of the controversy over the effect of weak doses of radiation (13)
If for a problem of this kind several decades of study are needed, several
million mice to carry out the necessary tests, how can one expect 'proofs'?
H. Novotny comes to this conclusion:

> Scientific judgements do not find solid ground in what is often only a
> basis, preliminary and incomplete data ... And if one must wait for
> decades to obtain solid bases for decisions then it is clear that one
> cannot find the ground for a non-ambiguous answer in the scientific field
> (13, p. 2).

To describe this situation which takes on particular importance when
major risks develop, Haefele has suggested using the qualifying term
'Hypothetical': we are entering fields where scientific examination proves
radically inconclusive. (14).

Along this line one can no longer consider oneself to be within the
strictly limited field of science and Weinberg has suggested speaking of
'transcientific' problems; they cause controversies which can no longer be
accommodated in the scientific field; one can no longer attain objective
proofs (15).

c) *Science invaded by politics*. One easily recognises that scientific
production, particularly in the field of risk, is not neutral towards the
social and political scene: a certain result will at the same time comfort
and weaken groups which are in conflict. One also recognises sometimes, as
Haefele and Weinberg have done, that the scientific approach cannot be a
goal in itself: the 'transcientific' problems lead to having to integrate
social analyses for orientation in the field of the 'hypothetical'.

There is still more however. One finds also that even the working out of
truly scientific results is intimately invested with politics. R. Johnston
and B. Gillespie spelled this point out in a concise way as far as the
question of risk is concerned which is particularly critical in this
connection (16):

Risk identification depends to a critical degree on the perception of risk
one has, on socially determined perception of damage. This immediately
affects the question of data selection and of the construction of the
scientific problem. R. Johnston says:

> If the construction of the issue is always a response to public or govern-
> ment pressure, then it is very unlikely that any coherent body of appro-
> priate scientific knowledge can emerge. Further still, if the risk is
> irremediably socially constructed, the attempt at building an exclusively
> objective basis to measure it reflects either scientific stupidity or
> political deception (16, p. 6).

The ambiguity of data, methods, results gives all kinds of political influences a grip. One cannot isolate such scientific practice from the power
complex in which it is located. Our two authors stress in this respect that
scientific production and interpretation may well reflect the social perception of one of the parties involved, and of one only:

The selection, the production, the interpretation of scientific knowledge may be modelled on a network of social and political forces (16, p. 7).

Adopting this framework of analysis, R. Johnston and B. Gillespie have studied the problems connected with the evaluations made in the field of toxicology and the one of carcinogenic effects of various substances.   H. Nowotny has looked into the examination of the controversy on weak doses of radiation.   She remarked in this case for instance (13, p. 27) that it had been explicitly admitted that there was an ethical problem to be taken into account, that in the USA one had been preoccupied with the double role (promoter+controller) of the Atomic Energy Commission; that the opposing scientist had clearly aimed at taking the debate beyond the usually closed scientific field which led to a controversy which she qualified as 'parascientific'*.

All this seriously complicates the presentation of science as being one and objective, particularly when it is a matter of major risk.   H. Nowotny and H. Hirsh spell out very clearly the problem thus created:

Historically speaking, the strategy which consists in keeping the political conflict outside the field of science, derived from the pretended political neutrality of science and from the rigid and very effective system set up so as to maintain the established frontiers by simply eliminating all questions and debates for which one did not seem to have a scientific solu- tion or which could not be separated from all non-scientific elements that attach to it.   While we must recognise the effectiveness of this strategy, its days are perhaps numbered (17, p. 18).

These various responses to the simple mechanistic approach, which is the most customary, because it is the easiest and most comfortable one, are just so many arguments developed in favour of the 'risk assessment' perspective. Without this political approach one falls victim to mystification; and this is difficult to justify and to keep up in a society which takes science as its essential point of reference.

This has been understood in some countries for several decades.   Little by little all industrial countries have taken to reexamining their practices in matters of decision making.

Whatever reluctance there may be, it is indicated to study the innovations, which the countries that are most attached to their democratic way of life, experience in this field.

2. MODALITIES FOR INTRODUCING AN OPENING-UP OF POLITICS

One must not ignore the vitality and determination of those who favour the scenario of firmness just described.   A chain of radical objections has therefore rapidly been raised against the explorations which would appear desirable where the exercise of democracy is concerned.

---

*The author does not use the term 'transcientific' which suggests that there is a limit which science cannot pass, a threshold of certainties which it can present and that from there on, social action must take the field. 'Parascientific' suggests that scientific and non-scientific debates mix, the latter leaning on the former.

The social body is divided into experts, who know, and 'the public', who are incompetent and irrational. It is up to the former to provide for the needs of the latter.

The solutions adopted by the experts can only be good solutions; it would be out of place to call them into question. The difficulty for the experts is to make rationality emerge sufficiently within the public, so that a broad consensus can establish itself on the options adopted which alone can provide for the welfare of the population.

Progress is a straight road on which many competitors face each other; success comes to the fastest. All hesitation, every step that is not taken, is a drop-back. All that one knows how to do, that one thinks one knows how to do, must be done.

Mankind has always extricated itself from difficulties; no technological risk must be rejected.

People have always accepted the price of progress. One cannot see why this should change.

We live in a world in which, no matter what we want, there is no longer a choice; 'necessity' is the order of the day.

An opening of the decision making process would not only lead to errors but would also unnecessarily worry the public.

The livelier these principles are, the less the modalities of a democracy which goes beyond simple, hostile non-reaction seem to be studied, tested, evaluated. The countries used to pluralism, to the confrontation of the interests at stake are by contrast more creative in this field. The record which G. Nichols has established in an OECD publication shows this very clearly (18).

1st. Information of the citizen

a) Access to information. The Scandinavian countries and the USA were the first to adopt legislation providing access to official documents. Other countries, such as France, have made arrangements in the same direction (Law No. 78 753 of July 17, 1978 — First section title: On the freedom of access to administrative documents) (19). This general development towards greater transparency of the actions of the public power is not achieved (or started) without difficulty: fear of impeding internal debate, of uncovering secrets, refusal to give up the classical model which assures the secrecy of ministerial deliberations; worry about having the habitual world in which the administration is considered to be neutral, impartial and objective vanish, fear also in terms of cost, loss of time and efficiency.

Nevertheless, the arguments in favour of greater openness have had some effect. G. Nichols recalls these lines from the reasons for the decision handed down in the USA to constrain the divulgence of a critical study on the project of civilian supersonic aircraft (SST):

The information requirement of the public is particularly important in the field of science and technology, because the ever growing expansion of scientific knowledge threatens to exceed our collective capacity to keep control of its effects on our existence ... It would run counter to the spirit of the law if one refused the public factual information on

scientific federal programs, the future of which are at the centre of
public debate (18, p. 26, 20).

Depending on the country, its administration and government tradition,
access to information has taken on different forms, the major factor being
the number of exceptions provided for withholding documents from disclosure
(18, pp. 24-26). In the USA for instance, documents that may be prejudicial
to the safety of the country are withheld, as are those prejudicial to defence or international relations. This is the country with the minimum of
exceptions. In Norway the exception is extended to documents, the communication of which might have a negative effect on the economic interests of
the country; by the same token minutes of cabinet meetings are exempt from
obligatory disclosure. The same situation exists in Sweden. In Denmark the
exemption extends to minutes of interministerial meetings as well as to
documents specially prepared for these meetings. American law contains no
exemptions of this kind. In France (Article 7 of the law) the administrations may refuse consultation or communication of a document which in
particular*might interfere with the secrecy of deliberations of the government and of the responsible authorities which have executive power.

It is still early days to draw up a balance sheet of these innovations.
G. Nichols remarks, however, that contrary to certain fears, heard earlier
on, the demands did not cause an exaggerated load of administrative work;
in addition, the possibility of bringing in the exemption has often permitted the service to avoid submitting to the duty of information; finally,
even when access to the information is possible, in fact and not only in
principle, the various interest groups do not enjoy equal possibilities of
profiting from such information: many citizens groups do not have the
necessary means to deal with the data obtained, whence the often heard
question whether or not to give public assistance to the less endowed
parties as concerns expertise (18, p. 16).

b) The production of information. Access to information touching on precise
projects is an essential demand by the citizen or by more or less organised
groups. Who decides? For when? How? What are the effects of the decision
etc., these are basic questions, followed immediately by the demand to
influence the choices made or to be made.

Interesting openings have been made to facilitate this quest for information concerning such and such project or such and such decision. One of

---

*More generally, the law says:
Article 6: The administrations mentioned in Article 2 may refuse permission
to consult or communicate an administrative document the consultation or
communication of which might interfere with:
  - the secrecy of the deliberations of the government and of the responsible
    authorities exercising executive power;
  - the secrecy of national defence or foreign policy;
  - the currency of public credit, the security of the state and public
    safety;
  - the execution of procedures started prior to jurisdiction or operations
    preliminary to such procedures, except if authorisation has been given
    by the competent authority;
  - the secrecy of private life, of personal and medical files;
  - the secrecy of commercial and industrial matters;
  - the search by the competent services for tax and customs offenders;
  - or generally any secrecy protected by the law.

the most important steps forward has been achieved by the legislation
providing the carrying out and publication of studies on the impact on the
environment. The Americans were again the pioneers in this field with their
law on national environmental policy which was completed for the first time
with a circular in 1973. This circular provided advance information for the
public, before any decision was taken, the preparation of a provisional
report, its discussion and the execution of a final study, which in an annex
takes up the essential elements of opinions voiced by the public; the
presentation of alternatives was of the kind to favour, even make possible,
discussion. It therefore went beyond simple information and proposed already
a certain form of intervention by the citizen.

In France, the law of July 10, 1976 is in a similar vein even if the text
remains short of the American legislation (particularly where the study of
alternatives is concerned) (21). The publication of the impact study in
France does not have to be made at the time the projects are subject to
study (as the Delmont report on "the participation of the French people in
the improvement of their living conditions", an unpublished report (22)
suggested) but at the time of the public enquiry*, i.e. rather late.

However, steps have been taken in this direction: a directive of May 14,
1976 has improved the procedure of public enquiry; a circular of October 12,
1977 has recalled the obligation to follow up the improvements provided in
the directive of 1976.

Arrangements permitting the giving of information to the public have also
been written into the law of July 19, 1976 concerning classified installations. As we have seen, the file given to the public enquiry must include
all documents submitted by the applicant (except those referring to industrial secrets). A circular of December 28, 1979 concerning information of
the public has recalled the importance of this publication particularly where
the putting up of notices and publication of the notice of enquiry in the
press is concerned (23, pp. 37-40).

The limits of this type of action are well known: the citizens are often
ill equipped to take advantage of such information; if they do not make a
great effort and take a close interest in the question raised they will not
be in a position to form an opinion on the problem. More fundamentally,
they know that the opinion they will express will have hardly any influence
on the actual principle of the decisions which are involved. This does not
encourage strong determination to gather the fruits of the information now
offered by the public authorities. To mitigate these difficulties other
complementary attempts have been made in various countries. They are aimed
foremost at the improvement of the comprehension of the subjects by the
public.

c) Assistance to citizens' comprehension. Various means have been used to
help the citizen understand better what is at stake with technological
choices. The nuclear question has given occasion for multiple initiatives
sometimes under the exclusive aegis of governments, sometimes very decentralised (18, pp. 29-44).

---

*For projects submitted to public enquiry. For the others (the less impor-
tant ones) publication is required only at the time the decision on
authorisation is taken. This has been improved by the circular of October
12, 1977: publication is required before the decision is made.

The 'study circles' in Sweden. The Swedish government decided at the end
of 1973 to make a major effort in the education of the public. The method
of 'study circles', small study groups directed by the associations for
adult education in connection with the political parties and the big social
organisations, was adopted: this practice has long been part of Swedish
tradition. The government invited the big social and political institutions
to organise study circles on energy; this effort would be assisted by the
public authorities who would provide technical and financial help. The
official documents supplied included arguments by those in charge who
favoured the development of a nuclear programme as well as those of its
opponents. A group of independent experts was available to the study circles.
The publicity campaign organised at the outset for the promotion of the
operation was given the name:

Learn more and you'll have more influence. Join a study circle on energy.

In this way some 10,000 circles were organised, comprising 80,000
participants. This was complemented by public hearings in 1974 and 1975.
The result was a greater awareness of the complexity of the problems concerning energy; the study circles permitted the opening of a much fuller debate
on technologies of a very large scale. If they have not simplified anything,
they have at least permitted to face the complex reality of the question of
technological choice in a modern society better.

The 'dialogue with the citizens' in West Germany. The 'Buergerdialog' was
launched in 1975 by the federal minister for Science and Technology. The
government first launched a big public relations campaign: press advertisements provided information material and prepared future discussions. Technical brochures were printed in more than a million copies. Since 1976,
seventeen public debates were held, in which 4,000 people took part. Efforts
were made to encourage social organisations to sponsor meetings and study
groups; this was developed further in 1977. The ministry let it be known
that it would supply these groups with technical and financial aid. In 1978
efforts were directed at the political parties as well as at employers'
associations and trade unions.

At the beginning there was much reticence; the attempts were denounced as
pro-nuclear propaganda or, by contrast, one worried about the dangers of
such a 'dialogue'. Later on judgements were more differentiated. As in
Sweden the operation did not lead to a simplification of the various
opinions. However, the campaign has no doubt permitted to channel the
discussion of certain questions like those on nuclear waste which in itself
is a non-negligible achievement.

The Austrian information campaign on energy. In 1975, Chancellor Bruno
Kreisky who had taken part in various public debates motivated by opposition
to projects for nuclear centres announced that such an important and controversial subject required very broad public discussion and could not be
dealt with by experts alone. In 1976, the Ministry for Industry prepared a
large information campaign which aimed at achieving broad agreement on government projects for the construction of three nuclear centres before 1990.
Experts for and against the programme drew up a list of controversial points
which were taken up in televised public debates.

These debates were used by the opponents of the nuclear industry to make
their views known. Like the preceding examples, the campaign contributed to
an awakening of public awareness and to the opening of a broader debate of

the country's energy policy. The Chancellor announced the holding of a
referendum in 1978*.

The Danish information campaign on energy. At the time of launching their
first reactor project the Danish public services prepared an information
campaign for the public to explain their choice. In the same month (April
1974) parliament opened its first great debate and decided to set up a
special commission for energy policy. The speakers for the party in power
suggested that it was perhaps necessary to launch a big information campaign
for the public. A committee was set up for this purpose in June 1974 by the
ministry for commerce and industry. It was a non-governmental committee,
given great freedom, financially assisted by the state. Here again, a well
known tool was adopted: big information campaigns had already been organised
previously (on joining the Common Market).

The approach adopted wanted to observe three principles:

 - getting beyond the purely technical framework of the questions raised in
   order better to understand the social and political aspects:
 - guaranteeing parity between positions in favour and against;
 - seeking throughout the campaign to improve the possibilities for public
   participation in the process of political decision making.

Financial aid was provided, not globally for the large organisations, but
to initiatives generated 'at the basis'. It was estimated that 150,000
people took part in this phase of highly decentralised self-education. To
help this development, the committee issued an important documentation of
a technical and political nature.

There was violent opposition from the protagonists of the nuclear programme
who were keen to keep the policy (the decisions) out of the discussion and
to limit information to the technical aspects of the problem. The approach
adopted did not satisfy them and they denounced the confusion of the debates
and their rather less than concrete nature. Despite these criticisms it
seems that the Danish people could in this way increase their understanding
of the complexity of the energy problems.

Information on the electro-nuclear industry in France. The council for
information on electro-nuclear energy was created by decree on November 10,
1977, following violent incidents at Creys-Malville during the summer of the
same year. The role of this council is not the information of the public
but giving the government its opinion on the conditions for the public's
access to information and proposing to the executive the forms and modalities of spreading the information. The council comes under the presidency
of Mme Simone Veil and consists of four mayors involved with the setting up
of nuclear centres, six representatives of associations, four members from
scientific academies and four specially qualified personalities. The council
was able to meet for the first time on April 4, 1978 and has held a dozen
sessions up to May 1979; the first year of operation produced an activity
report (24).

The report shows that during this first phase the council above all
informed itself about the general aspects of the problems of nuclear energy

---

*In the end the Austrian parliament, following the results of the referendum
 on November 5, 1978, which, however, was not binding on it, put an end to
 the country's nuclear programme.

and received several people in charge of large official services. In
addition the council was able to recommend and obtain publication of certain
documents, like the annual report of the central service against ionising
radiation (24, p. 8). It was able to record the government's intention to
publish the 'intervention plans' prepared for each centre, a publication
which it considered indispensible. (24, p. 8). It was further able to take
note of the intention by the ministry of industry to proceed with the updating of the report of 1976 concerning inter alia the problem of nuclear
waste and to publish subsequently a complete file on the subject (24, p. 8).
The council could also establish that considerable progress had been made in
the field of information on nuclear energy over the last five years (24,
p. 9). However, it took it upon itself to look for improvements that could
still be made in this field: initial announcement of projects at the time
of prospecting; communication of files to associations as soon as the
request for the declaration of public usefulness is made; issue of a condensed and simplified brochure to facilitate comprehension by everybody;
hearings during the public enquiry which must be chaired by a high-level
personality of recognised moral integrity; regular information of the
parliamentarians and the population on the funding of the centres in operation through the general council at annual and biannual intervals etc.
(24, pp. 9-12).

Summing up, the report says:

The council has thus tried to encourage the various people in charge of
the design and the execution of the electro-nuclear programme to make
available to everybody interested, in accessible form, the information
necessary to form a reasoned opinion on a subject which is essential for
the future of the country.

It is convinced that the abandonment of the traditional secrecy of the
administration, difficult and slow as it may be, is one of the necessary
conditions for the peaceful solution of the problem raised and, with
respect to procedures, without which there is not true democracy.

It therefore intends to continue and to develop an undertaking which is
often criticised because of the difficulties it has encountered, and the
modesty of the first results obtained, the failure of which, however,
could raise doubts about the ability of reconciling the needs of economic
and technical development with the rights of the citizens (24, p. 15).

A new effort in matters of nuclear information was decided on by the
French Government following the TMI accident as the official communique by
the Council of Ministers of Wednesday, November 7, 1979 emphasises:

The Minister of Industry has communicated information concerning electro-
nuclear energy.

The rapid progress of the sciences and techniques provokes a profound
change in French society. This requirement confers a special nature on
the electricity of nuclear origin, an account of the scope of the equipment
programme undertaken by France.

Very broad information is now available. It is, however, intended to
improve its content and to ensure its dissemination. A particular effort
will be made at the national level to facilitate the citizens' access to
this information. Better liaison will be established between the
authorities in charge of the nuclear programme and the council for electro-
nuclear information.

Moreover, information at the local level will be improved. Effective access to the public enquiry files will be facilitated when this is considered useful by publication of condensed files and then of responses given to the questions raised. The special emergency plans will be published from the start of 1980 for all centres in operation. Regular information will be given to the population and to the local parliamentarians on the functioning of the installations which are operating.

Thus, as the council of ministers held on April 4, 1979 has decided the report made by the Academy of Sciences on the accident at the electronuclear centre of TMI in the USA has been given to the competent commissions of parliament. This report which is deposited at the ministry of industry is available to the public (25).

2nd. Consultation of the citizen, strengthening of the information given to his representatives

a) Consultative bodies. This is not a new means. It is actually widely used. We have seen examples of it in the legislation that deals directly with the question of risk: British Consultative Committee on Major Risk, Higher Council for Classified Installations in France etc. The need is generally felt for the oldest structures to effect certain changes which would permit the citizen to exercise real influence. There arise questions of competence, time, inequality in the capacity of influencing the course of decision making. We shall notice one particular attempt: the American Food and Drug Administration indemnifies the consumer representatives who take part in their consultation groups. Some like G. Nichols feel that a number of consultative bodies often have only a figurative function and that their meetings and reports are manipulated by the officials in order to give the impression of support of public opinion for policies or decisions adopted beforehand. (18, p. 61).

b) The technical competence of the legislature. For some ten years the idea has been developed that the elected representatives of the citizens had seen their control power substantially weakened in matters of technological choice because of the increasing complexity of the question raised. The imbalance in favour of the executive branch kept growing. And at the same time multiple mechanisms of 'direct democracy' began to take shape.*

Parliaments have therefore tried to counteract these tendencies which are very damaging to their mission. The image which has become too classical in the circle of experts close to the executive branch, that the people's elect understand nothing about technical questions must be fought. The first country to innovate the capacity of the legislature in this direction was the USA by setting up a technical office to serve Congress. Great Britain, the FRG, Japan, Sweden etc. also instituted technical advice bodies in their parliaments. We shall examine here two opposite cases: the American attempt, the office for the evaluation of technological options established in 1972, and the French refusal to see such a system installed for the sake of balance between the institutions.

The control of technological options by the Congress of the USA. Following

---

*Among other innovations in the American Congress aimed at making its functioning more transparent and to get closer to the public one may note the televised sessions, circuits set up to permit citizens to participate in hearings without having to go to Washington etc.

up certain experiences in various American states, the creation of an office
for the control of technological options at federal level was envisaged in
1966/67. The idea was put into practice, with the law of October 13, 1972
which created the 'office of technology assessment', destined to help
Congress in its control task. The office permits Congress better to appreciate the nature and significance of government projects or parliamentary
proposals. The legislature would thus be able to exercise a function of
evaluating in advance the options under discussion.

The office depends exclusively on Congress so as to respond to the desire
to preserve its full influence on the executive branch. It is sponsored by
an administrative council of twelve members, each of the two houses del-
egating half of them. The same balance is ensured between the big political
parties in the choice for president and vice president. To ensure sufficient
independence for the body its director enjoys large powers and a long term
of office (6, p. 4).

As Ch. Brumter recalls the fundamental mission of the office is indicated
in the law that created it. To work in such a way that the consequences of
the application of technologies are studied in advance, defined and taken
into account during the working out of the policies concerning questions of
national importance, actual or potential, and this in the most exhaustive
manner (6. p. 6).

The office is therefore given charge of:

- defining the foreseeable and potential consequences of the adoption of
  new technologies;
- presenting, as the case may be, an account of cost-benefit analysis;
- identifying similar or close-by technologies that are susceptible to be
  used in the execution of the proposed programmes;
- putting into evidence equivalent programmes that are susceptible to
  achieve the same objectives;
- comparing the various programmes and methods;
- defining the areas in which further investigation or research into
  complementary data would prove to be necessary.

In practice, the office can intervene at the request of the chairman of
any group, any commission, even at the request of a small number of parliamentarians. The office can support parliamentarians. The office can support
parliamentarians in hearings; an annual activity report permits intervention
in the direction of the work of the legislature. The requirements in terms
of budget and staff for its body have grown rapidly; but it is thought that
it has now reached a level with an appropriation that has gone from 1.3 to
10 million dollars (in 1979) and staff from fifty to one hundred and fifty
people (compared with eight hundred employees of the research service of the
Congress Library for example). Congress can benefit, apart from these means,
from the support of a large number of outside collaborators. Between 1973
and 1977 (both years included) one hundred and ninety two studies have been
carried out. Some ten broad study subjects were adopted as priorities,
energy and transport in particular. A certain number of criteria such as
the importance of the problem for the nation, the irreversible nature of the
expected effects, the feasibility of the analysis to help further in
organising the research effort.

Without going into more detail of the modalities of work we shall mention
the fundamental aim of this innovation. Since 1972, the Congress of the
USA recognises (Section 2 of the law of October 13) that: technology

undergoing rapid changes and constant expansion, its applications which take place on an ever larger scale, cause more and more important and complex consequences, be they beneficial or ill-starred, on the natural or social environment (6, p. 5).

It deplored (Section 2, paragraphs 1 and 2, subparagraph c) the fact that: the federal administration, even though directly responsible to Congress, does not usually supply it with adequate information that is to the point and presented in an objective manner that would permit the evaluation of the consequences of the adoption of technologies (6, p. 6).

To respond to these difficulties, Congress gave itself an office to re-cover its capacity for evaluation. As Ch. Brumter writes:

By reestablishing a balance of the sources of information in relation to the means available to the executive branch nobody doubts that the work of the office will contribute to a strengthening of the legislative power and even accentuate the democratic character of the American institutions (6, p. 4).

The abortive attempt to create a committee for the evaluation of technologi-cal options at the French National Assembly. A first attempt to give the French parliament some permanent technical support was a draft law proposed by D. Julia and Cl. Labbe (No. 2495, extraordinary session 1975/76). We shall go from there directly to the last episode in the successive attempts that failed: the refusal by the government to adopt title 1A (proposing an evaluation committee for technical options which the commission for produc-tion and exchange wanted to attach to the project law concerning energy saving and the use of heating.)

In his report of May 21, 1980, P. Weisenhorn, recorder, clearly stated and defended the desired innovation (26).

The text. A committee for the evaluation of technical options is set up at the National Assembly.

This committee is made up of six competent scientific and technical personalities appointed for three years of which four are appointed by the National Assembly and two are coopted by the first four.

At the request of the president of the National Assembly, taken up by sixty members or by a competent commission, this committee formulates a reasoned opinion on all questions concerning the choice of production techniques, conversion or distribution of energy and the consequences of these choices for social and economic development and for the physical, biological and human environment. This opinion which is sent to the president of the National Assembly is published.

The president of the National Assembly is entitled, in order to enable the committee to accomplish its mission, to obtain from the administration and public establishments all official documents with the exception of those of a secret nature and those concerning national defence, foreign policy and the internal and external security of the state.

By the same token, the president of the National Assembly is entitled to summon, if need be, through a bailiff or a member of the forces of order, anybody who the committee considers to be useful to answer to a hearing. These persons are obliged to obey a summons served on them (26, p. 15).

Criticisms of the government and replies to them according to the recorder.
We have said that the Minister for Industry has presented mainly three
criticisms of a practical nature of the setting up of a committee for the
evaluation of energy choices.

1st: In the first place, according to the Minister, nothing is foreseen
to protect the secrecy of French inventions and the developments resulting
from them. Nor is anything foreseen to protect industrial secrets and the
secret of French enterprises including public enterprises, and nothing is
said about the nationality of the experts that might be called upon to take
a seat on the committee.

2nd: The Minister has actually indicated that he cannot see how six
experts could cover the whole field of the committee's competence. I am
asking myself, he said, how the Assembly will be able to find the six
miracle men who would be likely to supply all the services asked for, when
the execution of such work is only possible thanks to hundreds of competent
individuals who work in publishing establishments.

3rd: Finally, the Minister thought, and according to his own words "this
is the very crux of the matter", the committee would take over from the
national representative body which latter alone has the responsibility of
controlling the government.

These criticisms were the object of the following responses:

1st: As for the Minister's first criticism, it is obvious that all the
arrangements he foresees can be made the object of an addition to the internal regulations of the National Assembly.

It does not seem helpful to foresee such arrangements within the law.
Actually, when the law foresees the setting up of consultative committees in
the executive branch the guarantees demanded by the Minister for Industry
are not foreseen either (nationality of committee members, secrecy of the
work etc.).

2nd: As to the competence of the members of the committee for the evaluation of energy choices, one must go a bit further into the details of the
possible and probable operation of this body in order to answer the Minister.
There is obviously no question that these six individuals could work in
isolation on all the files with which the National Assembly would entrust
them. It is in the nature of things that they would have recourse to more
specifically competent experts for the needs of the study ...

3rd: The third argument of the Minister appears singularly weak. In fact,
there is no question of the National Assembly renouncing its role of controlling the executive branch and of transferring this role to some technocratic aeropagus. The committee for the evaluation of energy choices would
permit the National Assembly to have studies that would coincide with those
supplied to or by the government and would therefore be an additional element
for its reflection (26, pp. 12-14).

Fundamental justification of the proposal. To finalise, we would like to
insist on three points in support of the opinion held by the National
Assembly on Article 1A of the present draft law.

1st: It cannot be denied that the evaluation of technical choices and its
practice in France deserves reexamination. Administrative centralisation,

concentration of economic power within a very small number of companies,
public or private, the quasi-monopoly granted, particularly in this field,
to a string of university scholars of one-sided training require evidently
that counter-evaluations be made when required ...

2nd: It must be emphasised that the formula adopted which permits the
setting up of the committee for the evaluation of energy choices, be it by
the competent commissions, be it by sixty parliamentarians, gives the possibility to everyone of the political groups that make up the Assembly to
escape opportunist considerations which other groups or the majority might
rally against it. It must be stressed that our initiative would permit an
increase in democracy.

3rd: Finally, a certain number of choices, considered fundamental by the
authorities, hit non-negligible obstacles in their execution. This is
particularly the case with the electro-nuclear programme. The very recent
example of Plogoff shows that one should not underestimate this kind of
opposition. One of the strongest arguments put forward by the opponents of
the electro-nuclear programme is the lack of an evaluation of the advantages
and inconveniences of the formulae proposed to the population by a neutral
expert mechanism. It has been said that at present those who give the information are identical with those who decide and produce. In these circumstances their opinion is not considered credible or in any case tainted by
partiality. It is therefore indicated to diversify the sources of information and to facilitate the dissemination of the data by organisations which
are outside of the execution of the programme. The committee for the
evaluation of energy choices could be one of these organisations among others
(Academy of Sciences, Universities ...) (26, pp. 14-15).

The radical rejection of the proposal by the government. During the debate
which followed in the National Assembly on May 22, 1980 the recorder tried
to defend the article in question:

> Once more, it is not a question of the National Assembly transferring part
> of its control prerogatives to any technocratic committee just as, by the
> way, the government does not transfer its decision making powers to the
> commissions with which it surrounds itself as, for instance, the Planning
> Commission or the Peon Commission. I believe that this parliamentary
> initiative must be viewed in a completely objective manner and not be
> dumped into processes of intent or bad disputes and one must not make a
> travesty of the eventual intent to limit the role of parliament through
> concern for the conservation of its prerogatives. This is why the commis-
> sion for production recommends the adoption of the present amendment
> (27, p. 1234).

The recorder took up the whole of the arguments mentioned above. But the
government was inflexible.

The Minister for Industry: The government holds that the proposal to
create a committee for the evaluation of technical options is useless and
dangerous, and that it does not at all reflect reality nor the normal
balance between the functions of our institutions.

It is useless because parliament has all the necessary means of information,
investigation and expertise ...

What purpose, therefore, would this evaluation committee serve? Would it
be a matter of giving experts, who have no democratic mandate, investigation

powers which belong only to those whom the people have elected? If that
were the case we would affect the institutions of the Republic ...

This proposal presents dangers because it is seriously incomplete ...
nothing is provided in the setting up of the evaluation committee to protect
the industrial secrets of French inventions and the developments resulting
from them.

One can go further. I could draw the attention of the House to the fact
that nothing is said about the nationality of the experts who would sit on
the committee and the ties it must eventually entertain with one or the
other industrial group.

M. Emmanuel Hamel: Or with foreign powers ...

Minister for Industry: ... public or private.

M. E. Hamel: What thoughtlessness!

Minister for Industry: In addition, this proposal is unrealistic: it is
impracticable and generates irresponsibility.

Here is a committee made up of six experts, six competent, scientifically
and technically competent individuals. This would actually not be their
only characteristic since it would be required that four of them are appointed by the office of the National Assembly at the rate of one per parliamentary group. What will be the supplementary criteria other than the
scientific and technical ones that will be applied?

Actually, I declare that these six people, who are apparently omniscient,
would formulate a reasoned opinion on any question concerning the choice of
techniques of production, conversion and distribution of energy, on the
consequences of these choices for social and economic development as well
as for the physical, biological and human environment, and this apparently
for all kinds of energy, for all its forms of production and on all categories of eventual inconveniences they might cause.

Where are the six miracle men who will attempt to carry out such a vast
mission which is only possible thanks to thousands of experts who work in
public establishments, in the laboratories and institutions which France is
honoured to have as well as in companies and organisational bodies outside
the administration and in the public sector who can be made available to
contribute?

It is precisely in order to approach this technical evaluation better that
France has created, in the course of history, a certain number of scientific,
technical and administrative organisations of which it is proud.

This committee could, even if it managed to do as well as the thousands
of experts who are involved in these operations, only approach this technical
evaluation. In fact, how could one choose these experts in such a way that
they could give clarifying opinion without ambiguity about the choices
offered? So much technological information can be assembled quickly at a
given time and objectively, so much technological evaluation can lead to an
endless and contradictory sequence of studies and researches.

Have TV and the automobile in their time been the object of technological
evaluation? What would a technological evaluation of aviation, of the
automobile, of TV have produced if it had been carried out by these super-
experts chosen by parliament? We would not fail to notice in our country,

if such a committee for the evaluation of technical options were created, the
same perversions as the ones we see in the USA with an office of the same
kind.

What has happened? First of all a multiplication of systems and bureaux
for technological evaluation running parallel in the various ministries,
organisational bodies, public and private companies involved in the opinions
created by the committee which challenge just as easily the conclusions of
the committee's evaluations as the latter would naturally be eminently chal-
lengable. An inflation of the organisational body would rapidly require
that one surrounds oneself with all the competences which the six experts
would not have. Four years after its creation, the American office for
technical assessment had a budget of a billion centimes, two hundred and
thirty agents, five hundred temporary consultants and two hundred and thirty
outside contractors.

M. E. Hamel: What an inflation!

Minister for Industry: This creation of all the pieces of a technocracy
supposed to control one another did not lead to an improvement of American
knowledge of the consequences of technological choices, if one looks at the
results.

Technological evaluation belongs in the end in the field of responsibility
and choice, i.e. in the field of politics and this is just the major point
which to me seems to raise difficulties. It is absolutely necessary to take
up position on the broad choices for energy and this is only possible by
bringing about syntheses which include scientific, technical, economic,
social and diplomatic elements of all sorts. France's democratic and republican tradition demands that parliament be responsible for controlling government on the choices it makes through the synthesis of these various aspects.
Would parliament be tempted to abandon this essential prerogative to a committee of technocrats? What in fact are these six people who have no democratic mandate whatsoever? To whom are they responsible for the consequences
which the opinions they give may have? Is political control not better
exercised by people who have been elected by their fellow citizens?

I have heard that they would eventually have pronounced on the choice of
the nuclear site at Plogoff. I could not wish for a better illustration of
my opinion. Plogoff was chosen from among five other sites by the regional
council for Brittany, by the economic and social committee for Brittany, by
the province of Finistere.

Who are these six super-experts of whom one would not even know whether
they were French, let alone Britanny — people, and who would dare to formulate an opinion contrary to that of the elected representatives from Britanny?
(27, pp. 1235-1236).

The Minister for Industry announced that he would have recourse to a
blocking vote so as to throw out the arrangement aimed at the creation of the
committee for the evaluation of technical options. The executive branch
would use a suitable method to keep the balance of power in its favour. There
were strong regrets.

M. Julien Schwartz: I'll restrict myself to presenting some short observations.

In the first place, minister, when tabling the amendment in question I had
no desire to challenge the institutions; in fact, the committee which we would

like to see installed would only give opinions; it would have no decision
power. How could we be challenging the institutions by turning to a committee that would only give us information? I want to stress, as did the
recorder, that the information that we could gather in this way would not be
that of the deciders or executors.

Furthermore, you have indicated that the special or permanent commissions
could have all the experts it wanted to hear come before them. But, minister,
the experts we consult in this way don't give opinions. They are the members
of parliament who establish the reports on the basis of information which has
been given to them by those experts. And these reports do not carry the same
weight for the government as the opinions of experts.

Finally, when in 1975, you were not Minister for Industry then, and I
regret this today because we have appreciated the fruitful dialogue which you
have entered into with us, we did by means of an amendment enter the draft
law which we are discussing today, we were treated as if we were hare-
brained. We have been treated like idiots by the high-up people from the
company which today supplies our electricity, those people in charge, who
have opposed our project over the last five years. It took all your
authority, minister, and I thank you for it, to make them admit that there
is sense in our reasoning.

You see, minister, it is not altogether useless to be able to get information from several people. We have invited foreign experts to No. 101 rue
de l'Universite. The representative of the big national enterprise who took
part in the meeting still pretended that one could only tap water from the
centres at 30°C while the foreign experts confirmed that one could have water
at 100°C. It was a dialogue of the deaf.

In the circumstances, minister, where is the truth?

We cannot get objective information from certain decision makers or
executors. We have keenly felt this in all the actions we launched to arrive
at this draft law which has only seen the light of day thanks to you. This
is the reason why we are so obstinate with regard to this amendment (27,
p. 1237).

M. J. -P. *Cot:* Mr President, I feel obliged to raise a protest against the
use of the procedure of the blocking vote unless the Minister for Industry
desists from his intentions.

In this respect, I fully approve in fact the observations made by M. Julien
Schvartz. Concerning the arrangements in question, the Minister for Industry
has praised parliament, showing how necessary it is for the latter to assume
its responsibilities, particularly with regard to the creation of a technical
committee destined to help parliamentarians to legislate better. That he
should then have recourse in the same business to such a brutal method as the
blocking vote seems really very regrettable to me (27, p. 1238).

The blocking vote was used nevertheless and the proposal could be thrown out
by the government. For questions of national importance which require much
time and competence and also require close attention to the social groups
concerned, the mechanism of the enquiry commission has been used. This is an
old recourse which however has been broadened over the last few years in
certain countries as a recent OECD document remarks: They (the parliamentary
enquiry commissions) assume new and broader functions such as often opening
up controversial questions to broad public debate, supplying the politicians

in charge with a more representative opinion of the various needs and wishes
of the public, evaluating the implications of various possible actions before
a final choice is made (28, p. 108).

The double objective of better informing the decision maker and to permit
the groups concerned to express themselves has been apparent with the Berger
enquiry commission which is one of the most interesting achievements in this
respect. The document quoted above presents this attempt concisely:

> With the creation of the Berger enquiry commission for the study of the
> construction project of a gas pipeline of 5,000 km in the MacKenzie valley,
> between the arctic circle in the north and the border between Canada and
> the USA in the south, a certain number of innovatory participation mech-
> anisms have been set up with a view to the complete evaluation of the
> impact of this project on the environment. Four different groups of
> public hearings were instituted. Official hearings were organised to
> receive testimonies concerning the pipeline construction project and to
> examine the technical problem and the consequences of the project for the
> physical environment, living conditions and the human environment.
> Communal hearings were organised in parallel with official hearings to
> permit the residents of communities, located along the pipeline, to express
> their views on the project. A certain number of special hearings were
> also held to study the particular problems concerning prospecting and gas
> production activities in the northern strata. Finally, a series of hearings
> were organised in large cities all over the south of Canada to permit those
> Canadians who were not represented in the north to express their point of
> view.

The communal hearings in particular constituted an important innovation
with regard to earlier procedures in public enquries inasmuch as they were
essentially informal and as they permitted everybody who wanted to, and
the indigenous people to express their preoccupations openly and freely,
without having to submit themselves to contradictory interrogation. What
is more, the minutes of the testimonies received in the official hearings
were broadcast on the radio in six indigenous languages for the benefit of
northern communities which permitted the citizens to keep themselves informed about the conclusions of these official hearings and to respond to
them in the communal hearings.

A second great innovation of the enquiry of Judge Berger was the granting
of financial aid to the groups of participants. The Canadian Government
had allocated nearly 500,000 dollars for this enquiry to finance the researches and studies carried out by groups for the defence of the environment or other interest groups. An additional amount of a million dollars
was also granted by the government directly to indigenous groups to permit
them to undertake researches into certain still unsettled territorial
problems and to gather testimonies which would be presented to communal
hearings.

A third innovation was to request the legal counsel of the commission to
attend the hearings as an active and independent participant, authorised
to research any testimony that to him seemed of interest for the enquiry.
He also received instruction to present his arguments and opinions publicly,
the aim being to avoid giving the impression that the legal counsel of the
commission, or his fellow workers could constitute some sort of 'private
council' of the enquiry commission.

Finally, and this is perhaps the most important point, the enquiry lasted

as long as was required. Started at the end of spring 1974 with a series
of preliminary hearings it ended in autumn 1976. The final report on the
enquiry was completed a year later in November 1977 (28, pp. 109-110).

This kind of approach, which was taken up again in other Canadian projects,
is obviously not faultless; the cost, the time it takes, the organisational
difficulty are acute problems while even the ultimate effectiveness is debatable inasmuch as the public is only consulted; the enquiry bears only an
indirect relationship to the effective exercise of the decision power of
the government. Nevertheless this is a means among others which permit
building a bridge between government and the governed.

3rd. Mechanism for direct access by the citizen to the decision making process

a) The judiciary approach. This is a classical recourse approach but it has
been amended in recent years; the right to plead in court has been modified
in many countries in a more liberal direction. This has permitted an increasing number of individuals and associations to present complaints before
administrative tribunals and other appeal institutions. In addition, the
admitted reasons for complaints have been broadened; impacts on health, well-
being, aesthetics, safety are admissible besides effects on property for
example.

We have seen the recourse to the courts developing particularly in the USA
in connection with the legislation on impact studies: three hundred and
thirty two lawsuits were brought against federal agencies between 1970 and
1975 for insufficient studies. In other countries too, legal actions have
multiplied, particularly in connection with the setting up of nuclear
centres.

Like the previously mentioned means this one has been the object of criticism: costs, delays in getting satisfaction or getting authorities to go
ahead with a project; difficulty for the courts to pronounce on subjects
which do not strictly arise out of the domain of the law, when they involve
largely social and political considerations, this difficulty can create
loopholes in the procedure and transform an action into casuistic competition or, as an American lawyer remarked, the intervenor often becomes the
prisoner of his lawyer (28, pp. 112, 117-118).

b) The referendum. The recourse to the vote by the whole of the citizenry
is another means by which the public can get directly involved. There have
been various examples in connection with nuclear choices, in particular the
referendum in the state of California in June 1976 and the Austrian consultation of November 5, 1978. True, these votes do not necessarily bind the
legislators but they carry nevertheless non-negligible weight.

A great many objections have been raised against this means: risk of
manipulation, excessive simplification of the issue, manifestation of the
executive branch's shirking of duty, and of the incapacity of the political
parties, a means that often leads to maintaining the *status quo* etc. (28,
pp. 111-116).

c) 'Ad hoc' participation procedures. Local innovations, in a way tailor-
made, have been tried in various countries. In Canada for instance, in the
province of Ontario, the public electricity company created a certain number
of 'citizens' committees' and 'working groups', made up of participants from
community organisations and interest groups which helped the company choose

the layout for its high voltage power lines. In the USA 'citizen review boards' were set up. These are organisational bodies in which decision power is delegated to representatives of the population who have been either elected or appointed. They sit on this committee and are charged with examining the various possible plans and decide which one should be adopted. Similar initiatives (but less advanced) have been launched in Germany where 'planning cells', made up of non-elected citizens, work in close cooperation with the municipal authorities to come up with local decisions. The creativity of the citizen was also used to advantage in public gatherings (competitions such as we have seen in Denmark in 1975 for country planning on the island of Bornholm) (28, p. 123).

These are limited attempts at dealing with problems which are not as complex and of as much national importance as the issues connected with high risk technology. However, and despite all the criticism which one may have for them, these attempts at opening up the process of decision making must not be belittled.

REFERENCES

(1) V. T. COATES, Technology and Public Policy. Program of Policy Studies on Science and Technology, The George Washington University, 1972.

(2) O. GODARD, P. LAGADEC et S. PASSARIS, Environnement et planification de la science: quelques propositions méthodologiques. C.I.R.E.D., juin 1974.

(3) F. HETMAN, La sociéte et la maîtrise de la technologie. O.C.E.E., 1973.

(4) The Mitre Corporation, A Technology Assessment Methodology. Office of Science and Technology, Executive Office of the President, June 1971, Washington.

(5) A. SUGIER et E. THIRIET, Rapport de synthèse sur le concept d'évaluation des options techniques, sa mise en valeur et les principaux travaux de ce type réalisés dans le monde. Groupe Interministeriel d'Evaluation de l'Environnement, octobre 1973.

(6) Ch. BRUMTER, Le contrôle des options technologiques par le Congrès des Etats-Unis. *A.J.D.A.*, 20 décembre 1979, pp. 3-11.

(7) J. C. DERIAN et A. STAROPOLI, *La technologie incontrôlée*. P.U.F., Paris 1975.

(8) A. STAROPOLI, L'évaluation technologique. *La Recherche*, no 54, mars 1975, pp. 239-244.

(9) L. Mc GINTY and G. ATHERLEY, Acceptability versus Democracy. *New Scientist*, 12 May 1977, pp. 323-324.

(10) P. STREETEN, Cost-benefit and other problems of method. *Political Economy of Environment*, Ed. Mouton, La Hague-Paris 1972, p. 53

(11) K. W. KAPP, Social Costs, neoclassical economic, environmental planning: a reply. *Political Economy of Environment*, Ed. Mouton, La Hague-Paris 1972, pp. 113-124.

(12) H. J. OTWAY, Risk assessment and societal choices. International Institute for Applied Systems Analysis (I.I.A.S.A.), Research Memorandum no 75-2.

(13) H. NOWOTNY and H. HIRSCH, Ecologization within science: the nuclear power controversy and the case of the low dose effects. (Personal Communitcation).

(14) W. HAFELE, Hypotheticality and the new challenges: the pathfinder role of nuclear energy. *Minerva*, vol. XII, no 3, July 1974, pp. 303-322.

(15) A. WENBERG, Science and Trans-Science. *Minerva*, 10, 1972, pp. 209-222.

(16) R. JOHNSTON and B. GILLESPIE, Scientific advice and risk assessment: a sociological analysis Second annual Meeting of the Society for social Studies of Science, Harvard University, 14-16 October, 1977.

(17) H. NOWOTNY and H. HIRSCH, The consequence of dissent: the controversy of the low dose effects. (Personal Communication).

(18) G. NICHOLS, La technologie contestée. O.C.D.E., 1979 (132 pages).

(19) Loi no 78-753 du 17 juillet 1978. *Journal Officiel* du 18 juillet 1978.

(20) N. WADE, Freedom of Information Officials Thwart Public Rights to Know Science, 175, 4 février 1972, pp. 498-502.

(21) P. LAGADEC, Les études d'impact: l'attente et l'outil. *Aménagement et Nature*, no 50, 1977, pp. 2-4.

(22) P. DELMON *et al.* La participation des Francais à l'amélioration de leur cadre de vie janvier 1976. (Non publié).

(23) Circulaires et instructions d'ordre général intervenues depuis l'entrée en vigueur de la loi du 19 juillet 1976. Ministère de l'Environnement et du Cadre de Vie. Direction de la Prévention des Pollutions. Service de l'Environnement industriel. Mars 1980 (44 pages).

(24) Rapport annuel du Conseil de l'Information sur l'Energie Electronucléaire La Documentation Francaise, mai 1971 (171 pages).

(25) *Le Monde*, 8 novembre 1979.

(26) Rapport de la Commission de la Production et des Echanges de l'Assemblée Nationale sur le projet de loi modifié par le Sénat, relatif aux économies d'énergie et à l'utilisation de la chaleur. Présenté par P. Weisenhorn. no 1719, 21 mai 1980 (62 pages).

(27) Compte-rendu intégral des débats du jeudi 22 mai 1980 à l'Assemblée Nationale (2è séance). *Journal Officiel* du 23 mai 1980.

(28) *Politiques de l'environnement pour les années 1980.* O.C.D.E., 1980 (130 pages).

CONCLUSION: AN EXECUTIVE BRANCH DETERMINED TO FACE THE PROBLEM OF MAJOR TECHNOLOGICAL RISK

Urgency and necessity in this second scenario lead to paying serious and competent attention to the difficulties raised by major risk: to deal with them, not to brush them aside or deny them.

On the one hand, the safety function will be examined from scratch: from the status it enjoys to the means it is granted. It is a matter of doing general management work, not only work on the level of 'technical and maintenance' submanagement. The policies and strategies will be re-evaluated with regard to major risk and so will the great socio-technical options.

On the other hand, one is taken to consider that the citizen is not funda-mentally an internal powerful enemy who is hostile to the survival of the Western World. Like the Anglo-Saxons and the Scandinavians we may consider listening to him, training him, informing him, and not by subtle disinfor-mation, helping him by various means to develop his capacity for responsible intervention. In the same way we can try to help the legislature to in-crease its competence, not to be satisfied with sneering on every occasion at the fundamental ineptitude of the nation's representatives to deal with technological questions of any importance. Here again we can see the importance of having a parliament which is a real interlocutor (without being instantly afraid that it might render a service to a foreign power). There are, therefore, a whole lot of arrangements to be made and means to be developed. The possibilities are too numerous to dwell on. It remains to conduct these changes within a more organised project because a chain of isloated actions without a basic policy would soon succumb to dissolution and incoherence.

These background references for the first scenario (the approach marked by closure and authoritarianism) and for the second (the elements of which we have just reviewed) will be examined in the last chapter.

CHAPTER NINE

When risk calls the fundamentals of politics into question

I   Democracy brushed aside by major risk
II The project of a democracy facing the
    challenge of major risk

From the point of view of mental health the most satis-
factory solution for further peaceful use of atmoic energy
would be to see a new generation grow up that would have
learned to adapt to ignorance and uncertainty ...

World Health Organisation

Questions of mental health raised by the use of atomic
energy for peaceful purposes. Report by a study group.
Technical reports series No. 151, Geneva 1958, p. 49.

I. DEMOCRACY BRUSHED ASIDE BY MAJOR RISK

In the seventh chapter we have seen how the body politic might brush aside
the question of major risk by passing it on to a technical commission. It
was then only a 'tactical' response. (It is (now) a response, forged in
the same vein, but of a strategical order). This time the body politic
takes the measure of the problem and reviews its basic frame of reference.

The first strategic response that we shall examine in the following developments will start from the idea that the public is incompetent and quite
incapable of reasoning scientifically and serenely in these hightly technical
matters and quickly gives way to emotion. However, as the imperatives of
growth demand, one must go ahead. It remains therefore to plan political
operations accordingly, i.e. to relax the demands which come along with the
ideological framework of democracy.

For the general development of high risk techniques, such as the management
of disasters, the roles of the citizen and the (public) power will be redefined in order to achieve maximum efficiency ...

1. THE BASE LINE: A CULTURE OF IGNORANCE

Given the complexity of the problems and the risk of seeing resolute
opposition rise up, the choice is made to train the citizen in such a way he
will not search for a solution.

This is a complete reversal of the idea of education preached in the last
century. One feels that it is no longer a good idea to put the elements of
a decision on the table, that today's risks no longer permit other approaches.

This perspective has been sketched out since the second half of the
Fifties, when the problem of the development of civilian nuclear industry
was pondered. M. Damian*, in a publication he is preparing, goes back to
this period during which one notices the great perspicacity of international
organisations who were looking into the problem of acceptability of atomic
energy.

We shall dwell on this case which has been examined in depth already and
with transparency by international organisations, a little more than twenty
years ago. The technological turning point had been reached. The broad
lines of strategy were then thought over and drawn up. This first frame of
reference appears most interesting and instructive for the other great risks
which are nowadays more apparent.

Two main publications will be of particular interest to us. One is by
the World Health Organisation (WHO) and the other by UNESCO, and both raise
the problem of responding in depth to the question raised by the entry into
a new technical and cultural era. We shall add a complement to these two
precursors: the modalities which are at present drawn up concerning information of the public in case of a nuclear accident, and this for the case of
France.

1st. An 'educative redress', 'a process of conditioning'

In 1955 the WHO reported the following factors:

- the quantitative observations on the effects of radiation on man are
  altogether wrong (1, p. 16).
- The present rules concerning the radioactive effluents from (nuclear)
  centres are based on the hypothesis that an insignificant proportion
  of the world's population is exposed to danger; it will be necessary to
  make stricter rules when the exposed population grows. (1, p. 13).
- The installations facing most of the difficulties are the plants for the
  treatment of used nuclear fuel; nevertheless, it must be possible to
  eliminate the waste with an improved technique. (1, p. 15).
- A much more important problem of public safety is the one raised by the
  elimination of strongly radioactive waste; a large number of methods are
  under study but to this day no satisfactory solution has been found
  (1, p. 13).

However these uncertainties did not have a braking effect. The WHO
declared that the action on public health must not:

expose itself to the accusation of shackling the development of nuclear
energy and thus depriving the world of the benefits that go with it
(1, p. 13).

The logic, of the 'necessity' of technological development, leads to the
same development which was upheld in 1810 in France when the decree of
October 15 was prepared: one must not, the author said, impede the founding
of factories for chemical products which deserve all the protection and
benevolence of your majesty since they will supply us with products for
which we have before depended on foreign countries.

It remained therefore, one hundred and fifty years later and with high
risk, to prepare social acceptability. The WHO continues:

---

*University of Grenoble (IREP).

Even if it is difficult to refute the thesis which demands that nothing is
hidden from the public one must nevertheless study the psychological
principles to be respected when one wants to present the disquieting news,
taking into account the public's aptitude to put up with them ... Two
snags must be avoided: on the one hand, raising anxiety through publicity
and on the other, imposing precautions despite official declarations to
the effect that the risks are negligible (2, p. 48).

This difficulty makes some arrangement with the theoretical principles of
democracy necessary. How should one launch nuclear programmes. The WHO
analyses the British case in this respect:

One can get some indication of the true situation by examining the reality
which is hidden behind the procedure normally followed in the U.K. when
one wants to build a new atomic energy installation ... The national
energy production is insufficient with regard to the industrial and
domestic demand, and this deficit tends to get worse. Responsible citizens therefore consider it a national imperative to push ahead with the
production of nuclear energy. What is more, the recent experience of
World War II and its after effects has got them used to accepting as an
obvious duty certain sacrifices such as expropriation of land for 'enterprises of national importance' and being satisfied with the compensation
offered by the authorities. In addition a large number of public statements supported by figures which the man in the street could not judge at
all independently have affirmed that the atomic enterprises did not expose
the population to any danger (2, p. 23).

The WHO adds, as M. Damian points out, in order not to worry public
opinion unduly, the English atomic energy commission undertook an educative
redress of journalists (the wording is that of the WHO) to try and eliminate
completely the problem of sensational headlines (2, p. 49). There was in
fact a quite diffuse resistance that had to be fought and the WHO put forward
its suggestions as "positive measures destined to raise public confidence";
declaring in particular that the propaganda to restore public confidence
risked to end up in deadlock, the WHO indicated: one must rather look at the
problem from the angle of a process of conditioning (2, p. 47).

In the same publication the WHO again points out the task to be accomplished
and particularly what to expect from the behavioural sciences (psychiatry,
anthropology, social psychology). The function of these sciences is to give
mankind useful and concrete assistance for its adaptation to the atomic age
to make it as easy and painless as possible (2, pp. 6-7).

It is here a question of bringing about a climatization to change, i.e.
a culture in the bosom of which dangers and changes of orientation can take
place harmoniously (2, p. 43).

Thus, the need to create social conditions which would permit to regulate
the difficult problem of 'acceptability' more safely led it to go beyond the
simple tactical actions of which we have seen a well appointed bouquet in
chapter seven. It needed and still needs much more: a real strategy.

2nd. Managing the change-over to a new culture

In 1958 UNESCO was not unaware of this basic need. In 1959 a report by
the director general of the organisation clearly brought out the approach
to be adopted. The essential points of this document can be regrouped under
the following headings.

a) The civilian nuclear industry raises a problem in society.   This for
several reasons:

The inevitable intervention by the state (government).   A certain government
intervention appears inevitable and one may ask oneself what the effects of
such state control will be and how it will be ensured by the various organisations and competent individuals.   It does not look as if there was complete
agreement on the most efficient way and the best one to exercise it adminis-
tratively (3, p. 4).

The variety and complexity of social questions connected with nuclear
development.   It is often admitted that economic motives are sufficient to
explain why a government or a nation accepts to commit the enormous funds
which the creation of nuclear installations involve.   These funds are committed, it is said, when the discounted profit is sufficiently big.
Certainly, the influence of economic factors cannot be disputed; but there
are quite some other considerations which also intervene in decisions of
this kind.   One may mention particularly the symbolic value of nuclear research, the fact that the building of a reactor constitutes an achievement
in itself; the concern for national prestige; the possibility for the government to exercise greater influence on the national economy; the, manifestly
unjustified, belief that the construction of electro-nuclear centres would
permit the solution of all the problems of economically underdeveloped
countries etc.   The idea had been spread that the notion of social profit
could be useful here because it encompasses the notion of economic gain
without exclusion of the intervention of non-economic motives.   To sum up,
there is good reason to examine on what criteria the decision to use nuclear
energy for peaceful purposes is based (3, p. 4).

The problems of collective psychology.   It is possible that the situation is
characterised more by the psychological and social attitudes which result
from the use of nuclear energy than by the material and economic aspects of
this use.   (3, p. 5).

*b) Civilian nuclear industry calls for management in depth of collective
psychology*

Thorough knowledge of the basis of citizens' deepseated convictions.
Special research must therefore be undertaken into people's opinions, whether
they are clearly formulated or secret and difficult to elucidate (3, p. 5).

Education programmes to calm the anxieties of the citizen.   Knowledge of the
attitudes and opinions of the general public and of certain specialised
groups is an indispensible condition for the working out of any education
programme destined to calm the anxieties of those who do not know exactly
what to hope for and what legitiamtely to fear of nuclear energy, or from
whom they should get suitable information on these questions (3, p. 5).

As a first measure it is suggested to centralise the data on public
opinion already collected but largely unpublished and to examine them carefully (3, p. 5).

Education of those who play a part in shaping opinion but do not necessarily
dissipate the anxieties of the public.   Since it is a matter of public
opinion it is normal to think in the first place of those who determine the
content of publications and who on account of this have a particular role to
play in shaping opinion.   This is the case with journalists, newspaper
proprietors, editors-in-chief and in particular the 'headliners' who sometimes give articles a sensational character which the content does not

justify. The first preoccupation of journalists and of information specialists in general is to catch people's interest which does not necessarily
mean the dissipation of their worries or the enlargement of their knowledge.
Even with the best will in the world it is not easy to present the facts
about nuclear energy in terms which are simple and accessible to everybody.
Many attempts have been made, by the schools of journalism at certain universities, by the international press institute and other organisations, to
train scientific writers who specialise in these questions. However, there
remains much to be done in this respect.

It has been recommended that UNESCO should organise study courses for the
staff members of these information organisations whose task it is to dis-
seminate information concerning the use of nuclear energy for peaceful
purposes: professional journalists in charge of writing or editing articles
on these questions and important administrators. These courses would have
to be prepared with the cooperation of specialists in journalist training
(3, pp. 5-6).

*c) The nuclear industry raises a cultural problem and therefore calls for
management of cultural evolution*

Very vast consequences. It would seem appropriate to pay special attention
to the social, cultural and moral consequences of the use of nuclear energy
for peaceful purposes.

Beyond information: raising attitudes. As concerns action on public opinion,
the experts have stressed that it is specifically up to government and to
the various national organisations who have the authority to produce a proper
documentation for the information of the public about new facts and to watch
that this information actually reaches the public. This is a complex task
because it consists not only in distributing information but also in raising
attitudes which are an indispensible condition for the use of nuclear energy
for peaceful purposes in the interest of the population and taking into
account the social and moral consequences of this use for the whole country
(3, p. 6).

Raising values that suit a world in which nuclear energy plays an ever more
important part. It has been recommended that UNESCO should encourage and
help member states and non-governmental organisations to bring to the attention of the general public by means that are as effective and varied as
possible information material concerning the peaceful applications of nuclear
energy. This material must not only present exact and useful information but
must also contribute to the dissemination of healthier concepts from the
psychological and social points of view and to raise attitudes in the general
public and a sense of values which suit a world in which the peaceful use of
nuclear energy begins to play a more and more important part (3, p. 6).

*d) Acceptability of nuclear energy demands that one covers up the differences
of opinion among experts*

Differences of opinion which cause confusion of minds. It would be difficult
to find a question on the subject of which the public had more confused ideas
than on the question of nuclear energy. One fears the destruction of the
world, the degeneration of the human species, the pollution of water and food.
One is terror-stricken by the 'magical' aspect of nuclear phenomena. Or, by
contrast, one expects unlimited benefits for the future.

Some of these attitudes are rational, others are not. The fears are
obviously not without foundation but they are often exaggerated. If the

public only think this, it is largely because the specialists disagree and
the predictions of one scholar are contradicted by those of another. These
differences of opinion are often of limited significance but the public is
not in a position to judge. However difficult it may be to achieve unanimity
on some point among the men of science who are the authorities on the subject,
one must try at all cost to get there.

There will be disagreement on many points; in these cases it will be
necessary to spell out the nature and significance of the differences of
opinion. It is necessary that the public understands that most of the time
science is based on probability and one cannot ask it to confirm that there
is such a percentage of risk or that there is no risk at all. There is
doubtless more advantage than inconvenience in having the general public know
that there is a risk on condition that the true nature of the risk is known
and that one is not afraid of facing and evaluating it. In the same way, the
hopes built on nuclear energy need to be revised in a more realistic way. It
will no doubt be found encouraging in this respect that students who were
questioned in an opinion poll in which this was the first question declared
by a large majority that they had confidence in the information coming from
the UN and its specialised institutions (3, p. 7).

Bringing about agreement and drawing attention to these points. It is recommended that UNESCO researches the means for bringing about agreement among
experts on a certain number of facts concerning the use of nuclear energy for
peaceful purposes and to draw the public's attention to this agreement (3,
p. 7).

e) A general education programme to prepare for life in the nuclear era

A general but adapted action. The educative action must obviously cover
various aspects according to the category of public to which it is addressed.
In this respect one can distinguish a large number of categories of the
public: children, teachers for various levels and age groups, students of the
exact and natural sciences, leading elites and particularly parliamentarians
and politicians, industrialists, trade unionists, men of science and the
general public (3, p. 7).

Particular attention to the preparation of children. A general education
programme must obviously respond to the demands of all these categories but
it is especially recommended to UNESCO to undertake the educational work on
the general public mentioned above and to give particular attention to the
teachers whose role is obviously decisive as concerns the preparation of
children for life in the nuclear era (3, p. 7).

Education of the teaching staff. It is recommended that UNESCO consult the
competent non-governmental organisations in the field of education concerning
the most effective means for keeping the teaching staff constantly up to date
on the new facts about the use of nuclear energy for peaceful purposes.
Priority would be given to the arrangements to bring up to date the knowledge
of teachers, whose scientific training was somewhat dated. Particular
attention must be paid to countries in which the training of teaching staff
is the least organised (3, p. 7).

f) The basic perspective

Finally, the strongest words come from the WHO:

... from the point of view of mental health the most satisfactory solution
for the future peaceful use of nuclear energy would be to see a new

generation raised which would have learned to live with ignorance and uncertainty (3, p. 49).

This is a true cultural revolution. The ideal of the Third Republic,
education, the references to the culture of the twentieth century, learning
for enlightenment, collapse at the entry into force of complex and high risk
technology. It becomes imperative that the citizen accustoms himself to
ignorance and uncertainty, his own in the first place, perhaps also to that
of those who govern him and commit the society, within which he remains,
free to attend to his daily chores.

3rd. The exclusion of the citizen in times of crisis

In a crisis situation one does not just aim at setting up a General Staff
for Information; the right tactic is spelled out quite correctly by the
Augustin-Fauve report on TMI:

Setting up in the first hours of an accident with potential radiological
consequences under the responsibility of the police commissioner a General
Staff "Information of the Press and the Public" which may include:
- one representative of the operator,
- one representative of the Central Safety Service for Nuclear Installations or from its support organisation, the Institute for Nuclear
  Protection and Safety,
- one representative from the Central Service for the Protection against
  Ionising Radiation at the ministry of health.
- the director of Civile Protection of the province or his representative.

Led by an individual appointed by the police commissioner, this General
Staff would organise press conferences at fixed hours and would have to
reply to all questions from journalists. In addition it would publish at
predetermined intervals communiques on the situation.

Liaison to be established between the provincial command post and a Paris
command post which could have an identical structure and whose action vis-
à-vis the national media would be coordinated with the action at the local
level.

The persons appointed to assume this "Information" function at the national
level would have to be appointed in advance. Their guidance would be
entrusted to an individual appointed by the Prime Minister (4, pp. 74-75).

However, one aims still at a strict filtering of information as the un-
'expurged version of the document additionally indicates:

In case of a crisis all appointed persons would have to set up, for each
of the organisations which they represent, the obligatory screening of all
technical infromation, of draft decisions and of the latest assessments of
the situation so as to be able to ensure effectively their mission of informing the public (5, p. 31).

2. THE ECONOMIC WAR: JUSTIFICATION OF THESE COMPROMISES OF THE
THEORETICAL PRINCIPLES OF DEMOCRACY

Multiple difficulties, blockages, imperatives may justify, in the eyes of
the supporters of this policy, this exclusion of the citizen: the too great
complexity of the problems, their strongly emotional nature, the speed necessary for taking up options etc.

However, two final arguments are thrown into the balance: the general chaos in the absence of firm determination in the pursuance of present tendencies; the 'economic war' which leaves no other choice.

The idea of war, civil or military, is often invoked. Six British individuals*have referred to it to justify their support for a massive effort in the nuclear field.

Nowadays, the theme of the 'economic war' has become central. The conclusion of the article by H. Lewis on the warning of TMI takes up this key idea:

Now, every reader who has attentively read this article knows that there are better things to do than to ask the sempiternal senseless question: Are the reactors safe? You would perhaps, nevertheless, want to know if I believe that nuclear energy is safe enough to satisfy an important part of the nation's requirements. My answer is in the affirmative, not because there isn't any danger or any uncertainty in nuclear matters, or because coal is more or less dangerous but because the energy crisis which we are facing is the moral equivalent of a war. I do not see any reason for not fighting with all the means at our disposal; nuclear energy represents an essential weapon in this battle (7, p. 89).

One must clearly see all the implications of this reference to war: war means, among other things, the suspension of democracy. Between 1914 and 1918 Europe had lost six million of its citizens who were young and strong; this leaves a large margin for the gallantries which nuclear energy, chemical industry or liquified gas may cause.

3. THE ULTIMATE WAGER

The wager is the kernel, as *Le Monde* headlined its editorial at the time of TMI with these four words: This had to happen (8) — it will happen as late as possible. This means that at a time when irreversible commitments have been entered into, when the population will effectively have got used to ignorance and uncertainty.

A good speech would then be enough which if needed (as has been done in matters of the economy) would warn that the time of milk and honey is over; that the future belongs to brave and loyal citizens who respect the order which is necessary for proper functioning of a war economy.

If this discussion were not understood, the facade of sophism might give way to a second phase of the same policy: the show of force that would bring the citizen to Reason. The reason for power.

---

*Sir John Elstub, Chairman of IMI and former Chairman of the Institute of Mechanical Engineers; Sir John Atwell, Chairman (1978-79) of the Council of Engineering Institutions (CEI) and former Chairman of the Institute of Mechanical Engineers; Sir Charles Pringle, formerly of the CEI (1977-78) and former Chairman of the Royal Aeronautical Society; Professor R. C. Coates, vice Chairman of the Institute of Civil Engineers; G. Tony Dummet, Chairman of the Meeting of Chairmen of Professional Bodies, former Chairman of the CEI and former Chairman of the Institute of Chemical Engineers; C. Norman Thompson, Director General of Shell Research and Chairman of the Royal Institute of Chemistry.

Having the choice between repression and resignation, the citizen will prefer a sham; soundings can show how reasonable and trustful the citizen is. Acceptability will be saved.

Army, mourning, misery and bitterness: Courrieres and the great Clemenceau have already shown the approach. It remains only to adapt the response to the present difficulties, i.e. when 10,000, 100,000 or even a million people will be involved and not just 1,099 miners. One will then have to call upon all the modern technology to 'manage' social reaction. State of emergency on the one hand, substantial economic aid and development of ecological programmes on the other*, must in the end permit one to overcome the difficulty and get back to a stable situation, where the 'process of conditioning' that was started will be sufficient to guarantee social acceptability and the pursuit of the economic war by the same means.

REFERENCES

(1) Les problèmes généraux de la protection contre les radiations du point de vue de la santé publique, Organisation Mondiale de la Santé. Actes de la Conférence Internationale de Genève — Nations Unies. L'utilisation de l'énergie atomique à des fins pacifiques, août 1955, vol. XIII.

(2) Questions de santé mentale que pose l'utilisation de l'énergie atomique à des fins pacifiques, Organisation Mondiale de la Santé, Rapport d'un groupe d'étude, série de rapports techniques, no. 151, Genève 1958 (58 pages).

(3) Rapport du Directeur General de l'U.N.E.S.C.O. sur la réunion d'experts chargés d'examiner les conséquences sociales, morales et culturelles de l'utilisation de l'énergie atomique à des fins pacifiques. Maison de l'UNESCO 15 septembre 1958, U.N.E.S.C.O./SS/26, 16 avril 1959 (8 pages).

(4) B. AUGUSTIN et J. M. FAUVE, L'accident nucléaire de Harrisburg. Analyse d'une crise. Sofedir, Paris 1979 (83 pages).

(5) L'accident nucléaire de Three Mile Island, Rapport présenté par B. Augustin et J. M. Fauve, au nom de la mission d'étude sur le déroulement de la crise, 4 juin 1979. (55 pages).

(6) Supplément à Inter-Info no 42. Commissarait à l'Energie Atomique, ler décembre 1978.

(7) H. LEWIS, La sûreté des réacteurs nucléaires. *Pour la science*, mai 1980, pp. 73-39.

(8) *Le Monde*, ler-2 avril 1979.

---

*Return to the land, to the underprivileged areas for the most determined activists: which will alleviate so many problems of acceptability.

II. THE DEMOCRATIC SCHEME CONFRONTED BY THE
CHALLENGE OF MAJOR RISK

The eighth chapter has shown what directions one can take in responding to major risk: redefinition of the status of safety and its means; opening up the decision making process to the citizen to ensure a legitimacy which is often lacking in the political power.

One feels strongly that these are still immediate responses rather than fundamental ones, adjustments rather than changes. Well, major risk demands more. It confronts those in charge and the citizens with something unprecedented which is difficult to grasp, even to think ... but it requires responses in depth. Here again it is important to set up strategies which give direction to useful tactics.

In the following pages we attempt a first examination of this problem.

1. FACING THE UNPRECEDENTED

Never before have such risks emerged from the very core of human activity: from the industrious activity the aim of which is the improvement of the conditions of life and not destruction as in the case of the military.

Never before has the political power been confronted with this necessity to act urgently, it is about the economic and social life of the Western World, with such risks and such strong uncertainties about these risks.

Never before has the split been so neat between the need for massive support from the citizen, who alone can give sufficient legitimacy to inviting such colossal risks, and the poverty of procedures of legitimation offered by by the public powers.

Never before has the citizen had at the same time this, no doubt, ambiguous desire to weigh down on the decisions and the massive handicap of his incompetence.

Never before has the need for information been felt so strongly and at the same time the difficulty to achieve it and the dangers (at least in certain respects) of such information.

Etc. ...

Some points will permit us to assess the challenge.

1st. The technical challenge: high risk, urgency and uncertainty

The technical problem has been dealt with before. Let us just recall here the question of uncertainty with which one is immediately faced.

The technician finds himself facing two major difficulties. On the one hand, he sometimes hardly knows the theory of the physical phenomena which he is about to set in motion; on the other hand, he has no absolute assurance about the probabilities to assign to eventual accidents. Added to this is an imperfect knowledge of equipment reliability; unknown factors from the prototype stage to industrial application — and on top of it the difficulties which may occur in case of a change in the conditions of application.

The problem connected with the insufficient knowledge of the physical

phenomena is new inasmuch as one has often become incapable of experiencing
in full all of the systems involved. As concerns the probabilities one must
remember that probability is foremost a mathematical thing; its objective is
not to describe a reality but merely to approximate it. It does not at all
mean that the event will not happen shortly; the technician talks of low
probability; he does not say that the event will not happen.

A good illustration of these difficulties is given in the report on the
safety analysis of the Canvey Island site.

Canvey Island. On account of the difficulties inherent in risk assessment,
it is not possible to establish limits of confidence for the conclusions in
a clearly defined manner. The estimation of uncertainties is itself a
process subject to professional judgement (1, p. 14).

Uncertainties also exist in the estimation of the consequences of any
particular accident sequence (1, p. 14).

Probabilities could be assigned with reasonable confidence to certain
cases of accidents while with others considerable speculation was involved ...
The report by the investigation team indicates whether these estimations are
suitably founded or whether they are based on limited information or on an
arbitrary figure (1, p. 14).

The passage from technician to politician increases these uncertainties.
Sometimes the contempt of the specialist for the person in charge, the faith
of the expert in his technique, the fact that the technician is not prepared
to tell all (for very understandable reasons in terms of power) are just so
many elements which obscure communication.

In order to take his decision the politician will have to cut into the uncertainty and above all put aside the scientific approach. One moves from
the elementary probability of assessment to the probability of the wager.

This distance between the technician and the politician becomes a deep
trench when the disaster has happened, i.e. when there has been a sudden
eruption of a certain contingency in a programmed system. Science has much
to say about the functioning of the system. When there has been disruption,
i.e. *a posteriori*, policy is on its own. Uncertainty is the heavier to bear
when one must act and sometimes act with great speed. This is what happens
when big options have to be exercised or, worse still, when one has to face
a disaster. Let us recall the case of Seveso: must one evacuate one hundred
people, 5,000, 12,000, 220,000 or still more? Must one go along with the
NATO report which recommends a threshold four hundred times lower (in concentration of dioxin) than the measurable 'acceptable' one?

J. Kemeny had the same experience in 1979.

Three Mile Island.

Someone will have to make a choice for us between the various alternatives.
This is no doubt the role of the President or of Congress but what worries
me after my experience in Washington is that I don't see an adequate process
set up to take a decision.

Let me go a bit more into depth ... Consider for instance the technical
data. There was a time when, if someone had asked me for information on
science, I would have said: Ask a scientist. If they had asked me which

one, I'd probably have answered: As long as you have one who is good it
does not matter which one because if the question comes within his/her
field of competence and if he/she studies the problem long enough, scientific truth will come out. I still believe this, but it is not always
good practical value because certain technologies we use are on the very
border of scientific knowledge.

Our commission ecnountered this problem ... We have tried to clarify the
question: what would have happened if the accident had lasted longer than
it did in reality and if there had been a real melting of the core? Would
that melted core have been kept confined or not? We have turned to big
national laboratories to get the answer and they did all they could to find
it. Their answer was: We think*that it would have remained confined. This
is more reassuring than the opposite but we cannot be absolutely sure
because our experts quite simply don't have enough information to answer
the question with certainty (2, p. 70).

2nd. Democracy out of breath

a) A law that takes the backseat in the face of 'necessity'. H. Lavaill,
Section Chairman in the Council of State writes in the preface to the CEA
publication *'Nuclear Law'*:

Nuclear energy has at first been a fact: essentially scientific in its
aims, then military: two bombs at the end of a world war. The law came
afterwards (3, Preface).

But very quickly, the author adds. The 'but' does not eliminate the
problem that the law follows the fact.

In the same way J. Teillac, High Commissioner at the Atomic Energy writes
in the foreword of the same publication.

... all innovations that lead to applications which interfere in everyday
life invariably open up new chapters in the law taking into account the
immense possibilities they bring about as well as the potential inconveniences against which one must guard. (3, Foreword).

The most decisive elements in development first change the facts; then the
law brings about the desirable adjustments. Thus the problem of exercising
choice is taken very much upstream. Must the discussion always develop
'afterwards', i.e. always come too late?

Downstream the institutional and judicial problems are not slight either.
The setting up of centres for instance are also facts; legitimacy is obtained
or rather attached by purely formal procedures which deepen the gap still
further between the 'land of reality' and the 'land of the law'.

The case of Plogoff is symbolic of this. The police on the warpath come
to protect the 'town hall annexes' destined for the gathering of citizens'
observations from which one wants to take up the whispers but ignore the
shouts of despair, the indignation and frustration. This is the sad spectacle
of a blocked democracy which, however, one wants to 'turn around', the only
one good and right rule (cf. next page).

---

*Underlined in the original.

b) Information. Here we shall bring up a press notice published in 1979 by
the ministry of industry. The problem is difficult and complicated further
by the passionate climate and the irrational fears which surround the nuclear
question. There are in fact a limited number of secrets to be protected,
either of an industrial nature or having to do with national defence and
necessary protection against acts of malevolence which forbid for instance
that certain technical documents concerning nuclear installations be published in full.

One must guard against the belief that one could at once publish everything
without seriously prejudicing security: with the absolute desire to publish
everything at once one would inevitably create phenomena of self-censorship
by the experts, flight from responsibility, defensive chicanery in which
every expert would worry not about facts, but about rumours and one would
provoke the paralysis of the deliberating and decision making organisms. It
is therefore indispensible to ensure an area of calm so that the scientific
experts can breathe.

On the side of acceptance by the citizens, the problem is not easy either.
The multiplication of isolated information, which would often represent non-
verified hypotheses or observations that have not yet been interpreted, would
serve no purpose other than creating confusion and panic in minds which are
not prepared to analyse them. This would mean opening the doors to all kinds
of demagogy and discord.

It is important that the surveillance and information bodies are not cut
off from actual practice and detached from reality: this is the reason why
the solution of (creating) an independent agency will not be adopted. It is
necessary that the people in charge themselves, at the right time, give the
necessary information (4).

In the opinion of those in charge high risk makes information extremely
difficult: secrets to be protected, malevolence to be avoided, panic to be
prevented etc.

Thus one gives information only on points which are:

- not critical,
- not alarming,
- not urgent,
- not doubtful.

In short, the risk having been qualified as

- often critical,
- often alarming,
- often urgent,
- often uncertain,

one can say that information on the phenomenon is for the time being considered to be too delicate to be given in any other than a parsimonious way.
This is just what those in charge of the Brest oil project expressed so well:

The studies may well be communicated to those interested, always on
condition that they have been completed (5).

or the operator at Harrisburg:

I do not see why we should have to ... tell you in detail everything we are
doing (6, p. 120).

PLOGOFF: Democracy out of breath

Thus when the authorities take the most serious options, in a crisis situation, in a situation of uncertainty, there is no information. It is only given afterwards. High risk and information of the citizen are there-fore presented as mutually exclusive; risk will be a fact, information a slow and deceptive discourse responding to a simple psychological demand, not a political requirement.

Through this bias, yet again, democracy appears to be thrown to the wind.

c) Confidence in the people in charge. A modern democracy is no longer organised in the market place. There must be delegation, and therefore confidence. The discourse is one of the most used channels to ensure and enliven this confidence without which a democracy cannot function.

Seeing that information is presented as very difficult or even impossible seeing that the law gives way to force; seeing that words are used more to cover up than to show, to indicate the non-negotiable rather than to open the discussion ... confidence collapses.

In such a situation, at the first wrong move, the authorities lose their credibility.

> After consulting the population we therefore decided to build a nuclear reactor.

Drawing by PLANTU

Fig. 36: The conclusion by the enquiry commissioners favours the construction of a nuclear centre at Plogoff.
(Source: *Le Monde*, April 17, 1980)

The commissioners have examined the observations by one hundred and twenty one people and have declared that five hundred and forty two inhabitants went to consult the documents at the police commissioner's office and in the 'town hall annexes' installed at Plogoff between January 31 and March 14.

The report also states the energy situation in Britanny which produces only 6 per cent of its electricity consumption. It also takes into account the favourable opinions expressed in 1978 by the regional council chaired by M. Raymond Marcellin (R.l.) and the general council of Finistere chaired by M. Louis Orvoen (CDS). It recalls that the political parties of the majority and the communist party have pronounced in favour of the construction and that a move opposing the project was launched by the socialist party, the union democratique bretonne and the ecologist movements. This opinion which allows some reservations in respect of human problems will be sent to the Council of State which will declare public usefulness within eight months.

VIOLENT DISPUTE

In fact, the public enquiry took place in a climate of violent dispute. The mayors of the four communities concerned (Plogoff, Primelin, Goulien and Cleden-Cap Sizun) having refused to open their town halls, the police commissioner had vans installed which were referred to as 'town hall annexes' and guarded by mobile constabulary. Barricades were set up and violent clashes confronted the forces of order with the population. A daily protest demonstration, the 'Mass' was organised and several gatherings called involving large numbers of ecologists and the local population.

On several occasions the demonstrators were summoned and convicted. The court hearing on March 6 at Quimper where nine people were convicted, under the Vandalism Act in particular has been marked by incidents and specifically by the suspension of a lawyer while the court building was guarded by police. Subsequently, seven of the demonstrators who had been arrested were set free after a second ruling of 'appeasement' on March 17.

A big celebration was organised at the cape of Raz after the end of the enquiry and the socialist mayor of Plogoff, M. Jean-Marie Kerlock thought on March 16 that he had 'won a battle' and declared that the population would fight 'to the end'.

Just as a finance minister is not believed at the time of a monetary crisis, rather the contrary, if he proclaims that there is no question of a devaluation, so a person in charge will not be listened to when he affirms after an accident that 'everything is under control'.

A poll by EdF-Harris seemed to indicate elements of this kind. Carried out after the TMI accident it showed that 80 per cent of the neighbours of nuclear centres thought that the authorities did not tell the truth in case of an accident, 61 per cent thought even that a Harrisburg type accident might have happened already but the authorities had not said anything about it (!) (7, p. 5).

How, then, should one manage an accident situation? Distrust will oblige to extreme authoritarianism, more so as a strong resentment could then break forth from those who had given their trust yesterday by non-accepted submission.

d) The gap between 'experts' and 'politicians'. The Minister for Industry presented these difficulties clearly in the National Assembly on May 22, 1980:

At present we witness the development of complex technologies, be it in the field of energy or in the field of information (data processing), in medicine and biology. We are therefore all faced with a very delicate question: trying to reconcile the conduct of technological progress with the political responsibilities we take upon ourselves.

On the level of public opinion this finds expression in the complexity of information the difficulties of which some of you, whether you belong to the majority or to the opposition, often and justifiably recall.

As for the political authorities, government and parliament, they find it difficult to take their bearings given the complication of technical developments. In fact, those who have the most advanced knowledge or the most complete documentation are only a limited number. They have an effective privilege which is not easily reconciled with the proper exercise of democracy. I believe we are all agreed on this point (8, p. 1238).

From here on the control of options concerning risk largely escape the political powers, starting with the legislature to which, as we have seen in chapter eight, a useful means of improving its competence is refused. To have a legislature that is less 'competent' than one is usually prepared to admit seems unconstitutional. Here again, basic problems and contingent difficulties mix and sharpen the challenge.

e) A clear expression of these difficulties: the brute discourse between the most alert. We quote hereafter extracts from talks which members of the PEON commission (Consultative Commission for the Production of Electricity from Nuclear Sources) afforded Ph. Simmonot in 1976. These are the top people in French energy policy.*

---

*We take up, one after the other, extracts from talks with the top people from the following organisations:
1. Alsthom;
2. CEA;
3. EdF-
4. CEA
5. An enterprise involved in the oil industry;
6. Ministry of Industry;
7. One of the principal actors in the 'nuclear decision'.

Ph. Simmonot: What role has the Plan played?

1. Ah! the Plan! the Plan! the Plan is a hesitant body. It prepared political arbitration. On the whole it has been rather favourable because it was sensitive to economic arguments (9, p. 45).

Ph. Simmonot: Perhaps you can tell me then how the uranium prices will develop.

2. I give incorrect figures to the ministry of finance. If I gave the true figures they might look too low to our African suppliers and too high to the ministry (9, p. 132).

Ph. S.: The control by the environment ministry ...

3. When the environment ministry was created we were told: "Don't move too fast, don't build centres just anywhere". Still, they must be set up somewhere, these centres! The environment ministry therefore has a say in the choice of the sites we propose ...

    What seems serious to me are the precautions that are taken. We could do it ourselves. But we are normally subject to outside rules.

    Having said this, too much is imposed on us. It goes too far. True, this is a new technique and for the first time it is normal to take many precautions. But the pendulum has swung too far in one direction. Once people are accustomed certain rules will be softened.

    I have been told that in the USSR they have no containment walls*. This is perhaps not so stupid because such structures induce stress on the pipes passing through them, which consequently increases risk: there have recently been two deaths on account of this. Thus certain excessive precautions lead to additional risks (9, p. 113).

Ph. S.: Perhaps the nucleocrats did not express themselves very well?

4. Yes, that is possible. This is all a problem of communication. In this field the technicians are a bit like cripples, particularly when there are controversies. This is no reason to treat the people as mental defectives. I regret that the people are looked at in this way (9, p. 122).

Ph. S.: You must also take public opinion into account. Don't you think that the decision makers underestimate this?

5. How do you want them not to be underestimated? They are so stupid that a rational mind can only underestimate them. There are only politicians who can appreciate them (9, p. 95).

Ph. S.: The waste ...

1. This no acute problem. The storage of this waste nowadays presents no difficulty, it occupies only a few square metres. This will be a large-scale problem in the year 2000.

---

*The usefulness of the containment walls was clearly shown at TMI: one can imagine what would have happened if there had been no such wall.

Having said this, there is the long term question of waste. For the time being one cannot say that it has been brought under control but one cannot say either that the survival of the species is threatened by it ... So much nonsense is talked about this subject.

We are working on this problem. When something raises a problem, one beautiful day a solution will be found. If no solution has been found in fifty years time we shall make a package of those wastes, put them somewhere and stop nuclear energy; we shall have lived for fifty years (9, p. 87).

Ph. S.: But is it certain that a solution will be found?

5. One always finds, one always finds, one always finds. There is no limit to human malice ... (9, p. 96).

Ph. S.: The Cassandras ...

6. At the limit, if a radioactive cloud escaped, the small area affected would be evacuated.

What must be understood is that by contrast to Feyzin or to an earthquake there is no suddenness in case of a nuclear accident; one has time to react* (9, p. 106).

Ph. S.: You just said that one could evacuate a 'small area'. You realise the reaction of public opinion!

6. Oh, yes! The psychological consequences would be disastrous. This is why we redouble precautions. Thus we blow money into the air which could be better used to eliminate black spots on the motorways. But the psychological atmosphere is such that one must redouble the precautions (9, p. 106).

Having said this, I really lose no sleep on account of nuclear energy. What I fear most is the following scenario: for five years all will go well; then one will say: all these fears, it was ridiculous. And attention will be relaxed. And at that time one will have trouble (9, p. 107).

Ph. S.: The tension will have been such that ...

6. That's it. And then also the effects of routine; when one does the same thing for ten years, the same motions, the same controls ... Our task is to maintain the tension of alertness, if I may say so. One day the Minister of Finance will say: All these precautions are really very expensive; do they really serve any purpose?

The embarassing thing, in a way, is that one has not had any accident ... A real nuclear accident (9, p. 107).

Ph. S.: For how long have you (has your organisation) existed?

6. The central service has its third anniversary. It is therefore a recent service. Before — i.e. before Messmer — there was no central service

---

*This deserves more adequate spelling out: What time? For what reaction?

and therefore no represnetative from this service on the PEON commission (9, p. 99).

Ph. S.: And therefore safety was not represented?

6. The ruling idea at the creation of the service was that safety should be one of the elements to be taken into account in the decisions.   (9, p. 99).

Ph. S.: How is it that this did not occur earlier to the people in charge?

6. Because, as I have told you, the central service did not exist (9, p. 99).

Ph. S.: But the problem as such existed ...

6. One was not aware of it (9, p. 99).

Ph. S.: I thought the explanation by the absence of a central service was insufficient.

6. Exactly: one was not aware of it ... I sometimes tell challengers that their challenges are sometimes useful (9, p. 100).

The examination of the safety problems is done *a posteriori,* when the decisions have already been taken*. The main responsibility for safety lies with the nucelar operator. This is the idea. The operator is therefore given charge of designing safe installations.

Later on he justifies himself for them before the safety services. A kind of dichotomy is set up in the role of the state; a producer state takes the decisions, thinking implicityly that the technicians are safe. It's a wager. Subsequently the safety examinations permit verification whether the wager has been won (9, p. 100).

3rd. The solutions of the past

Faced with disquieting menaces, in an 'irrational' way, as one usually says, with infinite condescension, or in a perfectly well-founded way as many examples show, after the event, there is the temptation to go back to the candle. The temptation of the myth of the golden age which plays parts of which one must be aware.

The idea of the return to natural bliss provides no effective means for him who wants change: utopia is only dream and digression.

One may ask who really preaches such return. This attraction is perhaps used more by its detractors than wanted by the groups pointed out as being stuck in the past and dangerous. The idea of the return to the age of the candle permits those who refuse a new reflection about our development and who want to persevere along the same lines to win easy comfort for their tranquil assurance.

They even show themselves up sometimes as sponsors of the experiments of deviant communities in underprivileged areas which permits them to get rid

---

*Since this talk things have developed considerably as the central service follows up on the development of the project regularly and is liable to intervene before an irreversible decision is taken.

of opponents who threaten to call 'progress' into question, in other words the means of the 'economic war'. These are not laboratory views. Some renowned international organisation seems to work in this direction already.

4th. The problem of fear

The three preceding points show that there is no easy answer to the problem of major risk. Very often the body politic is on the crest of a wave. We must stop for a moment at a very difficult question in this respect: the problem of fear.

What can be the effect of information about high risks? One must consider our practice of democracy in order to examine the answer to this question. The citizen, here, seldom enacts his choices; misinformation is more current than information and the transmission of knowledge.

At a time when the body politic cannot sufficiently lean on knowledge, at a time when its choices involve the citizen en masse, too much to forget him altogether, the body politic may want to half-open its decision making process. At this point the citizen may refuse the offer of half-measures.

This is the rejection of information that is given only on the strict understanding that there is no power sharing.

Faced with very frightening information, in a disaster situation, the reaction threatens to be: we no longer want to know. We do not want to know. See how you sort the problems out, or more or less — so that we don't have to worry. If you drive us to fear we shall invent gods for ourselves, and the most accessible religious form is fascism.

Fear leads its victims most regularly to the strongest. Democracy, already out of breath, risks to be harmed still more by high risk. In a serious situation the strategic recourse to fear as the province of intervention by the executive branch would very seriously affect the health of our democracy.

2. OPERATION BASE FOR A START

1st. Renunciation

The unprecedented, by its very nature, leaves one perplexed, in some cases paralysed. The temptation is strong to renounce, to fall back on habitual behaviour.

A Maginot line of the mind, a discourse well rehearsed for the people who often only want to forget and to be protected (this atmosphere of anaesthesia which permits at least a vision) and a flight forwards may seduce (knowing also latently if not explicitly that in case of problems the public power can always have recourse to the means of legal violence).

This is the approach of renunciation.

It is not without risk. Risk of a socio-political nature as we have just mentioned: the cases of Plogoff and of the *Tanio**constitute warnings in this

---

*Cf. the sticker pictured on the next page which was sold by the hotel owners and shopkeepers of the north coast of Britanny in 1980 in memory of the *Torrey Canyon*, the *Amoco Cadiz* and the *Tanio*.

respect. Risk also, one will at least be sensitive to this aspect, for socio-
technical development; the experts who took part in the preparation of the
eighth plan did not hide it in their report:

> The risk of rejection of technological progress has been analysed as deeply
> as possible by the working group because this disruption, in the present
> state of events and frame of mind, cannot be considered as improbable (10,
> p. 13).

One must have the courage to say No when faced with certain situations and
to say like this expert who has given us long hours of conversation:

You are feeding the plague

Fig. 37: Thus was born the black flag of the rebels

2nd. The forces which raise the challenge

Those who cannot follow the approach of renunciation are led to look at the forces that are capable of supporting a different perspective. This is doubtless for those who can see. True, they do not seem to be much in evidence under such and such a banner: they are not necessarily marked out by a clear ideological reflection. This is a new issue for our societies: major technological risk creates new camps of socio-political forces.

Major risk is thus a new factor which is going to upset our political and social organisations which are modelled on the nineteenth century, the age of the machine. We join here the preoccupation of A. Touraine about social movements (11). But for lack of a theoretical bases and sufficient examination on our part to move ahead with assurance on this approach we shall come back in pragmatic fashion to our field of study by rephrasing the question: which forces oppose renunciation?

In the course of our exploration we have seen a good many motives which permit us to believe in a possible start. Among the active forces we count in the first place vigilant citizens who are determined not to succumb to infatuation, who are also careful not to delight in doubtful fashion in alarmism and disaster prophecy.

Without drawing an overall lesson from this, and one must guard against this temptation!, without wanting to say more than necessary about the phenomena (in order to put them back quickly into an old goatskin that serves as a recomforting theory), let us recall the following lessons:

The one from Dr. Reggiani, the man high up in the Hoffmann-la-Roche hierarchy who smashed the game of sham and impotence at Seveso:

The situation is very serious; the houses must be destroyed ...*

The one from Laura Conti, regional deputy of the Italian communist party, again at Seveso, who used all her energy to let a little truth emerge into a situation where everyting had become false:

I know the lies from the Right; I know the lies from the Left ...
... We have no figures ...
To reduce the danger one must first of all educate the population with concrete examples and not with circulars ...

Are we wrong or right? ... I shall do it again if it has to be done again but it is frightening because we know very well that a safety threshold does not exist. But if one day I shall be told that there is a child with leukaemia in zone A, then perhaps the painful feeling will grow inside me that I was wrong, terribly, irreparably wrong, a wrong that I shall carry with me for the rest of my life ... (12, pp. 47-48).

The one from Sir B. Braine, conservative MP for South Essex who relentlessly spoke out in parliament as he did during the night of July 23/24, 1974:

... The story I have told may be applicable to other communities in the U.K. What is at stake is not the quality of judgement of those who govern (us) but the whole of our democratic process. If day after day the resolutions

---

*Cf. the Seveso case, chapter one.

of the elected representatives of the population — on the local level as well as here at Westminster — are to be rejected and treated with distrust, if attempts are even made to impede the hearing of the case, what is our democracy worth? (13, p. 1497).

The same preoccupations exist with the people in charge of industries and administrations who have seen their acute sense of responsibility in action. We have quoted examples of interesting attempts. Let us remember here in particular the work done on Canvey Island, the safety analyses carried out in France on various installations which until then had been much less well studied, the efforts by certain people to improve the information of the citizen etc. On the level of international research we mention the efforts by the services of the EEC commission; in the introduction to the seminar on Technological Risk held in Berlin in April 1979 Dr G. Schuster*phrased the problem:

The hostility to the development and the introduction on a large scale of certain new technologies grows incessantly, just like the fear which the threat from invisible dangers inspires.

This, then, is the dilemma. The future, the survival of a free and autonomous Europe, depends on the setting up of new procedures and technologies which are often accompanied by considerable risks. But at the same time we should not impose these procedures and technologies on a society which does not want them, on a population that rejects them. We need a consensus by the majority**.

We need proof of acceptance. And this is an important problem which is added on to all our other difficulties (14, p. 3).

If one is ready to call for a new beginning and to show one's determination in action one has a heavy task ahead. We have previously mentioned a number of changes to be accomplished and adjustments to be made. True, this is work that takes a long breath but one can set oneself for the short term objectives that must imperatively be attained and within the shortest time possible. Very quickly (because we shall take this question up in the general conclusion) let us mention here some key points: redefinition of the status accorded to safety (on the level of particular organisations but also for the country in general); development of knowledge about major risks that exist or are in the process of being created; information of the citizens and their represen-tatives.

In addition, a certain number of innovations would have to be launched with the aim of creating a dynamism favourable to more responsible handling of the problem, and this to be done serenely, not under the reign of fear. Witness the case of the proposal of an open discussion of the setting of standards in the USA. It was a matter of establishing a directive in matters of protec-tion against ionising radiation. The risk was taken as a subject for debate as well as for an objective to be managed. The proposal was published in these terms in the Federal Register by the federal agency in charge of the environment (EPA):

---

*Director General of the Research, Science and Education Directorate.

**The notion of majority is no doubt a bit difficult to handle in matters of major risk.

PUBLIC ENQUIRY, INIQUITOUS ENQUIRY

Two hundred and six Bretons have written to the enquiry commissioners or have entered their observations in the enquiry registers open at the police commissioner's office at Quimper and in the vans called 'town hall annexes' which from February 4 to March 14 have been stationed in four communities in Cape Sizun.

Only two hundred and six out of 60,000 people who live in this corner of the hexagon; the boycott order for the procedure followed by the police commissioner has been followed by the population. In response, thousands of people demonstrated their hostility outside the legal procedures. Never has the split between an institutional mechanism of consultation and the citizens' desire for expression been so flagrant.

The strike against the public enquiry had started on the site of Brand-et-Saint Louis (Gironde); it grew in scale at Pellerin (Loire-Atlantique), was 'perfected' at Golfech (Tarn-et-Garonne) to attain its height in Brittany. Henceforth, will all consultations concerning nuclear centres have the same fate? For how much longer will the government be able to hang on to a procedure the paralysis of which spells out its inadequacy?

The public enquiry is already an old mechanism the purpose of which is to bring to the knowledge of the population the broad outlines of a project of general interest which necessitates the expropriation of land or buildings. The citizens involved are invited to formulate their eventual objections; then after the report by the enquiry commissioners and the opinion of the Council of State the government in a text published in the official journal makes a declaration of public usefulness. Convinced that its projects serve the community, it has,

for a long time, taken the procedure as a simple formality.

The enquiries had nothing public except the name. They took place stealthily not to say clandestinely, in any case at hours and seasons which did not permit the interested parties to go to the townhalls.

NO REFERENDUM

These faults became so flagrant and raised such protests that in 1976 M. Jacques Chirac, when he was Prime Minister, signed a decree reforming public enquiry. Time, duration and place of enquiry, the composition of the file, appointment of commissioners, publicity given to the consultation and to the conclusion by the commissioners were noticeably improved. These measures, commented M. Chirac, bear witness to the desire by the public powers to improve the participation by the inhabitants in the planning of their living conditions.

The word 'participation' awakened considerable hopes. They have been disappointed and quite particularly in the nuclear field because when it happened that town plannings were changed subsequent to public enquiries there was never an important project (motorways, rail tracks, port installations, nuclear centres) affected by one iota.

The citizens had imagined that they were being consulted about the public usefulness of a planning operation and that consequently if they had opposed it the authority would have to renounce it. For the common mortal this is what participation means. On the whole, some people likened a public enquiry to a kind of referendum which it has never been.

For the administration, by contrast, launching a public enquiry means informing the population and gathering eventually some observations on details. In fact, the decision is already taken; this is

so true that reservations by the enquiry commissioners and even negative opinions, however rarely, from the Council of State have never stopped the government. One has seen this in the affair of the motorway at Sologne which has been declared to be of public usefulness despite contrary opinion from the councillors of state. Ambiguity has, however, been entertained by the public power which in case of a challenge declared that the interested parties have been democratically consulted since there was a public enquiry. The misunderstanding had to come into the open sooner or later and this is the case with the most controversial and most impassioned subject: the nuclear file. It is true that the blows struck against democracy have been so numerous in this field that the exasperation of the citizens has reached its limit.

The French people involved are no longer content with saying No to the centres, they refuse to lend themselves to a sham of consultation, they want nothing to do with what they call tin-can-participation, they are defending in depth their understanding of democracy in the face of a public power (government) which has quite a different understanding of it.

The confrontations at Plogoff cannot be taken for folklore. They illustrate a conflict which goes far beyond the one always called forth between private and collective interests. It is here the question of a population affirming its right to decide its future itself in the face of a central power which, based on technocratic habits, wants to direct it otherwise.

MARC AMBROSE-RENCU.

Fig. 38:
(Source: *Le Monde*, March 16/17, 1980)

The recommendations for the directive aim at attaining adequate protection
of health for the small fraction of the total population who are subjected
to the greatest risk because of exposure to transuranic (elements) in the
environment and thus to offer much better protection to the vast majority
of the population who are exposed to a lesser risk. The risk, on the level
of the proposed directive, is estimated as being below one chance per one
million years and less than ten chances per hundreds of thousands because
in the course of a life an individual develops a cancer as a result of continued exposure to the established rate of doses. One must recognise that
these estimates are not precise and carry an uncertainty of at least a
factor of three for the risk of cancer. There may be differences between
estimable scientific opinions in the statements made but they represent the
the best judgement by the Agency. The social judgement of the fact of
knowing whether this risk level is acceptable will be subject to public
debate during the period foreseen for this purpose. The final recommendation by the Agency may be higher or lower depending on what will appear
to be appropriate.* (15, p. 60 956).

Resolution for immediate action, fixing minimum thresholds to be attained,
launching of procedures destined to give rise to different practices: this
is a first collection of demands in the face of major risk.

There remains the fundamental question to be asked. We shall see in what
terms we think it can be approached.

3. THE WESTERN WORLD FACED WITH MAJOR RISK: QUESTIONS ABOUT
   KNOWLEDGE, POWER AND DEMOCRACY

D. Nelkin recalls a formula from American industrial circles in 1933:
Science finds — Industry applies — Man conforms (16, p. 19).

Major technological risk renders such a shibboleth inapplicable.

One must recognise that there is today a discontinuity with the past,
mainly because of new conditions of socio-technical development. One aimed
at setting up a world of stability and certainty. During the nineteenth
century one worked with the idea: "tomorrow will be better". Since the
beginning of the twentieth century one endeavours to provoke change and catch
up with it. Today there is instability, insecurity, uncertainty (17). How
does one face it? No doubt not by accommodating oneself cheaply to the
difficulties but by approaching them with the greatest intelligence possible.

At the limit of what we can do there is above all the whole seriousness of
our thought (18, pp. 12-13).

Seriousness of thought and seriousness in its use. In the field of knowledge one must recognise that one finds in the very core of the exercise of
scientific reationality what has been a bit forgotten under the influence of
triumphant and arrogant attitudes: doubt. Doubt, the foundation of all
thought. In the domain of power one realises that one can no longer deceive
everybody all the time (19); that here again doubt, which does not necessarily
lead to hostility, founds an action which wants to survive the immediate
future.

_____

*Underlining by the author.

It is therefore of fundamental importance to relocate oneself in relation
to knowledge, power, the exercise of authority and responsibility at a time
when major technological risk has come to call the two terms and their relationship into question.

What do I know? would often be a good question to remember on the level of
knowledge.

What alternative? would be another question, whatever one has to say about
it, for the decision maker.

What place for the citizen? would be a fundamental question for the responsible politician. He is tempted by the closed shop of the decision making
process, convinced that without this closed shop decision making might become
impossible. If everybody knows it, the debate will never end, he thinks. He
may be seduced by false openings: The public must believe what the experts
know, it must have the impression that it participates. It is a different
perspective.

Nowadays risks must doubtless be run in certain fields, as long as there
still exists a large expanse between the feasible and the insensible. One
cannot remove all uncertainties which attach to them. One should know how
to keep the citizen out of it.

Is there a way out? We think so. However it requires a different relationship to knowledge and to power. The citizen must know consciously that he
does not know, that the people in charge do not know everything and know
nothing completely. There is doubt and uncertainty. The citizen must be
able to express his wishes and weigh down on choices. Information without
power is only nonsense and in the long run the ruin of democracy.

This presupposes a training of minds. A training that requires time and
also reflection, not the sort that limits itself to the acquisition of
formulae, not subtle and carefully developed conditioning, not a cultural
redress that is learnedly carried out. Trained for responsible doubt the
citizen can consider the complex and the uncertain without taking flight
into the camp where a mystical saviour is sought. It is a matter of having
experienced at the same time doubt, uncertainty, choice and responsibility.
This is not ignorance but lucid knowledge. This is not resignation to uncertainty but shared responsibility in well understood uncertainty.

This is perhaps the key difficulty raised by major technological risk. Not
only is democracy a project which postulates that the other party has a say,
this is the agora; it presupposes also a training of the mind. In recent
years it was believed that a mass of technical knowledge might be enough for
the contemporary *homo faber*.

With major technological risk one discovers that this is very dangerous,
that it is a blind alley, to dispossess the citizen of the faculty of judgement. The problems raised by the social acceptance of technology will only
find para-military answers outside this perspective of a society made up of
free, voluntary and responsible and therefore informed citizens. Only such
a society will be capable of choosing consciously, of managing, of regulating
and politically conducting public affairs in the era of major technological
risk.

It is part of modern culture to face doubt and uncertainty. To remain in
a world of certainties in order to impose choices which are strongly marked

by uncertainty (despite all assurances to the contrary); to remain in this world of stability in order to refuse all action stamped with uncertainty are two behaviours of another age, today they are no longer in fashion. Reconciling the people in charge and the citizens with uncertainty means taking up the means for making choices in full knowledge of the issue and also clearly pronouncing refusals, well-founded ones this time, when the degree of uncertainty becomes excessive, aberrant. This is the only way of avoiding the blind alley of the manichaeism which exacerbates major risk.

There is therefore need for a new culture, for a new training of the mind.

Without such reorientation which undertakes to review our concepts in matters of education (in its broadest sense) it is to be feared that the rational will only produce the irrational and that a very peculiar type of scientific rationality will come to doubt reason altogether.

If one remembers the disquieting aspects of our examination one may give thought to what P. Blanquart wrote in 1968:

It is a matter of letting man exist in the technological age, i.e. at a time when absurdity has taken rationality in its service (20, p. 116).

We have seen however, that many points permit hope, again without certainty, that intelligence will be able to carry off and at the same time save reason and the capacity of a society as a whole to conduct its own transformation freely, voluntarily and responsibly.

REFERENCES

(1) Health and Safety Executive, Canvey, an investigation of potential hazard from operations in the Canvey Island/Thurrock area. Health and Safety Commission, London, H.M.S.O., 1978.

(2) J. KEMENY, Saving American Democracy: the Lessons of Three Mile Island. *Technology Review*, June-July 1980, pp. 65-75.

(3) M. PASCAL, *Droit nucléaire*. Commissariat à l'Energie Atomique Eyrolles, octobre 1979 (462 pages).

(4) Regards ... sur l'information. *La Gazette Nucléaire*, no 31, décembre 1979-janvier 1980, pp. 3-5.

(5) Extrait du procès-verbal des délibérations. Conseil Général du Finistère, séance du 12 janvier 1973.

(6) *Report on the President's Commission on the accident at Three Mile Island*. Pergamon Press, New York, October 1979 (201 pages).

(7) *La lettre de l'Expansion*. no 462, lundi 7 mai 1979, p. 4.

(8) Assemblée Nationale, 2e séance du 22 mai 1980.

(9) Ph. SIMONOT, *Les Nucléocrates*. Presses Universitaires de Grenoble, 1978 (313 pages).

(10) Préparer l'avenir à long terme. La Société francaise et la technologie

Commissariat Général du Plan. Préparation du VIIIe plan 1981-1985 La Documentation Francaise, 1980.

(11) A. TOURAINE, Z. HEGDUS, F. DUBET et M. WIEVIORSKA, La prophétie anti-nucléaire. *Le Seuil* 1980.

(12) L. CONTI, Trop d'échéances manquées. *Survivre à Seveso?* Maspero/Presse Universitaire de Grenoble 1976, pp. 45-47.

(13) House of Commons, Parliamentary Debates, Official Report (Hansard). London, H.M.S.O., Vol. 877, no 82, 24 July 1974.

(14) G. SCHUSTER, Le risque technologique majeur: perception et évaluation dans la C.E.E. — Introduction. Berlin, 2-3avril 1979 (8 pages).

(15) Persons exposed to transuranium elements in the environment. Guidance on dose limits. *Federal Register,* Vol. 42, no 230, Wedn. November 30, 1977, pp. 60. 956-60. 969.

(16) D. NELKIN, Science as a source of political conflict. *Ethics for Science Policy* (T. Segestedt, editor) Report for a Nobel Symposium, Royal Swedish Academy, Pergamon Press, Oxford 1979, pp. 9-24.

(17) J. ELSTER, Risk uncertainty and nuclear power. *Social Science Information,* Vol. 18, no 3, 1979, pp. 371-400.

(18) K. JASPERS, *La bombe atomique et l'avenir de l'homme.* Buchet-Chasterl, Paris, 1963 (708 pages) (Die Atombombe und die Zukunft des Menschen, Piper, München, 1958).

(19) M. AMBROISE-RENDU, *Le Monde,* ler avril 1980.

(20) P. BLANQUART, Mai 68 en France, ou du rationnel encore irrationnel face à l'irrationnel sous forme rationnelle. *Recherches et débats,* no 63, mars 1969, pp. 103-117.

CONCLUSION: MAJOR TECHNOLOGICAL RISK — A PROBLEM
THAT SUBVERTS THE BODY POLITIC

Major technological risk does not only lead the body politic to stiffen up
and to throw out of its field an important stake. Nor does it just, and by
contrast, call for an opening of politics to other attitudes.

Major risk constrains the body politic to relocate itself in relation to
the development process, i.e. in relation to two of its main components:
knowledge and power, reason and democracy.

Two fundamentally strategic options may be adopted. First of all refusal
which takes up the attitude of closure which we examined in chapter seven by
systematising it and turning it into a project. A flight forwards into the
attack and a conscious process of conditioning which keeps the citizen in
ignorance will then be the fundamental choices.

By contrast, one seeks to pursue a different project. True, it is not yet
well defined if it is not defined by what it rejects. But can it be otherwise? By definition, the rationality of a situation which has not quite
arrived yet cannot be precisely known; it can only appear insufficiently
'rational' if it is judged by present rationality, which is already inadequate on its own terms.

Knowledge, power, delegation, authority, certainty, stability, consultation,
refusal ... all typical notions for the body politic are seen to be called
into question in their habitual meaning. Losing these references which
classically are its own, the body politic finds itself upset by the challenge
of major risk. At the limit, if one considers the temptation to brush aside
the reference to democracy or the need to reexamine today the conditions of
a democracy in the era of high risk technology one may find that risk
'subverts' the body politic.

We have shown that western societies have forces and means to settle down
to the task and come up with answers worthy of their most fundamental
cultural traditions. But, as Hegel said, is politics not the science of the
will?

The seriousness of our freedom

1. FACED WITH MAJOR RISK

Major technological risk: we have tried to explore the multiple facets of these three words, to grasp to the full the political problem raised by this challenge to the industrial societies of the western world.

In a first stage, a long one but the complexity of the issue demanded such developments, we examined the continuities and discontinuities we observed in the order of technological risk since the beginning of the industrial era. This view has permitted us better to appreciate the novel nature of contemporary risks: menaces of large scale and extreme seriousness, unknown in the past, which today are part of our world. True, no event has as yet shown this phenomenon clearly but serious warnings have already been given, we examined examples among the most significant ones in the first chapter.

From this first stage on, as we have actually announced in advance, the view was political as we have undertaken to examine the documents under the aspect of human responsibility. The conclusion is severe: the disaster must not be seen like the meteorite that falls out of the sky on an innocent world; the disaster, most often, is anticipated, and on multiple occasions. By force of deafness and blindness misfortune unfolds and throws a frequently cruel light on the often stupefying elements (at least for those who do not play ostrich and bypass the discussion of the circumstances). The history of disasters is the history of the irresponsibility of the public powers, of the vanity of their assurance, of the derisory nature of their speech.

Can this story continue in the same manner when the risks have become more serious? This raises a question when one passes from the mining disaster (1,099 deaths at Courrieres) to the menaces which weigh down on urban areas with several hundred thousand or several million inhabitants (Toulouse, Dunkirk, Grenoble, Lille, Lyons, Milan, Toronto ...).

One must therefore closely examine what our societies are capable of doing when faced with such a challenge.

In a second stage we then examined the institutional means and the scientific tools available for the management of major risk. A historical view

has shown the undeniable progress accomplished in this field: good adjustments have been made in matters of prevention, combat and reparation; better
still: significant breakthroughs have appeared on the level of tools.

Yet, here again, the body politic is challenged. Often it still has to
take notice of the very existence of these high risks and subsequently to
take into consideration the numerous insufficiencies of the means of management available. For each examined point we were able to produce evidence of
questions which require immediate determined actions. Even the best provided
sectors, such as the nuclear sector, demand significant progress. In other
cases, such as the chemical industry, the very first steps recently undertaken surprise (and this is a rather weak term) by their timidity and their
belated nature.

The body politic is challenged more seriously still when it accepts to
observe the very free limits of the means and tools available to the administrator and technician. The world's biggest insurance companies warn: if
development is not conducted in a reasonable fashion one cannot expect reparation in case of major disaster. The experts insist: in a disaster
situation at present there is hardly a chance for successful defence but only
for managing the rout as well as possible. There remains prevention. There
again the specialists sound doubtful: it is not possible to filter out, as
should be done, all defective elements of the existing systems.

Therefore, the body politic can, less than ever, afford to leave power to
the technician. Given the risks, it must decide the demands to be made on
the engineer; given the limits of technique it must choose well in advance
the approaches it is prepared to accept. Once the choices are made, the
possible adjustments in matters of safety will be only of limited effectiveness.

However there is more. All these insufficiencies and limits must be
examined *in situ*. The social and organisational realities in fact weigh
down heavily on the difficulty of controlling major risk. The third stage
of our exploration, devoted to the examination of the regulation of major
risk has shown the social scope of the problem. In a general way the brushing aside of the eventuality of the disaster, completed by a series of inauspicious, sometimes aberrant behaviours, increases the vulnerability of
the socio-economic systems, and this for a double reason. On the one hand,
every one of the agents tends to put the question of major risk outside his
frame of reference. On the other hand, complex social situations are not a
propitious atmosphere for the control of major risk. The disaster which is
made possible because of technical and institutional faults occurs because
of the irresponsibility of some people and because of very general social
situations which carry in themselves states of a responsibility which are no
longer adequate in the era of major risk.

Faced with all this, the body politic is more than just upset, it is
affected in depth. Nightmares haunt the responsible minds: has the Western
World gone mad? One can understand that such questions are sometimes quickly
brushed aside.

Our fourth and last stage has been devoted precisely to these brutal
challenges to the body politic. We have pursued three lines of reflection
and response.

The first is the one of refusal: defending progress, getting projects
passed, holding up in a disaster situation. The body politic stiffens up and

uses what tactics seem suitable, firmness and authority, to ensure the pursuit of the previous tendencies.

The second is the one of questioning: what reorganisations are necessary
to face the challenge? What status to give to the safety function? What
decision making process to adopt in order to give the citizen again a place
where he is very directly involved as soon as there is major risk?

The third one is different: it is here a question of deeper reflection on
the body politic. It does no longer separate risk from its key preoccupations
as in the first case; it does not only open up to questioning as in the
second. It senses the deepest issues to which major risk leads. Responses
of a strategic order are therefore given or sought. They are given when the
firmness of the first scenario is taken up, systematised and anchored in a
true project. This is mainly the project according to which the citizen will
no longer be able to act out his choices. He must, more than before, leave
the choice of development options to experts (the parliamentarians being
content with initialling the documents put before them). This represents a
new cultural era. In order to enter without too much shock into this new
culture one must have, as the international organisations stressed in 1958
for the case of nuclear energy, "educative redress" and a carefully managed
process of conditioning. It will be necessary that the citizen gets used to
ignorance. And in case of refusal, subsequent to a disaster or not, a discourse or if that does not work the use of force will have to set him right
again, this because of Reason, the reason of Power.

Side by side with this clear and simple approach there are doubtless other
approaches which are difficult to draw up and to follow but more innovating.
We have given some outlines of them but the difficulty here is very big. On
the one hand, social responses to such a challenge would not come from one
individual or even from several which is an evidence not to be forgotten.
On the other hand, the attribute of a real turning point is precisely that
one does not see what it hides; doubt and search are then the rule and are
actually more adequate than the vain assurance of the one who brandishes a
solution with the force of naivety, of shortcuts and doubtless distrust for
history. We maintain that much must be done to avoid the easy temptation of
authoritarianism which deceives itself by thinking that all can still be
solved "as before" or that the world must be set right in the end by the few
dozen most alert experts.

We prefer the approach of courageous humility which in the face of major
risk does not brush aside doubt and takes up the example set by Ilya
Prigogine and Isabelle Stengers calling for a 'new alliance':

The problem raised by the interaction of human and machine populations has
nothing in common with the relatively simple and controllable problem of
the construction of a specific machine. The technical world to the
creation of which classical science has contributed, in order to be understood needs concepts that are utterly different from those of that science
... The time has come for new alliances, always knotted, for a long time
misunderstood, between the history of men, their societies, their knowledge
and the adventure of exploring nature.*

New alliances to be woven between man and nature, between societies and

---

*Ilya Prigogine and Isabelle Stengers: *La Nouvelle Alliance — Metamorphose de
la Science*. Gallimard, Paris 1979, pp. 294 and 296.

their techniques, between men themselves, inserted into new relationships
bearing the stamp of their technological options. A new alliance to be
brought to light between knowledge and power, new realities to be brought to
life for the project of democracy, a project which could quickly and brutally
shrivel up if the body politic did not take up the challenge of major technological risk intelligently and with determination.

2. WITHOUT WAITING FOR THE TWILIGHT

May we be permitted at the end of this work a more personal implication
and not a classical aggregation on a problem which is in itself already
rather complex. May we be permitted to face what touches every one of us
and the whole world: the possibility of personal and collective death. Just
aggregation, it seems to us, would be a bit artificial in the present state
of our own reflection and would above all constitute, under the cover of
respecting the rules, an ultimate escape, a defence mechanism of great sub-
tlety, no doubt, because it would slip into the normal rule of the exercise,
but not legitimately in relation to the seriousness of the question raised.

Yet, it is indeed an ultimate aggregation that we did for a time foresee
these final pages. We want to see the philosopher, the one who is not
revolted by thinking out the human experiment.

What does the thinking of those with whom he is familiar teach us?

The philosopher treated us to Plato who wanted to found power on knowledge
and no longer on the false discussions of the sophists in the agora. This
is a difficult perspective to uphold when knowledge vacillates in the face
of uncertainty. He treated us to Descartes who wanted to reduce reality to
what is measurable, to space, to the calculable, by brushing aside everything
that would not fit this scheme which is operable but so impoverishing. ("The
silence of these infinite spaces frightens me", Pascal sighed.) A reduction
that is untenable when the measurable includes such incommensurable discontinuities. He treated us to Freud who clearly showed that there exist
rationalities other than the Cartesian rationality. How does one make a
world function when A may be $\bar{A}$, B, X or Z, parapet or umbrella?

He entertained us for a long time. That world was fascinating. On the
evidence, the problems which we raised were tailor-made for these great men
of thought.

But answers? None. And yet, does not the eventuality of death or of the
fate of several million people, perhaps more still, because of our use of
knowledge remain very real? The eventuality of the death of an area and the
fact that one does not want to know about it? That one cannot know much
about it?

The philosopher did not continue further except to come to the core of our
quest and to emphasise to us its illusory nature. There are times when
thought can no longer function. Thought cannot, at a time of discontinuity,
take precedence over history and experience. We found ourselves faced with
these well-known words by one of the greatest: Minerva's owl flies only when
night comes.

What was left to us? Having entered the philosopher's realm full of
illusory and naive confidence we went out naked, faced with ourselves and the
reality which we had studied. No magic key, not even at the thinker's.

Like Candide who lived through the Lisbon disaster and whose observations on the subject we quoted at the beginning, we were left with our 'garden'. Certainly not the one that is tended by resignation, by casting down one's eyes for fear that a single question might bring the sky down on our heads. But by free and voluntary work which, carried out together with others, aims at bringing forth from the soil some food, and rather grain than poisonous mushrooms, which requires knowledge, options and involvement in the task, both personally and collectively.

Would this not be a rather brief ending? An expert, the same one who told us at the beginning about the supreme value of our societies, the economic value in its narrowest sense, suggested to us as a whim the diversion: refuge in some gloomy office or at the Antipodes, under the sun of fantasy island. Much resignation and oblivion combined with the necessary hidden cynicism to ensure for oneself a fossil freedom in secret.

We have returned to Jaspers:

The historical events lead to the unexpected, they bring with them destruction but also salvation. The formula: he who does not believe in miracles is not a realist — is, however, true only if one joins to it the other one: He who relies on the miracle is a freak ... It is certain that reason cannot claim in principle that it rules the world; but it can affirm that it must be itself and that it must act without bringing about a restriction of the strength of its means.*

Ultimate escape? Unfounded confidence in human capacities? Or the ultimate act of will by which personal and collective creativity and freedom can still reaffirm themselves despite the possibility of colossal defeat?

The stakes call for determination. But what task to settle for as a priority? What does a Reason mean for us which must be itself and which acts without applying any restriction to the strength of its means?

Without falling for facile and illusory formulae one can take a few steps, knowing full well that the first one is the most costly but also the decisive one. Without expecting to have a clear idea of the extremely complex situation which we must face one can nevertheless try to awaken Minerva's owl within (us) before twilight comes and to act before it is too late.

A first snag is always denounced: the fall into the irrational. One will have to watch out in everything to keep sight of reality. There is certainly nothing inescapable; but one must know the price that must be paid for each degree of freedom which one wants to conquer or to reconquer.

A second snag is also difficult to avoid, and many of those who are thrashing the 'irrational' groups have not avoided it: it is the snag of the unreasonable**.

---

*K. Jaspers: *La bombe atomique et l'avenir de l'homme*. Buchet-Chastel, Paris 1963, pp. 677-678 (Die Atombombe und die Zukunft des Menschen, Piper, Munich 1958.)

**An action conforms to reason when it knows its preconditions, when it foresees its own consequences and when it is aware of the motivations which dictated its objectives. (G. Picht, *Reflections on the edge of the abyss*. R. Laffont Ed., Paris 1969, p. 119).

It is unreasonable to launch into innovations while hiding the risks from oneself; to count too much on the future to resolve the technical incapacities of the present and to commit the future massively on too fragile a basis.

It is unreasonable to launch oneself onto the ocean of major risk and its furies in skiffs which are obviously too frail; at the first real storm we shall drown. One cannot leave the various organisational bodies in charge of safety, and their political status in particular, in their present state; they keep — with means sometimes stingily counted out — lifeboats from another age while they should be on the bridge in order to choose together with others routes and speeds and put in their veto in case of choices which are contrary to reason. Today they are far from being pilots. They are received rather like spoil-sports that must be satisfied at minimum expense.

It is unreasonable always to smile after accidents, which are immediately qualified as 'incidents', while claiming that there has been more fear than fault and pollution only of the minds. Our examination has shown only too well: the disaster is most often anticipated. The defeat, the rout, the collapse of yesterday's smooth-talkers, the heroism of some people could have been avoided, often without much difficulty.

It is unreasonable therefore to pursue our development with this sovereign and haughty ingeniousness which one sees sometimes. It is time to ask oneself like the British expert who suggested to us as a theme for a future conference: what industrial policy after the first civilian holocaust? This not in order to succumb to fear but in order better to size up the stakes: the eventuality also of very 'strange' defeats, so big is the vain assurance of our day; the eventuality of the collapse of the organisations that should be in charge 'afterwards'; the possibility of seeing the citizen panic-stricken and abandoned (yesterday's sophists shrivelling up in culpability and impotence) in search of an Imam or a Fuehrer.

To counter these tendencies must be the first task. We shall not set forth any formulae here; we shall not line up recommendations. Only one perspective must be affirmed: suitably rearranging politics, risk and the development process, each of these terms having to be reexamined in itself and in its relationships with the other two, and this according to what is at stake.

Redefining this presupposes a first act of will.

What are the collective conditions of such an act of will? It presupposes at least a general awareness. A free and responsible awareness, open to doubt, which is the foundation of thought and intelligence.

The difficulty lies in the fact that a community, a civilisation affected to its foundations — may refuse to take notice of the means that lie in wait for it. When ignorance is bliss, 'tis folly to be wise* ... And ignorance entertains unreason. It may also, at a moment's notice, succumb to fear and precipitate (itself) headlong into such unreason.

Making headway is possible. The first step is to give information while at the same time arousing the citizen's sense of responsibility. Information will therefore not at all be the last annex to a decision — deceptive wrapping for some non-negotiable present; information will be one of the first elements of real involvement of the citizen.

---

*Thomas Gray, 1716-1771.

We are very far from establishing this political effort to train free,
voluntary and responsible citizens.

For the time being, more or less, one must bring out a little awareness in
order to let some willingness well up, on the level of those in charge as
well as on the level of the citizen.

This was the aim of our contribution: to offer an outline of knowledge for
an outline of will.

3. THE SURVIVAL OF THE BODY POLITIC

A very last observation must be made: it will recall the massive and
aberrant nature of major technological risk, the colossal challenge which
originates not in Nature but in our decisions and on account of this provokes
a direct and brutal challenge to the body politic.

One cannot hide it from oneself: when one passes the 'limit' (the watershed
on which our whole effort here is situated) and this would be the case if
a very large scale disaster struck, one truly realises the scope of the
challenge. Major risk subverts all knowledge and all power. From there it
perverts the body politic. In the face of the event, the body politic dis-
embarks with this haggard astonishment of the man in charge paralysed in
front of the 'unforeseeable' rout. In the collapse, the body politic would
realise that it has controlled major risk only on a misunderstanding: by
assuming that this risk was neither major nor real.

Major risk is, however, today's reality. It has been committed. How can
one live under its shadow? What must be asked of the body politic?

One can no longer ask it to brush the menace radically aside. Irreversi-
bilities have been committed. Zero probability does not exist.

All energy may be dissipated in the denial of the split, in the brushing
aside of the question which is intellectually uncontrollable because major
risk presents itself at the limit of the body politic as an "altérité" (the
total opposite). An "altérité" which only an experience, as terrible as
death, and in this case: collective death, would permit one to know.

Shall we say, as with defeat in the military field, that these are forbidden
questions because they are too serious and there is no answer to them?

At the limit of our thinking ability there is above all, according to us,
the seriousness of our freedom.

A freedom which demands responsibility: a responsibility which at least
requires immediate actions towards vigorous responses in the face of the
challenge, fundamental reflections on our development and its means, new
dynamism in our approach and in our use of knowledge and power.

In the era of major risk the survival of the body politic is at this price.

Annex

LIST OF ACCIDENTS IN THE CHEMICAL INDUSTRY

by

R. ANDURAND

in "The safety report and its application in industry", *Annales des Mines*, 7-8, July/August 1979, 115-38

Location	Country	Date	Product involved	Damages	Causes
Hull	GB	1921	Hydrogen	Windows shattered within 3 km radius. Pressure felt within a 7 km radius and tremors up to 70 km.	Explosion of confined gas
Cleveland	USA	1944	LNG	136 deaths. Nearby roads swept by burning gas. Windows shattered, pavements ripped up, drain covers blown across rooftops. One fire engine blown into the air.	Explosion of confined gas, fireball
Manhattan District Project	USA	1946	Uranium hexafluoride	Two deaths, three people seriously injured, thirteen slightly injured. Explosion of UF_6 and very hot water in a laboratory. HF aerosol carried up to 100 m by the wind.	Explosion
Ludwigshafen	FRG	1948	Dimethyl-ether	245 deaths, 2,500 injured. Wagon ruptured near a dimethyl-ether factory followed by explosion and fire (Cost: 80 million FF).	Explosion of non-confined vapor cloud
Newark (Warren Oil Port)	USA	1951	Not specified	No record	Explosion of non-confined gas cloud
Wilsum	Germany	1952	Chlorine	Seven deaths in an escape of 15 tonnes, coming from a storage tank.	Toxic product
Whiting, Indiana	USA	1955	Not specified	Two deaths, thirty injured following a detonation in a pressurised container. Storage tanks pierced by the burst burned for eight days (Cost: 80 million FF).	Detonation
New York	USA	1956	Ethylene	1,100 m^3 of ethylene escaped into the atmosphere causing an explosion in the air.	Explosion of non-confined vapour cloud

503

Location	Country	Date	Product involved	Damages	Causes
Niagra Falls, N.Y. State	USA	1958	Nitro-methane	Two hundred injured when a tank wagon detonated and caused a big crater (Cost: 5 million FF).	Detonation
Signal, California	USA	1958	Not specified	Two dead when vapor coming from an overflowing tank ignited and ravaged 70 per cent of the installation.	Fireball
La Barre, Los Angeles	USA	1961	Chlorine	One dead in a cloud of 27.5 tonnes, released by a tank wagon.	Toxic product
Kentucky	USA	1962	Ethylene oxide	One dead, nine injured. Explosion 'equivalent' to 18 t of TNT.	Explosion of non-confined vapour cloud
Berlin, NY State	USA	1962	Propane	No records.	Explosion of non-confined gas cloud
Louisiana	USA	1963	Ethylene	Long lasting fire	Explosion of confined vapour
Texas	USA	1963	Propylene	Fire and explosion in a low-pressure polymerisation unit for polypropylene (Cost: 30 million FF).	Explosion of confined vapour
Texas	USA	1964	Ethylene	Fire explosion subsequent to an escape of gaseuous ethylene (Cost: 15 million FF).	Explosion of confined gas, fireball
Texas	USA	1964	Ethylene	Two dead in a fire following rupture of a high-pressure ethylene pipeline (Cost: 20 million FF).	Explosion of non-confined vapour
Massachusetts	USA	1964	Vinyl monomere chloride	Seven dead, 40 injured. Rupture of an observation window under tension and pressure. The escaping gas caught fire and exploded (Cost: 25 million FF).	Explosion of confined vapour
Pierrelatte	France	1965	Uranium Hexafluoride + fluorising agent	No deaths of injuries or evacuations. Fire subsequent to escape of 300 kg of a mixture caused by chemical corrosion of a sleeve of a distilling column at the CEA pilot plant. Pilot run stopped for eight days. Aerosol of hydrofluoric acid carried over 200 metres by the wind. No external contamination: uranium remained confined in the building natural uranium).	Product of of only chemical toxicity
Louisiana	USA	1965	Ethylene	12 injured by fire and explosion of ethylene from ruptured pipes (Cost: 15 million FF).	Explosion of confined gas
Texas	USA	1965	Propylene	Explosion and fire subsequent to pipe break in a propylene polymerisation plant (polypropylene) (Cost: 30 million FF).	Explosion of confined gas and fireball

Location	Country	Year	Substance	Description	Type of event
Feyzin	France	1966	Propane	16 dead and 63 injured. A valve blocked by freezing during sample-taking from a storage sphere permitted formation of a gas cloud which exploded near a motorway.	Explosion of non-confined gas cloud
La Salle	Canada	1966	Styrene	11 dead after explosion following the breaking of an observation window (Cost: 20 million FF).	Explosion of confined vapour
Not specified	FRG	1966	Methane	Three dead, eighty three. injured. Circumstances not specified.	Explosion of confined gas
Fernhald, Ohio	USA	1966	Uranium Hexafluoride	One injured (Six days in hospital). Eight people under observation. Human error: the maintenance man thought he was unscrewing the top of a valve but unscrewed the valve itself. Escape of 1.7 t of UF_6 in vapour. Duration forty minutes (Lead Company of Ohio).	Toxic gas + dangerous derivatives (HF)
Santos	Brazil	1967	Coal gas	300 injured, eighty buildings of various sizes within a radius of 2 km either destroyed or damaged.	Explosion of confined gas
Hawthorne, N.J.	USA	1967	Not specified	Two dead, sixteen injured. Building explosion.	Not specified
Buenos Aires	Argentina	1967	Propane	100 injured. Fire destroyed four hundred houses in the neighbourhood.	Fire
Antwerp	Belgium	1967	Vinyl monomere chloride	Four dead, 33 injured. Fire lasted three days.	Fire
Lake Charles Louisiana	USA	1967	Isobutane	Seven dead. A leaking 10 inch shutter valve released a cloud which exploded. Secondary fires and explosions continued for two weeks.	Explosion of a non confined gas cloud
Bankstown, New South Wales	Australia	1967	Chlorine	Five people intoxicated by evaporations. Evacuation of a large part of the town.	Toxic product
Perris	Holland	1968	Light hydrocarbons	Two dead, Seventy five injured. Pressure wave broken windows 2 km away.	Explosion of non-confined vapour cloud
Not specified	GDR	1968	Vinyl chloride monomere	24 four dead.	Toxic product
Paris	France	1968	Petro-chemical products	400 people evacuated. Explosion shook houses in the neighbourhood.	Explosion of confined vapours
Hull	GB	1968	Acetic acid	Two dead, thirteen injured.	Explosion of con-fined vapours
Rjukan	Norway	1968	Gas	Windows of cars and shops shattered. No statement on type of gas or circumstances.	Explosion of confined vapours

Location	Country	Date	Product involved	Damages	Causes
Soldatna, Alaska	USA	1968	Pressurised liquified gas	Two people seriously injured.	Not specified
Tamytown	USA	1968	Propane	3,500 people evacuated.	Not specified
Lievin	France	1968	Ammonia	Explosion of road tanker in the process of unloading. Escape of 19 tonnes. Six dead, twenty people living in the neighbourhood hospitalised for poisoning.	Formation of a toxic aerosol
Grandes Armoises	France	1969	Ammonia	During transfer of NH_3 from a fixed to a mobile tank a hose ruptured. Escape of 4 tonnes.	Formation of a aerosol
Teeside	GB	1969	Cyclohexane	Vegetation burned over a surface of $2,000 \times 450$ metres. Sixteen cows, one dog and various chicken killed near living quarters.	Fireball
Not specified	Libya	1969	LNG	Two dead, twenty three injured.	Not specified
Puerto la Cruz	Venezuela	1969	Light hydrocarbon	12 injured.	Not specified
Long Beach, California	USA	1969	Mineral Oil	Five dead. Considerable damage to windows and ceilings in town.	Explosion of confined vapours
Escombreas	USA	1969	Petroleum	One dead, eighty three injured. The cover of a 2,600 t tank was blown off in a suburban area.	Explosion of non-confined vapours
Repesa	Spain	1969	Pressurised liquified gas	Four dead, three injured, 5,000 people evacuated. The shock wave broke windows within a radius of several km.	Fireball
Crete, Nebraska	USA	1969	Ammonia	An escape of liquified propylene gas caused a refinery fire that burned for six days.	Toxic product
Basle	Switzerland	1969	Liquified nitric product	Six dead. Escape of 64 t of ammonia from a wagon.	Detonation
Philadelphia	USA	1970	Petrol products	Three dead, twenty eight injured. The pressure shook windows up to 1 km away.	Detonation
Osaka	Japan	1970	Gas	Five dead, twenty seven injured. Explosion in an oil refinery.	Detonation of confined gas
				92 dead. Gas explosion on a subway construction site in Osaka.	
Mitcham, Surrey	GB	1970	Propane,	Substantial destruction of private property in the neighbourhood: roofs cracked, windows broken, fences overturned, dwellings destroyed by fire, two cars destroyed.	Explosion of confined gas

Location	Country	Year	Substance	Description	Explosion of
St. Thomas	Virgin Islands	1970	Natural gas	25 injured. The explosion shook practically the whole island.	Explosion of
New Jersey	USA	1970	Petroleum products	40 injured. The shock waves shook windows in an area of 150 km².	Explosion of non-confined vapours
Port Hudson	USA	1970	Propane	No human casualties. Windows were broken up to 18 km away. The derivative cloud was ignited by an electric motor at a cold storag eunit in its trajectory.	Explosion of non-confined gas cloud
Blair, Nebraska	USA	1970	Ammonia	Overflow of a dryogenic tank of 32,000 t for two and a half hours. Escape of 145 tonnes. Animals and fish killed. Three foliage burned over 40 hectares of woodland. Low cloud of 2.50 to 9 metres thickness stretching over 365 hectares at 2,500 meters from the tank. Affected area: one house, one farm; two dogs killed at 1,770 metres distance.	Toxic product
Crescent City	USA	1970	Propane	Derailment of a wagon; the commercial centre of the town was destroyed.	Explosion of vapours emitted by flash fire and boiling liquified gas
Emmerich	FRG	1971	Not specified	Four dead, four injured, many buildings in the area damaged.	Explosion of confined gas
Not specified	Holland	1971	Butadiene	38 dead, seventy five injured, five hundred people evacuated; window frames dislocated within a radius of 15 km. Accident in a 20-story tower.	Not specified
Not specified	Holland	1972	Hydrogen	230 injured; windows damaged up to 3 km from the site of the accident which occurred during shunting of a wagon.	Explosion of confined gas
Sao Paolo	Brazil	1972	Gas	21 dead, twenty injured, an island put out of action completely.	Explosion of confined vapours
St. Louis	USA	1972	Propylene	350 injured. Thirteen tank wagons of butane exploded.	Explosion of non-confined vapours
Virginia	USA	1972	Gas	One dead, sixteen injured.	Explosion of confined gas and fire
Not specified	Mexico	1972	Butane	Seven dead. Vapours released from a reactor exploded. Evacuation of hundreds of people within a radius of several hundred meters (Cost: 10 million FF).	Explosion
Not specified	Japan	1973	Vinyl monomer chloride	1,000 people evacuated.	Explosion of non-confined vapours
Lodi	USA	1973	Methanol	Four dead, twenty four injured. The explosion damaged buildings over a large area and blew in hundreds of windows. Cars were covered with a shower of debris and crashed by enormous pieces of concrete.	Explosion of non-confined vapours

Location	Country	Date	Product involved	Damages	Causes
Gladbeck/Ruhr	FRG	1973	Cumene	Four dead, two missing, thirty seven injured following the overturning of a truck carrying liquified presurised gas.	Not specified
Sheffield	GB	1973	Gas from	One dead, four injured.	Explosion of non-confined gas
St. Amand les	France	1973	Propane		Explosion of non-confined gas
Tokuyama	Japan	1973	Ethylene		Explosion of non-confined gas
California	USA	1973	Vinyl monomor chloride	200 litre steel containers of chemical products where hurled on top of houses, fields and into the bay.. Thousands of windows were broken and at least eight small houses seriously damaged. The shock wave was felt 80 km away.	Unknown
Cologne	FRG	1973	Vinyl monomor chloride	Rupture of a joint caused escape of 10 t of product.	Explosion of non-confined gas
New York	USA	1973	Pressurised liquified gas	40 dead.	Explosion of non-confined gas
Western Bohemia	Czecho-slovakia	1973	Gas	47 dead in a factory.	Explosion of non-confined gas
Potchefstroom	South Africa	1973	Ammonia	18 dead of which six were outside the factory. Sixty five injured. Release of 38 t of ammonia. The aerosol cloud of 20 metres thickness and a diameter of 150 metres drifted on to the neighbouring town.	Aerosol of toxic product
Falkirk	GB	1973	Inflammable liquid	Destruction of a tar factory.	Fireball
Texas	USA	1974	Isoprene	Twelve injured. Windows broken over a large area.	Explosion of non-confined gas cloud
Los Angeles	USA	1974	Organic peroxides	Run-away road tanker carrying organic peroxides exploded, causing 250 million FF damge.	Detonation
Beaumont, Texas	USA	1974	Isoprene	Two dead, ten injured: explosion of a vapour cloud which followed a big spillage of isoprene (Cost: 80 million FF).	Explosion of non-confined gas cloud
Not specified	Czecho-slovakia	1974	Ethylene	14 dead, 79 injured.	Explosion of non-confined gas

Place	Country	Year	Substance	Description	Type of explosion
Flixborough	GB	1974	Cyclohexane	28 dead, 104 injured, 3,000 people evacuated, 10 houses damaged; fishing in river Trent banned.	Explosion of non-confined vapour
Rotterdam	Holland	1974	Petro-chemical products	Enormous fire.	Fireball
Not specified	Romania	1974	Ethylene	One dead, 50 injured.	Explosion of non-confined gas cloud
Nebraska	USA	1974	Chlorine	500 people evacuated. Toxic vapour clouds spread about.	Toxic product
Floride	USA	1974	Propane	Two storages destroyed. Cars crushed and windows broken within an area of four blocks of buildings.	Explosion of non-confined gas cloud
Wenatchee	USA	1974	Monomethyl aminonitrate	Two dead, 66 injured in the explosion of a wagon.	Detonation
Not specified	Holland	1975	Ethylene	Four dead, 35 injured.	Explosion of non-confined gas cloud
Marseille	France	1975	Petro-chemical products	One dead, three injured; the explosion broke windows in a large area around the complex.	Explosion in confined area
Not specified	South Africa	1975	Methane	Seven dead, seven injured. Whole town gas supply was cut for two days.	Unknown
Antwerp	Belgium	1975	Ethylene	Six dead, 13 injured. Ethylene escaping from a compressor exploded, causing extensive damage to buildings and at the plant.	Explosion of non-confined gas cloud
Philadelphia	USA	1975	Crude oil	Eight dead, two injured. Vapours from a storage tank exploded in a boiler house when a marine lighter was refuelled. Cost: 50 million FF.	Explosion of confined vapour
Not specified	Holland	1975	Propylene	14 dead, 104 injured.	Explosion of non-confined gas cloud
Oak Ridge	USA	1976	Uranium hexafluoride	Reaction of UF_6 and oil from a vacuum pump in a type 30 (2 tonnes) container during transport: liquid UF_6 under pressure. Two injured. Eight days' stoppage.	Explosion of UF_6 + hydro-carbon product
Seveso	Italy	1976	TCDD	Complete evacuation of the area until now (1979). Abortions authorised exceptionally. Decontamination made very difficult because of the non-soluble nature of the product.	Aerosol of solid toxic product

Location	Country	Date	Product involved	Damages	Causes
Beek	Holland	1976	Naphta	14 dead, 30 injured when an escape caught fire. The explosion damaged windows of shops and houses. Cost: 100 million FF.	Explosion of non-confined vapour cloud
Baton Rouge, Louisiana	USA	1976	Chlorine	10,000 people evacuated. Mississipi banned for navigation over 80 km to the north.	Toxic product
Sandelfjord	Norway	1976	Inflammable liquid	Rupture of underground piping; fire and explosion killing six people and causing 100 million FF damage.	Fireball, explosion of confined vapour
Pierre Benite	France	1976	Acrolin	Escape from container of a wagon in the Rhone following human error (21 tonnes(. River fauna destroyed from Pierre Benite to Vienne (320 tonnes).	Toxic product
Brachead	GB	1977	Sodium chloride	Fire and explosion. Circumstances not specified.	Detonation
Mexico City	Mexico	1977	Ammonia	Two dead, 102 people treated for poisoning. Gas entered the drainage system.	Toxic product
Umm Said	Quatar	1977	Pressurised liquified gas	Seven dead, many injured. The explosion superficially burned the villages up to 2 km around. The international airport of Doha was closed for two days.	Fireball
Mexico City	Mexico	1977	Vinyl monomor chloride	90 injured. No details.	Unknown
Not specified	Taiwan	1977	Vinyl monomor chloride	Six dead, 10 injured.	Unknown
Pierrelatte	France	1977	Uranium hexafluoride + hydro-fluoric acid	One dead, nine injured.	Chemically toxic product: fluorhydric acid
Jacksonville	USA	1977	Pressurised liquified gas	Neither deaths nor injuries nor poisoning. Comhurex factory. Subsequent to human error rupture of a valve in "6 o'clock" position on a type 48 container. Expulsion of 7.1 t of UF_6 liquified under pressure.	Not specified
Rockingham North Carolina	USA	1977	Uranium hexafluoride	2,000 people evacuated.	Derailment

Derailment of a 29 train. Four type 48 (12 tonnes) containers of UF_6 involved in the accident. Fire of ammonium nitrate, fertiliser and ground nuts. The containers held: no escape of UF_6.

Location	Country	Year	Substance	Details	Type of event
Gela	Italy	1977	Ethylene oxide	One dead, two injured	Explosion of confined vapour
Not specified	India	1977	Hydrogen	20 injured. The explosion shook a fertiliser factory, an oil refinery and a village.	Explosion of confined gas
Not specified	Italy	1977	Ethylene	Three dead, 22 injured. Shop windows and doors smashed. Car blown several metres up into the air.	Explosion of confined gas cloud
Pierrelatte	France	1977	Uranium hexafluoride + hydrofluoric acid	Neither deaths nor injuries nor poisoning.f Break of a shutter clamp on a tank that was overfilled with UF_6 under hydrostatic pressure, in the course of warming up. The passage from the solid to the liquid state causes a volume increase in the order of 25-30 per cent. Release of 1,200 kg of "natural" UF_6, confined in the building.	Toxic product: gaseous hydrofluoric acid
Pasacabalo near Cartagena	Columbia	1977	Ammonia	30 dead, 22 injured. The villagers in the neighbourhood suffered the effect of the gas. The installations of the state factory Acobal were destroyed at the time of shift change (some hundred workers were then present). It has not been proved that NH_3 was the cause of the accident but rather a fire.	Toxic product
Cadarache	France	1977	UF_6 + hydrofluoric acid	Neither deaths nor injuries nor poisoning. During warm-up of a thermic trap which was overfilled with UF_6 following human error a crack developed in the partioning wall through hydrostatic pressure which put UF_6 in contact with cooling fluid and produced an aerosol of hydrofluoric acid and UO_2F_2 which moved the filters of the extractor fan: two neighbouring workshops, becoming depressurised, were invaded by HF aerosol. Restart of the workshops after one week. UF_6 released: 20 kg. Natural uranium.	Toxic product: gaseous acid hydrofluoric
Seoul	South Korea	1977	Explosives	58 dead, 1,300 injured. Explosion of train. Circumstances not specified.	Explosions
Los Alfaques, South Tarragona	Spain	1978	Propylene	216 dead, many disappeared, several hundred injured following an escape of liquified propylene under pressure, following road accident of a tanker near a camping site in summer.	Explosion of non-confined gas
Not specified	USA	1978	Grain dust	Explosion followed by large fire. Number of dead and injured not specified. Criminal attack not excluded.	Dust explosion
Portsmouth	USA	1978	UF_6 + hydrofluoric acid	Number of people poisoned not specified. No deaths, no injuries. Rupture in the piping of a hydraulic jack of a lorry which carried a 48C type container (thin partioning wall, not used in France) which held 9.6 t of liquified UF_6 under pressure. In the fall of the container a cylinder clamp cuased a 19 cm long fissure in the partioning wall when it struck the ground violently and expelled the whole UF_6 content.	Toxic product: aerosol of hydrofluoric acid
Waverley, Tennessee	USA	1978	Propane	12 dead and at least 50 injured when a tank wagon which had derailed exploded.	Explosion of vapour emitted by flash of pressurised liquid that was suddenly brought to atmospheric pressure

Location	Country	Date	Product involved	Damages	Causes
Youngstown, Florida	USA	1978	Chlorine	Eight dead, 10 injured, evacuation of 3,500 people in an area of 10 km² following spread of chlorine escaping from a derailed tank wagon. The enquiry concluded that a criminal attempt was likely.	Toxic product
Baltimore	USA	1978	Sulphur trioxide	Toxic fumes drifting up to 15 km. Mor ethan 100 people treated fro nausea.	Toxic product
Paris(Passy)	France	1978	Gas	13 dead, 13 injured, 60 flats destroyed. Cars damaged by flying debris. Series of explosions in a building and in underground piping after rupture of a gas pipe.	Explosion of confined gas
New York	USA	1978	Not specified	130 injured: explosion of a deep-freeze lorry near Wall Street.	Not specified
Pierre Benite	France	1978	Acrolein	Escape of some 100 kg of acrolein into the atmosphere. Inconvenience for several thousand people at Pierre Benite and Oullins (tear gas and nauseous gas).	Toxic product
Xilatopec	Mexico	1978	Presurised liquified gas	100 dead, 150 injured. Explosion of a lorry carrying 10,000 litres of LPG in a collision of 12 vehicles on a motorway 85 km north of Mexico City. 85 people died within minutes from the explosion.	Explosion of non-confined gas
Bantry Bay	Ireland	1979	Hydro-carbon vapours	48 disappeared (41 sailors and seven workers) in the explosion of the oil tanker Betelgeuse at the quayside. The oil tanker was not equipped with an inering system to inject inert gas as the tanks are emptied. An efficient inverting system only became obligatory when the "convention of safeguarding human life at sea" of 1974 came into force (only 15 countries, France among them, out of 25 ratified the convention).	Explosion of gas in confined volume, perhaps preceded by fire
Warsaw	Poland	1979	Not specified	41 dead, 77 injured, several hundred people evacuated. Under the effect of the explosion most of the windows of the neighbouring buildings were sent flying in splinters within a radius of 200-300 metres. The accident occurred in the basement of the savings bank building. There were no gas pipes in the building. Welding work was going on in the basement.	Explosion of confined gas probably
Islamabad	Pakistan	1979	Not specified	26 dead, 50 injured, several buildings affected by the pressure. Explosions of an artisan shop in the Raja Bazar at Rawalpindi.	Explosion of instable solid chemical products
Crestview, Florida	USA	1979	Ammonia, Chlorine	4,500 people evacuated within a radius of 2,500 metres. Derailment of a convoy of 28 tank wagons (NH_3Cl_2) on leaving the bridge over the Yellow River.	Toxic gases

INDEX OF CASES

LIST OF ILLUSTRATIONS

FIGURES

TABLES

PLATES AND PHOTOS